LAPLACE TRANSFORMS AND CONTROL SYSTEMS THEORY FOR TECHNOLOGY

INCLUDING MICROPROCESSOR-BASED CONTROL SYSTEMS

Theodore F. Bogart, Jr., PE
University of Southern Mississippi

JOHN WILEY & SONS

NEW YORK   CHICHESTER   BRISBANE   TORONTO   SINGAPORE

Copyright © 1982, by John Wiley & Sons, Inc.

All rights reserved. Published simultaneously in Canada.

Reproduction or translation of any part of
this work beyond that permitted by Sections
107 and 108 of the 1976 United States Copyright
Act without the permission of the copyright
owner is unlawful. Requests for permission
or further information should be addressed to
the Permissions Department, John Wiley & Sons.

Library of Congress Cataloging in Publication Data:

Bogart, Theodore F.
   Laplace transforms and control systems theory
for technology.

   Includes index.
   1. Control theory. 2. Laplace transformation.
I. Title.

QA402.3.B63   629.8'312   81-14708
ISBN 0-471-09044-1   AACR2

Printed in the United States of America

10 9 8 7 6 5 4 3 2 1

ELECTRONIC TECHNOLOGY SERIES

Santokh S. Basi
SEMICONDUCTOR PULSE AND SWITCHING CIRCUITS (1980)

Theodore Bogart
LAPLACE TRANSFORMS AND CONTROL SYSTEMS THEORY FOR TECHNOLOGY (1982)

Rodney B. Faber
APPLIED ELECTRICITY AND ELECTRONICS FOR TECHNOLOGY, 2nd edition (1982)

Luces M. Faulkenberry
AN INTRODUCTION TO OPERATIONAL AMPLIFIERS, 2nd edition (1982)

Joseph D. Greenfield
PRACTICAL DIGITAL DESIGN USING ICs (1977)

Joseph D. Greenfield and William C. Wray
USING MICROPROCESSORS AND MICROCOMPUTERS: THE 6800 FAMILY (1981)

Curtis Johnson
PROCESS CONTROL INSTRUMENTATION TECHNOLOGY, 2nd edition (1982)

Irving L. Kosow
STUDY GUIDE IN DIRECT CURRENT CIRCUITS: A PERSONALIZED SYSTEM OF INSTRUCTION (1977)

Irving L. Kosow
STUDY GUIDE IN ALTERNATING CURRENT CIRCUITS: A PERSONALIZED SYSTEM OF INSTRUCTION (1977)

Sol Lapatine
ELECTRONICS IN COMMUNICATION (1978)

Ulises M. Lopez and George E. Warrin
ELECTRONIC DRAWING AND TECHNOLOGY (1978)

Dan I. Porat and Arpad Barna
INTRODUCTION TO DIGITAL TECHNIQUES (1979)

Herbert W. Richter
ELECTRICAL AND ELECTRONIC DRAFTING (1977)

Thomas Young
LINEAR INTEGRATED CIRCUITS (1981)

LAPLACE TRANSFORMS AND CONTROL SYSTEMS THEORY FOR TECHNOLOGY

INCLUDING MICROPROCESSOR-BASED CONTROL SYSTEMS

TO MY PARENTS
ON THEIR GOLDEN WEDDING ANNIVERSARY

PREFACE

LaPlace Transforms and Control Systems Theory for Technology is intended for one- or two-semester courses in network analysis and control-systems theory. The material covered in Chapters 1 through 5 and the Appendix form the basis for a one-semester course in LaPlace transformation and its application to network analysis. The remainder of the book can be used for a course that covers the basics of continuous control-systems theory, analog- and digital-analysis techniques, and microprocessor-based control systems. Also, an accelerated single-semester course, which would touch on a majority of topics, could be designed around the material in Chapters 3, 4, 5, and 6 (through Section 6.9), 7 (through Section 7.5), and 8 (through Section 8.5).

The material covered and its level of treatment are appropriate for any two- or four-year program that includes technical calculus. Students beginning the two-course sequence at the University of Southern Mississippi have completed two semesters of calculus, although conscientious students with only one semester of calculus, including an introduction to integration, have successfully completed the sequence. Students should also have completed the traditional courses in dc- and ac-circuit analysis and electronic devices and circuits. For the material in Chapter 10, students should have completed at least one course in digital electronics, but no prior knowledge of microprocessors is assumed.

The intent of this book is to expose students to as broad a range of topics as possible, from traditional continuous-systems analysis, to the most recent developments in digital and microprocessor control techniques. The breadth of coverage precludes an exhaustive treatment of any one topic, though it is hoped that the depth is sufficient to stimulate the interest of readers and to make them feel comfortable with new concepts in *every* area covered. In any event, the level of mathematical abstraction that would be required to treat many topics in greater detail is not considered appropriate for the technology student. Thus, for example, the mathematical procedure for finding an inverse LaPlace transform and the rigorous justification of the Nyquist stability criterion, as well as its application to generalized systems, are omitted, as they might not be in a traditional engineering curriculum. A knowledge of calculus of the complex plane is not an expected background for this course, nor is it discussed in the text.

The profusion of topics that today's electronics technologist should know something about (even if only the jargon) is so great that a single course in, say, LaPlace transforms, or continous-type control-systems theory, may not be justified in a technology program simply because such specialty courses must necessarily displace instruction in other equally vital areas of the curriculum. Nevertheless, the application of LaPlace transforms to both circuit and systems analysis receives thorough coverage in this text. Topics that are traditionally relevant to this area are also covered: control-systems stability analysis, compensation, operational amplifiers, and analog computers, though not in the same depth. A substantial part of the text is devoted to digital techniques, in recognition of the incursion that the digital computer has made in recent years both as a tool for systems analysis and as an integral part of control systems. Therefore, one entire chapter deals with digital-computer analysis of continuous systems, and another chapter introduces the microprocessor-based control system. Appendix A also deals with the application of LaPlace transforms to the analysis of circuits driven by pulse-type waveforms.

Wherever possible and appropriate, the theory is illustrated by examples from mechanical as well as electrical systems. The analogy between mechanical and electrical networks, and the equations that describe their behavior, is stressed throughout. Portions of this book are appropriate for use in some mechanical or manufacturing curriculums as well as electronics and electrical-technology programs, particularly in view of the increasingly important role of the microprocessor in automation.

A large number of worked-out examples and student exercises reinforce the theory; this is especially true in those chapters that introduce new ideas or make greater demands on the students' ability to reason quantitatively with abstract concepts. Answers to exercises are provided, except to those for which a large number of correct solutions is possible or whose solutions are quite long or require extensive diagrams.

Many illustrative examples are based on manufacturers' product literature and specifications. Specification sheets and illustrations of commercially available servo-system components, both analog and digital, are reproduced in Appendix B. Data taken from this product literature are used in examples to give students a feeling for the orders of magnitude of variables used in the theory and to motivate and reinforce learning by confirming the existence of real hardware that behaves according to theory.

The SI system of units is used throughout the book, except in illustrations where actual product specifications are used. Lamentably, most American manufacturers' specifications are still given in English or CGS units; but in the majority of the examples that use these units, the conversions to SI are shown and used.

I wish to thank Professor C. Howard Heiden, Chairman of the Department of Industrial Technology at the University of Southern Mississippi, for the encouragement and departmental support given me during the many months required to prepare this book. Thanks also to Professor Jack Lipscomb, University of Southern Mississippi, for reviewing portions of the manuscript and for his assistance in the section dealing with CSMP. I am grateful to those instructors from other institutions, who reviewed the manuscript for John Wiley & Sons, and made many constructive suggestions. A large measure of thanks is due my students, who suffered through the first drafts of the book in class, and who were very diligent in seeking out errors.

T. F. Bogart, Jr.

CONTENTS

1. INTRODUCTION 1

   1.1 What Is a LaPlace Transform? 1
   1.2 What Is a Control System? 2
   1.3 Analog Versus Digital Systems 3
   1.4 Mathematical Models 5

2. REVIEW OF COMPLEX ALGEBRA AND PHASORS 7

   2.1 Complex Numbers 7
   2.2 The Complex Plane 9
   2.3 The Arithmetic of Complex Numbers 12
   2.4 Complex Conjugates 14
   2.5 Rationalizing 16
   2.6 Properties of j 17
   2.7 The Exponential Form 19
   2.8 Phasors 20
   2.9 Impedance and the Phasor Form 24
   2.10 Review of Some Important Facts from Calculus 29

3. TRANSFER FUNCTIONS 38

   3.1 Gain 38
   3.2 Gain as a Function of Frequency 41

xii CONTENTS

3.3	Transfer Functions	42
3.4	Calculating the Transfer Function of Passive Networks	43
3.5	Bode Plots	46
3.6	Frequency Response in Decibels	51
3.7	Asymptotic Plots	54

4. LAPLACE TRANSFORMATION 63

4.1	Differential Equations	63
4.2	Differential Equations in Electronics and Mechanics	65
4.3	LaPlace Transforms	72
4.4	Using the Table of Transform Pairs	79
4.5	Transforms of Derivatives and Integrals	82
4.6	Transfer Functions in the Frequency Domain	84
4.7	Inverse Transformation by Partial Fractions	87

5. NETWORK ANALYSIS USING LAPLACE TRANSFORMS 104

5.1	Review of Basic Network Analysis Techniques	104
5.2	Sources and Initial Conditions in the Frequency Domain	119
5.3	Kirchoff's Laws using LaPlace Transforms	123
5.4	Mesh Analysis using LaPlace Transforms	128
5.5	Nodal Analysis using LaPlace Transforms	134
5.6	Superposition using LaPlace Transforms	135
5.7	Thevenin's and Norton's Theorems using LaPlace Transforms	137
5.8	The Effects of Damping on Circuit Response	141
5.9	Analysis of Mechanical Networks	148

6. CONTROL SYSTEMS THEORY 163

6.1	Open and Closed Loop Control Systems	163
6.2	Block Diagrams	165
6.3	Integrators	173
6.4	The Servomotor in a Closed Loop	174
6.5	Bode Plot for the Simple Position Control System	178
6.6	The Dynamics of Rotating Systems	181
6.7	Gear Trains	184
6.8	The Second Approximation of a Servomotor	189
6.9	A Velocity Control System	195
6.10	A Comprehensive Analysis of the Servomotor in a Control System with Mechanical Components	196
6.11	Speed-Torque Curves	202
6.12	Error Coefficients	205

7. STABILITY AND COMPENSATION 221

7.1	Stable and Unstable Control Systems	221
7.2	Stability and the Open Loop Transfer Function	222
7.3	Stability Analysis Using Bode Plots	227
7.4	Stability Analysis of the System with Complex Conjugate Poles	230
7.5	Gain and Phase Margins	234
7.6	Polar Plots	235
7.7	The Nyquist Stability Criterion	243
7.8	The Nichols Chart	246
7.9	Root Locus	252
7.10	Compensation	260

8. ANALOG COMPUTATION AND SIMULATION 276

8.1	Analog Computers	276
8.2	Operational Amplifiers	277
8.3	Operational Amplifier Circuits	283
8.4	Analog Solution of Differential Equations	290
8.5	Analog Computer Modes	299
8.6	Magnitude Scaling	304
8.7	Nonlinear Differential Equations	309
8.8	Hybrid Computers	313

9. DIGITAL COMPUTER SIMULATION 325

9.1	Role of the Digital Computer	325
9.2	Computer Analysis of Transfer Functions	326
9.3	Computing Time-Domain Responses	344
9.4	Optimization by Automatic Iterative Computation	356
9.5	Continuous System Modeling Program (CSMP)	363

10. MICROPROCESSOR-BASED CONTROL SYSTEMS 381

10.1	Role of the Microprocessor	381
10.2	The Analog/Digital Interface	382
10.3	Microprocessor System Design Considerations	389
10.4	Microprocessor Architecture and Programming for Control Systems Applications	403
10.5	Interrupts	423
10.6	Design Example—A Sorting Robot	435

APPENDIX A ADVANCED TOPICS IN LAPLACE TRANSFORMS 459

A.1	Representation of Delayed Time Functions	459
A.2	Representation of Nonlinear Periodic Functions	468

	A.3	LaPlace Transforms of Delayed and Nonlinear Periodic Waves	471
	A.4	LaPlace Transform Analysis of Circuits with Delayed and Nonlinear Periodic Waves	475
APPENDIX B		SPECIFICATIONS AND DATA SHEETS FOR PRODUCTS CITED IN THE EXAMPLES	486
INDEX			535

LAPLACE TRANSFORMS AND CONTROL SYSTEMS THEORY FOR TECHNOLOGY

INCLUDING MICROPROCESSOR-BASED CONTROL SYSTEMS

CHAPTER 1
INTRODUCTION

1.1 WHAT IS A LAPLACE TRANSFORM?

Since a major portion of this text deals with the rather formidable sounding term *LaPlace transform*, we wish to reassure the reader at the outset that it is a concept designed to *simplify* rather than complicate the analysis of circuits and systems. Although differential and integral calculus are the mathematical prerequisites for the material in this book, the student who has weathered these subjects will be pleased to learn that LaPlace transforms effectively convert many calculus problems into ordinary algebra problems. Indeed, a good command of algebra, particularly the ability to handle algebraic fractions, may be a more important skill in using LaPlace transforms than is a thorough knowledge of calculus.

Calculus, as we all know, is the mathematical tool that we must have in order to analyze, predict, and understand *dynamic* systems: systems whose variables change with time. A derivative, after all, represents a *rate of change* and is, therefore, the fundamental means we have for describing the behavior of time-varying quantities. For example, the familiar relation between the charge and voltage on a capacitor $Q = CV$ is useful when the charge and voltage are constant with respect to time, that is, in the dc case, but it is more useful when we differentiate both sides of the equation to obtain

$$\frac{dQ}{dt} = C\frac{dv}{dt}$$

or equivalently,

$$i = C\frac{dv}{dt}$$

2 INTRODUCTION

This allows us to relate the current through and voltage across a capacitor when these two quantities are changing with time. When we attempt to relate voltage and charge, or current, in a circuit containing capacitors, inductors, and resistors, we find that we suddenly have a number of derivatives to deal with and that we, in fact, have a *differential equation* to solve. We intend to cover the techniques for setting up the differential equations of electrical as well as mechanical networks in a later chapter, but suffice it for now to say that LaPlace transformations of these equations will allow us to solve them by simply solving ordinary algebra equations.

A *transformation* is a new name for an already familiar procedure. In simplest terms, we say that we transform a function whenever we change or modify what that function does. We usually think of a function as a rule we follow to change the value of a number; similarly, a transformation is a rule we follow to change what a function does. For example, the transformation "multiply by 2" would transform the function $y = x^2$ into $y = 2x^2$, the function $y = \sin x$ into $y = 2 \sin x$, and so forth. Thus, a transformation operates on functions to change *them*, just as a function operates on numbers to produce new numbers. For this reason, transformations are sometimes called *operators*. When the operator "multiply by 2" is applied to the function $y = x^2$, we obtain the transform $y = 2x^2$.

The LaPlace transformation is an *integral operator*, so called because it changes a given function into a new function by the process of integration. In practice, we rarely need to perform that integration when we wish to obtain the LaPlace transform of a function. We rely instead on *tables* of LaPlace transforms and upon our ability to manipulate algebraic expressions so that they fit the format of the tables. The major advantage of the LaPlace transformation is that it transforms the mathematical operations of differentiation and integration into multiplication and division. Thus, a differential equation that has been transformed in this manner has no derivatives present, and its solution may therefore be obtained using purely algebraic methods (and a table of transforms). Those who have struggled with the classical methods for solving differential equations will appreciate what this means and will reap considerable dividends from investing a little more time and effort to learn a little more mathematics.

So rather than feeling that vaguely threatening and familiar apprehension often experienced on a first encounter with a new mathematical concept, the student will, we trust, welcome the transform method and embrace it with confidence as simply a useful tool for simplifying complicated problems.

1.2 WHAT IS A CONTROL SYSTEM?

The subject we call control systems pervades all of technology. In fact, we may go so far as to say that at least one aspect or another of control-systems theory is appropriate to every human activity. In its most general sense, a control system is simply an aggregate of components that, working together, generate a response to some stimulus.

The human body, which responds to internal and extenal stimuli, is certainly a control system. Internal stimuli (thoughts, ideas, physiological changes) stimulate human responses, as do such external stimuli as temperature changes, threatening events, social interactions, and a host of others. The very act of reaching out to pick up a pencil is an example of a human control system in operation. It is an example of a closed-loop control system where a visual stimulus (the relative position of the hand with respect to the pencil as perceived by the eye) is continually fed back to the brain and where the response, that is, the movement of the hand, is continually adjusted until the hand successfully reaches the pencil.

We are surrounded by man-made control systems, both open-loop and closed-loop. All manner of electrical appliances and mechanical devices perform useful functions, that is, *respond* to all manner of stimuli. In control-systems parlance, we refer to a stimulus as an input, while the response is called an output. Thus, turning a knob on an oven represents an input, and the oven responds by producing an output that causes an increase or decrease in temperature. More sophisticated examples include aircraft autopilots, guided missiles, automobile-speed controllers, nuclear-reactor safety systems, scanning radio receivers, dc power supplies, oil refineries and chemical-processing plants, robots, and many, many others.

Perhaps the very broadness of scope has led us to characterize the subject as a "system"—a word nearly as vague and as often misused as "thing." However, in the evolution of this subject as a specialty area of mechanical and electrical engineering and technology, control-systems theory has come to mean the study of certain resonably well-defined techniques for predicting, analyzing, and improving the behavior of devices (traditionally electromechanical devices) that perform some useful function in response to electrical or mechanical inputs. These techniques have been successfully applied in other disciplines including the human control system. Control-systems theory is sufficiently general, so that its concepts and results, though frequently expressed in electrical or mechanical terms, have their counterparts in a wide variety of scientific studies.

1.3 ANALOG VERSUS DIGITAL SYSTEMS

An analog system is one whose variables, for example, input and output voltages, can assume *any* values within their permissible range. Such variables are said to be *continuous*. For example, a sinusoidal voltage with 10-volt (V) peak value is a continuous or analog variable, since it can have *any* value between -10 V and $+10$ V. A motor-driven shaft is an analog device if we regard its output as some angular rotation, since the shaft can assume any angular position we wish.

A digital system, on the other hand, is one whose variables can assume only certain *discrete* values. Typically, the magnitudes of the variables are represented by binary numbers, which, in turn, are represented by a sequence of "on" or "off" levels; these levels correspond to distinct voltages. For example, a system may produce only the four binary outputs 00, 01, 10, and 11, where a 0 is represented by, say, -5 V and a 1 is represented by $+5$ V. Processing these signals is done by digital-logic circuitry or by a digital computer. A stepping motor whose shaft can only be rotated in fixed increments of, say, ten degrees is an example of a digital device.

4 INTRODUCTION

A control system that employs analog devices and processes analog signals is an analog, or continuous, control system, while one that utilizes digital devices and discrete-voltage levels is a digital-control system. Frequently, a control system uses both types of devices and performs both analog- and digital-signal processing; such a system is said to be *hybrid*. For example, while the input signal to a system may be analog in nature, it is often converted to digital form by an analog-to-digital (A/D) converter. The digital signals are processed as necessary, and the result is then converted to an analog output by a digital-to-analog (D/A) converter.

Because the development of analog systems preceded the widespread use of digital systems, much of the classical theory of control systems is based on mathematical relationships between continuous variables. Control-systems theory is, for example, very much concerned with the solution of differential equations. These equations represent continuous, analog-type relations between variables and their rates of change, about which we will have more to say in subsequent chapters.

In recent years, the trend has been to replace analog-signal processing with its digital counterpart. Digital-signal processing has some significant advantages, most notably a greater immunity to error caused by noise and circuit simplicity. Furthermore, the advent of the microprocessor, a compact, inexpensive means of handling large quantities of digital data rapidly and efficiently, has resulted in its appearance in many control systems that have been traditionally analog in nature. Automobile ignition systems, microwave ovens, automated production lines, and navigation systems are but a few examples.

It is nonetheless true that our world is essentially analog in nature, since we deal with continuous variables (our position in a coordinate system, a shaft rotation, light intensity, and the like). The mathematical relations between these continuous variables and their rates of change are still of interest to us; analyzing analog systems in terms of continuous variables is, therefore, still an appropriate endeavor. In a sense, we can think of a digital system as one that operates by approximating the actual analog variables that we experience and observe. Thus, analog solutions represent the true state of affairs with which we must certainly be acquainted in order to implement reasonable and efficient approximations. This is not to imply that digital systems and digital-control signals are any less valid, useful, or efficient than their analog counterparts. Indeed, as we have already suggested, digital-signal processing enjoys significant practical advantages. Accuracy is limited only by the resolution (number of binary digits) that the user is willing to employ. A case in point: A differential equation may be solved using an analog computer, which yields a continuously time-varying function, or by a digital computer, which yields a finite set of values of the function at specific instants of time. The accuracy of the digital solution may be as good or better than a practical analog solution, since the user may choose as many instants of time as he wishes and use numbers with as many significant digits as he wishes.

In view of these considerations, this text treats the subject of control-systems theory in the classic manner, that is, by analog methods, after developing the mathematical tools necessary for such a treatment. Then attention focuses on analysis techniques using a digital computer and, finally, on digital-control techniques, most notably the microprocessor and its role as a system controller.

1.4 MATHEMATICAL MODELS

A meaningful, quantitative analysis and understanding of a system's behavior can be achieved only through the use of mathematics. Through the years, a body of mathematics and certain mathematical tools have been developed for this purpose. These tools have proved to yield an efficient and systematic approach to the analysis problem. In many practical problems, it is only necessary to bring these tried and true methods to bear on the problem, turn the crank, and grind out a solution. There is, however, a certain amount of art involved in applying mathematics to other more significant problems. This art is called mathematical modeling, and, in essence, it is the process of deciding which variables in the problem have truly significant effects on the nature of the solution. A mathematical model may be nothing more than an equation that governs the behavior of a certain variable we are interested in: the unknown, whose solution we desire. For example, suppose we are interested in the velocity as a function of time of a rock dropped from the top of a cliff. It is well known that $v(t) = v_0 + gt$, where v_0 is the initial velocity (zero in this case), g is the gravitational constant, and t is time. This equation is a mathematical model with which we can predict the velocity of the rock at any instant of time. It is, however, only *one* possible model for the situation.

Our model overlooks a host of other variables that may affect the velocity of the rock and should or should not be taken into account, depending on how accurate a solution we want. This is where the art of mathematical modeling becomes important. For example, if the rock is falling through the atmosphere, we may have to take into account the drag forces on it due to air friction. And what of the air temperature and humidity? These affect the density of the air and, hence, the amount of friction force on the rock. Furthermore, the temperature and humidity are themselves changing as a function of time, due to the rock changing altitude as it falls. We may even wish to take into account the fact that the gravitational constant g is itself changing very slightly as the rock loses altitude. We see that a relatively simple problem can become very complex if *all* the variables influencing the behavior of the solution are accounted for. Indeed, in nearly every situation of this sort, it is impossible to account for exactly each and every variable, since there are inevitably random factors that influence the behavior and, therefore, produce effects that we are not capable of predicting with 100 percent confidence.

The art of mathematical modeling, then, is the art of being able to extract those variables that have significant influence on the solution to a problem and to ignore the others. If the rock in our example is dropped from a 20-foot cliff and we are only interested in knowing its velocity ± 1 foot/second at any instant of time, then the original model is wholly adequate.

In modeling a control system, we attempt to write equations relating ouput to input. Factors that affect the input signal as it is processed through the system to the output may include such variables as inertia, amplifier gain, frequency response, gear ratios, inductive and capacitive reactance, friction, and others. We then decide to what extent each of these fractors must be considered in developing the model. Since the magnitude of many of these quantities depends on signal rates of change, we must also take this factor into account. Thus, we are confronted with a problem whose solution (a function of time) we must have before we can estimate the significance of the variables we have included or excluded! In short, it may be necessary to solve numerous differential equations before we have a feel for the effect that different combinations of the variables have on the system.

Fortunately, there is an alternative technique for acquiring this sense of the appropriateness of a variable. As we will see, we can transform our equations by using the LaPlace transform into another domain where it is much easier to visualize the contribution of any given variable to the behavior of the system. This domain is, in fact, the *frequency domain*, and with some practice, we will be able to predict the degree to which any component in the system contributes to the overall frequency response of the system. As we will see, the behavior of a control system with respect to the frequency at which it is driven is really all we need to know in order to evaluate it, analyze it, and predict its response to any stimulus.

CHAPTER

2

REVIEW OF COMPLEX ALGEBRA AND PHASORS

2.1 COMPLEX NUMBERS

Complex numbers find wide application in the study of control-systems theory, electronic circuits, network analysis, and many other branches of science and technology. It is imperative for the student who wishes to understand, analyze, or synthesize electrical networks and control systems to be well versed in complex number theory, for it forms the basis of many advanced mathematical techniques used in these fields, including phasors and LaPlace transforms.

The choice of the word complex to describe these numbers is perhaps unfortunate. To the student with no previous background in this subject, "complex" probably implies complicated, which is not at all the case. In fact, any student who understands and is able to apply simple vector analysis should be able to comprehend complex number theory with no difficulty whatsoever; the concepts are quite similar. A complex number is not a complicated, abstract idea, but rather an alternative way of describing a vector in a two-dimensional space.

What makes a complex number distinctive is the concept of the imaginary number $\sqrt{-1}$. Again, imaginary may be a poor choice of terminology, since the quantity $\sqrt{-1}$ can be treated in many ordinary, down-to-earth ways that other numbers are. For example $(\sqrt{-1})^2 = -1$, just as $(\sqrt{5})^2 = 5$. The use of such abstract terms as complex and imaginary has probably caused much undue apprehension for students beginning their study of this subject.

The imaginary number $\sqrt{-1}$ is frequently denoted by j. Other texts, particularly

8 REVIEW OF COMPLEX ALGEBRA AND PHASORS

mathematics texts, use $i = \sqrt{-1}$, but electrical/electronics theory has adopted $j = \sqrt{-1}$ because of the possible confusion with the symbol i used for current. It should be noted that the square root of *any* negative number can be expressed as the product of a real number with j. Such a number is also said to be imaginary.

EXAMPLE 2.1

(a) $\sqrt{-4} = \sqrt{(+4)(-1)} = \sqrt{+4}\sqrt{-1} = \pm 2j$

(b) If $x > 1$, then $\sqrt{1-x^2}$ is the square root of a negative number, and $\sqrt{1-x^2} = \sqrt{-1(x^2-1)} = \pm j\sqrt{x^2-1}$, since x^2-1 is positive.

Note that the notation jx and xj mean the same thing and are both commonly used; hereafter, we will generally use jx.

A complex number consists of a *real* part and an *imaginary* part. The complex number is represented in the general form $a + jb$, where a is the real part and b is the imaginary part. A real number is the familiar number that we have all used and grown up with and can be plotted on the real-number line, between $-\infty$ and $+\infty$. Note that in the form $a + jb$, jb is an imaginary number, while b itself (the imaginary *part*) is a real number.

EXAMPLE 2.2

Find the real and imaginary parts of the following complex numbers:

(a) $2 + j4$
(b) $-5 + j7$
(c) 4
(d) $-j6$

SOLUTION

(a) Real part = 2, imaginary part = 4.
(b) Real part = -5, imaginary part = 7.
(c) Real part = 4, imaginary part = 0 (Note that all real numbers have imaginary part zero: $a = a + j0$).
(d) Real part = 0, imaginary part = -6. (Note that all imaginary numbers have real part zero: $jb = 0 + jb$).

It should be noted that the plus (+) symbol joining the real and imaginary parts of a complex number does *not* mean addition in the ordinary sense that we use in adding real numbers. As will be seen in the next paragraph, we cannot add a real and imaginary number except in a vector sense. The plus symbol is merely a convenient way of separating the real and imaginary parts of a complex number. We could have used a comma or a semicolon or any other symbol to set off the real from the imaginary part.

2.2 THE COMPLEX PLANE

Since the real and imaginary parts of a complex number are quite distinct, it is convenient to imagine them as points in two different dimensions. That is, we can visualize the real part a as a point on the usual real-number line, while the imaginary part b can be regarded as a point on a vertical axis. This is exactly the same idea we use when plotting a two-dimensional point (x, y) in the plane, where x is the abscissa and y is the ordinate. In the case of a complex number $a + jb$, a is the abscissa and b is the ordinate. The plane formed by all possible complex numbers $a + jb$ is called the complex plane and is quite analogous to the xy-plane. We merely use j to designate the ordinate when plotting $a + jb$ in this particular plane. Any complex number can be located (plotted) in the complex plane simply by specifying its real and imaginary parts. The axis on which the real part is located (the x-axis) is called the *real axis*, while the imaginary part is located on the *imaginary axis* (the y-axis).

EXAMPLE 2.3

Plot the following complex numbers in the complex plane:
(a) $2 + j4$
(b) $-1 - j2$
(c) 3
(d) j

SOLUTION

See Figure 2.1.

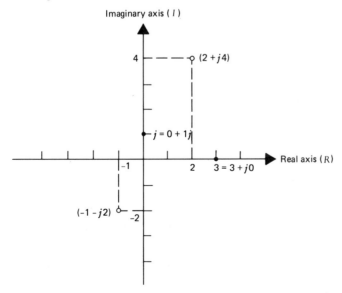

FIGURE 2.1 Plotting numbers in the complex plane (Example 2.3).

The complex number $a + jb$ is said to be in *rectangular form*, since a and b are rectangular coordinates just as x and y are the rectangular coordinates of the number (x, y) in the xy-plane. Given the coordinates a and b of any complex number, there is another way we can specify exactly (uniquely) the location of the number of the complex plane. If we imagine a line drawn from the origin $(0 + j0)$ to the number $a + jb$, we can locate the number by stating the length of this line and the angle the line makes with respect to the positive real axis. Let M represent the length of the line and θ the angle it makes measured (counterclockwise) from the real axis. Then $M \angle \theta$ represents the *polar form* of the complex number.

EXAMPLE 2.4

Find the polar form of the complex number $3 + j4$. (See Figure 2.2.)

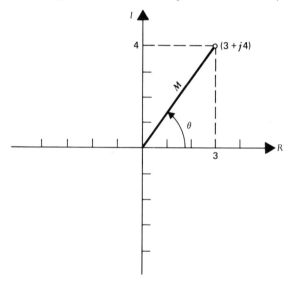

FIGURE 2.2 Polar form of $3 + j4$ (Example 2.4).

SOLUTION

By the Pythagorean theorem, $M = \sqrt{3^2 + 4^2} = 5$. Also, $\theta = \arctan(4/3) = \angle 53.1°$; thus, $3 + j4 = 5 \angle 53.1°$.

Example 2.4 illustrates the general method for converting a complex number from rectangular to polar form. The complex number $a + jb$ has polar form.

$$M \angle \theta, \qquad (1)$$

where

$$M = \sqrt{a^2 + b^2}$$

$$\theta = \angle(a + jb) = \arctan b/a \qquad (2)$$

The quantity M is called the *magnitude*, or *absolute value*, of the complex number. Note that it is always a positive real number. The absolute value sign is frequently used to denote the magnitude of a complex number, for example, $|9 + j12| = \sqrt{9^2 + 12^2} = 15 = M$.

Care must be exercised when finding the angle θ, particularly with regard to the signs of the real and imaginary parts. For example, the complex number $-3 - j4$ has angle tan $-4/-3$ and though $-4/-3$ is algebraically equal to 4/3, the angle of $-3 - j4$ is *not* the same as the angle of $3 + j4$. In fact, arc tan $-4/-3$ is 233.1°, while arc tan 4/3 is 53.1°. This distinction is especially important when using a calculator to find the angle of a complex number, since the preliminary calculation $-4/-3$ will yield $+1.333$ and the subsequent calculation arc tan (1.333) will yield the incorrect result of 53.1°. When converting from rectangular to polar form, a sketch should *always* be drawn in the complex plane to verify that the calculated angle is in the correct quadrant.

EXAMPLE 2.5

Convert $2 - j2$ to polar form. (See Figure 2.3.)

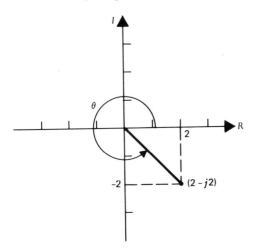

FIGURE 2.3 Polar form of $2 - j2$ (Example 2.5).

SOLUTION

$M = \sqrt{2^2 + 2^2} = \sqrt{8} = 2.828$
$\theta = $ arc tan $-2/2 = 315°$
$M \angle \theta = 2.828 \angle 315°$

Note that θ is *not* the same as arc tan $2/-2$ (even though $-2/2 = 2/-2$), since arc tan $2/-2 = 225°$, which is in the third quadrant.

An angle that is measured counterclockwise from the positive real axis is considered *positive*, while the same angle measured in a clockwise direction from the positive real axis is considered negative. Thus, any complex number can be equally

12 REVIEW OF COMPLEX ALGEBRA AND PHASORS

specified by either a positive or negative angle. In Example 2.4, $5 \angle 53.1° = 5 \angle -306.9$ $(53.1 - 360 = -306.9)$, while in Example 2.5, $2.828 \angle 315° = 2.828 \angle -45°$.

To convert from polar to rectangular form, we may use the basic trigonometric relations between hypotenuse, adjacent, and opposite sides of a right triangle. (See Figure 2.4.)

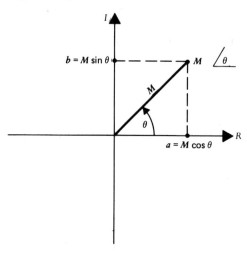

FIGURE 2.4 Trigonometric relations between polar and rectangular forms.

Thus, $M \angle \theta = M \cos \theta + jM \sin \theta$. (3)

EXAMPLE 2.6

Convert $10 \angle 225°$ to rectangular form.

SOLUTION

$$10 \angle 225° = 10 \cos 225° + j10 \sin 225°$$
$$= -7.07 - j7.07$$

2.3 THE ARITHMETIC OF COMPLEX NUMBERS

To add (or subtract) two complex numbers, it is merely necessary to add (or subtract) the real and imaginary parts separately. Thus,

$$(a + jb) + (c + jd) = (a + c) + j(b + d) \qquad (4)$$

and

$$(a + jb) - (c + jd) = (a - c) + j(b - d) \qquad (5)$$

2.3 THE ARITHMETIC OF COMPLEX NUMBERS

EXAMPLE 2.7

Perform the following operations:
 (a) $(3 + j2) + (-1 + j4)$
 (b) $(-2 + j1) - (1 - j2)$

SOLUTION

 (a) $(3 + j2) + (-1 + j4) = (3 - 1) + j(2 + 4) = 2 + j6$
 (b) $(-2 + j1) - (1 - j2) = (-2 - 1) + j(1 + 2) = -3 + j3$

Multiplication of complex numbers can be performed using the same technique used to multiply two algebraic binomials. Recall that

$$(a + b)(c + d) = ac + ad + bc + bd$$

Similarly,

$$(a + jb)(c + jd) = ac + jad + jbc + j^2bd$$
$$= (ac - bd) + j(ad + bc) \qquad (6)$$

(Recall from section 2.1 that $j^2 = (\sqrt{-1})^2 = -1$, so $j^2bd = -bd$.)

EXAMPLE 2.8

Perform the following operations:
 (a) $(1 + j2)(3 + j4)$
 (b) $(2 - j1)(-2 + j5)$

SOLUTION

 (a) $(1 + j2)(3 + j4) = 3 + j4 + j6 + j^28$
 $= (3 - 8) + j(4 + 6)$
 $= -5 + j10$
 (b) $(2 - j1)(-2 + j5) = -4 + j10 + j2 - j^25$
 $= (-4 + 5) + j(10 + 2)$
 $= 1 + j12$
 [Note that $-j^2 = -(-1) = +1$.]

Complex numbers can also be multiplied in polar form. This operation is much simpler than multiplication in rectangular form, provided the numbers are available in polar form to begin with. The rule is as follows:

$$(M_1 \angle \theta_1)(M_2 \angle \theta_2) = M_1M_2 \angle \theta_1 + \theta_2 \qquad (7)$$

As can be seen in Equation 7, the magnitudes are multiplied and the angles are added in order to form the product. Division of complex numbers in polar form is performed as follows:

14 REVIEW OF COMPLEX ALGEBRA AND PHASORS

$$\frac{M_1 \angle \theta_1}{M_2 \angle \theta_2} = \left(\frac{M_1}{M_2}\right) \angle \theta_1 - \theta_2 \qquad (8)$$

The magnitudes are divided and the denominator angle is subtracted from the numerator angle. If the complex numbers to be multiplied or divided are in rectangular form, it is, of course, necessary to convert them first to polar form if it is desired to use the rules just given.

EXAMPLE 2.9

Perform the following operations in polar form:
(a) $(4 \angle 30°)(0.5 \angle 65°)$
(b) $(2 \angle -30)(1 \angle 70°)$
(c) $\dfrac{5 \angle 120°}{20 \angle -50°}$
(d) $(2 + 2j)(3 + j4)$

SOLUTION

(a) $(4 \angle 30)(0.5 \angle 65°) = (4)(0.5) \angle 30° + 65° = 2 \angle 95°$
(b) $(2 \angle -30)(1 \angle 70) = 2 \times 1 \angle -30° + 70° = 2 \angle 40°$
(c) $\dfrac{5 \angle 120°}{20 \angle -50°} = \dfrac{5}{20} \angle 120° - (-50°) = 0.25 \angle 170°$
(d) $2 + 2j = 2.828 \angle 45°$
  $3 + j4 = 5 \angle 53.10°$
  $(2 + 2j)(3 + j4) = (2.828 \angle 45°)(5 \angle 53.1°)$
  $= 14.14 \angle 98.1°$

2.4 COMPLEX CONJUGATES

The complex conjugate of the number $a + jb$ is defined to be $a - jb$. Thus, to form the complex conjugate of any complex number, it is only necessary to change the sign between the real and imaginary parts. In polar form, the complex conjugate is formed by changing the sign of the angle.

EXAMPLE 2.10

(a) The complex conjugate of the number $-4 + j2$ is $-4 - j2$.
(b) The complex conjugate of $0.5 - j7$ is $0.5 + j7$.
(c) The complex conjugate of $13 \angle 57°$ is $13 \angle -57°$ and the complex conjugate of $3 \angle -120°$ is $3 \angle 120°$.

In the complex plane, the complex conjugate is simply the mirror image of the original number, as reflected on the other side of the real axis. This fact is illustrated in Example 2.11.

EXAMPLE 2.11

Plot the complex conjugates of $-3 - j4$ and $5\ \angle 60°$.

SOLUTION

See Figure 2.5.

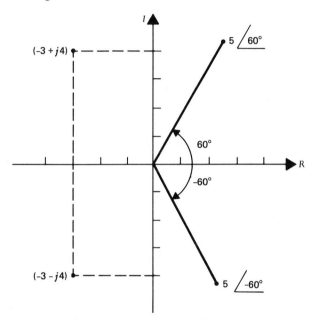

FIGURE 2.5 Plotting complex conjugates (Example 2.11).

When a number is multiplied by its complex conjugate, the result is a purely *real*, positive number. This can be shown as follows:

$$(a + jb)(a - jb) = a^2 - jab + jab - j^2b^2$$
$$= a^2 + b^2 \tag{9}$$
$$(M\ \angle\theta)(M\ \angle -\theta) = M^2\ \angle 0 \tag{10}$$

In Equation 9, the result is real, since $a^2 + b^2 = (a^2 + b^2) + j0$, and in Equation 10, $M^2\ \angle 0$ is real, since any number with angle zero must lie on the positive real axis. To form the product of a complex number and its conjugate, it is only necessary to find the sum of squares of the real and imaginary parts, *regardless* of the sign joining them. Note that the product of a complex number and its conjugate is the square of the magnitude of the complex number

16 REVIEW OF COMPLEX ALGEBRA AND PHASORS

$$(a + jb)(a - jb) = a^2 + b^2$$

while $M = \sqrt{a^2 + b^2}$ so $(a + jb)(a - jb) = M^2$

EXAMPLE 2.12

(a) Find the product of $2 - j5$ and its complex conjugate.
(b) Find the magnitude of $6 - j8$ by using complex conjugates.

SOLUTION

(a) $(2 - j5)(2 + j5) = 2^2 + 5^2 = 29$
(b) $M^2 = (6 - j8)(6 + j8) = 36 + 64 = 100$
   $M = \sqrt{100} = 10$

2.5 RATIONALIZING

In complex number computations, the form $(u + jv)/(x + jy)$ frequently arises. For many applications, it is necessary to express this in the *rationalized* form $a + jb$ so that the real and imaginary parts can be readily identified. One approach to this problem would be to convert numerator $(u + jv)$ and denominator $(x + jy)$ to polar forms, divide using the rule for division of polar forms, and then convert the resulting polar number to rectangular form.

Another, perhaps easier, method involves multiplying numerator and denominator of the fraction by the complex conjugate of the denominator. This operation is called *rationalizing*, and it is perfectly legitimate, since it is equivalent to multiplying the fraction by one. The advantage of this approach stems from the fact that multiplying the denominator by its complex conjugate yields a real number. The following example will illustrate these ideas.

EXAMPLE 2.13

Find the real and imaginary parts of $(3 + j4)/(-1 - j1)$ by (a) converting to polar forms and (b) rationalizing.

SOLUTION

(a) $3 + j4 = 5 \angle 53.1°$
   $-1 - j1 = 1.414 \angle -135°$

$$\frac{3 + j4}{-1 - j1} = \frac{5 \angle 53.1°}{1.414 \angle -135°} = 3.535 \angle 188.1°$$

$a + jb = 3.535(\cos 188.1) + j3.535(\sin 188.1)$
$= -3.5 - j.5$

(b) $\dfrac{3+j4}{-1-j1} \dfrac{-1+j1}{-1+j1} = \dfrac{-3+j3-j4-4}{1^2+1^2}$

$= \dfrac{-7-j1}{2} = \dfrac{-7}{2} - j\dfrac{1}{2}$

$= -3.5 - j.5$

Thus, the real part of $(3+j4)/(-1-j1)$ is -3.5, and the imaginary part is -0.5.

Factoring a constant from both the real and imaginary parts of a complex number is permissible. For example, $40 + j80 = 40(1 + j2)$.

EXAMPLE 2.14

Rationalize $(2000 + j2000)/(50 - j50)$.

SOLUTION

$\dfrac{2000 + j2000}{50 - j50} = \dfrac{2000(1+j1)}{50(1-j1)}$

$= 40\dfrac{(1+j1)(1+j1)}{(1-j1)(1+j1)} = \dfrac{40(0+j2)}{1+1}$

$= 20(0+j2) = j40 = 40\angle 90°$

2.6 PROPERTIES OF j

There are certain properties of j that prove very useful in complex-number computations. We have already seen that $j^2 = -1$. Furthermore,

$$j^3 = (j^2)(j) = -j$$
$$j^4 = (j^2)(j^2) = (-1)(-1) = +1$$
$$j^5 = (j^4)(j) = (+1)j = j$$

and so forth.

Note that the polar form of j is $1\angle 90°$ (the magnitude of $j1$ is 1, and since $j = 0 + 1j$ is located directly on the positive imaginary axis, its angle must be $+90°$). Similarly, $-j = 1\angle -90°$, $1 = 1\angle 0°$, and $-1 = 1\angle 180° = 1\angle -180°$. (See Figure 2.6.)

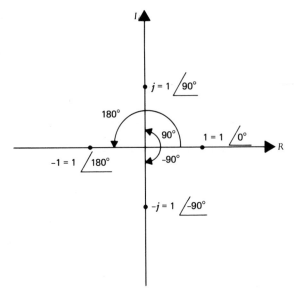

FIGURE 2.6 Plots of numbers with unity magnitude.

These ideas can, of course, be extended to any multiple of j. Thus, $-4j = 4 \angle -90°$ and $0.5j = 0.5 \angle 90°$. Also, $-25 = 25 \angle -180°$.

Another useful property of j is that it can be shifted from the numerator to the denominator of a fraction simply by changing its sign. This can be verified as follows:

$$\frac{1}{j} = \frac{1}{j}\left(\frac{j}{j}\right) = \frac{j}{j^2} = \frac{j}{-1} = -j \tag{11}$$

and

$$j = j\left(\frac{j}{j}\right) = \frac{j^2}{j} = \frac{-1}{j} = \frac{1}{-j} \tag{12}$$

EXAMPLE 2.15

Convert $(4 + j3)/5j$ to polar form.

SOLUTION

$$\frac{4 + j3}{5j} = -j\left(\frac{4}{5} + j\frac{3}{5}\right)$$

$$= -j\,(1 \angle 36.87°)$$

$$= (1 \angle -90°)(1 \angle 36.87°)$$

$$= 1 \angle -53.1°$$

2.7 THE EXPONENTIAL FORM

There is still a third way to represent a complex number, the so-called *exponential form*. Given a complex number $M \angle \theta$, the exponential form is defined as follows:

$$M \angle \theta = Me^{j\theta} \tag{13}$$

where e is the natural logarithm base and θ is usually expressed in radians.

EXAMPLE 2.16

Express (a) $10 \angle 180°$ and (b) $1.73 - j1$ in exponential form and plot them in the complex plane.

SOLUTION

(a) $10 \angle 180° = 10 \angle \pi = 10e^{j\pi}$
(b) $1.73 - j1 = 2 \angle -30° = 2e^{-j\pi/6}$

See Figure 2.7.

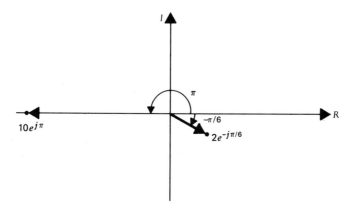

FIGURE 2.7 Plots of complex numbers in exponential form (Example 2.16).

Note that any positive real number a when expressed in exponential form becomes $ae^{j0} = a$, while any negative real number $-a$ becomes $ae^{j\pi}$.

When complex numbers are in exponential form, we can perform arithmetic on them by following the usual algebraic rules of exponents. Thus, for example,

$$(M_1 e^{j\theta_1})(M_2 e^{j\theta_2}) = M_1 M_2 e^{j(\theta_1 + \theta_2)}$$

$$\frac{M_1 e^{j\theta_1}}{M_2 e^{j\theta_2}} = \frac{M_1}{M_2} e^{j(\theta_1 - \theta_2)}$$

and

$$(Me^{j\theta})^m = M^m e^{jm\theta}$$

20 REVIEW OF COMPLEX ALGEBRA AND PHASORS

There are two very useful identities involving complex numbers in exponential form; they are:

$$\sin \omega t = \frac{e^{j\omega t} - e^{-j\omega t}}{2j} \tag{14}$$

$$\cos \omega t = \frac{e^{j\omega t} + e^{-j\omega t}}{2} \tag{15}$$

EXAMPLE 2.17

Express $10 \sin(5t + \pi/3)$ as a function of exponential forms.

SOLUTION

Since the argument of the sine function (the term in parentheses in this example) becomes the angle in the exponent terms of Equation 14 and since 10 is simply a multiplying factor, we have:

$$10 \sin(5t + \pi/3) = 10 \left[\frac{e^{j(5t + \pi/3)} - e^{-j(5t + \pi/3)}}{2j} \right] = \frac{5}{j} \left[e^{j(5t + \pi/3)} - e^{-j(5t + \pi/3)} \right]$$

2.8 PHASORS

Complex numbers can be effectively used to represent electrical voltages, currents, and impedances. When the complex number is used in this fashion, it is said to be a *phasor*. The phasor form of an electrical quantity usually corresponds to the polar form of the associated complex number, though in applications, it is often converted to rectangular or exponential form to simplify computations.

A sinusoidal ac voltage or current is converted to phasor form by writing a complex number that in polar form has magnitude equal to the peak value and has angle equal to the phase of the voltage or current. (Note that some authors define the phasor in terms of the RMS rather than peak value of the ac wave. Since for sine waves the RMS and peak values are a fixed multiple of each other, it makes no difference which value is used, so long as we are consistent in all computations.) A bold-faced symbol will be used to designate a phasor form.

EXAMPLE 2.18

Express the following in phasor form:
 (a) $e = 120 \sin(377t + 60°)$ V
 (b) $i = 25 \sin(10^6 t - \pi/6)$ milliamps (mA)

SOLUTION

  (a) **e** = 120 ∠60° V

(b) $i = 25 \angle -\pi/6$ mA
$= 25 \times 10^{-3} \angle -30°$ amps (A)

Note that the frequency of the sinusoidal voltage or current does not appear in its phasor representation.

It should be emphasized that the angle used in the phasor form is the phase angle of the voltage or current when expressed in *sine form*. It may be necessary to convert a given voltage or current to an equivalent sine-wave representation before writing the phasor form, in order to ensure that the correct angle is used.

EXAMPLE 2.19

Express the following in phasor form:
(a) $e = 40 \cos(\omega t - 20°)$ V
(b) $i = -16 \sin(50t + 40°)$ A

SOLUTION

(a) Since $\cos x = \sin(x + 90°)$, we have:
$$e = 40 \cos(\omega t - 20°) = 40 \sin(\omega t - 20° + 90°)$$
$$= 40 \sin(\omega t + 70°)$$
$$\mathbf{e} = 40 \angle 70° \text{ V}$$

(b) Since $-\sin x = \sin(x \pm 180°)$, we have:
$$i = -16 \sin(50t + 40°) = 16 \sin(50t + 40° \pm 180°)$$
$$= 16 \sin(50t + 220°) = 16 \sin(50t - 140°)$$
$$\mathbf{i} = 16 \angle 220° = 16 \angle -140° \text{ A}$$

It is desirable to express the angle of a phasor as an angle between 0 and 180° or between 0 and $-180°$, so the form $16 \angle -140°$ is preferred.

Having noted the correspondence between a phasor and the polar form of a complex number, we can legitimately manipulate phasors in precisely the same way we do complex numbers. That is, we can locate the phasor in the complex plane, find its real and imaginary parts, rationalize, and do arithmetic computations just as if it were any other complex number. The only distinction to be aware of is that certain mathematical operations on two or more phasors (addition and subtraction) require that all phasors involved in the process represent voltages or currents of the *same* frequency. More will be said on this point subsequently.

It is important to recognize the distinction between a phasor and Equation 14. A phasor is a *representation* of a sine wave and, strictly speaking, is not *equal* to the sine wave; after all, the phasor says nothing about the frequency of the sine wave. Equation 14, on the other hand, is a strict equality. In spite of this distinction, we may occasionally place an equals sign between a sinusoidal function and its phasor representation, with the understanding that equality holds in the sense we have described.

One advantage of phasor representation is that it provides a simple visual concept of lead-lag relations among ac voltages and currents when the phasors are plotted in the complex plane. Recall that a voltage or current is said to lead another when it has greater positive phase angle. In the complex plane, this corresponds to a greater counterclockwise rotation of the phasor that is leading.

EXAMPLE 2.20

Plot the following as phasors in the complex plane and find the rectangular form of each:
(a) $e_1 = 100 \sin(150t - 30°)$ V
(b) $e_2 = 50 \sin(150t + 60°)$ V

SOLUTION

See Figure 2.8.

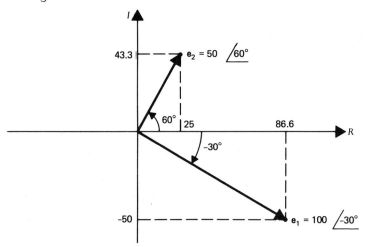

FIGURE 2.8 Rectangular forms of phasors (Example 2.20).

$$e_1 = 100 \angle -30° = 100 \cos(-30°) + j100 \sin(-30°)$$
$$= 86.6 - j50 \text{ V}$$
$$e_2 = 50 \angle 60° = 50 \cos(60°) + j50 \sin 60°$$
$$= 25 + j43.3 \text{ V}$$

Note that in Example 2.20 e_2 leads e_1 by 90°.

By this time, the reader may have begun to notice the similarity between phasors (or complex numbers, for that matter) and the familiar concept of *vectors*; indeed, the two are synonymous in their application. Just as two vector forces acting in different directions may be added (vectorially) to obtain a resultant force, so may two phasors representing voltages with different phase angles be added to obtain a resultant voltage. The vector technique of finding the horizontal and vertical components of each force vector corresponds exactly to finding the rectangular form of each phasor. The only stipulation is that each phasor must represent voltages or currents of the same frequency. This is the case because only then is the sum of two or more sinusoidal voltages or currents equal to another sinusoidal voltage or current of the same frequency. This is an extremely important result that is often overlooked by beginning students. We cannot, for example, add $e_1 = 10 \sin 50t$ and $e_2 = 10 \sin 100t$ and expect to obtain a sinusoidal result. On the other hand, *any* two sinusoidal waveforms of the same frequency, regardless of their individual peak values or phase angles, will yield a third sine wave with that same frequency when added. The only problem at hand is to determine the peak value and phase angle of the resultant. The frequency is known to be the same.

EXAMPLE 2.21

Find the sum of $e_1 = 20 \sin(\omega t + 27°)$ and $e_2 = 30 \sin(\omega t - 50°)$. Express the sum in sinusoidal form.

SOLUTION

We first express e_1 and e_2 as phasors and convert each to rectangular form

$$e_1 = 20 \angle 27 = 17.82 + j9.08$$

$$e_2 = 30 \angle -50 = 19.28 - j22.98$$

Then

$$e_1 = 17.82 + j9.08$$

$$e_2 = 19.28 - j22.98$$

$$e_1 + e_2 = 37.10 - j13.9$$

Converting $e_1 + e_2$ to polar form

$$M = |e_1 + e_2| = \sqrt{(37.10)^2 + (13.9)^2} = 39.62$$

$$\theta = \arctan\left(\frac{-13.9}{37.1}\right) = \arctan(-0.375) = -20.54°$$

So

$$(e_1 + e_2) = 39.62 \angle -20.54°$$

and, therefore,

$$e_1 + e_2 = 39.62 \sin(\omega t - 20.54°) \text{ V}$$

See Figure 2.9.

24 REVIEW OF COMPLEX ALGEBRA AND PHASORS

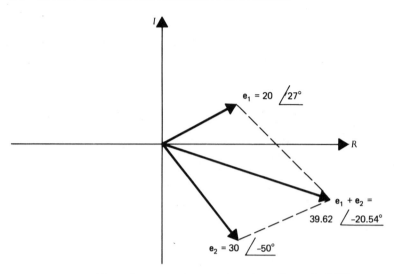

FIGURE 2.9 Adding phasors in the complex plane (Example 2.21).

When Example 2.21 is sketched in the complex plane as shown in Figure 2.9, the similarity between phasor addition and the process of finding a resultant vector is quite evident. In this process, we found the horizontal and vertical components of each phasor (by converting each to rectangular form) and then added the horizontal components (real parts) together as well as the vertical components (imaginary parts). We then converted the resultant back to polar form and finally to sinusoidal form. A similar process may be used when subtracting two voltages or currents. For example, if $\mathbf{i}_1 = 12\angle 20°$ and $\mathbf{i}_2 = 5\angle 95°$ and we wish to find $(\mathbf{i}_1 - \mathbf{i}_2)$, we write

$$\mathbf{i}_1 - \mathbf{i}_2 = 12\angle 20° - 5\angle 95°$$
$$= 12\angle 20° + 5\angle -85°$$

(recall that a minus sign is equivalent to a $\pm 180°$ phase shift) and proceed in the same way as before.

2.9 IMPEDANCE AND THE PHASOR FORM

Just as an ac voltage or current may be represented by the polar form of a complex number, so may an impedance. Recall that an impedance Z may, in general, consist of resistance R, inductive reactance X_L and/or capacitive reactance X_C.

2.9 IMPEDANCE AND THE PHASOR FORM

Resistance as a complex quantity is real and positive; that is, the phasor form of a pure resistance of, say, 60 ohms (Ω) is $\mathbf{R} = 60 + j0 = 60\angle 0°$ and is located on the positive real axis.

Inductive reactance is defined by $jX_L = j\omega L$, where ω is the angular frequency in rad/sec and L is the inductance in henries (H), and it is located on the positive imaginary axis. For example, a 20-Ω inductive reactance expressed as a complex quantity is $jX_L = 0 + j20 = 20\angle 90°$.

Capacitive reactance on the other hand is defined by

$$-jX_c = \frac{1}{j\omega C} = \frac{-j}{\omega C}$$

where C is in farads and can be seen to lie on the negative imaginary axis. Fifty Ω of capacitive reactance would be written as $-jX_c = 0 - j50 = 50\angle -90°$. For a series combination of resistance, inductive reactance, and capacitive reactance, we find that $\mathbf{Z} = R + j(X_L - X_c)$. (Note that phasor impedances in *series* add just as series resistors add). Since Z is a complex quantity, we can find its magnitude and angle:

$$|Z| = \sqrt{R^2 + (X_L - X_c)^2}$$

$$\angle Z = \arctan\frac{(X_L - X_c)}{R}$$

EXAMPLE 2.22

Find the magnitude and angle of the total impedance of a series circuit consisting of 20 Ω of resistance, 40 Ω of inductive reactance, and 20 Ω of capacitive reactance. Sketch the impedance and its components in the complex plane.

SOLUTION

$$\mathbf{Z} = R + j(X_L - X_c) = 20 + j(40 - 20)$$

$$= 20 + j20 \; \Omega$$

$$|Z| = \sqrt{(20)^2 + (20)^2} = 28.25$$

$$\angle Z = \arctan\frac{20}{20} = 45°$$

Thus,

$$\mathbf{Z} = 28.25\angle 45° \; \Omega$$

See Figure 2.10.

26 REVIEW OF COMPLEX ALGEBRA AND PHASORS

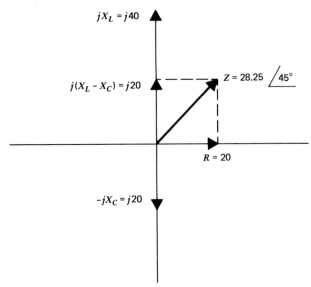

FIGURE 2.10 Phasor form of an impedance (Example 2.22).

When impedance is expressed in phasor form, we may use the phasor form of voltage and current to apply Ohm's law for ac circuits

$$\mathbf{E} = \mathbf{IZ}$$

All of the usual extensions of this law are valid in phasor form. Thus, $\mathbf{I} = \dfrac{\mathbf{E}}{\mathbf{Z}}$ and $\mathbf{Z} = \dfrac{\mathbf{E}}{\mathbf{I}}$ may both be calculated in phasor form.

EXAMPLE 2.23

The voltage across $\mathbf{Z} = 60\angle 30°$ is $e = 30 \sin(1000t + 58°)$. Find the current through $\mathbf{Z}$.

SOLUTION

$$\mathbf{I} = \frac{\mathbf{E}}{\mathbf{Z}} = \frac{30\angle 58°}{60\angle 30°} = 0.5\angle 28°$$

$$i = 0.5 \sin(1000t + 28°) \text{ A}$$

EXAMPLE 2.24

The voltage across an unknown impedance is $0.34\angle 50°$, while the current through it is $2\angle -40°$. Find the impedance in phasor form.

2.9 IMPEDANCE AND THE PHASOR FORM

SOLUTION

$$Z = \frac{E}{I} = \frac{0.34 \angle 50°}{2 \angle -40°} = 0.17 \angle 90° \; \Omega$$

The impedance in this example is seen to be purely inductive reactance.

Impedances in parallel or series parallel may be combined to find a total equivalent impedance using exactly the same techniques that are used to combine resistors in a circuit. The only difference is that all multiplication, addition, and so forth, performed must conform to the rules of phasor arithmetic. Remember that each impedance element has magnitude and *angle* and so must be treated as a phasor (vector) when combining it with other impedances. In a purely resistive circuit, this was never a problem, since all resistors have angle zero. Furthermore, all of the rules that apply to resistive networks are applicable to impedance, recalling again that arithmetic operations must account for magnitude and angle. For example, the equivalent resistance of two resistors in parallel is known to be $R = R_1 R_2/(R_1 + R_2)$; similarly, $Z = Z_1 Z_2/(Z_1 + Z_2)$. In the latter relation, it may be necessary to convert Z_1 and Z_2 to polar form in order to perform the multiplication in the numerator and to rectangular form in order to perform the addition in the denominator. As another example, consider the *voltage-divider rule* illustrated in Figure 2.11.

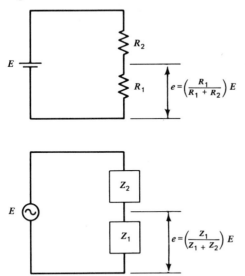

FIGURE 2.11 The voltage-divider rule in resistive and reactive networks.

EXAMPLE 2.25

Find the total equivalent impedance of the network in Figure 2.12. Express the result in polar and rectangular forms.

28 REVIEW OF COMPLEX ALGEBRA AND PHASORS

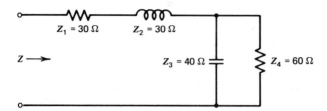

FIGURE 2.12 Find the total phasor impedance (Example 2.25).

SOLUTION

$$\frac{Z_3 Z_4}{Z_3 + Z_4} = \frac{(40\angle -90)(60\angle 0)}{(0 - j40) + (60 + j0)}$$

$$= \frac{2400\angle -90}{60 - j40} = \frac{2400\angle -90}{72.11\angle -33.7°}$$

$$= 33.28\angle -56.3° \; \Omega$$

$$= 18.46 - j27.69 \; \Omega$$

$$Z_1 + Z_2 = 30 + j30$$

$$Z = (30 + j30) + (18.46 - j27.69) = 48.46 + j2.31 \; \Omega$$

Recall that *susceptance B* is defined to be the reciprocal of reactance. In phasor form, inductive susceptance B_L has angle *minus* 90°, since

$$\frac{1}{jX_L} = \frac{-j}{X_L} = \frac{1}{X_L}\angle -90° = \frac{1}{\omega L}\angle -90°$$

$$= 0 - j\left(\frac{1}{\omega L}\right) \text{ Siemens (S)}$$

(16)

Also capacitive susceptance B_C is found from

$$\frac{1}{-jX_c} = \frac{j}{X_c} = \frac{1}{X_c}\angle 90° = \omega C \angle 90°$$

$$= 0 + j\omega C \; \text{S}$$

(17)

and has angle *plus* 90°.

The reciprocal of resistance is *conductance G* and like resistance, has angle zero

$$\frac{1}{R\angle 0°} = \left(\frac{1}{R}\right)\angle 0° = \frac{1}{R} + j0 \; \text{S}$$

Admittance Y is the general term applied to a combination of susceptance and conductance, just as impedance is the general term applied to combinations of

2.10 REVIEW OF SOME IMPORTANT FACTS FROM CALCULUS

resistance and reactance. For a *parallel* combination of conductance and susceptance, we have:

$$Y = G + j(B_c - B_L) \text{ S} \tag{18}$$

EXAMPLE 2.26

Find the total equivalent admittance and the total equivalent impedance of the network in Figure 2.13. Verify that $Y = 1/Z$.

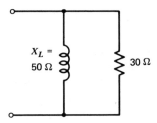

FIGURE 2.13 Find the total phasor impedance and admittance (Example 2.26).

SOLUTION

(a) $B_L = \dfrac{1}{50} \angle -90 = .02 \angle -90 = 0 - j.02$

$G = \dfrac{1}{30} \angle 0° = .0333 \angle 0° = .0333 + j0$

$Y = G - B_L = .0333 - j.02$ S

(b) $Z = \dfrac{(50\angle 90)(30\angle 0)}{30 + j50} = \dfrac{1500\angle 90}{58.3\angle 59°} = 25.73\angle 31° \; \Omega$

(c) $Y = \dfrac{1}{Z} = \dfrac{1}{25.73}\angle -31°$

$= .0389 \cos(-31°) + j(0.0389)\sin -31°$

$= .033 - j.02$ S

2.10 REVIEW OF SOME IMPORTANT FACTS FROM CALCULUS

As was indicated in Chapter 1, a comprehensive knowledge of differential and integral calculus is not altogether necessary for a student wishing to learn how to apply LaPlace transforms to analyze circuit and control systems. There are, however, certain important results from calculus, mainly from the calculus of *transcendental* (nonalgebraic) functions, that are used repeatedly. The student should be familiar with both the mechanics and interpretations of these results.

In the examples and problems worked by students in most calculus courses, the

30 REVIEW OF COMPLEX ALGEBRA AND PHASORS

dependent and independent variables are usually designated y and x, respectively. Thus, for example, most students will recall that if

$$y = x^n \quad \text{then} \quad \frac{dy}{dx} = nx^{n-1}$$

the so-called power rule. Let us emphasize that in our use of calculus hereafter, the dependent variable will almost always be time (t) instead of x. The independent variable may be designated by a number of different symbols: y, θ, or even x. We might, for example, express the power rule as follows:

$$\text{If } x = t^n \quad \text{then} \quad \frac{dx}{dt} = nt^{n-1}$$

We recognize that most students fresh from calculus have a strong mental attachment to the symbols they are familiar with and rely heavily on these symbols (x and y) to understand and apply the results of calculus. Now is the time to switch that allegiance, to begin the mental processes that lead to a confortable feeling with the symbol t in place of x and to learn to interpret results as functions of *time*.

Let us first review the differentiation and integration of sinusoidal functions. Most students remember that when u is a function of x, that is, $u(x)$ then

$$\frac{d(\sin u)}{dx} = (\cos u) \frac{du}{dx}$$

Let us apply this fundamental result to a situation that is frequently encountered in applied analysis:

$$\frac{d[\sin(\omega t + \theta)]}{dt} = \omega \cos(\omega t + \theta] \tag{19}$$

Here we regard $u(t)$ as $(\omega t + \theta)$ and use the fact that

$$\frac{du}{dt} = \frac{d(\omega t + \theta)}{dt} = \omega$$

Similarly,

$$\frac{d \cos(\omega t + \theta)}{dt} = -\omega \sin(\omega t + \theta) \tag{20}$$

Recall also that

$$\int \sin u \, du = -\cos(u) + C$$

where C is an arbitrary constant. If $u = (\omega t + \theta)$, then $du = \omega \, dt$, and we can write the integral as

$$\frac{1}{\omega} \int \sin \underbrace{(\omega t + \theta)}_{u} \underbrace{\omega \, dt}_{du}$$

Thus,

$$\int \sin(\omega t + \theta) dt = -\frac{1}{\omega} \cos(\omega t + \theta) + C \tag{21}$$

2.10 REVIEW OF SOME IMPORTANT FACTS FROM CALCULUS

Similarly,

$$\int \cos(\omega t + \theta)dt = \frac{1}{\omega}\sin(\omega t + \theta) + C \tag{22}$$

Recall that a *definite* integral has *limits of integration* and that we evaluate a definite integral by substituting these limits in a certain way, a way that eliminates the arbitrary constant of integration C from the expression. For example,

$$\int_1^2 x\,dx = \left.\frac{x^2}{2}\right]_1^2 = \frac{2^2}{2} - \frac{1^2}{2} = \frac{3}{2}$$

In many practical problems, we will eliminate the constant of integration by using 0 and t as limits on a definite integral. Thus, our results will still be functions of time t, since we substitute t for the upper limit of t in the expression. As examples,

$$\int_0^t \sin\omega t\,dt = -\frac{1}{\omega}\cos\omega t\bigg]_0^t$$

$$= -\frac{1}{\omega}\left(\cos\omega t - \cos 0\right)$$

$$= -\frac{1}{\omega}\left[\cos(\omega t) - 1\right]$$

$$= \frac{1}{\omega} - \frac{1}{\omega}\cos\omega t$$

Also,

$$\int_0^t \cos\omega t\,dt = \frac{1}{\omega}\sin\omega t\bigg]_0^t$$

$$= \frac{1}{\omega}\left(\sin\omega t - \sin 0\right)$$

$$= \frac{1}{\omega}\sin\omega t$$

Consider now the *exponential function*. Recall that

$$\frac{de^u}{dx} = e^u\frac{du}{dx}$$

and

$$\int e^u du = e^u + C$$

where u is a function of x, $u(x)$. We will be primarily concerned with cases where u is the function of time: $u(t) = at$, where a is a constant. Thus,

$$\frac{de^{at}}{dt} = ae^{at} \tag{23}$$

32 REVIEW OF COMPLEX ALGEBRA AND PHASORS

and

$$\int e^{at} dt = \frac{1}{a}\int e^{at} d(at) = \frac{1}{a} e^{at} + C \qquad (24)$$

As before, we will frequently use 0 and t as the limits of integration for the integral and so obtain

$$\int_0^t e^{at} dt = \frac{1}{a} e^{at}\Big]_0^t = \frac{1}{a}(e^{at} - e^0) = \frac{1}{a}(e^{at} - 1) \qquad (25)$$

EXAMPLE 2.27

Evaluate the following expressions:

(a) $\dfrac{d[\sin(50t + \pi/4)]}{dt}$

(b) $\dfrac{d[.02\cos(2\pi \times 10^4 t)]}{dt}$

(c) $\int_0^t 400 \sin(100t - 60°) dt$

(d) $\dfrac{d(60e^{-t/4})}{dt}$

(e) $\int_0^t 2t + 12e^{-4(t+1)} dt$

SOLUTION

(a) $\dfrac{d[\sin(50t + \pi/4)]}{dt} = 50 \cos(50t + \pi/4)$

(b) $\dfrac{d[.02\cos(2\pi \times 10^4 t)]}{dt} = -.02 \times 2\pi \times 10^4 \sin(2\pi \times 10^4 t)$

$$= -400\pi \sin(2\pi \times 10^4 t)$$

(c) $\int_0^t 400 \sin(100t - 60°) dt = \dfrac{-400}{100}\{[\cos(100t - 60°)]\}_0^t$

$$= -4 [\cos(100t - 60°) - \cos(-60°)]$$
$$= -4 [\cos(100t - 60°) - .5]$$
$$= 2 - 4\cos(100t - 60°)$$

(d) $\dfrac{d(60e^{-t/4})}{dt} = -\dfrac{1}{4}(60)e^{-t/4} = -15e^{-t/4}$

(e) $\int_0^t 2t + 12e^{-4(t+1)}dt = \left[\dfrac{2t^2}{2} + \dfrac{12}{-4}e^{-4(t+1)}\right]_0^t$

$= t^2 - 3e^{-4(t+1)} - [0 - 3e^{-4}]$

$= t^2 + 3e^{-4} - 3e^{-4(t+1)}$

REFERENCES

1. **Boylestad, Robert L.** *Introductory Circuit Analysis.* 3d ed. Charles E. Merrill, 1977.
2. **Cooke, N. M.,** and **Herbert Adams.** *Basic Mathematics for Electronics.* McGraw-Hill, 1960.
3. **Romanek, Richard J.** *Introduction to Electronic Technology.* Prentice-Hall, 1975.
4. **Thomson, Charles M.** *Mathematics for Electronics.* Prentice-Hall, 1976.
5. **Washington, Allyn J.** *Basic Technical Mathematics with Calculus.* Benjamin/Cummings, 1978.

EXERCISES

2.1 Find the real and imaginary parts of the following numbers and sketch each number in the complex plane:

(a) $4 - j2$
(b) $-1 - j.35$
(c) 5
(d) $j3$
(e) $\sqrt{-25}$
(f) $-0.5 + \sqrt{-0.09}$

2.2 Find the polar form of each of the numbers in Exercise 2.1.

2.3 Convert each of the following to rectangular form:

(a) $10\angle 30°$
(b) $0.2\angle 300°$
(c) $-4\angle 45°$
(d) $4.7\angle -90°$
(e) $150\angle 180°$
(f) $8\angle 150°$

2.4 Perform the following operations:

(a) $(3 + j4) + (2 - j2)$
(b) $(-7 - j1.5) + (6.2 - j4.1)$
(c) $(12 - j13) - (13 - j12)$
(d) $(2 + j5)(3 - j2)$
(e) $(-1.1 + j4)(6 - j1.8)$
(f) $(11\angle -20°)(4\angle 83°)$
(g) $(-5\angle 30°)(1.2\angle -50°)$
(h) $\dfrac{124\angle -75°}{6\angle 32°}$
(i) $\dfrac{32\angle 91.5°}{(6\angle 13°)(4.5\angle -25°)}$
(j) $4.4\angle 45° + 4.4\angle -45°$
(k) $\dfrac{-3 - j3}{12\angle 75° - 8\angle -60°}$

34 REVIEW OF COMPLEX ALGEBRA AND PHASORS

2.5 Find the complex conjugate of each of the following:
(a) $14 - j2$
(b) $-1 + j2$
(c) $30\angle 25°$
(d) $-5\angle 121.5°$

2.6 Rationalize each of the following expressions and identify the real and imaginary parts of each.

(a) $\dfrac{1}{3 - j3}$
(b) $\dfrac{2 + j5}{2 - j1}$
(c) $\dfrac{3 - j2.5}{5\angle 30°}$
(d) $\dfrac{12 + j3}{j}$

2.7 Using Equations 14 and 15 verify that $(\sin \omega t)^2 + (\cos \omega t)^2 = 1$.

2.8 Express each of the following in exponential form:
(a) $.01\angle 175°$
(b) $14 + j14$

2.9 Write the phasor form of each of the following:
(a) $15 \sin (40t + \pi/3)$ V
(b) $20 \sin (300t - 17°)$ microamps (μA)
(c) $18 \cos \omega t$ V
(d) $-42 \cos 10^6 t$ A

2.10 Sketch each of the phasors found in Exercise 2.9 in the complex plane, and identify the real and imaginary parts of each.

2.11 Perform each of the following operations using phasor algebra. Express each result as a phasor, and illustrate work performed by means of a sketch in the complex plane. Also express each result in sine wave form:
(a) $50 \sin (377t) + 40 \sin (377t + 100°)$
(b) $.04 \sin (\omega t - 45°) - 0.02 \sin (\omega t + 45°)$
(c) $\cos \omega t + \sin \omega t$

2.12 An ac voltage, $e = 12 \sin 10^5 t$ is applied to two series impedances. The voltage drop across one of the impedances is $8 \sin (10^5 t + 60°)$. Find the voltage drop across the other impedance. (Apply Kirchoff's voltage law).

2.13 Find the total equivalent phasor impedance in both polar and rectangular form for each of the networks in Figure 2.14.

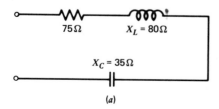

(a)

FIGURE 2.14

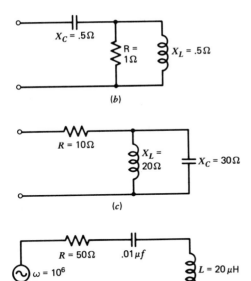

FIGURE 2.14—continued

2.14 A voltage $e = 40 \sin(10^4 t + \pi/4)$ V is measured across a 10-millihenry (mH) inductor. Use phasors to find the current through the inductor. Express the current in phasor and sinusoidal form.

2.15 An impedance consists of 25 Ω of resistance in series with 60 Ω of capacitive reactance. The current through the impedance is $i = 70\angle 65°$ mA. Use phasors to find the voltage across the entire impedance. Express the voltage in phasor form.

2.16 Find the total admittance of each of the networks in Exercise 2.13. Express each in phasor and rectangular form.

2.17 Evaluate each of the following:

(a) $\dfrac{d[0.16 \sin(6t - 75°)]}{dt}$

(b) $\dfrac{d(\omega \cos \omega t)}{dt}$

(c) $\dfrac{d(e^{-t}\sin 2t)}{dt}$

(d) $\displaystyle\int_0^t 10^3 e^{-10^3 t} dt$

(e) $\displaystyle\int_0^t 2 \sin(2t + \pi/3) dt$

(f) $\displaystyle\int_0^t \omega \cos \omega t \, dt$

2.18 Differentiate Equation 14 to prove that

$$\dfrac{d(\sin \omega t)}{dt} = \omega \cos \omega t$$

2.19 Integrate Equation 15 to prove that

$$\int \cos \omega t = \dfrac{1}{\omega}\sin \omega(t) + C$$

36 REVIEW OF COMPLEX ALGEBRA AND PHASORS

ANSWERS TO EXERCISES

2.1 (a) real = 4, imaginary = -2
    (b) real = -1, imaginary = -0.35
    (c) real = 5, imaginary = 0
    (d) real = 0, imaginary = 3
    (e) real = 0, imaginary = ± 5
    (f) real = -0.5, imaginary = ± 0.3

2.2 (a) $4.47\angle -26.6°$
    (b) $1.06\angle -160.7°$
    (c) $5\angle 0°$
    (d) $3\angle 90°$
    (e) $5\angle 90°, 5\angle -90°$
    (f) $0.583\angle 149°, 0.583\angle -149°$

2.3 (a) $8.67 + j5$
    (b) $0.10 - j.173$
    (c) $-2.83 - j2.83$
    (d) $0 - j4.7$
    (e) $-150 + j0$
    (f) $-6.93 + j4.0$

2.4 (a) $5 + j2$
    (b) $-0.8 - j5.6$
    (c) $-1 - j1$
    (d) $16 + j11$
    (e) $0.59 + j25.98$
    (f) $44\angle 63°$
    (g) $-6\angle -20° = 6\angle 160°$
    (h) $20.67\angle -107°$
    (i) $1.18\angle 103.5°$
    (j) $6.22\angle 0°$
    (k) $0.229\angle 132.2°$

2.5 (a) $14 + j2$
    (b) $-1 - j2$
    (c) $30\angle -25°$
    (d) $-5\angle -121.5° = 5\angle 58.5°$

2.6 (a) $0.167 + j.167$
    (b) $-0.20 + j2.4$
    (c) $0.270 - j.734$
    (d) $3 - j12$

2.8 (a) $.01 e^{j3.05}$
    (b) $19.8 e^{j.785}$

2.9 (a) $15\angle \pi/3$ V
     (b) $20\angle -17°$ μA $= 20 \times 10^{-6}\angle -17°$ A
     (c) $18\angle 90°$ V
     (d) $42\angle -90°$ A

2.10 (a) $7.5 + j12.99$ V
     (b) $(19.13 - j5.85) \times 10^{-6}$ A
     (c) $0 + j18$ V
     (d) $0 - j42$ A

2.11 (a) $43.04 + j39.39 = 58.35\angle 42.46°$
         $= 58.35 \sin(377t + 42.46°)$
     (b) $.01414 - j.04242 = .0447\angle -71.56°$
         $= .0447 \sin(\omega t - 71.56°)$
     (c) $1 + j1 = 1.414\angle 45°$
         $= 1.414 \sin(\omega t + 45°)$

2.12 $8 - j6.93 = 10.58\angle -40.9°$
     $= 10.58 \sin(10^5 t - 40.9°)$ V

2.13 (a) $75 + j45 = 87.46\angle 30.96°$ Ω
     (b) $0.2 - j.1 = 0.224\angle -26.56°$ Ω
     (c) $10 + j60 = 60.83\angle 80.54$ Ω
     (d) $50 - j80 = 94.34\angle -58°$

2.14 $0.4\angle -\pi/4$ A
     $= 0.4 \sin(10^4 t - \pi/4)$ A

2.15 $4.55\angle -2.38°$ V

2.16 (a) $.0114\angle -30.96° = (9.8 - j5.88) \times 10^{-3}$ S
     (b) $4.46\angle 26.56°$ S
     (c) $.0164\angle -80.54° = (2.7 - j16.18) \times 10^{-3}$ S
     (d) $.011\angle 58° = (5.62 - j8.99) \times 10^{-3}$ S

2.17 (a) $0.96 \cos(6t - 75°)$
     (b) $-\omega^2 \sin \omega t$
     (c) $e^{-t}(2 \cos 2t - \sin 2t)$
     (d) $1 - e^{-10^3 t}$
     (e) $\frac{1}{2} - \cos\left(2t + \frac{\pi}{3}\right)$
     (f) $\sin \omega t$

CHAPTER 3
TRANSFER FUNCTIONS

3.1 GAIN

Most students of electronics are familiar with the concept we call *gain*. The voltage gain of an amplifier, for example, is known to be the ratio of the output voltage to input voltage. Stated differently, gain is the number we multiply the input voltage by in order to obtain the output voltage.

If the input voltage to an amplifier with gain 10 is 1 V, then we would expect the output voltage to be $10 \times 1 = 10$ V. Assuming (as we generally will) that the amplifier is *linear*, a 2-V input would produce a 20-V output, and so on.

In this chapter, we will generalize the concept of gain to see how it may be applied to a broader range of situations as well as how it may be used to predict outputs more precisely (e.g., phase angle). To begin our generalization, let us understand that if a device has a certain gain, it does not necessarily imply that the output is greater than the input. Although use of the word *gain* suggests that this is the case, there are, in fact, many devices that we wish to study whose outputs are *smaller* than their inputs. When this is the case, it simply means that the gain of the device is less than 1. Indeed, the gain of some devices (such as the voltage gain of a short circuit) may be zero. Consider, for example, a voltage divider consisting of two series resistors, as shown in Figure 3.1.

Using the voltage divider rule,

$$e_o = \left(\frac{3K}{3K + 1K}\right)e_{in} = 0.75 e_{in}$$

So $e_o/e_{in} = 0.75$, and we say the gain of this device is 0.75.

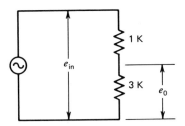

FIGURE 3.1 Voltage divider.

In many applications of the gain concept, gain is itself a dimensionless quantity. The gain of the voltage divider in the previous example is certainly dimensionless, since it is computed by dividing output *volts* by input *volts*. In our study of control systems and the various electromechanical devices of which they are composed, we will find that gain may have specific units, depending upon the type of device under study. This leads to a second generalization of gain. When the term gain is applied to any device, the units of gain will always be the same as the units of the output divided by the units of the input. It is very important to keep track of the units of gain as they apply to each device in a control system, for only then can we be certain that we understand the overall gain of the system; that is, we must know precisely how the gain of each component affects the quantity we are attempting to control.

As an example, consider the simple electromechanical device known as a potentiometer, which finds wide use in control systems; see Figure 3.2.

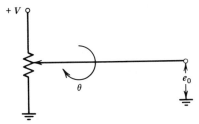

FIGURE 3.2 Potentiometer with input θ and output e_o.

In this case, the input is a certain angular *rotation* θ of the potentiometer shaft, while the output is the voltage appearing at the wiper arm. Therefore the gain of the potentiometer has units V/rad (or V/deg). We can calculate the magnitude of this gain by assuming that a full 360° of rotation (for a single-turn potentiometer) will cause the output voltage to vary from zero to its maximum (+V in Figure 3.2).

EXAMPLE 3.1

Find the gain of the potentiometer shown in Figure 3.3 in V/rad and in V/deg.

40 TRANSFER FUNCTIONS

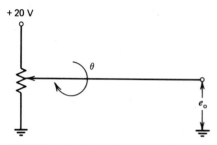

FIGURE 3.3   Potentiometer in Example 3.1.

SOLUTION

A shaft rotation of Θ degrees corresponds to a fractional revolution equal to $\Theta/360$. Thus,

$$e_o = \frac{\Theta}{360} \, 20 \text{ V}$$

or

$$\frac{e_o}{\Theta} = \frac{20}{360} = \frac{1}{18} \text{ V/deg}$$

Similarly,

$$\frac{e_o}{\Theta} = \frac{20 \text{ V}}{2\pi \text{ rad}} = 3.18 \text{ V/rad}$$

A practical potentiometer cannot generally be rotated through a full 360°, so the manufacturer's specifications should be consulted to determine how many degrees constitute one full rotation. Many precision potentiometers are of the multiturn type: More than one 360° rotation is required for full excursion of the wiper from minimum to maximum output. Five- and ten-turn potentiometers are typical examples. In general,

$$e_o = \frac{V}{2n(360)} \text{ V/deg} \quad \text{or} \quad \frac{V}{2n\pi} \text{ V/rad} \tag{1}$$

where n = the number of 360° turns required for full rotation.

Rotary potentiometers are examples of *transducers*: Components that generate an electrical quantity (generally voltage) that is in direct proportion to the magnitude of some physical quantity. Thus, rotary potentiometers produce a voltage proportional to the physical quantity of angular rotation. Every control system has at least one transducer of some type, and care must be exercised to express the magnitude and units of its gain correctly.

EXAMPLE 3.2

Find the gain, including units, of the following transducers:

(a) The Vernitech Model 112 linear-motion transducer whose specifications are given in Appendix B. Assume a 4-in. stroke, a 250-K total resistance, and an applied voltage equal to 1/10 the maximum allowed by the specifications.

(b) The Vernitech Model 1000 pressure transducer whose specifications are given in Appendix B. Assume a pressure range of 0 to 100 lb/in^2 and a 10-V supply voltage.

SOLUTION

(a) From the manufacturer's specifications, the rated power dissipation is 1 watt/in. Hence a 4-in. unit is rated 4 watts (W). Assuming a negligible electrical load on the wiper arm, the maximum applied voltage could then be found from

$$\frac{E^2_{MAX}}{250 \times 10^3} = 4 \text{ W}$$

$$E_{MAX} = \sqrt{10^6} = 1000 \text{ V}$$

Hence, the applied voltage in this example will be

$$E = (1/10)E_{MAX} = 100 \text{ V}$$

The gain is therefore,

$$G = \frac{100 \text{ V}}{4 \text{ in.}} = 25 \text{ V/in.}$$

(b) We see from the manufacturer's specifications that the Model 1000 pressure transducer displaces the wiper arm of a potentiometer in direct proportion to the applied pressure. Hence with a 10-V supply voltage and a 0 to 100 psi range, we have

$$G = \frac{10 \text{ V}}{100 \text{ psi}} = 0.1 \text{ V/lb/in}^2$$

3.2 GAIN AS A FUNCTION OF FREQUENCY

When we speak of the gain of an ac amplifier as being, say, 10, it is usually understood that the output voltage is ten times the input voltage over some frequency range, the so-called pass band of the amplifier. We know that if the input frequency is much higher than the so-called upper cutoff frequency of the amplifier, then the gain is likely to be much less than 10. That is to say, gain is a function of frequency.

This frequency dependence of gain applies to *every* device. (It can be shown mathematically that no physically realizable device can have infinite bandwidth.) Therefore, we can further generalize our notion of gain to include its variation with frequency. The manner in which gain varies with frequency (whether gain increases or decreases) will, of course, depend on the nature of the device itself. The gain of a low-pass filter, for example, increases as frequency decreases.

The symbol G is often used to designate the gain of a device. If we wish to emphasize the fact that G is a function of frequency, we write $G(\omega)$ or $G(j\omega)$. It

42 TRANSFER FUNCTIONS

is often the case that the gain of a particular device is constant over such a wide bandwidth that we can ignore the fact that it must eventually decrease at some very high frequency. This, of course, depends on the frequency range over which we are interested in analyzing the behavior of the device. If we are studying an amplifier with pass-band 0–500 Hz, we would not need to take into account the fact that a potentiometer in the circuit begins to cause addditional attenuation due to its shunt capacitance in the vicinity of several megahertz.

Consider the RC network shown in Figure 3.4.

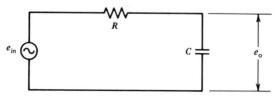

FIGURE 3.4   A circuit whose gain depends on frequency.

The gain of this circuit will certainly depend upon the frequency of the input signal. Since capacitive reactance is inversely proportional to frequency, we would expect the output to decrease as the input frequency is increased. Conversely, at lower frequencies, X_c would be large, and a greater proportion of the input signal would be dropped across the capacitor, resulting in a higher output voltage. This is an example of a low-pass filter.

3.3 TRANSFER FUNCTIONS

Having established that gain is a function of frequency, we are now in a position to define the transfer function of a device: It is simply an expression for the gain where frequency is a variable. Thus the transfer function allows us to calculate the gain of a device *at any frequency* we choose. As mentioned in the previous paragraph, when gain is stated as a function of frequency, we may use the notation $G(\omega)$ to emphasize this fact.

We may further refine and improve our concept of a transfer function by including some provision by which it is possible to calculate the *phase shift* of the output relative to the phase angle of the input. This may be accomplished by expressing the gain, or transfer function, as a phasor. Suppose, for example, we knew that at a particular frequency, the transfer function of a device is given by $5\angle 30°$. Suppose also that the input to the device is $e_{in} = 0.6 \sin(\omega t - 60°)$. Then,

$$\frac{e_o}{e_{in}} = 5\angle 30°$$

$$\frac{e_o}{0.6\angle -60°} = 5\angle 30°$$

$$e_o = (5\angle 30°)(0.6\angle -60°)$$

$$e_o = 3 \sin(\omega t - 30°)$$

3.4 CALCULATING THE TRANSFER FUNCTION OF PASSIVE NETWORKS

EXAMPLE 3.3

The input signal to an amplifier is a 4-V peak sine wave at a frequency of 1 kHz. At that same frequency, the output is a 12-V peak sine wave with phase angle 75°. Find the transfer function of the device at 1 kHz.

SOLUTION

$$e_{in} = 4\angle 0°$$

$$e_o = 12\angle 75°$$

Therefore, at 1 kHz

$$\frac{e_o}{e_{in}} = \frac{12\angle 75°}{4\angle 0°} = 3\angle 75°$$

In Example 3.3, it should be clearly understood that a change in frequency above or below 1 kHz might well change both the magnitude and angle of the output and therefore change the transfer function. Again, we emphasize the dependency of transfer function upon frequency.

3.4 CALCULATING THE TRANSFER FUNCTION OF PASSIVE NETWORKS

One way to determine the transfer function of a particular device would be to vary the frequency of the input and measure the magnitude and phase of the output at a number of different frequencies. In this way, we could calculate e_o/e_{in} in phasor form at each frequency (as in Example 2.2 at 1 kHz) and thus obtain the transfer function versus frequency. In many cases, this is not practical, nor does this approach give us an exact value for the transfer function at those frequencies where measurements were not made.

For passive networks (composed of resistance, inductance, and capacitance), we can obtain an exact expression for the transfer function by expressing impedance in phasor form and performing conventional circuit and analysis. Consider for example the low-pass filter discussed in Section 3.2. (See Figure 3.5.)

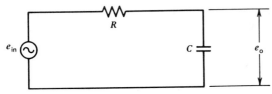

FIGURE 3.5   Low-pass filter.

Application of the voltage divider rule (Section 2.8) shows that

$$e_o = \left(\frac{-jX_c}{R - jX_c}\right) e_{in}$$

44 TRANSFER FUNCTIONS

or

$$\frac{e_o}{e_{in}} = \frac{-jX_c}{R - jX_c} \qquad (2)$$

Since $X_c = 1/\omega c$, we have

$$\frac{e_o}{e_{in}} = G(j\omega) = \frac{-j/\omega C}{R - j/\omega C} = \frac{-j}{\omega RC - j}$$

Now $G(j\omega)$ is a phasor with both magnitude and angle. However, it is not possible to identify the magnitude and angle in the form that it is written.

One approach is to rationalize $G(j\omega)$ to obtain the form $a + jb$, so that we may find $M = \sqrt{a^2 + b^2}$ and $\theta = \arctan b/a$ in the usual way.

$$G(j\omega) = \frac{-j}{\omega RC - j}\left(\frac{\omega RC + j}{\omega RC + j}\right) = \frac{-j^2 - j\omega RC}{(\omega RC)^2 + (1)^2}$$

$$= \frac{1 - j\omega RC}{1 + (\omega RC)^2} = \frac{1}{1 + (\omega RC)^2} - j\frac{\omega RC}{1 + (\omega RC)^2} \qquad (3)$$

It can be seen now that

$$a = \frac{1}{1 + (\omega RC)^2} \qquad b = \frac{-\omega RC}{1 + (\omega RC)^2}$$

Therefore,

$$|G(j\omega)| = M = \sqrt{a^2 + b^2} = \sqrt{\left[\frac{1}{1 + (\omega RC)^2}\right]^2 + \left[\frac{-\omega RC}{1 + (\omega RC)^2}\right]^2}$$

$$= \frac{\sqrt{1 + (\omega RC)^2}}{1 + (\omega RC)^2} = \frac{1}{\sqrt{1 + (\omega RC)^2}} \qquad (4)$$

In this case, a simpler approach would be to recognize that

$$|G| = \frac{|-j|}{|\omega RC - j|} = \frac{1}{\sqrt{(\omega RC)^2 + 1}}$$

directly.

This result tells us that the *magnitude* of the gain (disregarding phase) decreases as ω increases, a result that we would certainly expect from our intuitive analysis of the network as a low-pass filter. In fact, as we approach zero frequency (dc) we see that

$$\lim_{\omega \to 0} |G(j\omega)| = \lim_{\omega \to 0} \frac{1}{\sqrt{1 + (\omega RC^2)}}$$

$$= \frac{1}{\sqrt{1}} = 1$$

This result agrees with our intuitive understanding of the network, since at dc the

3.4 CALCULATING THE TRANSFER FUNCTION OF PASSIVE NETWORKS

capacitor is an open circuit (with infinite impedance), and consequently, *all* of the input signal appears at the output (none is dropped across the series resistor).

Furthermore, as we aproach infinite frequency,

$$\lim_{\omega \to \infty} |G(j\omega)| = \lim_{\omega \to \infty} \frac{1}{\sqrt{1 + (\omega RC)^2}} = 0$$

This result again agrees with our intuitive understanding of the circuit, since at infinite frequency, $X_c = 1/\omega C$ is zero (a short circuit), and the output voltage must therefore be 0 V.

As for the phase of $G(j\omega)$,

$$\theta = \angle G(j\omega) = \arctan \frac{b}{a} = \arctan \left(\frac{-\omega RC/[1 + (\omega RC)^2]}{1/[1 + (\omega RC)^2]} \right) \quad (5)$$

$$= \arctan(-\omega RC)$$

$$\lim_{\omega \to 0} \angle G(j\omega) = \lim_{\omega \to 0} \arctan(-\omega RC)$$

$$= \arctan 0 = 0°$$

$$\lim_{\omega \to \infty} \angle G(j\omega) = \lim_{\omega \to \infty} \arctan(-\omega RC)$$

$$= \arctan(-\infty) = -90°$$

Since the transfer function has phase between $0°$ and $-90°$, we can conclude that the output will *lag* the input at any frequency between dc and infinity. (In control systems parlance, this is often referred to as a *lag network*.)

EXAMPLE 3.4

In the low-pass filter just described, suppose $C = 1$ microfarad (μF) and $R = 1.59$ K. Find (a) the transfer function at 100 Hz and (b) the output voltage when driven by a 100-Hz sine wave with 5-V peak value.

SOLUTION

(a) From Equation 3,

$$G(j\omega) = \frac{1}{1 + (\omega RC)^2} - j\frac{\omega RC}{1 + (\omega RC)^2}$$

$$= \frac{1}{1 + (2\pi \times 10^2 \times 1.59 \times 10^3 \times 10^{-6})^2}$$

$$- j\frac{(2\pi \times 10^2)(1.59k)(10^{-5})}{1 + (2\pi \times 10^2 \times 1.59 \times 10^3 \times 10^{-6})^2}$$

$$= \frac{1}{2} - j\frac{1}{2} = 0.707\angle -45°$$

(b) $e_{in} = 5\angle 0°$

$$e_o = e_{in}G(j\omega) = (5\angle 0°)(0.707\angle -45°)$$
$$= 3.535\angle -45°$$

In this example, 100 Hz is tthe *cutoff* frequency of the network (by definition, the frequency at which the output magnitude is 0.707 times the input magnitude). Notice that the output lags the input by 45° at 100 Hz.

When amplifiers are connected in cascade (series), we know that the overall gain of the system is the product of the gains of each amplifier. Consider, for example, three amplifiers with gains $G_1 = 5$, $G_2 = 1.2$, and $G_3 = 12$ (see Figure 3.6). Since $e_2 = 5e_1$, $e_3 = (5e_1)(1.2) = 6e_1$, and $e_4 = (6e_1)(12) = 72e_1$, we have $e_4/e_1 = 72$.

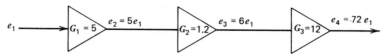

FIGURE 3.6 Three amplifiers connected in cascade.

In a similar way, the transfer functions of devices connected in cascade can be multiplied to obtain the overall transfer function of the system. If

$$G_1 = 5\angle 0°,\ G_2 = 1.2\angle -45°,\ \text{and}\ G_3 = 12\angle 60°$$

then

$$\frac{e_4}{e_1} = (5\angle 0°)(1.2\angle -45°)(12\angle 60°) = 72\angle 15°$$

The peak value of the output would be 72 times the peak value of the input, and the output would lead the input by 15°.

At this point, a word of caution is in order. When finding the gain or transfer function of a device such as an amplifier, it is usually assumed that the output of the device is *open-circuited*, that is, it is driving a load with infinite input impedance. The load on an amplifier may very well affect its gain. Therefore, unless the input impedances of the devices connected in cascade are very large, so that their loading effects can be neglected, it would not be correct to find the overall system gain by multiplying the individual component gains. This holds true for transfer functions as well. For example, we have seen (Equation 2) that the transfer function of the low-pass filter is $G(j\omega) = -jX_c/(R - jX_c)$. If two identical filters were connected in series, it would *not* be correct to state that the overall transfer function is $[-jX_c/(R - jX_c)]^2$.

3.5 BODE PLOTS

The behavior of the output of a particular device (its magnitude and phase) when the input frequency is varied is called the *frequency response* of the device. If we know the transfer function $G(j\omega)$ of the device, then, as we have seen, we can predict the output magnitude and phase at any frequency. We can, therefore, plot the magnitude of the gain |G| as well as the phase ∠G versus frequency to obtain a graphical description of the frequency response. When the logarithm of the magnitude of G is plotted versus the logarithm of frequency, the result is called a *Bode plot*. Phase angle is also plotted versus log frequency.

It would be rather tedious to compute all the logarithms of gain magnitude and frequency necessary to obtain a complete frequency response. For this reason, log-log graph paper is used. The lines on this type of paper are spaced logarithmically on both horizontal and vertical axes. When points are plotted on log-log paper, the result is exactly the same as if the logarithms of gain magnitude and frequency were computed and plotted on conventional linear graph paper. It should be emphasized that when using log-log paper, it is *not* necessary to compute logarithms of the quantities plotted. So-called semi-log paper is also available, which is graph paper with one axis logarithmically scaled and the other axis linear. It takes some practice to use log-log or semi-log paper correctly, particularly when assigning values to the scales. Students are urged to obtain such paper, study the scales, and practice plotting some points until they are convinced that they understand the technique. Log-log paper is described by the number of *cycles* it has on each axis. One cycle corresponds to one *decade*, that is, a one-to-ten range. Log-log paper with three cycles on the vertical axis and three cycles on the horizontal axis is said to be 3 by 3. With this paper, we could plot data on each axis with a 1000-to-1 range (three decades).

EXAMPLE 3.5

In order to obtain the frequency response of a device, its gain is measured over a frequency range from 15 Hz to 900 Hz. The gain magnitude was found to vary from 25 to .07. What type of log-log paper is required to make a Bode plot?

SOLUTION

Two decades will be required to plot the frequencies used (10 to 100 and 100 to 1000). Hence, two cycles will be required on the horizontal axis. Similarly, four decades will be required for the vertical axis (.01 to 0.1, 0.1 to 1, 1 to 10, and 10 to 100). Therefore, 2-by-4 graph paper is required.

Most devices that are used in control systems have nonlinear frequency responses; that is, the plot of gain magnitude versus frequency is not a straight line. For example, the gain magnitude versus frequency plot of any device whose output is inversely proportional to frequency is hyperbolic. The advantage of a Bode plot is that it generally transforms substantial portions of nonlinear plots into straight lines, making it much easier to sketch the frequency response.

An example of such a device is the *integrator*. An integrator is any device whose output is proportional to the integral of the input. In practice, integration is almost always with respect to time. Suppose, for example, that the input to an electronic integrator (one whose output voltage is proportional to the integral of the input voltage) is $e_{in} = A \sin(\omega t + \theta)$. (See Figure 3.7.)

48 TRANSFER FUNCTIONS

FIGURE 3.7 An electronic integrator.

Then

$$e_o = \int A \sin(\omega t + \theta) dt$$

$$= -\frac{A}{\omega} \cos(\omega t + \theta)$$

$$= -\frac{A}{\omega} \sin(\omega t + \theta + 90°)$$

Expressing e_{in} and e_o in phasor form,

$$e_{in} = A \angle \theta \tag{6}$$

$$e_o = \frac{-A}{\omega} \angle \theta + 90°$$

Therefore, the transfer function of the integrator is

$$G = \frac{e_o}{e_{in}} = \frac{-A/\omega \angle \theta + 90°}{A \angle \theta} = -\frac{1}{\omega} \angle 90°$$

$$= \frac{1}{\omega} \angle -90° = -j/\omega = 1/j\omega \tag{7}$$

Then

$$|G| = \left| \frac{1}{\omega} \angle -90° \right| = \frac{1}{\omega} \tag{8}$$

and

$$\angle G = -90° \tag{9}$$

Thus, the gain magnitude of an integrator is $1/\omega$, that is, inversely proportional to frequency. If $|G|$ were plotted on linear axes, we would obtain the *hyperbola* $|G| = \frac{1}{\omega}$. (See Figure 3.8.)

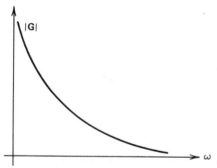

FIGURE 3.8 Gain magnitude of an integrator versus frequency (hyperbola).

It would be a tedious and time-consuming procedure to plot this graph. On the other hand, suppose we take the logarithm of each side

$$\log |G| = \log \left(\frac{1}{\omega}\right)$$

Then

$$\log |G| = \log 1 - \log \omega$$

and

$$\log |G| = -\log \omega$$

since $\log 1 = 0$.

Recall that the general equation of a straight line in the xy-plane is $y = mx + b$, where m is the slope and b is the y-intercept. If we let $y = \log |G|$ and $x = \log \omega$, we see that $\log |G| = -\log \omega$ is a straight line with slope -1 and intercept zero when plotted on log-log paper. (See Figure 3.9.)

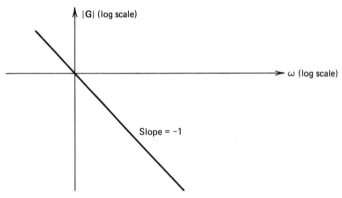

FIGURE 3.9 Plot of $\log |G| = -\log \omega$.

If the integrator incorporates a certain frequency-independent gain K (for example, if the electronic device performing integration also amplifies by K over a very wide bandwidth), then the transfer function would become

$$G(j\omega) = \frac{K}{j\omega} \quad (10)$$

and

$$|G(j\omega)| = \left|\frac{K}{j\omega}\right| = \frac{K}{\omega} \quad (11)$$

Taking the logarithm of both sides, $\log |G| = \log (K/\omega) = \log K - \log \omega$. The Bode plot is then modified to the extent that the straight line is now shifted upwards until it intercepts the y-axis at $\log K$. (The y-intercept is now $\log K$). (See Figure 3.10.)

50 TRANSFER FUNCTIONS

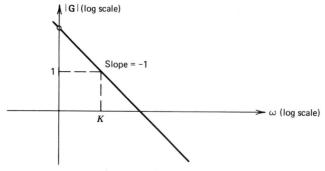

FIGURE 3.10 Plot of log |G| = −log K − log ω.

The result illustrates another advantage of Bode plots: When the gain of a device is increased or decreased, the Bode plot is simply shifted up or down by a corresponding amount. As we will see, much of control-systems design and analysis is concerned with the effect of gain variations, so the Bode plot provides a very effective tool for studies of this nature.

Note that the gain of the integrator is unity at frequency ω = K

$$|G| = 1 = \frac{K}{\omega} \Rightarrow \omega = K \tag{12}$$

This is an important result that will be used extensively in our future development of control-systems theory.

EXAMPLE 3.6

An integrator has transfer function $G(j\omega) = 40/j\omega$. How much additional frequency-independent gain would have to be included in cascade with this integrator in order for the combination to have gain magnitude equal to unity at ω = 500 rad/sec? See Figure 3.11.

FIGURE 3.11 Block diagram for Example 3.5.

SOLUTION

$$\frac{e_o}{e_{in}} = \left(\frac{40}{j\omega}\right)(K) \qquad \left|\frac{e_o}{e_{in}}\right| = \frac{40K}{\omega}$$

In order that

$$\left|\frac{e_o}{e_{in}}\right| = 1 \text{ at } \omega = 500 \text{ rad/sec we require that}$$

$$\left|\frac{e_o}{e_{in}}\right| = 1 = \frac{40K}{400}$$

so

$$K = \frac{500}{40} = 12.5$$

Integration and differentiation are two of the most important operations performed on electrical signals when conditioning them to achieve certain desirable control-systems responses. It can be shown that an electronic differentiator with frequency-independent gain K has transfer function $G(j\omega) = jK\omega$. (See Exercise 3.6.) For the differentiator, $|G| = K\omega$ and, therefore, $\log |G| = \log K\omega = \log K + \log \omega$. By comparison with the standard form of a straight line $y = mx + b$, it can be seen that the Bode plot for a differentiator has slope $+1$ and y-intercept $\log K$, as shown in Figure 3.12.

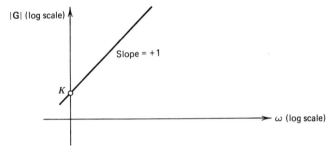

FIGURE 3.12 Gain-magnitude bode plot for a differentiator.

If two integrators are connected in cascade, the resulting Bode plot has slope -2, while two differentiators connected in cascade produce a Bode plot with slope $+2$. Generalizing, n integrators in cascade produce a Bode plot slope of $-n$ and n differentiators in cascade produce a Bode plot slope of $+n$.

3.6 FREQUENCY RESPONSE IN DECIBELS

Decibels (dB) are widely used to specify power gain (or attenuation). Recall that decibel-power gain is defined to be ten times the logarithm (base 10) of the ratio of two powers. Thus, if the power at one point in a system is P_1 and the power at another point is P_2, we say that the gain (or attenuation) in decibels between the second and first points is

$$dB = 10 \log_{10} \frac{P_2}{P_1} \qquad (13)$$

If P_2 is greater than P_1, then $\log_{10}(P_2/P_1)$ is a *positive* number, indicating an increase in power; if P_2 is less than P_1, $\log_{10}(P_2/P_1)$ is *negative*, indicating an attenuation.

EXAMPLE 3.7

An amplifier with input resistance 4 K draws a current of 2 mA from a certain source. It delivers 5 V to a 100-Ω load. Calculate the power gain of the amplifier in decibels.

SOLUTION

$$P_{in} = I^2R = (2 \times 10^{-3})^2(4 \times 10^3)$$
$$= (4 \times 10^{-6})(4 \times 10^3)$$
$$= 16 \times 10^{-3} = 16 \text{ mW}$$

$$P_{out} = \frac{E^2}{R} = \frac{5^2}{100} = \frac{25}{100} = 0.25 \text{ W} = 250 \text{ mW}$$

$$\text{Power gain} = 10 \log_{10} \frac{250 \times 10^{-3}}{16 \times 10^{-3}}$$
$$= 10 \log_{10}(15.625)$$
$$= 11.94 \text{ dB}$$

Voltage gain is also frequently specified in decibels. Strictly speaking, the specification of voltage gain in decibels is only valid when the voltages compared are taken across the *same* value of resistance. Suppose that power P_2 is delivered across resistance R at the output of an amplifier, and power P_1 is developed across an equal resistance R at the input to the amplifier. Then the gain in dB may be found from

$$10 \log_{10} \frac{P_2}{P_1} = 10 \log \frac{E_2^2/R}{E_1^2/R}$$
$$= 10 \log_{10} \left(\frac{E_2}{E_1}\right)^2 = 20 \log_{10} \left(\frac{E_2}{E_1}\right) \tag{14}$$

where E_1 is the input voltage and E_2 is the output voltage. The expression $20 \log_{10}(E_2/E_1)$ is called the dB voltage gain and is often defined this way in practice, whether or not the voltages are across equal-valued resistors.

Consider an integrator with transfer function $G(j\omega) = 1/j\omega$. Then

$$\left|\frac{e_o}{e_{in}}\right| = |G| = \frac{1}{\omega}$$

and

3.6 FREQUENCY RESPONSE IN DECIBELS

$$20 \log_{10} \left| \frac{e_o}{e_{in}} \right| = -20 \log \omega \qquad (15)$$

We see that the voltage attenuation in decibels is equal to $-20 \log \omega$. Suppose we wish to determine the attenuation of dB when the frequency is changed from 1 rad/sec to 10 rad/sec (i.e., one decade). Then

at $\omega = 1$ dB = $-20 \log (1) = 0$

at $\omega = 10$ dB = $-20 \log (10) = -20$

There is a total attenuation of 20 dB over this frequency decade. In fact, any device whose Bode plot has a slope of -1 will exhibit 20 dB of attenuation over every decade. For the integrator, we would find 20 dB of attenuation between $\omega = 10$ and $\omega = 100$, between $\omega = 4$ and $\omega = 40$, between $\omega = 200$ and $\omega = 2000$, or between the limits of any other decade of frequency.

EXAMPLE 3.8

An integrator has frequency-independent gain of 50. Find the voltage attenuation in dB when the frequency is changed from 2 Hz to 4 Hz.

SOLUTION

$$|G| = \frac{50}{\omega}$$

dB = $20 \log_{10} |G| = 20 \log_{10} 50 - 20 \log_{10} \omega$

at $f = 2$ Hz   dB = $20 \log_{10} 50 - 20 \log_{10} (2\pi \times 2)$
= $33.98 - 21.98 = 12$ dB

at $f = 4$ Hz   dB = $20 \log_{10} 50 - 20 \log_{10} (2\pi \times 4)$
= $33.98 - 28.00 = 5.98$ dB

Attenuation = 12 dB $-$ 5.98 dB
= 6.02 dB

It can be seen from Example 3.7 that the integrator output experienced an attenuation of 6.02 dB over the octave 2 Hz to 4 Hz. We would find this same integrator attenuation over *any* octave, regardless of the amount of frequency-independent gain. In practice, the attenuation over an octave is usually rounded off and referred to as 6 dB/octave. Thus, an integrator causes attenuation at the rate of -6 dB/octave and -20 dB/decade.

A differentiator with transfer function $G(j\omega) = jK\omega$ will produce an increase in gain magnitude at the rate of $+6$ dB/octave or $+20$ dB/decade. Two integrators in cascade will produce an overall attenuation of -12 dB/octave or -40 dB/decade.

54 TRANSFER FUNCTIONS

In general, if the Bode plot slope of the gain magnitude is $\pm n$, the gain magnitude changes at the rate $\pm 6n$ dB/octave or $\pm 20n$ dB/decade.

3.7 ASYMPTOTIC PLOTS

In many practical situations involving the design or analysis of a control system, it is not necessary to know the exact values of gain magnitude or phase at every frequency of interest. Frequently, an approximation is adequate. In fact, the complexity of some systems composed of many different devices with a wide variety of transfer functions may dictate an approximation of the overall transfer function.

The method most commonly used to approximate the gain magnitude of a device over a certain frequency range is called the *asymptotic* approximation. Recall that an asymptote is a line that a function approaches but never reaches except in a limiting sense. The straight line $y = 1$ is, for example, an asymptote of the function $y = 1 - e^{-x}$, since $\lim_{x \to \infty} (1 - e^{-x}) = 1$.

We have seen that gain-magnitude lines on a Bode plot are straight lines for some types of devices. For some other devices, we may use the straight-line asymptotes of the gain magnitude as approximations. Consider the low-pass filter discussed in Section 2.4. (See Figure 3.13.)

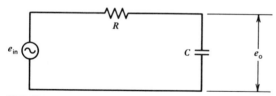

FIGURE 3.13 Low-pass filter.

We saw (Equation 4) that

$$|G| = \frac{1}{\sqrt{1 + (\omega RC)^2}}$$

If this function were plotted exactly on log-log graph paper, we would obtain a curve resembling Figure 3.14.

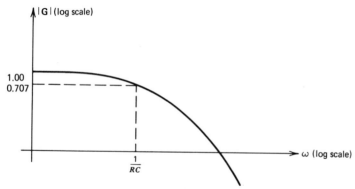

FIGURE 3.14 Log-log plot of gain magnitude for the low-pass filter.

3.7 ASYMPTOTIC PLOTS

Notice that the cutoff frequency (or break frequency) where $|G| = 0.707$ occurs at $\omega = 1/RC$. At frequencies lower than $\omega = 1/RC$, $|G|$ is essentially flat and approaches 1. A horizontal line through 1.0 is, therefore, an asymptote for the gain magnitude as frequency approaches zero. On the other hand, at frequencies above the break frequency, the gain magnitude approaches a straight line with slope -1. This can be verified as follows:

$$\log |G| = \log \frac{1}{\sqrt{1 + (\omega RC)^2}}$$

$$= \log 1 - \log [1 + (\omega RC)^2]^{1/2}$$

$$= 0 - \frac{1}{2} \log [1 + (\omega RC)^2]$$

As $\omega \to \infty$, $[1 + (\omega RC)^2] \approx (\omega RC)^2$ [since $(\omega RC)^2$ is much larger than 1 for large ω]. Thus, for large ω

$$\log |G| = -\frac{1}{2} \log (\omega RC)^2$$

$$= -\log \omega RC$$

which is a line with slope -1 on log-log graph paper. This line is, therefore, the asymptote for log $|G|$ at frequencies above the break frequency. Figure 3.15 shows the actual gain magnitude and its asymptotic approximation on the same graph. We see that the greatest error in using the asymptotic approximation occurs at the break frequency, where the actual gain magnitude is 0.707 times the asymptotic approximation. This is equivalent to a maximum error of 3 dB and is not considered excessive for most purposes. (In some systems, the gain may vary randomly over a greater range than this.)

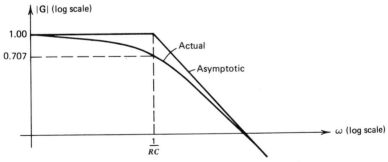

FIGURE 3.15 Asymptotic gain-magnitude plot for the low-pass filter.

Asymptotic approximations are also used to sketch phase angle versus log frequency. However, in this case, the approximation is not so accurate, and care must be exercised in using the approximation if phase angle is critical at a certain frequency. As will be discussed in a subsequent chapter, the frequency at which the phase angle is $-180°$ is a critical one in control-systems theory.

The actual and asymptotic phase-angle plot for the low-pass network is shown in Figure 3.16 as it would appear on semi-log graph paper.

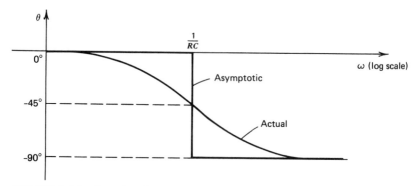

FIGURE 3.16 Actual and asymptotic phase-angle plots for the low-pass filter.

Mechanical components in a system have frequency-dependent transfer functions just as electrical networks do. These transfer functions may be interpreted in the same way as network transfer functions: They relate magnitude and phase of the output to magnitude and phase of the input. In the case of a transducer, for example, the transfer function is the ratio of the output phasor (voltage) to the input quantity (pressure, displacement, etc.) expressed as a phasor. Manufacturers typically specify a *transducer time constant* from which we can calculate a cutoff frequency in the same way we do for an RC network. A transducer behaves like a low-pass filter in a system, since its output voltage can only follow input variations up to a certain frequency.

EXAMPLE 3.9

Find the frequency dependent transfer function $G(j\omega)$ for the Vernitech Model 1000 pressure transducer of Example 3.2(b). Consult the specification sheet in Appendix B.

SOLUTION

From the manufacturer's specifications, we find that the response time is 20 milliseconds (msec). Thus, the break frequency is

$$\omega = \frac{1}{20 \times 10^{-3}} = 50 \text{ rad/sec}$$

In Example 3.2(b), we found the gain of the transducer to be 0.1 V/lb/in². This figure actually represents the dc gain (the gain at $\omega = 0$). Thus, the transfer function of the pressure transducer is

$$G(j\omega) = \frac{(0.1)(50)}{j\omega + 50} = \frac{5}{j\omega + 50} \text{ V/lb/in}^2$$

Note that at $\omega = 0$, $|G| = 0.1$, as required.

3.7 ASYMPTOTIC PLOTS

EXAMPLE 3.10

Sketch the asymptotic approximation of the Bode plot for the gain magnitude of the network in Figure 3.17.

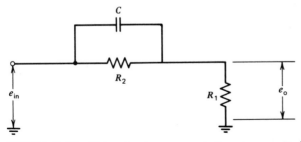

FIGURE 3.17 Network whose asymptotic gain magnitude is required (Example 3.9).

SOLUTION

We will first find the equivalent phasor impedance of the parallel combination of C and R_2

$$Z = \frac{R_2(-j/\omega C)}{R_2 - j/\omega C} = \frac{-jR_2}{\omega R_2 C - j} \tag{16}$$

Then applying the voltage divider rule yields

$$e_o = \left(\frac{R_1}{Z + R_1}\right) e_{in} = \frac{R_1}{[-jR_2/(\omega R_2 C - j) + R_1]} e_{in}$$

$$G = \frac{e_o}{e_{in}} = \frac{R_1}{(-jR_2 + \omega R_1 R_2 C - jR_1)/(\omega R_2 C - j)}$$

$$= \frac{\omega R_1 R_2 C - jR_1}{\omega R_1 R_2 C - j(R_1 + R_2)} \left(\frac{j}{j}\right) = \frac{R_1 + j\omega R_1 R_2 C}{(R_1 + R_2) + j\omega R_1 R_2 C}$$

$$= \frac{R_1 R_2 C[1/(R_2 C) + j\omega]}{R_1 R_2 C[(R_1 + R_2)/(R_1 R_2 C) + j\omega]} = \frac{1/(R_2 C) + j\omega}{1 / \left(\frac{R_1}{R_1 + R_2}\right) R_2 C + j\omega}$$

let

$$T = R_2 C$$

and

$$\alpha = \frac{R_1}{R_1 + R_2}$$

58 TRANSFER FUNCTIONS

Then,

$$G = \left[\frac{1/T + j\omega}{1/\alpha T + j\omega}\right] \tag{17}$$

At $\omega = 0$ (dc), $G = \alpha$. Clearly at dc, the capacitor acts as an open circuit and the network, therefore, behaves as a simple resistive voltage divider with gain $\alpha = R_1/(R_1 + R_2)$.

As frequency increases, the capacitive reactance decreases until it becomes equal in magnitude to R_2. This is the break frequency $\omega_1 = 1/R_2C = 1/T$ beyond which the output voltage begins to increase in magnitude due to the decreasing impedance of the R_2C combination. When the frequency is high enough, the capacitive reactance becomes so small that virtually all of the signal is dropped across R_1. Thus, $|G| = 1$ is an asymptote for higher frequencies. The break frequency here is $\omega_2 = 1/\alpha R_2C$. Note that α is always less than one (and, therefore, ω_2 is greater than ω_1). The asymptotic plot is sketched in Figure 3.18.

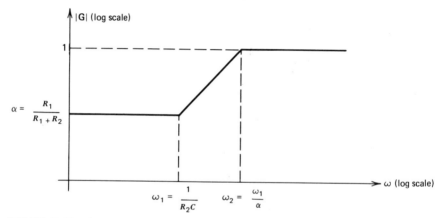

FIGURE 3.18 Asymptotic gain magnitude of the network in Figure 3.17 (Example 3.9).

This network is used in many control systems to provide a signal whose output leads the input over the range ω_1 to ω_2; it is called a *lead network*.

REFERENCES

1. **Belove, Charles,** and **Melvyn M. Drossman.** *Systems and Circuits for Electrical Engineering Technology.* McGraw-Hill, 1976.
2. **Boylestad, Robert L.** *Introductory Circuit Analysis.* 3d ed. Charles E. Merrill, 1977.
3. **Romanek, Richard J.** *Introduction to Electronic Technology.* Prentice-Hall, 1975.
4. **Wojslaw, Charles F.** *Electronic Concepts, Principles, & Circuits.* Reston, 1980.

EXERCISES

3.1 Find the gain of the circuits in each of the following figures (Figure 3.19) (assume that the gain in each case is independent of frequency):

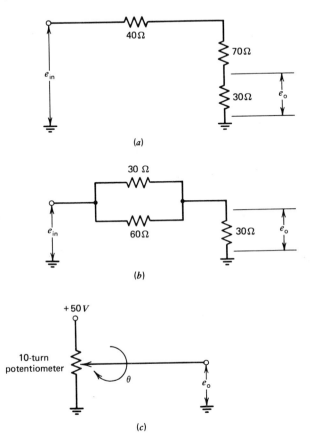

FIGURE 3.19

3.2 A network has transfer function $1.5\angle 30°$. The input is a 5-V peak sine wave with phase angle $-45°$. Find the output in phasor form. Does the output lead or lag the input?

3.3 Given the passive network in Figure 3.20:

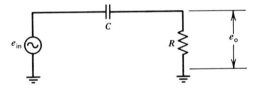

FIGURE 3.20

60 TRANSFER FUNCTIONS

(a) Find $G(j\omega)$ in rectangular form.
(b) Find an expression for $|G|$.
(c) Find an expression for $\angle G$.
(d) Find $\lim_{\omega \to \infty} |G|$ and $\lim_{\omega \to 0} |G|$.
(e) Find $\lim_{\omega \to \infty} \angle G$ and $\lim_{\omega \to 0} \angle G$.

3.4 In Exercise 3.3, suppose $C = 1 \text{ } \mu F$ and $R = 1.59 \text{ K}$:

(a) Find the transfer function $G(j\omega)$ at 100 Hz.
(b) Find the output voltage (peak value and phase) when driven by a 100-Hz sine wave with 5-V peak value.

3.5 The transfer function of the following network (Figure 3.21) is $0.5\angle 0$:

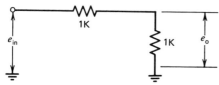

FIGURE 3.21

Find the transfer function of the network in Figure 3.22:

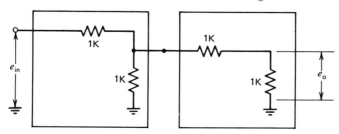

FIGURE 3.22

Why is it true in this case that the transfer function of two networks with identical transfer functions connected in cascade does not equal the product of the transfer functions $(0.25\angle 0)$?

3.6 Show that the transfer function of an electronic differentiator with frequency-independent gain K is $G(j\omega) = jK\omega$.
[Hint: Recall that the derivative of $A \sin (\omega t + \theta)$ is $A\omega \cos (\omega t + \theta)$.]

3.7 An integrator has frequency independent gain $K = 100$. How much frequency-independent gain would have to be included in cascade with this integrator in order for the combination to have:

(a) Gain magnitude equal to unity at $\omega = 250$ rad/sec?
(b) Gain magnitude equal to unity at $\omega = 50$ rad/sec?
(c) Gain magnitude equal to unity at $f = 400$ Hz?

3.8 The gain magnitude |G| of an integrator is 47.8 at $f = 20$ Hz. At what frequency is the gain magnitude equal to 1?

3.9 What type of log-log paper would be required to construct the Bode plot of a device whose gain magnitude varies from 87 to .025 over a frequency range from 0.5 Hz to 670 Hz?

3.10 The voltage at the input to an amplifier with input resistance 2 K is 4.2 VRMS. The amplifier produces 10 VRMS across a 500-Ω load. What is the power gain in dB?

3.11 The input to an electronic device is 6 V across 1 K, and its output is 1 V across 1 K. Find the voltage gain of this device in dB.

3.12 An integrator has a frequency-independent gain of 200. Find the attenuation in dB when the frequency of the input signal is changed from 8 Hz to 32 Hz. How many octaves does this frequency range represent?

3.13 Sketch the Bode plot of an integrator whose transfer function is $G(j\omega) = 400/j\omega$. Be certain to label axes and intercepts and include phase as well as gain magnitude.

3.14 Sketch the asymptotic Bode plot for the network in Exercise 3.4. What is the break frequency of this network?

3.15 Design an RC network whose gain magnitude on an asymptotic Bode plot is flat (horizontal) up to a frequency of 48 Hz and decreases at a rate of -20 dB/decade for frequencies beyond 48 Hz. Find the phase angle of the transfer function at 144 Hz.

3.16 Design a lead network with transfer function

$$G(j\omega) = 0.1\left(\frac{1 + j100\,\omega}{1 + j10\,\omega}\right)$$

Sketch the Bode plot of the gain magnitude of your design.

ANSWERS TO EXERCISES

3.1 (a) 0.214
  (b) 0.60
  (c) 0.796 V/rad; or .0139 V/deg

3.2 $7.5\angle -15°$; leads

3.3 (a) $G = \dfrac{R^2 + jRX_C}{R^2 + X_C^2} = \dfrac{R^2}{R^2 + (1/\omega C)^2} + j\dfrac{R/\omega C}{R^2 + (1/\omega C)^2}$

  (b) $|G| = \dfrac{R}{\sqrt{R^2 + (1/\omega C)^2}}$

62 TRANSFER FUNCTIONS

    (c) $\angle G = \arctan\left(\dfrac{1}{\omega RC}\right)$

    (d) $\lim\limits_{\omega \to \infty} |G| = 1; \quad \lim\limits_{\omega \to 0} |G| = 0$

    (e) $\lim\limits_{\omega \to \infty} \angle G = 0°; \quad \lim\limits_{\omega \to 0} \angle G = 90°$

3.4 (a) $0.5 + j.5 = 0.707 \angle 45°$
    (b) $3.54 \angle 45°$

3.5 0.2

3.7 (a) 2.5
    (b) 0.5
    (c) 25.13

3.8 956 Hz

3.9 4 × 4

3.10 13.56 dB

3.11 −15.56 dB

3.12 −12.04 dB, 2 octaves

3.14 Break frequency = 100 Hz

3.15 $2\pi \times 48 = \dfrac{1}{RC}; \quad \theta = -71.56°$

CHAPTER 4
LAPLACE TRANSFORMS

4.1 DIFFERENTIAL EQUATIONS

In science and technology, we frequently study the relationship between a variable and its *rate of change*. Examples are the relations between distance, velocity (which is the rate of change of displacement), and acceleration (the rate of change of velocity), the rate of change of voltage and current in capacitors and inductors, and many others. Accordingly, we frequently encounter equations that contain *derivatives*. The derivative of a variable with respect to time is, of course, the rate of change of that variable (with respect to time). An equation that contains one or more derivatives is called a *differential equation*. A *solution* to a differential equation is a function that satisfies the differential equation.

EXAMPLE 4.1

$dy/dt - y = 0$ is a differential equation. The function $y(t) = e^t$ is a solution, since $dy/dt = d(e^t)/dt = e^t$, and when we substitute $y = e^t$ and $dy/dt = e^t$ in the differential equation, we see that the equation is satisfied:

$$e^t - e^t = 0$$

Of course, not all differential equations involve derivatives with respect to time. However, in our study of control systems and networks, we will only be concerned with the time rate of change of variables and so will confine our discussion to derivatives with respect to time.

64 LAPLACE TRANSFORMS

A differential equation may also contain derivatives of higher *order* (second derivatives, third derivatives, etc.). For example, we know that acceleration is the second derivative of distance (or displacement) with respect to time: $a = d^2s/dt^2$, where s is linear displacement. The order of a differential equation is by definition the same as the order of the highest order derivative it contains. The *degree* of a differential equation is the same as the degree of its highest order derivative. For example, the differential equation in Example 4.1 is first order and first degree, since the highest order derivative is the first derivative, and this derivative is raised to unity power.

EXAMPLE 4.2

$(d^2y/dt^2)^3 - 2(dy/dt)^4 + y = 5$ is a second-order, third-degree equation. We will be primarily concerned with first- and second-order differential equations of first degree.

It is conventional to use dot notations when writing derivatives with respect to time. In this notation, a derivative is represented by writing the variable symbol and a number of dots over it equal to the order of the derivative. For example, the differential equations in Examples 4.1 and 4.2 would be written, respectively, as: $\dot{y} - y = 0$ and $\ddot{y}^3 - 2\dot{y}^4 + y = 5$. We will adopt this notation hereafter.

A differential equation may have an infinite number of solutions. For example, the differential equation in Example 4.1, $\dot{y} - y = 0$, is satisfied by $y = e^t$, $y = 2e^t$, $y = -e^t$, $y = .01e^t$, and, in fact, *any* function of the form $y = ce^t$, where c is an arbitrary constant. (The student should verify this fact.) A solution that contains one or more arbitrary constants is called a *general* solution, while a solution in which the arbitrary constant(s) have a particular value is called a *particular* solution. Thus, $y = e^t$ is a particular solution to the equation in Example 4.1 with arbitrary constant $c = 1$, while $y = 2e^t$ is a particular solution with $c = 2$. The number of arbitrary constants in the general solution to a differential equation is equal to the order of the differential equation. The general solution to the differential equation in Example 4.2 would contain two arbitrary constants.

A particular solution is obtained from a general solution by specifying certain *initial conditions*; that is, a value specified for a variable (or its derivative) at $t = 0$.

EXAMPLE 4.3

Find a solution to the differential equation $y + \dot{y} = 0$ subject to the initial condition $y(0) = -5$.

SOLUTION

$y = ce^{-t}$ is a general solution, since $\dot{y} = -ce^{-t}$ and $y + \dot{y} = ce^{-t} - ce^{-t} = 0$. Invoking the initial condition, $y(0) = ce^{-0} = -5$, $c(1) = -5$, or $c = -5$, we obtain the particular solution $y = -5e^{-t}$.

Since the general solution to a second-order differential equation would contain two arbitrary constants, we would have to specify initial conditions on both y and $\dot{y}$ in order to obtain a particular solution; that is, we would have to state values for $y(0)$ and $\dot{y}(0)$.

EXAMPLE 4.4

Given that $y = c_1 e^{-t} + c_2 e^{4t}$ is a general solution to the differential equation $\ddot{y} - 3\dot{y} = 4y$, find the particular solution corresponding to $y(0) = 0$ and $\dot{y}(0) = 1$.

SOLUTION

$$y(0) = 0 = c_1 e^{-0} + c_2 e^{0} \tag{1}$$

$$0 = c_1 + c_2$$

also

$$\dot{y} = \frac{dy}{dt} = -c_1 e^{-t} + 4c_2 e^{4t}$$

$$\dot{y}(0) = 1 = -c_1 e^{-0} + 4c_2 e^{0} \tag{2}$$

$$1 = -c_1 + 4c_2$$

Solving Equations 1 and 2 simultaneously for c_1 and c_2, we find $c_1 = -1/5$ and $c_2 = 1/5$. Thus, $y = -1/5(e^{-t}) + 1/5(e^{4t})$ is the particular solution.

4.2 DIFFERENTIAL EQUATIONS IN ELECTRONICS AND MECHANICS

Since the study of control systems theory involves the analysis of *electromechanical* systems, we may expect to encounter differential equations that relate both electrical and mechanical variables to their derivatives. This section reviews some of the fundamental realtionships in electrical networks and in mechanical systems and illustrates how these relationships may be stated in the form of differential equations.

As a first example, recall that the voltage across an inductor L is related to the time rate of change of the current through it by

$$e_L = L \frac{di_L}{dt} \tag{3}$$

where L is the inductance in henries. We may apply this relationship when writing Kirchoff's voltage law around a closed loop consisting, for example, of an inductor in series with a resistor.

EXAMPLE 4.5

In the network shown in Figure 4.1, the switch is closed at $t = 0$. There is no initial current in the inductor. Write the differential equation for the current in the loop and verify that its particular solution is

$$i_L(t) = \frac{E}{R} - \frac{E}{R}e^{-(R/L)t} \qquad (4)$$

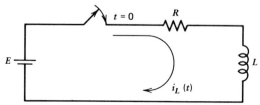

FIGURE 4.1 RL network (Example 4.5).

SOLUTION

Applying Kirchoff's law after the switch is closed, we have

$$E = i_L R + L\frac{di_L}{dt}$$

To verify that

$$i_L(t) = \frac{E}{R} - \frac{E}{R}e^{-(R/L)t}$$

is a solution, we must first find di_L/dt

$$\frac{di_L}{dt} = \frac{d[E/R - E/R\, e^{-(R/L)t}]}{dt}$$

$$= 0 - \frac{R}{L}\left[-\frac{E}{R}e^{-(R/L)t}\right] = \frac{E}{L}e^{-(R/L)t}$$

Then, substituting i_L and di_L/dt in the original equation, we obtain

$$E = \left[\frac{E}{R} - \frac{E}{R}e^{-(R/L)t}\right]R + L\left[\frac{E}{L}e^{-(R/L)t}\right]$$

$$= E - Ee^{-(R/L)t} + Ee^{-(R/L)t}$$

$$= E$$

and we see that the differential equation is satisfied. The fact that there was no initial current in the inductor corresponds to the initial condition $i(0) = 0$. Note that from Equation 4, $i_L(0) = 0 = E/R - E/R$.

The current through a capacitor is related to the rate of change of the voltage across it by

4.2 DIFFERENTIAL EQUATIONS IN ELECTRONICS AND MECHANICS

$$i_c = C \frac{dv_c}{dt} \qquad (5)$$

That is, current through a capacitor is proportional to the rate of change of the voltage across it, the constant of proportionality being the capacitance C in farads.* One well-known result of this fact is that the current through a capacitor is zero when the voltage is dc, which follows from the fact that the rate of change of a dc voltage (being constant with respect to time) is zero. Since $i_c = C\, dv_c/dt$, it is also true that

$$v_c = \frac{1}{C} \int_0^t i_c\, dt \qquad (6)$$

This follows from integrating both sides of Equation 5 or simply from the fact that if one variable is the derivative of another, then the second is by definition the integral of the first. This fact is frequently used to convert an equation involving an integral into a differential equation, as illustrated by Example 4.6.

EXAMPLE 4.6

A 12-V dc source is applied to a series RC circuit at $t = 0$. If $R = 10$ K and $C = 0.1$ μf, find the differential equation for the current in the circuit, and show that $i = 1.2 \times 10^{-3} e^{-10^3 t}$ is the (*particular*) solution. Assume the capacitor is initially uncharged. (See Figure 4.2.)

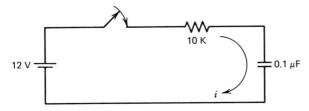

FIGURE 4.2 RC network (Example 4.6).

SOLUTION

Application of Kirchoff's voltage law around the loop shows that

$$12 = i(10^4) + \frac{1}{0.1 \times 10^{-6}} \int_0^t i\, dt$$

Differentiating both sides of this equation in order to obtain a differential equation,

---

*This fact can be derived from the fundamental relationship of charge, voltage, and capacitance, namely, $Q = CV$. Taking the derivative of both sides, we have $dQ/dt = C\, dv/dt$. But dQ/dt, the time rate of change of charge in Cs/sec is by definition equal to current. Thus, $i = C\, dv/dt$.

we find

$$\frac{d(12)}{dt} = 0 = 10^4 \frac{di}{dt} + 10^7(i)$$

A *general* solution to this equation is $i = ce^{-10^3 t}$, where c is an arbitrary constant. This can be verified as follows:

$$\frac{di}{dt} = -10^3 ce^{-10^3 t}$$

Substituting into the differential equation,

$$10^4(-10^3 ce^{-10^3 t}) + 10^7(ce^{-10^3 t})$$
$$= -10^7 ce^{-10^3 t} + 10^7 ce^{-10^3 t} = 0$$

and the equation is satisfied. To find the particular solution, we note that when the switch is first closed, the current is $E/R = 12/10^4 = 1.2$ mA, since the initial voltage on the capacitor is assumed to be zero. Hence,

$$i(0) = 1.2 \times 10^{-3} = ce^{-10^3(0)} = c$$

So $i = 1.2 \times 10^{-3} e^{-10^3 t}$ is the required particular solution.

A *translational* mechanical system is one where displacement is along a straight line, as opposed to a *rotational* mechanical system where displacement is an angular rotation. Although control systems are most often concerned with the angular position or angular velocity of a shaft (such as a motor shaft that is positioning a mechanical load or rotating it at a certain angular velocity), we will illustrate the application of differential equations to the more familiar translational systems for the time being and reserve our discussion of angular systems for a later chapter.

The three parameters of every mechanical system are *spring constant, damping,* and *mass*. Each of these parameters is responsible for a *reaction force* whose magnitude may depend upon displacement, velocity, or acceleration. The force exerted by a spring is directly proportional to the displacement of the spring, the constant of proportionality being the spring constant K (see Figure 4.3).

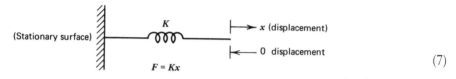

(7)

FIGURE 4.3 The mechanical spring.

For example, the force exerted by a spring with spring constant, say, 15 N/m, is 15 Newton (N) for every meter of displacement x (or, in more practical terms, 0.15 N for every cm of displacement). That the units of spring constant are force/

displacement can be verified by solving Equation 7 for K

$$K = \frac{F}{x} = \frac{\text{force}}{\text{displacement}}$$

For example, the force exerted by a spring with spring constant 5 N/m when it is stretched 4.2 cm is $F = Kx = (5 \text{ N/m})(.042 \text{ m}) = 0.21$ N.

Damping in a mechanical system, usually represented by a *dashpot*, exerts a force proportional to *velocity*. A good example of a dashpot is a shock absorber on an automobile. The force that resists the motion of the automobile when it strikes a chuckhole is proportional to the (vertical) velocity of the vehicle. If this force is large (due to good shock absorbers with large damping), then there will be very little vibration or displacement of the vehicle in a vertical direction. One method used to test for worn shock absorbers is to jump on the bumper of the automobile being tested. If the vehicle vibrates excessively, it is an indication that the retarding force exerted by the shock absorbers is too small, that is, the *damping constant* has deteriorated. The relationship between force and damping is given by

$$F = B\dot{x} \qquad (8)$$

where B is the damping constant. Since the units of $\dot{x}$ are displacement over time (for example, m/sec), we see that the units of B are

$$B = \frac{F}{\dot{x}} = \frac{\text{N}}{\text{m/sec}} = \frac{\text{N-sec}}{\text{m}}$$

Thus, a dashpot with damping constant $12 \frac{\text{N/sec}}{\text{m}}$ would exert a force of 12 N for every m/sec of velocity. The symbol for damping is shown in Figure 4.4.

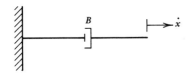

FIGURE 4.4 The mechanical damper (dashpot).

The force required to accelerate a mass m is given by Newton's familiar second law:

$$F = ma \qquad (9)$$

Recall that mass has units kilograms (kg), which are equivalent to N/m/sec², and can be found from weight W by dividing the weight by the gravitational constant g. On the surface of the earth, g is approximately 9.8 m/sec².

EXAMPLE 4.7

A 1-N force is applied to a body with weight 2 N. What will be the acceleration of this body?

SOLUTION

$$F = ma \qquad m = F/a$$

(Note that weight is a force.)

$$m = \frac{2 \text{ N}}{9.8 \text{ m/sec}^2} = 0.20408 \frac{\text{N-sec}^2}{m}$$

Then

$$a = \frac{F}{m} = \frac{1 \text{ N}}{0.20408 (\text{N/sec}^2)/m} = 4.90 \text{ m/sec}^2$$

Since the forces developed by the various parameters of a mechanical system have been shown to depend on displacement as well as velocity and acceleration, we might expect that an equation describing the behavior of a mechanical system (its displacement versus time) would be a differential equation. This is indeed the case, as illustrated in Example 4.8.

EXAMPLE 4.8

A mass of 4 kg connected to a spring with spring constant 12 N/m is subjected to a sudden (step) force of 24 N. Write the differential equation of motion (in terms of the displacement x) and show that $x = 2(1 - \cos \sqrt{3} t)$ is the particular solution. Assume zero initial displacement. (See Figure 4.5.)

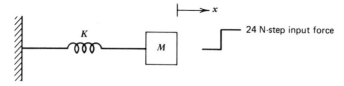

FIGURE 4.5 A spring-mass combination (Example 4.8).

SOLUTION

The applied force must equal the sum of the mechanical forces:

$$F = M\ddot{x} + Kx \tag{10}$$

Thus

$$24 = 4\ddot{x} + 12x \tag{11}$$

4.2 DIFFERENTIAL EQUATIONS IN ELECTRONICS AND MECHANICS

To show that $x = 2(1 - \cos \sqrt{3}t)$ is a solution, we find $\ddot{x}$

$$\dot{x} = \frac{d}{dt}[2(1 - \cos \sqrt{3}t)] = 2\sqrt{3} \sin \sqrt{3}t$$

$$\ddot{x} = 2\sqrt{3}(\sqrt{3}) \cos \sqrt{3}t = 6 \cos \sqrt{3}t$$

Substituting into Equation 11

$$24 = 24 \cos \sqrt{3}t + 12[2(1 - \cos \sqrt{3}t)]$$
$$= 24 \cos \sqrt{3}t + 24 - 24 \cos \sqrt{3}t$$
$$= 24$$

and the equation is satisfied. Note that $x(0) = 2(1 - \cos 0) = 2(1 - 1) = 0$, as required by the initial condition.

EXAMPLE 4.9

A spring is connected to a dashpot as shown in Figure 4.6. A step force F is applied to the system. Find the differential equation of motion and show that a solution is $x = (F/K)(1 - e^{-(K/B)t})$.

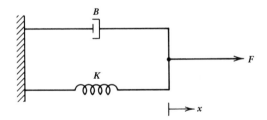

FIGURE 4.6 A spring-dashpot combination (Example 4.9).

SOLUTION

Again, we use the fact that the applied force must equal the sum of the forces due to each element in the system

$$F = B\dot{x} + Kx \qquad (12)$$

To verify the solution

$$\dot{x} = (-F/K)(-K/B)e^{-(K/B)t} = \frac{F}{B}e^{-(K/B)t}$$

$$F = B\left[\frac{F}{B}e^{-(K/B)t}\right] + K\left[\frac{F}{K}(1 - e^{-(K/B)t})\right]$$

$$= Fe^{-(K/B)t} + F - Fe^{-(K/B)t} = F$$

The solution is, in fact, the particular solution corresponding to $x(0) = 0$. (Verify this.)

4.3 LAPLACE TRANSFORMS

At this point, the student may legitimately wonder how to go about *finding* the solution to a differential equation. In all of the previous examples, the solutions to the differential equations have been presented in advance as if conjured up by magic, and it has only been necessary to verify that they are, in fact, solutions. There are many so-called classical methods that may be applied to differential equations in order to find solutions; these methods are very effective provided the equations appear in certain standard forms. Examples of these techniques include the methods called separation of variables, integrable combinations, and others. The utility of a particular technique depends upon the form of the equation to which a solution is desired, and determining which technique is applicable is often a matter of trial and error. Indeed, there are many differential equations to which no solution exists!

In engineering and technology, we are only concerned with differential equations that represent physically realizable systems and that, therefore, have solutions. We also rely on a systematic technique for finding the solution to a differential equation, namely, the technique called *LaPlace transformation*.

The classical methods used to find a solution to a differential equation are based on certain relationships between the variables in the equation and their derivatives and integrals. Consequently, these methods rely heavily on the investigator's knowledge of differential and integral calculus. LaPlace transformation, on the other hand, converts differentiation and integration to multiplication and division, so it is only necessary to be able to perform and apply purely algebraic operations (and use tables) in order to find the solution to a differential equation.

A *transformation* is merely a restatement of a functional relationship in terms of a different variable. For example, $y = 2t$ in a transformation from the set of all t-values to the set of y-values that are twice the value of t in magnitude. Thus, if $t = 1$, t is transformed to $y = 2$ by the transformation $y = 2t$. A somewhat more sophisticated example is found in the theory of logarithms. Two numbers x and y may be transformed to the "logarithm domain" by taking the logarithm of each number. In this case, multiplication of numbers is transformed to addition of numbers (in the logarithm domain), since the log of a product is the sum of the logarithms. In a similar fashion, LaPlace transformation of variables results in the conversion of differentiation and integration to multiplication and division, respectively. This greatly simplifies the manipulation of variables in a differential equation by converting differentiation and integration into simple algebraic operations.

These concepts will become clearer once we have defined and developed the theory of LaPlace transformation. It is recommended that the reader review these remarks once he or she has mastered the concepts of LaPlace transformation and understands the utility of this technique when searching for solutions to differential equations.

When a function $f(t)$ has been transformed by the LaPlace method, we use the notation $\mathcal{L}[f(t)]$ to denote this fact. $\mathcal{L}[f(t)]$ can be thought of as having been transformed to a different domain, just as taking the logarithm of a number transforms it to the "logarithm domain."

4.3 LAPLACE TRANSFORMS

The LaPlace transform of a function $f(t)$ is defined by

$$\mathcal{L}[f(t)] = \int_0^\infty f(t)e^{-st}\, dt \qquad (13)$$

This definition requires a few explanatory remarks. First, we must discuss the meaning of an integral whose limits are zero and infinity. This is called an improper integral, terminology that is, perhaps, misleading, since the integral can be readily evaluated in all practical situations and is, therefore, perfectly respectable for our purposes. If we wish to be rigorous in our mathematics, we would say that such an integral is defined by a *limit*, namely, the limit approached when the upper limit of the integral approaches infinity. That is

$$\int_0^\infty f(t)e^{-st}\, dt = \lim_{a \to \infty} \int_0^a f(t)e^{-st}\, dt$$

In all practical situations, this limit *exists* just as $\lim_{t \to \infty} e^{-t} = 0$ exists.

Another observation we may make about the definition of a LaPlace transform is that it will definitely be a function of the variable s. This is true because the integral is a *definite* integral with respect to t, that is, we substitute the limits $t = 0$ and $t = \infty$ once we have performed the integration, so it is certainly true that the variable t will not appear in the result. In fact, this integration constitutes transformation to the s-domain (more commonly called the *frequency* domain, as will be explained later), since the result is a function of s rather than t.

EXAMPLE 4.10

Find the LaPlace transform of $f(t) = e^{-at}$.

SOLUTION

Applying the definition of the LaPlace transform of $f(t) = e^{-at}$, we simply substitute $f(t) = e^{-at}$ in Equation 13 and proceed to evaluate the integral

$$\mathcal{L}(e^{-at}) = \int_0^\infty e^{-at} e^{-st}\, dt$$

$$= \int_0^\infty e^{-t(s+a)}\, dt = \frac{-1}{s+a}\left[e^{-t(s+a)}\right]_{t=0}^{t=\infty}$$

$$= \frac{-1}{s+a}\left[e^{-\infty(s+a)} - e^{-0(s+a)}\right]$$

$$= \frac{-1}{s+a}\left[0 - 1\right] = \frac{1}{s+a}$$

It is a very important fact illustrated in Example 4.10 that $\lim_{t \to \infty} e^{-t(s+a)} = 0$. This is simply an assertion that an exponentially decaying function becomes zero after the elapse of a sufficiently long time, a fact often used in the study of control systems. This limit is evident when the function $f(t) = e^{-KT}$ is plotted versus time (K is *any* positive constant); see Figure 4.7.

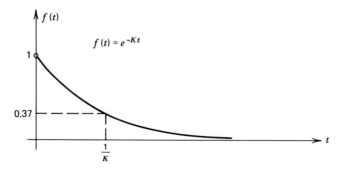

FIGURE 4.7 Plot of e^{-Kt} versus t.

Recall that $1/K$ is called the time constant and is the length of time it takes for $f(t)$ to decrease to a value equal to 0.37 times its maximum value.

EXAMPLE 4.11

Find the LaPlace transform of a function $f(t)$ that is zero for $t<0$ and 1 for $t>0$. This is an example of a *step function*, in this case a *unit* step function. A step function corresponds to a dc voltage that is switched into a circuit at $t = 0$ and is represented by the notation $u(t)$. $5u(t)$ represents a five-V dc source switched into a circuit at $t = 0$.

SOLUTION

$$\mathcal{L}[f(t)] = \int_0^\infty u(t)e^{-st}\, dt = \int_0^\infty e^{-st}\, dt$$

$$= -\frac{1}{s}e^{-st}\Big]_0^\infty = -\frac{1}{s}(0 - 1) = \frac{1}{s}$$

If $f(t) = 5u(t)$, then the corresponding LaPlace transform would be $5/s$.

The notation $F(s)$ is frequently used to denote the fact that the LaPlace transform of a function $f(t)$ has been transformed to the s-domain and is now a function of the variable s. Thus, in the previous example, we would write

$$\mathcal{L}(e^{-at}) = F(s) = \frac{1}{s + a}$$

It would *not* be correct to write $e^{-at} = 1/(s + a)$, since the left-hand side is in the so-called *time domain*, while the right-hand side is in the s-domain. In transforming the number 100 into the logarithm domain, we know that $\log_{10} 100 = 2$, but it would certainly not be correct to write $100 = 2$.

4.3 LAPLACE TRANSFORMS

An *impulse* function δ(t), sometimes called a *Dirac delta function*, is a function whose value is zero everywhere except at $t = 0$. The impulse function has the curious mathematical property of being infinite in magnitude and zero in width yet has a finite area called the *weight* of the function. The weight of δ(t) is 1, while the weight of Kδ(t) is K. Most rigorous mathematicians would deny the existence of such a function, but it plays a useful role as an approximation of many physical functions whose nonzero time durations are very short. The voltage spikes resulting from differentiating a square wave are a good example. If it were possible to produce a perfect square-wave voltage (one whose rise and fall time were zero), then differentiating this square wave would yield a series of positive and negative impulse functions, that is, functions of infinite magnitude occurring at points in time where the voltage changed value instantaneously (in zero time). Clearly, any voltage that changes value in zero time must have an infinite rate of change. Although such a voltage change is not possible, we frequently represent and approximate voltage changes that occur in very short-time intervals as impulse functions.

Differentiating the positive-going (leading) edge of a square wave yields a positive impulse function, while differentiating the trailing edge yields a negative impulse function, since it represents a negative rate of change. The weight of an impulse function is equal to the total change undergone by the square wave when the impulse is produced. These ideas are illustrated in Figure 4.8 where arrows are drawn to represent impulse functions.

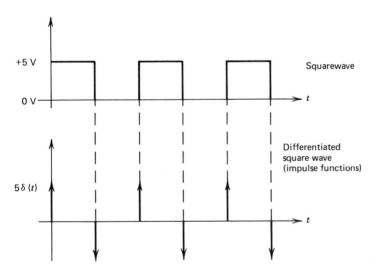

FIGURE 4.8 Differentiation of a square wave produces impulse functions.

EXAMPLE 4.12

Find the LaPlace transform of $f(t) = 10\delta(t)$.

$$F(s) = \mathscr{L}[10\delta(t)] = \int_0^\infty 10\delta(t)e^{-st}\, dt$$

76 LAPLACE TRANSFORMS

SOLUTION

Since $10\delta(t)$ is zero everywhere except at $t = 0$, the function being integrated need only be considered at $t = 0$. Since integration corresponds to finding the area under a function between the limits of integration and the area or weight of $10\delta(t)$ is 10, we have $F(s) = 10$.

EXAMPLE 4.13

Find the LaPlace transform of $\sin \omega t$.

SOLUTION

$$F(s) = \int_0^\infty (\sin \omega t) e^{-st}\, dt$$

Using the exponential identity in Equation 14 in Chapter 2, we may write

$$F(s) = \int_0^\infty \left(\frac{e^{j\omega t} - e^{-j\omega t}}{2j}\right) e^{-st}\, dt$$

$$= \frac{1}{2j} \int_0^\infty \left[e^{-t(s-j\omega)} - e^{-t(s+j\omega)}\right] dt$$

$$= \frac{1}{2j} \left[\frac{e^{-t(s-j\omega)}}{-(s-j\omega)} - \frac{e^{-t(s+j\omega)}}{-(s+j\omega)}\right]_0^\infty$$

$$= \frac{1}{2j} \left[0 - 0 + \frac{1}{s-j\omega} - \frac{1}{s+j\omega}\right]$$

$$= \frac{1}{2j} \left[\frac{(s+j\omega) - (s-j\omega)}{(s+j\omega)(s-j\omega)}\right]$$

$$= \frac{1}{2j} \left[\frac{s+j\omega - s + j\omega}{s^2 + \omega^2}\right] = \frac{\omega}{s^2 + \omega^2}$$

LaPlace transforms of time-domain functions can often be combined. This is done in order to find the LaPlace transform of algebraic combinations of time-domain functions. For example,

$$\mathcal{L}[f_1(t) \pm f_2(t)] = \mathcal{L}[f_1(t)] \pm \mathcal{L}[f_2(t)] \tag{14}$$

The transform of a sum is the sum of the transforms; this can be extended to three or more functions. It is also true that for any constant k,

$$\mathcal{L}[kf(t)] = k\, \mathcal{L}[f(t)] \tag{15}$$

EXAMPLE 4.14

Find the LaPlace transform of

(a) 20 sin 100t
(b) $u(t) - \delta(t) + e^{-4t}$
(c) $6\, u(t) - \dfrac{e^{-t}}{10}$

SOLUTION

(a) $\mathcal{L}[20 \sin 100t] = 20\, \mathcal{L}[\sin 100t]$

$$= \frac{2000}{s^2 + (100)^2}$$

(b) $\mathcal{L}[u(t) - \delta(t) + e^{-4t}]$

$= \mathcal{L}[u(t)] - \mathcal{L}[\delta(t)] + \mathcal{L}[e^{-4t}]$

$= \dfrac{1}{s} - 1 + \dfrac{1}{s+4}$

(c) $\mathcal{L}\left[6u(t) - \dfrac{e^{-t}}{10} \right]$

$= 6\, \mathcal{L}[u(t)] - \dfrac{1}{10} \mathcal{L}[e^{-t}]$

$= \dfrac{6}{s} - \dfrac{1/10}{s+t}$

We have begun to build a repertoire of LaPlace transforms of various time-domain functions. A time-domain function and its corresponding LaPlace transform is called a *transform pair*. Transform pairs are usually listed in tables to facilitate finding transforms as well as finding the time-domain function corresponding to a particular transform. These tables are similar to a table of numbers and their logarithms—we may find the logarithm of a given number or the number that corresponds to a given logarithm. Table 4-1 presents the transform pairs that we will refer to frequently in the future; note that this table contains LaPlace transforms for many functions that we have not yet derived.

TABLE 4-1 LAPLACE TRANSFORM PAIRS

f(t)	F(s)
1. $\delta(t)$ unit impulse	1
2. 1 or unit step $u(t)$	$\dfrac{1}{s}$

TABLE 4-1   LAPLACE TRANSFORM PAIRS (Continued)

f(t)	F(s)
3. t	$\dfrac{1}{s^2}$
4. $\dfrac{1}{(n-1)!} t^{n-1}$	$\dfrac{1}{s^n}$
5. e^{-at}	$\dfrac{1}{s+a}$
6. $\dfrac{1}{(n-1)!} t^{n-1} e^{-at}$	$\dfrac{1}{(s+a)^n}$
7. $\dfrac{1}{a}(1 - e^{-at})$	$\dfrac{1}{s(s+a)}$
8. $\dfrac{1}{ab}\left[1 - \left(\dfrac{b}{b-a}\right)e^{-at} + \left(\dfrac{a}{b-a}\right)e^{-bt}\right]$	$\dfrac{1}{s(s+a)(s+b)}$
9. $\dfrac{1}{ab}\left\{\alpha - \left[\dfrac{b(\alpha-a)}{b-a}\right]e^{-at} + \left[\dfrac{a(\alpha-b)}{b-a}\right]e^{-bt}\right\}$	$\dfrac{s+\alpha}{s(s+a)(s+b)}$
10. $\dfrac{1}{b-a}(e^{-at} - e^{-bt})$	$\dfrac{1}{(s+a)(s+b)}$
11. $\dfrac{1}{a-b}(ae^{-at} - be^{-bt})$	$\dfrac{s}{(s+a)(s+b)}$
12. $\dfrac{1}{b-a}\left[(\alpha-a)e^{-at} - (\alpha-b)e^{-bt}\right]$	$\dfrac{s+\alpha}{(s+a)(s+b)}$
13. $\sin \omega t$	$\dfrac{\omega}{s^2+\omega^2}$
14. $\cos \omega t$	$\dfrac{s}{s^2+\omega^2}$
15. $\dfrac{\sqrt{\alpha^2+\omega^2}}{\omega} \sin(\omega t + \theta)$ where $\theta = \arctan \dfrac{\omega}{\alpha}$	$\dfrac{s+\alpha}{s^2+\omega^2}$
16. $\dfrac{1}{\omega^2}(1 - \cos \omega t)$	$\dfrac{1}{s(s^2+\omega^2)}$
17. $\dfrac{\alpha}{\omega^2} - \dfrac{\sqrt{\alpha^2-\omega^2}}{\omega^2} \cos(\omega t + \theta)$ where $\theta = \arctan \dfrac{\omega}{\alpha}$	$\dfrac{s+\alpha}{s(s^2+\omega^2)}$
18. $\dfrac{1}{b} e^{-at} \sin bt$	$\dfrac{1}{(s+a)^2+b^2}$

TABLE 4-1 LAPLACE TRANSFORM PAIRS (Continued)

f(t)	F(s)
19. $e^{-at} \cos bt$	$\dfrac{s+a}{(s+a)^2 + b^2}$
20. For $\zeta < 1$, $\dfrac{1}{\omega_n \sqrt{1-\zeta^2}} e^{-\zeta\omega_n t} \sin(\omega_n\sqrt{1-\zeta^2}\, t)$	$\dfrac{1}{s^2 + 2\zeta\omega_n s + \omega_n^2}$
21. $\dfrac{1}{a^2}(at - 1 + e^{-at})$	$\dfrac{1}{s^2(s+a)}$
22. $\dfrac{1}{a^2}(1 - e^{-at} - ate^{-at})$	$\dfrac{1}{s(s+a)^2}$
23. $\dfrac{\sqrt{(\alpha-a)^2 + b^2}}{b} e^{-at} \sin(bt + \theta)$ where $\theta = \arctan \dfrac{b}{\alpha - a}$	$\dfrac{s+\alpha}{(s+a)^2 + b^2}$

It should be noted that Table 4-1 does not list *every* transform pair that has been worked out by investigators. In fact, entire books containing nothing but transform pairs have been published. Our table contains only the pairs that we will use most frequently in later work.

4.4 USING THE TABLE OF TRANSFORM PAIRS

The process of finding a time-domain function that corresponds to a given transform is called *inverse transformation*. The symbol $\mathcal{L}^{-1}[F(s)]$ is called the inverse transform of $F(s)$. Finding the inverse transform is analogous to finding an antilogarithm. For example, since

$$\mathcal{L}(e^{-at}) = \frac{1}{s+a}$$

it is also true that

$$\mathcal{L}^{-1}\left(\frac{1}{s+a}\right) = e^{-at}$$

Also

$$\mathcal{L}^{-1}(5) = 5\delta(t) \qquad \mathcal{L}^{-1}\left(\frac{1}{s^2 + \omega^2}\right) = \sin \omega t$$

and so on. It is often necessary to manipulate a transform algebraically in order to make it fit a form in the table.

EXAMPLE 4.15

Find the inverse LaPlace transforms of:

(a) $F(s) = \dfrac{12}{4 + 2s}$

(b) $F(s) = \dfrac{40}{s^2 + 25}$

SOLUTION

(a) Dividing numerator and denominator by 2, we obtain

$$F(s) = \frac{6}{s + 2} = 6\left(\frac{1}{s + 2}\right)$$

which fits the form of transform pair 5 in Table 4-1. Hence, $f(t) = 6e^{-2t}$.

(b) $F(s) = \dfrac{40}{s^2 + 25} = 8\left(\dfrac{5}{s^2 + 5^2}\right)$

therefore, from pair 13 in Table 4-1, $f(t) = 8 \sin 5t$.

In some published tables of LaPlace transform pairs, the form $F(s) = 1/(1 + Ts)$ is used. Since

$$\frac{1}{1 + Ts} = \frac{1/T}{s + 1/T}$$

we see that

$$f(t) = \frac{1}{T}e^{-t/T}$$

Notice also that we can algebraically manipulate some pairs in the table to obtain equivalent pairs. Consider pair 4 for example. Since multiplication of a function $f(t)$ by a constant results in multiplication of the corresponding $F(s)$ by the same constant, we may write pair 4 in the equivalent form

$$\mathcal{L}[t^{n-1}] = \frac{(n - 1)!}{s^n} \quad \text{or alternatively} \quad \mathcal{L}[t^n] = \frac{n!}{s^{n+1}}$$

It is frequently necessary to find the inverse transform of a function that has a quadratic in s (second-degree polynomial) in the denominator. Two examples are (1) $1/(s^2 + 5s + 6)$ and (2) $1/(s^2 + 4s + 7)$. In order to find the inverse transform, we search the table of transform pairs for functions $F(s)$ that have second-degree denominators. We discover that there are at least two possible candidates: pairs 10 and 18, since the denominator of the former may be written $s^2 + (a + b)s + ab$ and the denominator of the latter as $s^2 + 2as + (a^2 + b^2)$. The question is, Which of these forms do we use for each of the cases (1) and (2)? The solution lies in our

4.4 USING THE TABLE OF TRANSFORM PAIRS

ability, or lack thereof, to factor the denominators of (1) and (2) into factors containing only *real* numbers. We note in case (1) that

$$\frac{1}{s^2 + 5s + 6} = \frac{1}{(s + 3)(s + 2)}$$

and, therefore, transform pair 10 applies

$$\mathcal{L}^{-1}\left[\frac{1}{(s + 3)(s + 2)}\right] = \frac{1}{(2 - 3)}(e^{-3t} - e^{-2t}) = e^{-2t} - e^{-3t}$$

On the other hand, the denominator in case 2 *cannot* be factored into factors containing real numbers (the factors are complex), and in cases such as this, we use pair 18. It is first necessary to manipulate the denominator in case 2 so that it fits the form of pair 18, which is accomplished by completing the square. Recall that to complete the square, we add and subtract the square of one-half of the coefficient of s to the quadratic

$$\frac{1}{s^2 + 4s + 5} = \frac{1}{s^2 + 4s + (2)^2 + 7 - (2)^2} = \frac{1}{(s + 2)^2 + 3}$$

Then, applying pair 18, we have

$$\mathcal{L}^{-1}\left[\frac{1}{(s + 2)^2 + (\sqrt{3})^2}\right] = \frac{1}{\sqrt{3}}e^{-2t} \sin\sqrt{3}\, t$$

One last observation on using the table relates to finding the transform of sinusoidal time functions. Suppose, for example, we wished to find the LaPlace transform of 10 sin (50t + 30°). We observe that pair 15 contains a term of the form sin (ωt + θ), and with considerable algebraic manipulation, we may be able to force our example to fit this case. However, a more direct technique, which is also frequently more desirable, is to resolve our sinusoidal function into a sine component with zero phase angle and a cosine component with zero phase angle. This may be accomplished using a phasor diagram as illustrated in Figure 4.9.

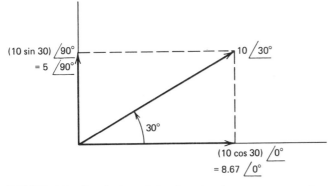

FIGURE 4.9 Resolution of sine function into components for ease of transformation.

Thus, $10 \sin(50t + 30°) = 8.67 \sin 50t + 5 \cos 50t$, and we may use pairs 13 and 14 to find the transform

$$\mathcal{L}\left[8.67 \sin 50t + 5 \cos 50t\right] = \frac{(8.67)(50)}{s^2 + 50^2} + \frac{5}{s^2 + 50^2} = \frac{5s + 433.5}{s^2 + 2500}$$

4.5 TRANSFORMS OF DERIVATIVES AND INTEGRALS

Suppose we know the transform $F(s)$ of a particular time function $f(t)$ and we wish to find the transform of the *derivative* of $f(t)$. That is, we know $\mathcal{L}[f(t)] = F(s)$, and we want to know $\mathcal{L}[df/dt]$, which may be found by using the following relationship:

$$\mathcal{L}\left[\frac{df}{dt}\right] = sF(s) - f(0) \tag{16}$$

Equation 16 tells us to multiply the transform of $f(t)$ by s and subtract $f(t)$ evaluated at $t = 0$ in order to obtain the transform of df/dt.

EXAMPLE 4.16

From Table 4-1, we know that $\mathcal{L}[e^{-at}]$ is $1/(s + a)$. Use Equation 16 to find the LaPlace transform $\mathcal{L}[de^{-at}/dt]$.

SOLUTION

$$\mathcal{L}\left[\frac{de^{-at}}{dt}\right] = s\left(\frac{1}{s + a}\right) - e^{-at}\bigg|_{t=0} = \frac{s}{s + a} - 1$$

$$= \frac{s - (s + a)}{s + a} = \frac{-a}{s + a}$$

We can verify this fact by observing that

$$\frac{d(e^{-at})}{dt} = -ae^{-at} \quad \text{and} \quad \mathcal{L}[-ae^{-at}] = \frac{-a}{s + a}$$

It is also true that if $\mathcal{L}[f(t)] = F(s)$, then

$$\mathcal{L}\left[\frac{d^2f(t)}{dt^2}\right] = s^2F(s) - sf(0) - \frac{df}{dt}\bigg|_{t=0} \tag{17}$$

Continuing Example 4.16,

$$\mathcal{L}\left[\frac{d^2 e^{-at}}{dt^2}\right] = s^2\left(\frac{1}{s+a}\right) - s(1) - \left[-ae^{-at}\right]_{t=0}$$

$$= \frac{s^2}{s+a} - s + a$$

$$= \frac{s^2 - s(s+a) + a(s+a)}{s+a}$$

$$= \frac{a^2}{s+a}$$

a result which is verified by noting that

$$\frac{d^2 e^{-at}}{dt^2} = a^2 e^{-at}$$

Suppose that we once again know the transform $F(s)$ corresponding to some $f(t)$ and now wish to find the transform of the integral of $f(t)$, namely, $\mathcal{L}\int_0^t f(t)dt$. (The upper limit of this integral is an artibrary t—recall that the limits are values of t—because otherwise, we would not have a time-domain function after integrating.) The rule for finding this transform is

$$\mathcal{L}\left[\int_0^t f(t)\,dt\right] = \frac{F(s)}{s} \qquad (18)$$

In other words, divide the original transform by s in order to obtain the transform of the integral.

EXAMPLE 4.17

By using Equation 18, derive transform pair 3 from transform pair 2.

SOLUTION

In this case, the known relation is assumed to be

$$\mathcal{L}[f(t)] = \mathcal{L}[1] = \frac{1}{s} = F(s)$$

Since $\int_0^t 1\,dt = t$, we can find $\mathcal{L}[t]$ by writing

$$\frac{F(s)}{s} = \frac{1/s}{s} = \frac{1}{s^2}$$

Equations 16, 17, and 18 illustrate a point that was made earlier in regard to the utility of LaPlace transformation. We see that differentiation and integration in the time domain are converted to *multiplication* and *division* (by s) in the s-domain. This fact greatly simplifies manipulating differential equations when searching for solutions once the equations have been transformed to the s-domain. Example 4.18 illustrates the point.

EXAMPLE 4.18

Find a solution to the differential equation $\dot{y} + y = 0$ subject to the initial condition $y(0) = 2$.

SOLUTION

We take the LaPlace transform of the equation and obtain

$$sY(s) - y(0) + Y(s) = 0$$

where $Y(s)$ represents the transform of $y(t)$. (It is conventional to use capital letters to designate the transform of a function.) Note that our ultimate task is to find $y(t)$, but our immediate task is to find its transform $Y(s)$. Thus, $Y(s)$ is the "unknown" in the equation. Note also that the transform of $\dot{y}(t)$ is $sY(s) - 2$, since $\dot{y}$ is the first derivative of y and $y(0) = 2$. Solving for $Y(s)$, we find

$$Y(s)(s + 1) = 2$$

$$Y(s) = \frac{2}{s + 1}$$

We can now find $y(t)$ by taking the inverse transform of $Y(s)$

$$y(t) = \mathscr{L}^{-1}\left[\left(\frac{2}{s + 1}\right)\right] = 2e^{-t}$$

Example 4.18 should be studied very carefully and thoroughly understood. It is our first example of the most important practical use of LaPlace transforms, a use that we will see often again. We have solved a differential equation using ordinary algebra and a table of LaPlace transform pairs. Many beginners ask, What is s, anyway? For our purposes, it doesn't really matter. (It is, in fact, a complex variable.) Note that we never actually solve for s; we are interested in finding $Y(s)$ and $y(t)$, and s itself is simply a transformed version of t, reminding us that we are in a different domain.

4.6 TRANSFER FUNCTIONS IN THE FREQUENCY DOMAIN

Since electrical voltages and currents are represented by time-domain functions, we can certainly find the LaPlace transforms of such functions. For example, the

4.6 TRANSFER FUNCTIONS IN THE FREQUENCY DOMAIN

ac current $i(t) = .02 \cos 10^3 t$ has transform $I(s) = .02s/(s^2 + 10^6)$ from transform pair 14.

Likewise, an impedance may be transformed to the s-domain. This transformation is very simple. When the impedance is expressed in phasor form, that is, as a function of $j\omega$, simply replace $j\omega$ by s. This is an important and useful result, and justifies using the term frequency domain when referring to the s-domain. Capacitive reactance $X_c = 1/j\omega C$ thus becomes $X_c = 1/sC$, while inductive reactance $X_L = j\omega L$ becomes sL. These results follow from the voltage and current relations in inductors and capacitors: Since

$$V_L = L \, di/dt \quad \text{we have} \quad V_L(s) = Lsi(s)$$

or

$$\frac{V_L(s)}{I(s)} = Ls \tag{19}$$

and

$$V_c = \frac{1}{C} \int_0^t i \, dt$$

implies

$$V_c(s) = \frac{1}{C}\left(\frac{I(s)}{s}\right)$$

so

$$\frac{V_c(s)}{I(s)} = \frac{1}{Cs} \tag{20}$$

Any algebraic operations that can be legitimately performed to combine impedances in phasor form can also be performed on their transforms.

EXAMPLE 4.19

Find the equivalent impedance of the network shown in Figure 4.10 in the frequency domain.

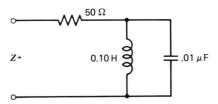

FIGURE 4.10 Network for Example 4.19.

SOLUTION

$$Z = 50 + \frac{(0.1s)(1/10^{-8}s)}{0.1s + (1/10^{-8}s)}$$

$$= 50 + \frac{10^7}{(10^{-9}s^2 + 1)/10^{-8}s} = 50 + \frac{10^{-1}s}{10^{-9}s^2 + 1}$$

$$= 50 + \frac{10^8 s}{s^2 + 10^9} = \frac{50(s^2 + 10^9) + 10^8 s}{s^2 + 10^9}$$

$$= \frac{50s^2 + 10^8 s + 50 \times 10^9}{s^2 + 10^9}$$

Since a transfer function G is by definition the ratio of an output function, whose transform may be found, to an input function that also has a transform, it is reasonable to expect that G may be also expressed as a transform. This is, indeed, the case. For example, if the output of a device has transform $s/(s^2 + \omega^2)$ while its input has transform $1/(s^2 + \omega^2)$, then the transfer function is

$$G(s) = \frac{s/(s^2 + \omega^2)}{(s^2 + \omega^2)} = s$$

(The device is a differentiator; why?)

As we have seen, transfer functions may often be found by applying the voltage-divider rule. In these cases, the transfer function is expressed as the ratio of impedances; and if these impedances are expressed in the frequency domain, so again will be the transfer function.

EXAMPLE 4.20

Find the transfer function of the network shown in Figure 4.11.

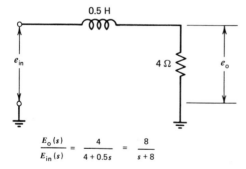

$$\frac{E_o(s)}{E_{in}(s)} = \frac{4}{4 + 0.5s} = \frac{8}{s + 8}$$

FIGURE 4.11 Network for Example 4.20.

When performing calculations in the frequency domain, we often switch notation ($s = j\omega$) in order to find magnitudes and angles at particular frequencies.

EXAMPLE 4.21

The input to a device with transfer function $G(s) = 1/(1 + s)$ is $e_{in} = 40\angle 0°$. Find the magnitude and angle of the output at $\omega = 2$ rad/sec.

SOLUTION

Substituting $s = j2$ into $G(s)$,

$$G(j2) = \frac{1}{1 + j2}$$

$$= \frac{1 - j2}{1^2 + 2^2} = \frac{1}{5} - j\frac{2}{5} = \sqrt{0.2} \angle -63.4°$$

Then

$$e_o = e_{in} G(j2)$$

$$= (40\angle 0°)(\sqrt{0.2}\angle -63.4)° = 17.89\angle -63.4°$$

Thus, the output has magnitude 17.89 and lags the input by 63.4°.

4.7 INVERSE TRANSFORMATION BY PARTIAL FRACTIONS

When using the LaPlace transform method to find solutions to differential equations, we often obtain a solution $Y(s)$ whose inverse is not listed in the table of transform pairs. Suppose, for example, that

$$Y(s) = \frac{s + 1}{s(s + 2)(s + 3)}$$

In order to find the inverse transform $y(t)$ corresponding to $Y(s)$, it is necessary to convert $Y(s)$ to some algebraically equivalent form that appears in the table.

One method for converting $Y(s)$ is called the method of *partial fractions*. The method is best explained by way of an example.

First write $Y(s)$ as the sum of simple fractions with unknown coefficients A, B, C

$$Y(s) = \frac{s^2 + 1}{s(s + 2)(s + 3)} = \frac{A}{s} + \frac{B}{s + 2} + \frac{C}{s + 3}$$

Note that there is one fraction for each *factor* in the denominator of $Y(s)$. To find A, multiply $Y(s)$ by the denominator of the first fraction and evaluate the result at a value of s that makes the denominator of the first fraction equal zero.

$$A = sY(s)\bigg]_{s=0} = \frac{s(s^2 + 1)}{s(s + 2)(s + 3)}\bigg]_{s=0} = \frac{1}{6}$$

Similarly, to find B

$$B = (s+2)Y(s)\Big]_{s=-2} = \frac{(s+2)(s^2+1)}{s(s+2)(s+3)}\Big]_{s=-2} = \frac{-5}{2}$$

Note that in this case $s = -2$ is the value of s that makes the denominator of the second fraction equal to zero.

Finally,

$$C = (s+3)Y(s)\Big]_{s=-3} = \frac{(s+3)(s^2+1)}{s(s+2)(s+3)}\Big]_{s=-3} = \frac{(s^2+1)}{s(s+2)}\Big]_{s=-3} = \frac{10}{3}$$

We now have:

$$Y(s) = \frac{s^2+1}{s(s+2)(s+3)} = \frac{1/6}{s} - \frac{5/2}{s+2} + \frac{10/3}{s+3}$$

and, therefore, we can use Table 4-1 to find

$$y(t) = \frac{1}{6} - \frac{5}{2}e^{-2t} + \frac{10}{3}e^{-3t}$$

The general rule for expressing a function $F(s)$ as a sum of partial fractions may be stated as follows:

If

$$F(s) = \frac{P(s)}{(s+a)(s+b)\ldots(s+z)}$$

where $P(s)$ is an arbitrary polynomial in s and $a, b, \ldots, z$ are all distinct, write

$$F(s) = \frac{A}{s+a} + \frac{B}{s+b} + \ldots + \frac{Z}{s+z} \tag{21}$$

Then

$$A = (s+a)F(s)\big|_{s=-a}$$
$$B = (s+b)F(s)\big|_{s=-b}$$
$$\vdots \quad \vdots \quad \vdots \tag{22}$$
$$Z = (s+z)F(s)\big|_{s=-z}$$

Note that one of the a, b, and so forth, may be zero.

Once the coefficients of a partial fraction expansion have been determined, it is a good idea to check them by combining the fractions and verifying that their

4.7 INVERSE TRANSFORMATION BY PARTIAL FRACTIONS

sum is equal to the original Y(s). Checking the above partial fraction expansion,

$$\frac{1/6}{s} - \frac{5/2}{s+2} + \frac{10/3}{s+3}$$

$$= \frac{1/6(s+2)(s+3) - 5/2(s)(s+3) + 10/3(s)(s+2)}{s(s+2)(s+3)}$$

$$= \frac{1/6(s^2 + 5s + 6) - 5/2(s^2 + 3s) + 10/3(s^2 + 2s)}{s(s+2)(s+3)}$$

$$= \frac{(1/6)s^2 - (5/2)s^2 + (10/3)s^2 + (5/6)s - (15/2)s + (20/3)s + 1}{s(s+2)(s+3)}$$

$$= \frac{s^2 + 1}{s(s+2)(s+3)}$$

as required.

The inverse transform in this example could also have been found using Table 4-1 and the fact that F(s) may be written

$$F(s) = \frac{s}{(s+2)(s+3)} + \frac{1}{s(s+2)(s+3)}$$

The reader is invited to show by using Table 4-1 that the inverse transformation of this expression is equivalent to the inverse transformation found by using partial fractions.

EXAMPLE 4.22

Find the solution of the differential equation in Example 4.5 using LaPlace transforms and partial fractions.

$$E = i_L R + L \frac{di_L}{dt}$$

SOLUTION

Note that in this example the dc voltage E is a *constant*—actually a step function Eu(t) applied to the circuit at $t = 0$—while i_L is a *variable*—the unknown whose solution we desire. Transforming the equation,

$$\frac{E}{s} = I(s)R + LsI(s) - i(0)$$

Since $i(0) = 0$ (no initial current in the inductor),

$$\frac{E}{s} = I(s)R + Ls$$

$$I(s) = \frac{E/s}{R + Ls} = \frac{E/L}{s(s + R/L)} = \frac{A}{s} + \frac{B}{(s + R/L)}$$

$$A = s\left[\frac{E/L}{s(s + R/L)}\right]_{s=0} = \frac{E/L}{R/L} = E/R$$

$$B = (s + R/L)\left[\frac{E/L}{s(s + R/L)}\right]_{s=-R/L} = \frac{E/L}{-R/L} = \frac{-E}{R}$$

So

$$I(s) = \frac{E/R}{s} - \frac{E/R}{s + R/L}$$

and, therefore,

$$i_L(t) = \frac{E}{R} - \frac{E}{R}e^{-Rt/L}$$

$$= \frac{E}{R}(1 - e^{Rt/-L})$$

Example 4.22 was chosen to illustrate the method of partial fractions, although the inverse transform $i(t)$ could have been found directly from Table 4-1. (The reader should verify that the same result is obtained when using the table.)

It is frequently instructive to plot the time-domain function versus time. This function, after all, represents what is "really" happening in the system or circuit. In Example 4.22, $i_L(t)$ specifies what the actual current in the circuit will be at any time t. It is plotted in Figure 4.12.

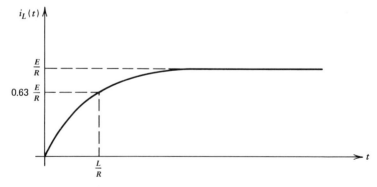

FIGURE 4.12 Plot of $i_L(t)$ versus time (Example 4.22).

4.7 INVERSE TRANSFORMATION BY PARTIAL FRACTIONS

Figure 4.12 illustrates the familiar rise of current in a series LR-circuit. It is well known that after a long period of time, the inductor will have no effect on the current in the circuit; that is, as $t \to \infty$, the current is determined only by the resistance R, $i = E/R$. Note that the time constant in this case is L/R, and it is the length of time required for the current to rise to 0.63 times its maximum value.

Example 4.22 illustrates once again that the time-domain response of a physical system, that is, the behavior of a current, voltage, or some other variable as a function of time is represented by a particular solution to a differential equation. It is convenient to regard this solution as being comprised of two parts: the *transient* part and the *steady-state* part, often referred to as the transient solution and the steady-state solution. The steady-state solution represents the condition that ultimately prevails, for example, the final voltage or current, the state of affairs as t approaches infinity. In Example 4.22, the steady-state solution is $I = E/R$, the ultimate current in the circuit. The transient solution, on the other hand, is that short-lived part of the response that represents how the system is changing from $t = 0$ until steady-state conditions are reached. The rising portion of the current in Figure 4.12 is the transient part of the solution in Example 4.22.

The steady-state solution for a given problem is not always a constant but depends rather upon the nature of the input to the system. In Example 4.22, if a sinusoidal voltage had been applied to the network at $t = 0$, then the steady-state solution would have been a sinusoidal current, a current with constant *peak* value, but a current changing with time, nonetheless. The transient part of the solution in that case would have been an oscillatory current with gradually rising amplitude. While the nature of the steady-state solution depends on the type of input, the nature of the transient solution depends on the system parameters. In Example 4.21, the time constant associated with the transient rise in current depends on the magnitudes of the system parameters L and R. LaPlace transform analysis is a particularly valuable way of investigating the transient behavior of systems. Students who have observed and measured ac-circuit behavior in elementary circuit-analysis classes have probably never appreciated the fact that there is always a transient buildup of voltage and current when the signal generator is first connected to an RC or RL circuit. By the time the waveform is observed on an oscilloscope, only the steady-state conditions are apparent. But in many practical systems, especially control systems, the transient behavior after the sudden application of an input is the most important characteristic of the system.

In developing the theory of partial fractions, we have referred to "the value of s that makes the denominator zero." Such a value of s is called a *pole*. A pole of any function $F(s)$ is a value of s that makes the denominator of $F(s)$ zero and, hence, makes $F(s)$ itself *infinite*. Thus, $s = 0$, $s = -2$, and $s = -3$ are all poles of $F(s) = (s + 1)/s(s + 2)(s + 3)$. A *zero* of a function $F(s)$ is a value of s that make $F(s)$ equal to zero; in the latter example, $s = -1$ is a zero of $F(s)$.

It can be seen that poles are the algebraic *roots* of the denominator. If the denominator is written as a product of factors, as in our example, it is only necessary to set each factor equal to zero and solve each of the resulting equations for s in order to find the roots [poles of $F(s)$]. Thus,

$$s = 0$$
$$s + 2 = 0 \Rightarrow s = -2$$
$$s + 3 = 0 \Rightarrow s = -3$$

If, however, the denominator is expressed as an algebraic polynomial in s, it may be necessary to factor it in order to find the roots. Also, it may well be that the roots are complex rather than real, as illustrated in Example 4.23.

EXAMPLE 4.23

Find the poles of $F(s) = 4s/(s^2 + 2s + 10)$.

SOLUTION

The quadratic formula may be used to find the roots (and factors) of the denominator. The equation $s^2 + 2s + 10 = 0$ has solutions

$$s = \frac{-2 \pm \sqrt{2^2 - 4 \times 10}}{2}$$

$$= -1 \pm \frac{\sqrt{-36}}{2} = -1 \pm j3$$

The poles of $F(s)$ are, therefore, the complex conjugates $(-1 + j3)$ and $(-1 - j3)$. Complex poles *always* occur in complex conjugate pairs. We could write $F(s)$ in factored form as

$$F(s) = \frac{4s}{(s + 1 - j3)(s + 1 + j3)}$$

Note that the factors always have opposite signs from the poles. This follows from the fact that we set the factors equal to zero in order to find the poles. For example, the factor $(s + 1 - j3) = 0 \Rightarrow s = -1 + j3$ is a pole.

An important theorem called the *fundamental theorem of algebra* states that the total number of roots of a polynomial is equal to the *degree* of the polynomial. For example, $s^4 + 3s^2 - 2s + 1$ is a fourth-degree polynomial and, therefore, has four roots. If any of these roots are complex, there will be an even number of them (two or four in this case), since *complex roots always occur in conjugate pairs*.

There are several methods of finding the inverse transform of $F(s)$ when the poles are complex, including the method of partial fractions. When $F(s)$ is expressed as a sum of partial fractions, there must be one fraction for each complex conjugate pole. The same technique as previously outlined may be used to evaluate the coefficients A, B, C, and so forth, of the fractions. It usually requires a considerable amount of rather tedious computation involving complex quantities to evaluate these coefficients. It can be shown that the result of these computations leads to an inverse transform that can be found from a "formula" illustrated in Example 4.24.

EXAMPLE 4.24

Find the inverse transform of

$$F(s) = \frac{s + 2}{s(s^2 + 6s + 25)} = \frac{s + 2}{s(s + 3 + j4)(s + 3 - j4)}$$

SOLUTION

Expanding $F(s)$ in partial fraction form, we would write

$$F(s) = \frac{s + 2}{s(s^2 + 6s + 25)} = \frac{A}{s} + \frac{B}{s + 3 + j4} + \frac{C}{s + 3 - j4}$$

The coefficient A is found in the usual way

$$A = s\left[\frac{s + 2}{s(s^2 + 6s + 25)}\right]_{s=0} = \frac{2}{25}$$

The inverse transform corresponding to the term $(2/25)/s$ is then $2/25$. Now, let the general form of the complex poles be $\alpha \pm j\omega$. In this example, $\alpha = 3$ and $\omega = 4$. Further, let $K(s)$ be what remains of $F(s)$ after the quadratic has been removed. In this case, $K(s) = (s + 2)/s$ [the quadratic $s^2 + 6s + 25$ having been removed from the denominator of $F(s)$]. Finally, define

$$M \angle \theta = K(-\alpha + j\omega)$$

That is, evaluate $K(s)$ at $s = (-\alpha + j\omega)$ and express the result in phasor form. In this example,

$$M \angle \theta = K(s)\Big]_{s = -\alpha + j\omega} = \frac{s + 2}{s}\Big]_{s = -3 + j4}$$

$$= \frac{-3 + j4 + 2}{-3 + j4} = \frac{-1 + j4}{-3 + j4} = 0.825 \angle -22.83°$$

So

$$M = 0.825 \text{ and } \theta = -22.83°$$

The formula that gives the inverse transform corresponding to the *pair* of complex poles is

$$f(t) = \frac{M}{\omega} e^{-\alpha t} \sin(\omega t + \theta) \tag{23}$$

In Example 4.24, the inverse transform corresponding to the pair of poles $(-3 \pm j4)$—that is, corresponding to $s^2 + 6s = 25$—is, therefore,

$$\frac{M}{\omega} e^{-\omega t} \sin(\omega t + \theta) = \frac{0.825}{4} e^{-3t} \sin(4t - 22.83°)$$

$$= 0.206 e^{-3t} \sin(4t - 22.83°)$$

Therefore, the *complete* inverse transform of $F(s)$ is the inverse transform due to the pole at $s = 0$, $2/25$, *plus* the inverse transform due to the complex conjugate poles

$$f(t) = \frac{2}{25} + 0.206e^{-3t} \sin(4t - 22.83°)$$

The time-domain function is a constant $2/25$ added to a *damped sinusoid*. In Figure 4.13, this function is sketched versus time.

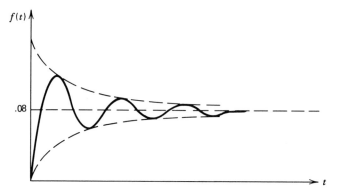

FIGURE 4.13 Time-domain plot of a damped sinusoid (Example 4.24).

Note that $f(0) = .08 + .206e° \sin(-22.83°)$
$= .08 + (0.206)(-0.388)$
$= 0$

and $\lim_{t \to \infty} f(t) = 0.08$, since $\lim_{t \to \infty} e^{-3t} = 0$. The transient solution is $-0.206e^{-3t} \sin(4t - 22.83°)$, while the steady-state solution is 0.08.

In some cases, roots may be *repeated*. For example, the polynomial $F(s) = s^2 + 4s + 4 = (s + 2)^2$ is said to have the root $s = 2$ *repeated twice* (or to be a *second-order* root) in order to satisfy the fundamental theorem. The partial fraction technique is somewhat different when $F(s)$ has repeated poles (i.e., repeated roots in the denominator). The general rule for evaluating coefficients in this case is as follows:

Suppose

$$F(s) = \frac{P(s)}{(s + a)(s + b) \ldots (s + x)(s + y)^2}$$

Then write

$$F(s) = \frac{A}{s + a} + \frac{B}{s + b} + \ldots + \frac{X}{s + x} + \frac{Y}{(s + y)^2} + \frac{Z}{(s + y)}$$

4.7 INVERSE TRANSFORMATION BY PARTIAL FRACTIONS

Find A, B, ..., X in the usual way (Equation 22). However, for the repeated pole $s = -y$

$$Z = \left[\frac{d}{ds}(s + y)^2 F(s) \right]_{s=-y} \tag{24}$$

and

$$Y = (s + y)^2 F(s) \Big|_{s=-y} \tag{25}$$

Example 4.25, in which $F(s)$ has a pole at $s = 0$ and a repeated pole at $s = -5$, illustrates the method.

EXAMPLE 4.25

Use partial fractions to find the inverse LaPlace transform of

$$F(s) = \frac{20}{s(s + 5)^2}$$

SOLUTION

As already shown, when a pole is repeated twice, we must write one fraction for *each* power (first and second) of the corresponding factor as well as one fraction for each of any other factors

$$\frac{20}{s(s + 5)^2} = \frac{A}{s} + \frac{B}{(s + 5)^2} + \frac{C}{s + 5}$$

Note that there is a fraction $B/(s + 5)^2$ for the second power of $(s + 5)$ as well as a fraction $C/(s + 5)$ for the first power of $(s + 5)$. To evaluate the coefficient C in this case, we must find the derivative of $(s + 5)^2 F(s)$ with respect to s and then evaluate the result at the pole $s = -5$ (Equation 24)

$$C = \left[\frac{d}{ds}\left((s+5)^2 \frac{20}{s(s+5)^2} \right) \right]_{s=-5}$$

$$= \left[\frac{d}{ds}\left(\frac{20}{s} \right) \right]_{s=-5}$$

$$= \frac{-20}{s^2} \Big|_{s=-5} = \frac{-20}{25} = -\frac{4}{5}$$

To find B, evaluate $(s + 5)^2 F(s)$ at $s = -5$ (Equation 25)

$$B = \left[(s+5)^2 \left(\frac{20}{s(s+5)^2} \right) \right]_{s=-5} = -4$$

Find A in the usual way

$$A = \left[\frac{d}{ds}\left(\frac{20}{(s+5)^2}\right)\right]_{s=0} = \frac{20}{25} = \frac{4}{5}$$

Checking

$$\frac{(4/5)}{s} - \frac{4}{(s+5)^2} - \frac{4/5}{s+5}$$

$$= \frac{(4/5)(s+5)^2 - 4s - 4/5(s+5)s}{s(s+5)^2}$$

$$= \frac{\frac{4}{5}(s^2 + 10s + 25) - 4s - (4/5)s^2 - 4s}{s(s+5)^2}$$

$$= \frac{20}{s(s+5)^2} = F(s)$$

This technique may be extended to poles that are repeated three, four, or more times, but we will not be concerned with these cases in the practical examples to be considered hereafter.

EXAMPLE 4.26

A 5-kg mass is connected to a dashpot with damping $B = 40$ N/m/sec and to a spring with spring constant $K = 80$ N/m as shown in Figure 4.14.

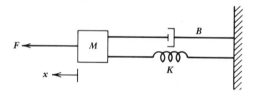

FIGURE 4.14 A spring-mass damper system (Example 4.26).

A step force of 60 N is applied to the system. Find (a) the differential equation of motion in the time domain; (b) the equation of motion in the frequency domain; (c) the solutions to these equations $X(s)$ and $x(t)$. Assume $x(0) = 0$.

SOLUTION

(a) $F = M\ddot{x} + B\dot{x} + Kx$

$60 = 5\ddot{x} + 40\dot{x} + 80x$

$12 = \ddot{x} + 8\dot{x} + 16x$

(b) $\dfrac{12}{s} = Xs^2 + 8sX + 16X$

(c) $\dfrac{12}{s} = X(s^2 + 8s + 16)$

$\dfrac{12}{s} = X(s + 4)^2$

$X(s) = \dfrac{12}{s(s + 4)^2}$

$X(s) = \dfrac{12}{s(s + 4)^2} = \dfrac{A}{s} + \dfrac{B}{(s + 4)^2} + \dfrac{C}{s + 4}$

$A = \left[s\left(\dfrac{12}{s(s + 4)^2} \right) \right]_{s=0} = \dfrac{12}{16} = 0.75$

$B = \left[(s + 4)^2 \left(\dfrac{12}{s(s + 4)^2} \right) \right]_{s=-4} = -3$

$C = \left[\dfrac{d}{ds} \left((s + 4)^2 \dfrac{12}{s(s + 4)^2} \right) \right]_{s=-4}$

$= \left[\dfrac{-12}{s^2} \right]_{s=-4} = \dfrac{-12}{16} = -0.75$

$X(s) = \dfrac{0.75}{s} - \dfrac{3}{(s + 4)^2} - \dfrac{0.75}{s + 4}$

$x(t) = 0.75 - 3te^{-4t} - 0.75e^{-4t}$

Again, the inverse transform of Example 4.26 could have been taken directly from the table.

REFERENCES

1. **Aidala, Joseph B.**, and **Leon Katz**. *Transients in Electronic Circuits*. Prentice-Hall, 1980.
2. **Churchill, Ruel V.** *Operational Mathematics*. 3d ed. McGraw-Hill, 1972.
3. **D'Azzo, John J.**, and **Constantine Houpis**. *Feedback Control-System Analysis and Synthesis*. McGraw-Hill, 1960.
4. **Franklin, Philip.** *An Introduction to Fourier Methods and the Laplace Transformation*. Dover Publications, 1949.
5. **Gardner, Murray F.**, and **John L. Barnes**. *Transients in Linear Systems*. John Wiley & Sons, 1956.

6. **Nixon, Floyd E.** *Handbook of Laplace Transformation: Tables and Examples.* Prentice-Hall, 1960.
7. **Stanley, William D.** *Transform Circuit Analysis for Engineering and Technology.* Prentice-Hall, 1968.

EXERCISES

4.1 Show that $y(t) = 3(e^{-t} - e^{-5t})$ is a solution of the differential equation $\ddot{y} + 6\dot{y} + 5y = 0$.

4.2 Show that $y(t) = c_1 \sin(\omega t + c_2)$ is a general solution to the differential equation $\ddot{y} + \omega^2 y = 0$.

4.3 Find the particular solution in Exercise 4.2 when $\omega = 1$ and the initial conditions are $y(0) = \dot{y}(0) = 1$.

4.4 Find the voltage $V_L(t)$ as a function of time across the inductor after the switch is closed at $t = 0$ (Figure 4.15).

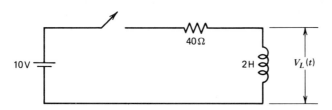

FIGURE 4.15

Hint: use Example 4.5 and the fact that $V_L = L\, di_L/dt$.

4.5 Write a differential equation for the current in the following circuit after the switch is closed at $t = 0$ (Figure 4.16). Show that $i(t) = te^{-2t}$ is a solution.

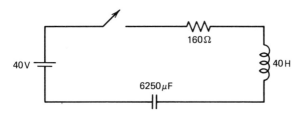

FIGURE 4.16

4.6 Express the voltage across the 0.1 μF capacitor in Example 4.6 as a function of time. (Use Equation 6 and integrate between the limits 0 and t.)

4.7 Write the differential equation of motion for the following mechanical system

(Figure 4.17). A 48-N step force is applied at $t = 0$. Show that $x(t) = 2/3(3t - 1 + e^{-3t})$ is a solution.

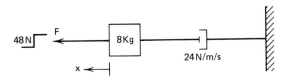

FIGURE 4.17

4.8 Find the LaPlace transform of $f(t) = t$ directly from the definition (Equation 13).

4.9 Find the LaPlace transform of $f(t) = \cos \omega t$ directly from the definition (Equation 13).

4.10 Use the table of LaPlace transform pairs to find the LaPlace transform of each of the following:

(a) $14e^{-6t}$

(b) $\dfrac{t^2}{4}$

(c) $\dfrac{1}{2}e^{-3t} - .01\delta(t)$

(d) $4.2 \sin 4t + 0.8 \cos 5t$

(e) $0.3e^{-4t} \cos t$

(f) $120e^{-t} \sin 10t$

(g) $3 \sin (20t + 45°)$

(h) $\cos (100t + 60°)$

(i) $12 \sin (\omega t - 20°)$

4.11 (a) Use Equation 16 to derive transform pair 14 from pair 13 in Table 4-1.
(b) Use Equation 18 to derive transform pair 2 from pair 1 in Table 4-1.

4.12 Use the table of LaPlace transform pairs to find the inverse LaPlace transform of each of the following:

(a) $\dfrac{12}{s^2} - 3$

(b) $\dfrac{2}{s(4s + 2)}$

(c) $\dfrac{3}{(2s + 6)^2}$

(d) $\dfrac{3s}{(s + 4)(2s + 1)}$

(e) $\dfrac{12s}{4s^2 + 16}$

(f) $\dfrac{7}{2s^2(s + 5)} - \dfrac{7}{s(4s + 3)}$

(g) $\dfrac{2}{s^2 + 0.5s}$

(h) $\dfrac{15}{s^2 + 10s + 25}$

(i) $\dfrac{.01}{s^2 + 8s + 7}$

(j) $\dfrac{1}{s^2 + s + 1}$

4.13 Use the table of LaPlace transform pairs to find the inverse transform of Example 4.21 directly.

4.14 Find G(s) for each of the following (Figure 4.18):

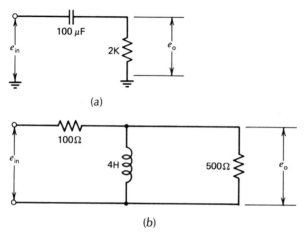

FIGURE 4.18

4.15 For each of the following transfer functions find $|G|$ and $\angle G$ at $\omega = 20$ rad/sec:

(a) $G(s) = \dfrac{s}{s + 20}$

(b) $G(s) = \dfrac{s}{s^2 + 60s + 1000}$

4.16 Write the poles and zeroes of each expression in Exercise 4.12.

4.17 Find the inverse transform of each of the following using partial fractions:

(a) $\dfrac{10s + 2}{(s + 5)(s + 1)}$

(b) $\dfrac{s^2 + 4}{s(s + 3)(s + 6)}$

(c) $\dfrac{2s}{s^2 + 3s + 2}$

(d) $\dfrac{1}{s^2(s + 1)}$

(e) $\dfrac{3s + 2}{s(s^2 + 18s + 81)}$

4.18 Find the inverse LaPlace transform of each of the following:

(a) $\dfrac{2s + 5}{(s + 2 - j4)(s + 2 + j4)}$

(b) $\dfrac{4}{s(s^2 + s + 2)}$

4.19 Find the current after the switch is closed at $t = 0$ (Figure 4.19). (Write Kirchoff's voltage law and differentiate both sides of the equation.)

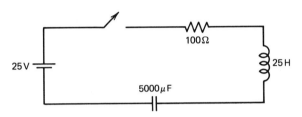

FIGURE 4.19

4.20 Write the LaPlace transform of the differential equation for $x(t)$ in Exercise 4.7. Use LaPlace transforms to find the solution to this equation.

ANSWERS TO EXERCISES

4.3 $y(t) = \sqrt{2} \sin\left(t + \dfrac{\pi}{4}\right)$

4.4 $V_L(t) = 10e^{-20t}$ V

4.5 $40 \dfrac{d^2i}{dt^2} + 160 \dfrac{di}{dt} + 160i = 40\delta(t)$

4.6 $V_c = 12(1 - e^{-10^3 t})$ V

4.7 $8\ddot{x} + 24\dot{x} = 48$

4.8 $F(s) = \dfrac{1}{s^2}$

4.9 $F(s) = \dfrac{s}{s^2 + \omega^2}$

4.10 (a) $\dfrac{14}{s+6}$

(b) $\dfrac{1}{2s^3}$

(c) $\dfrac{1/2}{s+3} - .01$

(d) $\dfrac{16.8}{s^2+16} + \dfrac{0.8s}{s^2+25}$

(e) $\dfrac{0.3(s+4)}{(s+4)^2+1} = \dfrac{0.3s+1.2}{s^2+8s+17}$

(f) $\dfrac{1200}{(s+1)^2+100} = \dfrac{1200}{s^2+2s+101}$

(g) $\dfrac{2.12s+42.4}{s^2+400}$

(h) $\dfrac{0.5s-86.7}{s^2+10^4}$

(i) $\dfrac{-4.1s+11.28\omega}{s^2+\omega^2}$

4.12 (a) $12t - 3\delta(t)$
(b) $1 - e^{-t/2}$
(c) $\dfrac{3}{4} t e^{-3t}$
(d) $\dfrac{3}{7}\left(4e^{-4t} - \dfrac{1}{2}e^{-1/2t}\right)$
(e) $3\cos 2t$
(f) $\dfrac{7}{50}(5t - 1 + e^{-5t}) - \dfrac{7}{3}(1 - e^{-3/4t})$
(g) $4(1 - e^{-0.5t})$
(h) $15te^{-5t}$
(i) $.00166\,(e^{-t} - e^{-7t})$
(j) $1.15e^{-t/2}\sin 0.867t$

4.14 (a) $\dfrac{s}{s+5}$ (b) $\dfrac{(5/6)s}{s+(125/6)}$

4.15 (a) $0.707\angle 45°$ (b) $.015\angle 26.6°$

4.16 Poles:
(a) $s = 0$
(b) $s = 0, -\frac{1}{2}$
(c) $s = -3$, repeated twice
(d) $s = -4, -\frac{1}{2}$
(e) $s = \pm j2$
(f) $s = 0, -5, -\frac{3}{4}$
(g) $s = 0, -0.5$
(h) $s = -5$, repeated twice

Zeroes: (a) $s = \pm 2$ (b) none (c) none (d) $s = 0$ (e) $s = 0$ (f), (g), (h), (i), (j) none

4.17 (a) $12e^{-5t} - 2e^{-t}$

(b) $\frac{2}{9} - \frac{13}{9}e^{-3t} + \frac{20}{9}e^{-6t}$

(c) $4e^{-2t} - 2e^{-t}$

(d) $t - 1 + e^{-t}$

(e) $\frac{2}{81} - \frac{2}{81}e^{-9t} + \frac{25}{9}te^{-9t}$

4.18 (a) $2.02e^{-2t}\sin(4t + 82.87°)$

(b) $2 + 2.14e^{-t/2}\sin(1.32t - 110.7°)$

4.19 $i = \frac{1}{2}e^{-2t}(\sin 2t)$

4.20 $\frac{48}{s} = 8s^2 X(s) + 24s X(s)$

$x(t) = \frac{2}{3}(3t - 1 + e^{-3t})$

CHAPTER

5
NETWORK ANALYSIS USING LAPLACE TRANSFORMS

5.1 REVIEW OF BASIC NETWORK ANALYSIS TECHNIQUES

The fundamental techniques of network analysis that we will use in this chapter are Kirchoff's voltage and current laws, the voltage and current divider rules, mesh and nodal analysis, Thevenin's and Norton's theorems, and superposition. We will apply these techniques in the s- (frequency) domain to obtain solutions of network variables in much the same way as they are used to obtain solutions in the time domain. Although it is assumed that the reader is acquainted with these techniques, a brief review is appropriate at this time in order to establish and define the notation and conventions that we will use in later work.

Voltage and Current Sources

An *ideal* voltage source is one that will maintain the same voltage at its terminals regardless of the load connected to it; that is, regardless of the current drawn from it. Of course, no such source exists, but it is convenient to describe the *real* voltage source in terms of an ideal voltage source. A real voltage source has some internal impedance that causes the terminal voltage to decrease as the current drawn from it increases. A real voltage source can then be described (and analyzed mathematically—which is our real goal) as an ideal voltage source in series with the series-equivalent internal impedance of the real voltage source. If the internal impedance is small compared to the load impedance, then the real source approximates an ideal source, as illustrated in Example 5.1.

EXAMPLE 5.1

A 12-V dc-power supply has a series-equivalent internal resistance of 5 Ω. Find the terminal voltage when loads of 4.995K, 495 Ω, and 5 Ω are connected to it.

SOLUTION

The real power supply may be represented as in Figure 5.1. When $R_L = 4.995K$, $I = 12 \text{ V}/5K = 2.4$ mA, and the terminal voltage V_T is

$$V_T = (2.4 \text{ mA})(4.995 \text{ K}) = 11.988 \text{ V}$$

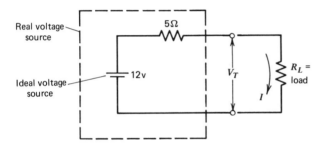

FIGURE 5.1 A real voltage source (Example 5.1).

When $R_L = 495$ Ω, $I = 25$ mA, and $V_T = 11.88$ V. When $R_L = 5$ Ω, $I = 1.2$ A, and $V_T = 6$ V.

We see that for resistance loads much greater than the internal resistance (small load currents), the real voltage source is a close approximation of the ideal source, since the terminal voltage remains fairly constant. But for loads approaching the value of the internal resistance, the terminal voltage falls off considerably.

The series-equivalent internal impedance we have referred to in defining the real source is in fact the Thevenin equivalent impedance of the source, a concept that we will review subsequently.

An ideal current source produces the same current to any size load connected at its terminals. The real current source may be represented by an ideal current source in parallel with an internal resistance.

EXAMPLE 5.2

A real current source is represented by a 10-mA ideal current source in parallel with internal resistance R_{INT}. The real source produces 9 mA in a 10-Ω load. (a)

What is the value of R_{INT}? (b) How much current would be produced in a 100-Ω load? See Figure 5.2.

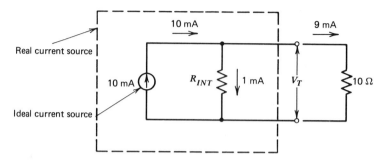

FIGURE 5.2 A real current source (Example 5.2).

SOLUTION

(a) Since there are 9 mA in the load, there must be 1 mA in R_{INT}. Since the voltage across the load is the same as the voltage across R_{INT}, we have

$$V_T = (9 \text{ mA})(10 \text{ Ω}) = 90 \text{ mV}$$

Then

$$R_{INT} = \frac{90 \text{ mV}}{1 \text{ mA}} = 90 \text{ Ω}$$

(b) With a 100-Ω load, 90/190 × 10 mA = 4.47 mA will be produced in the load.

Note that the real current source behaves more like an ideal current source with *smaller* load resistance; that is, with greater load currents in contrast to the real voltage source. Note also that the terminal voltage of the current source may vary considerably while the current is substantially constant, just as the current from a voltage source varies considerably while maintaining substantially constant terminal voltage.

A real voltage source may always be converted to an *equivalent* real current source, and vice versa. The conversion is quite straightforward. Given a voltage source consisting of an ideal source E in series with Z, the equivalent current source has ideal current E/Z in parallel with Z, as shown in Figure 5.3.

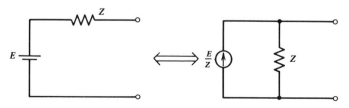

FIGURE 5.3 Conversions to equivalent sources.

5.1 REVIEW OF BASIC NETWORK ANALYSIS TECHNIQUES

Note especially that the polarity of the equivalent current source must be consistent with that of the original voltage source, in the sense that both must produce current in the same direction through any load connected across the terminals. These sources are equivalent in every sense of the word. Given a "black-box" source with only the terminals available to the user, it would be impossible by any measurement or test to determine whether the black box contained a voltage source or an equivalent current source.

The method for converting a current source to an equivalent voltage source should be apparent, and it is illustrated in Example 5.3.

EXAMPLE 5.3

(a) Convert the current source in Figure 5.4 to an equivalent voltage source.
(b) Show that each source produces the same open-circuit voltage and short-circuit current.

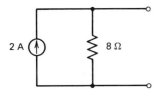

FIGURE 5.4 Convert to an equivalent voltage source (Example 5.3).

SOLUTION

(a) The equivalent voltage source has $E = (2\text{ A})(8\text{ }\Omega) = 16$ V and series resistance 8 Ω. See Figure 5.5. (This can be checked by converting to an equivalent current source.)

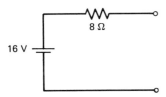

FIGURE 5.5 The voltage-source equivalent to Figure 5.4 (Example 5.3).

(b) The open-circuit voltage of the current source is the drop across the 8-Ω resistor: $(2\text{ A})(8\text{ }\Omega) = 16$ V, since all current must flow through the 8 Ω. The open-circuit terminal voltage of the voltage source is also 16 V, since there is no drop across the series 8 Ω. The short-circuit current of the current source is 2 A, since all current must flow in the short, while the short-circuit current of the voltage source is 16 V $\div$ 8 Ω = 2 A.

Kirchoff's Voltage Law

Kirchoff's voltage law states that the sum of the voltage drops around any closed loop is equal to the sum of the applied voltages in that loop. Some authors use the term "voltage rises" instead of applied voltages. In some ways, the former is better usage, since it is often the case when writing a loop equation that a voltage source in the loop actually constitutes a voltage drop, due to its polarity with respect to the direction of the loop. Recall that the direction of a loop (clockwise or counterclockwise) is purely arbitrary. In any event, we summarize the procedure for finding the current in a closed loop by writing a loop equation, as follows:

1. Convert any current sources in the circuit to equivalent voltage sources..
2. Be certain that each source has its polarity clearly indicated. For an ac source, positive and negative signs may be placed in accordance with its polarity at $t = 0$.
3. Assume a loop direction, clockwise or counterclockwise.. (As a matter of practice, we will always assume clockwise loops in subsequent examples.)
4. Write Kirchoff's voltage law in equation form around the loop. It is good practice to place applied voltages, or voltage rises, on one side of the equation and voltage drops on the other side. An important rule: When traveling around the loop in the assumed direction, if we go through a source from minus to plus, it is a voltage rise; if we pass through it from plus to minus, it is a drop.
5. Solve the equation for the unknown current I.

EXAMPLE 5.4

Find the voltage drop across the four-Ω resistor by using Kirchoff's voltage law. (See Figure 5.6.)

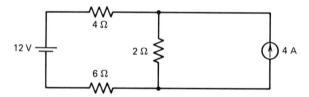

FIGURE 5.6 Find the voltage drop across the four-Ω resistor (Example 5.4).

SOLUTION

Converting the current source to an equivalent voltage source and assuming a clockwise loop, we have the circuit shown in Figure 5.7.

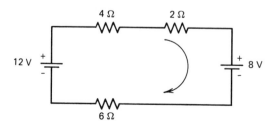

FIGURE 5.7 Circuit of Figure 5.6 redrawn for solution (Example 5.4).

5.1 REVIEW OF BASIC NETWORK ANALYSIS TECHNIQUES 109

Then
$$12 = 4I + 2I + 8 + 6I$$

Note than the 8-V source constitutes a drop, since we pass through it from plus to minus. Solving for I,

$$12 - 8 = 12I$$
$$I = (1/3)A$$

Thus, the drop across the 4-Ω resistor is $(1/3)A(4\,\Omega) = 1.33$ V.

Kirchoff's Current Law

According to Kirchoff's current law, the sum of the currents entering any junction is equal to the sum of the currents leaving that junction.

EXAMPLE 5.5

Use Kirchoff's current law to find the current (including direction) through the 1-K resistor in Figure 5.8.

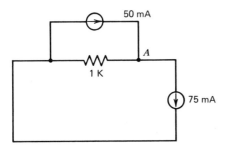

FIGURE 5.8   Find the current in the 1-K resistor (Example 5.5).

SOLUTION

At the junction labeled A, we have the situation shown in the following figure.

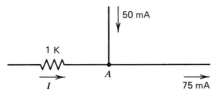

If we assume I in the direction indicated, then

$$\text{current entering} = I + 50 \text{ mA} = 75 \text{ mA} = \text{current leaving}$$

and $I = 25$ mA entering A.

(If we had assumed I in the opposite direction, we would have found $I = -25$ mA, which has the same meaning as our first solution.)

The Voltage and Current Divider Rules

The voltage-divider rule was discussed in Chapter 2, Section 2.8. The current divider rule is used to find the current in one branch of two parallel branches when the total current entering the junction of the branches is known. The rule may be formulated as shown in Figure 5.9.

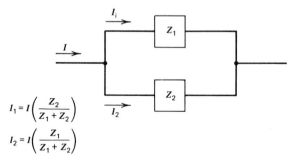

$$I_1 = I\left(\frac{Z_2}{Z_1 + Z_2}\right)$$

$$I_2 = I\left(\frac{Z_1}{Z_1 + Z_2}\right)$$

FIGURE 5.9 The current-divider rule.

EXAMPLE 5.6

What should be the value of R in Figure 5.10 in order for 1.5 A to flow through it?

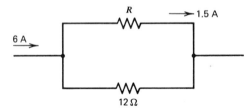

FIGURE 5.10 Find the value of R (Example 5.6).

SOLUTION

Let the current through R be I_1. Then

$$I_1 = 1.5 = 6\left(\frac{12}{R + 12}\right)$$

and

$$1.5R + 18 = 72$$

$$R = 36 \, \Omega$$

Mesh Analysis

Mesh analysis is the application of Kirchoff's voltage law to two or more currents in a network. The equations that result from writing the loop equations are solved

5.1 REVIEW OF BASIC NETWORK ANALYSIS TECHNIQUES

simultaneously for the unknown currents. The same procedures and conventions described in the section on Kirchoff's voltage law apply here, with one additional consideration. The voltage drop across an element that is common to two or more loops is computed by using the sum (or difference) of the loop currents through the element. The direction of positive current is the same as the direction of the assumed loop current in the loop for which Kirchoff's voltage law is being written.

EXAMPLE 5.7

Find the current in the four-Ω resistor in the network of Figure 5.11.

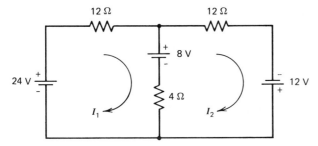

FIGURE 5.11   Find the current in the four-Ω resistor (Example 5.7).

SOLUTION

For loop 1, we have

$$24 = 12I_1 + 8 + 4(I_1 - I_2) \quad (1)$$
$$16 = 16I_1 - 4I_2$$

Note that the drop in the 4-Ω resistor is computed as if the net current in it were $I_1 - I_2$, since I_1 is the positve direction in loop 1, while I_2 opposes it. Note also that in loop 1, the 8-V source is a drop. For loop 2,

$$8 + 12 = 12I_2 + 4(I_2 - I_1) \quad (2)$$
$$20 = -4I_1 + 16I_2$$

In this case, I_2 is the assumed positive direction, so the drop in the 4-Ω resistor is $4(I_2 - I_1)$. Also, the 8-V source constitutes a voltage rise in loop 2. Solving Equations 1 and 2 simultaneously, we find that $I_1 = 1.4$ A and $I_2 = 1.6$ A. The actual current in the 4-Ω resistor may be calculated using either loop 1 or loop 2 as a reference. In loop 1 where clockwise current is positive, we have: net $I = 1.4 - 1.6 = -0.2$ A, signifying that the actual current flows *up* through the 4 Ω. If loop 2 had been chosen as the reference, we would write: net $I = 1.6 - 1.4 = +0.2$ A, which has the same interpretation. Of course, the direction and magnitude of the current in each of the 12-Ω resistors is the same as the assumed loop currents through them, 1.4 and 1.6 A, since each calculated current was positive.

A negative result would simply mean that the actual direction of current is opposite to the assumed loop direction.

Nodal Analysis

Nodal analysis is a method for finding unknown voltages at the nodes of a network. The simplest way to identify a network node is to visualize it as a junction where we would connect or solder component leads together. We summarize the procedure followed in nodal analysis.

1. Convert any voltage sources in the network to equivalent current sources.
2. Identify all nodes. It is a good idea to label them by assigning voltage designations to each: V_1, V_2, and so forth. These are the node voltages, and they constitute the unknowns in this analysis.
3. Choose a reference node V_0. This will usually be the node to which most of the components are connected. It may be thought of as ground or "common" and will be assumed to be at 0 V. When the unknowns V_1, V_2, and so forth, are found, the values will be the same as we would measure at the corresponding nodes with respect to V_0.
4. Make an assumption about the relative magnitudes of V_1, V_2, and so forth, with the proviso that V_0 is the smallest. (The assumption is arbitrary as is the assumption of loop directions in mesh analysis.) In subsequent examples, we will make the assumption $V_1 > V_2 > \ldots > V_0 = 0$.
5. Draw arrows indicating the current directions through each component of the network. These directions should be consistent with the assumed magnitudes of the node voltages (current flowing from a node of higher voltage). It is also a good idea to label each arrow with a current magnitude in terms of the node voltages and component values (e.g., $I = V/R$).
6. Write Kirchoff's current law at each node. Solve the resulting equations simultaneously for the unknown node voltages.

EXAMPLE 5.8

Find the voltages with respect to ground at points A and B in the network shown in Figure 5.12.

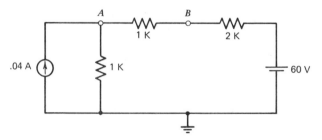

FIGURE 5.12 Find the node voltages at A and B (Example 5.8).

5.1 REVIEW OF BASIC NETWORK ANALYSIS TECHNIQUES

SOLUTION

The circuit is redrawn with the voltage source converted to an equivalent current source, the nodes labeled, and the currents identified according to the assumption $V_1 > V_2 > V_0$. (Each heavy line represents an entire node.) See Figure 5.13.

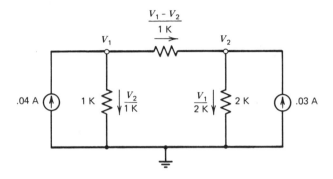

FIGURE 5.13 Circuit of Figure 5.12 redrawn for nodal analysis.

Writing Kirchoff's current law at each node,

Node 1: current entering $= .04 = \dfrac{V_1}{1\text{ K}} + \dfrac{V_1 - V_2}{1\text{ K}} =$ Current leaving

Node 2: current entering $= .03 + \dfrac{V_1 - V_2}{1\text{ K}} = \dfrac{V_2}{2\text{ K}} =$ Current leaving

Solving for V_1 and V_2 simultaneously, we find $V_1 = 45$ V and $V_2 = 50$ V, which correspond to the voltages at A and B in the original network.

Superposition

Superposition is a powerful technique for finding voltages and currents in a network containing multiple sources and linear, bilateral components (the only kind we will encounter in this book). The principle of superposition allows us to solve for a voltage or current by considering the effects of *one* source at a time. After we have found the voltage or current due to each source acting alone, we simply add all of these individual contributions to obtain the actual voltage or current due to all sources acting simultaneously. When finding the voltage or current due to a single source acting alone, all other *voltage sources must be replaced by a short circuit*, and all other *current sources must be replaced by an open circuit*. Example 5.9 illustrates the method.

EXAMPLE 5.9

Use the principle of superposition to find the current in, and voltage across, the 4-Ω resistor in Figure 5.14.

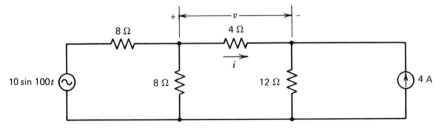

FIGURE 5.14 Find v and i using superposition (Example 5.9).

SOLUTION

To find the current due first to the ac source acting alone, we open circuit the current source and redraw the resulting equivalent circuit as in Figure 5.15.

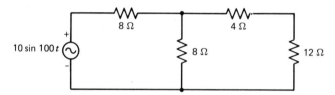

FIGURE 5.15 Figure 5.14 with current source removed (Example 5.9).

The series combination of 4 Ω and 12 Ω is equivalent to 16 Ω, which is in parallel with 8 Ω. Hence, we have Figure 5.16.

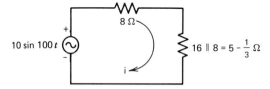

FIGURE 5.16 Circuit equivalent to Figure 5.15 (Example 5.9).

$$i = \frac{10 \sin 100t}{13\frac{1}{3} \ \Omega} = 0.75 \sin 100t$$

Using the current-divider rule, the portion of $0.75 \sin 100t$ that flows in the branch containing the 4-Ω resistor is

$$i = (75 \sin 100t)\left(\frac{8}{8 + 16}\right) = 0.25 \sin 100t$$

5.1 REVIEW OF BASIC NETWORK ANALYSIS TECHNIQUES 115

and the resulting voltage across this resistor is

$$e = 4(0.25 \sin 100t) = \sin 100t$$

Note especially the polarity of this voltage and current (see Figure 5.17).

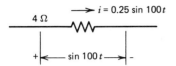

FIGURE 5.17 Polarity of voltage and current due to voltage source (Example 5.9).

To find the current due to the current source acting alone, we replace the voltage source with a short and redraw the resulting equivalent circuit (see Figure 5.18).

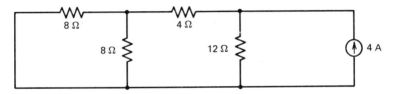

FIGURE 5.18 Figure 5.14 with voltage source removed (Example 5.9).

The parallel combination of the 8-Ω resistors is equivalent to 4 Ω, which is in series with the 4-Ω resistor. Hence, we have the circuit of Figure 5.19.

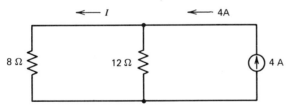

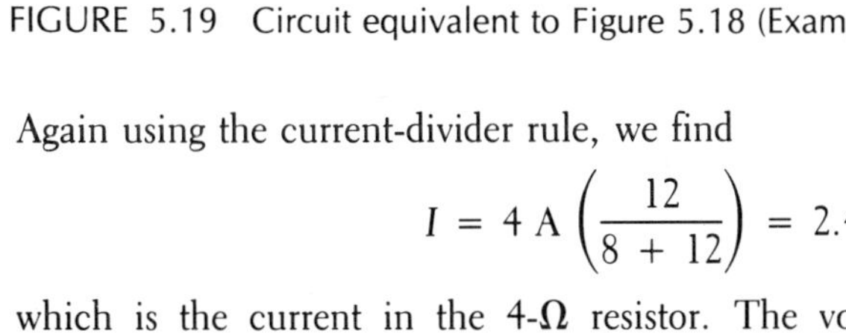

FIGURE 5.19 Circuit equivalent to Figure 5.18 (Example 5.9).

Again using the current-divider rule, we find

$$I = 4 \text{ A} \left(\frac{12}{8 + 12} \right) = 2.4 \text{ A}$$

which is the current in the 4-Ω resistor. The voltage across it is, therefore, (2.4 A)(4 Ω) = 9.6 V. Again, note the polarities (see Figure 5.20).

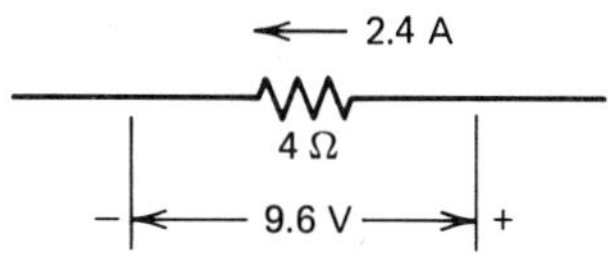

FIGURE 5.20 Polarity of voltage and current due to current source (Example 5.9).

NETWORK ANALYSIS USING LAPLACE TRANSFORMS

We now combine results from each source acting alone, observing that we must, in fact, subtract these due to opposite polarities. With respect to the assumed positive (reference) polarities shown on the original network, we write

$$\text{Actual } i = (0.25 \sin 100t - 2.4)\text{A}$$

$$\text{Actual } v = (\sin 100\,t - 9.6)\text{V}$$

Thevenin's and Norton's Theorems

Thevenin's theorem states that any network containing one or more sources and linear, bilateral components may be replaced by an equivalent circuit consisting of a single source and a single series impedance. In effect, any network is equivalent to a single (real) voltage source. This result may be applied to any portion of a given network as well, a fact that often simplifies the task of finding voltage or current in another portion of the network.

The procedure for finding a Thevenin equivalent circuit is summarized as follows:

1. Isolate that portion of the network for which a Thevenin equivalent circuit is desired. That is, open circuit the terminals with respect to which the equivalent circuit is desired by removing all other components connected to them.
2. Replace all voltage sources by short circuits and all current sources by open circuits in that portion of the network for which the Thevenin circuit is desired.
3. Calculate the total equivalent impedance looking into the open-circuited Thevenin terminals [with the sources removed, as described in (2)]. This impedance is the Thevenin equivalent impedance Z_{TH}.
4. Restore all sources to the circuit and calculate the voltage that would appear across the open-circuited terminals. If more than one source is present, mesh analysis or superposition may be used to find this voltage. The result is the Thevenin equivalent voltage E_{TH}.

EXAMPLE 5.10

Find the Thevenin equivalent circuit with respect to the terminals A–B in Figure 5.21. Find the current in the 2-K resistor.

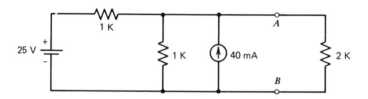

FIGURE 5.21 Find the Thevenin equivalent circuit with respect to A–B (Example 5.10).

SOLUTION

Removing the 2-K resistor, shorting the voltage source, and open circuiting the current source, we have the equivalent circuit in Figure 5.22.

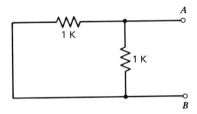

FIGURE 5.22 Circuit of Figure 5.21 with voltage source shorted (Example 5.10).

Clearly, $Z_{TH} = 1\text{ K} \parallel 1\text{ K} = 500\ \Omega$.
To find the open-circuit voltage across A–B, we will use the principle of superposition and first open circuit the current source (see Figure 5.23). By the voltage-divider rule,

$$V_{AB} = \frac{1\text{ K}}{2\text{ K}}(25) = 12.5\text{ V}$$

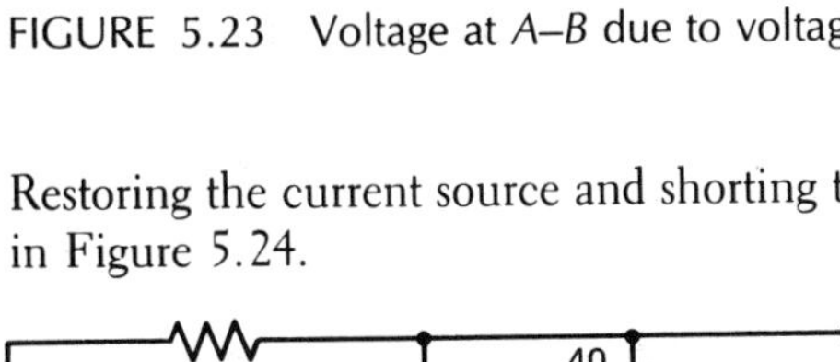

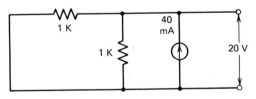

FIGURE 5.23 Voltage at A–B due to voltage source (Example 5.10).

Restoring the current source and shorting the voltage source, we obtain the circuit in Figure 5.24.

FIGURE 5.24 Voltage at A–B due to current source (Example 5.10).

$$V_{AB} = (40\text{ mA})(1\text{ K} \parallel 1\text{ K}) = 20\text{ V}$$

Thus, E_{TH} = 12.5 V + 20 V = 32.5 V, and the Thevenin equivalent circuit is shown in Figure 5.25.

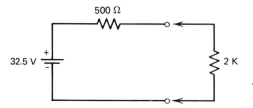

FIGURE 5.25 The Thevenin equivalent circuit for Example 5.10.

When the 2-K resistor is restored to the terminals A–B, we see that the current through it will be

$$I = \frac{32.5}{2.5\ K} = 13\ mA$$

This is, of course, the same current we would find in the 2-K resistor had we analyzed the original network without finding the Thevenin equivalent circuit. The Thevenin approach is particularly useful in situations where the 2-K load resistor is subject to change; given the Thevenin circuit, it is a simple matter to find the current and voltage across the load when its value is changed.

A Norton equivalent circuit may be used to replace a network by a single equivalent current source I_N in parallel with a single impedance Z_N. A Norton equivalent circuit may be found by converting the Thevenin equivalent circuit for a network to its equivalent current source. Thus, $I_N = E_{TH}/Z_{TH}$, and $Z_N = Z_{TH}$.

EXAMPLE 5.11

Find the Norton equivalent circuit with respect to the terminals A–B in Example 5.10.

SOLUTION

See Figure 5.26.

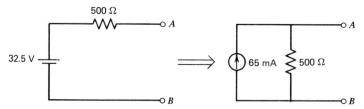

FIGURE 5.26 The Norton equivalent circuit (Example 5.11).

In all of the preceding examples, we have used resistive networks to illustrate basic network analysis techniques. Of course each technique is applicable to net-

5.2 SOURCES AND INITIAL CONDITIONS IN THE FREQUENCY DOMAIN

works containing reactive components (X_L and X_C) and ac sources. It is only necessary to modify the calculations to the extent that all arithmetic is performed in phasor form, taking into account the real and imaginary components of each variable.

EXAMPLE 5.12

Find the Thevenin equivalent circuit with respect to the terminals A–B in Figure 5.27.

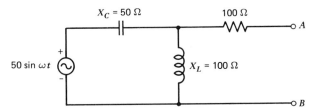

FIGURE 5.27 Find the Thevenin equivalent circuit at terminals A–B (Example 5.12).

SOLUTION

The Thevenin equivalent impedance is

$$Z_{TH} = R + \frac{(jX_L)(-jX_C)}{jX_L - jX_C} = 100 + \frac{(j100)(-j50)}{j100 - j50}$$

$$= 100 + \frac{5000}{j50} = 100 - j100 = 141.4 \angle -45°$$

Since there is no drop across the 100-Ω resistor, the Thevenin voltage may be found from the voltage-divider rule

$$e_{TH} = 50\angle 0° \left(\frac{jX_L}{jX_L - jX_C} \right)$$

$$= 50\angle 0° \left(\frac{j100}{j50} \right) = 100 \angle 0°$$

$$= 100 \sin \omega t$$

5.2 SOURCES AND INITIAL CONDITIONS IN THE FREQUENCY DOMAIN

In Chapter 4, we indicated that it is possible to find the LaPlace transform of time-domain voltages and currents. Thus, we may represent voltage and current sources in the frequency domain as well as convert one to an equivalent form of the other.

EXAMPLE 5.13

Find the frequency domain representation of the voltage source, and convert it to an equivalent current source in the frequency domain. (See Figure 5.28.)

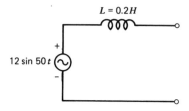

FIGURE 5.28 Find the frequency-domain representation (Example 5.13).

SOLUTION

$$\mathcal{L}[12 \sin 50\ t] = \frac{(50)(12)}{s^2 + 50^2}$$

$$Ls = 0.2s$$

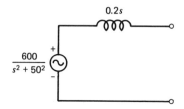

FIGURE 5.29 Solution to Example 5.13.

The equivalent current source is found in the usual way

$$I(s) = \frac{E(s)}{Z(s)} = \frac{600/(s^2 + 50^2)}{0.2s} = \frac{3000}{s(s^2 + 50^2)}$$

The equivalent current source is shown in Figure 5.30.

FIGURE 5.30 Current-source equivalent of Figure 5.29.

5.2 SOURCES AND INITIAL CONDITIONS IN THE FREQUENCY DOMAIN

When analyzing networks, there is frequently an initial voltage on a capacitor or an initial current in an inductor. These are conditions which may exist at $t = 0$. In order to include the effects of these initial conditions on the analysis, we must replace them by equivalent voltage or current sources in the frequency domain. If we are performing a mesh analysis where all sources must be voltage sources, the Thevenin equivalent circuits of the components with initial values must be used, while nodal analysis requires the use of Norton equivalent circuits.

Suppose a capacitor C is charged to an initial voltage V_o. Then its equivalent time-domain and frequency-domain Thevenin and Norton circuits are as shown in Figure 5.31.

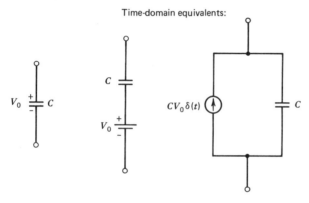

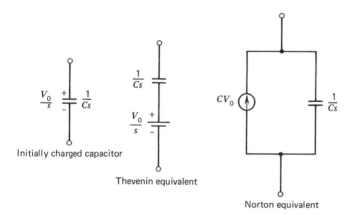

FIGURE 5.31 Time- and frequency-domain equivalent circuits of a charged capacitor.

An inductor L with initial current I_0 may be represented as shown in Figure 5.32.

122 NETWORK ANALYSIS USING LAPLACE TRANSFORMS

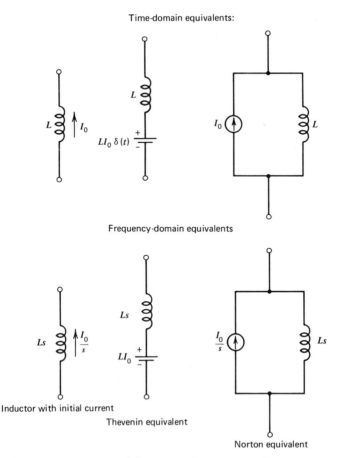

FIGURE 5.32 Time- and frequency-domain equivalent circuits of an inductor with initial current.

EXAMPLE 5.14

It is desired to perform mesh analysis of a network that contains a 100-μF capacitor with an initial voltage of 16 V and a 60-mH inductor with an initial current of 8 A. Find the frequency-domain equivalents of these components that would be appropriate for the analysis.

SOLUTION

Since all sources in a mesh analysis must be voltage sources, we will use the Thevenin equivalent circuits of each component. For the capacitor, we therefore have Figure 5.33.

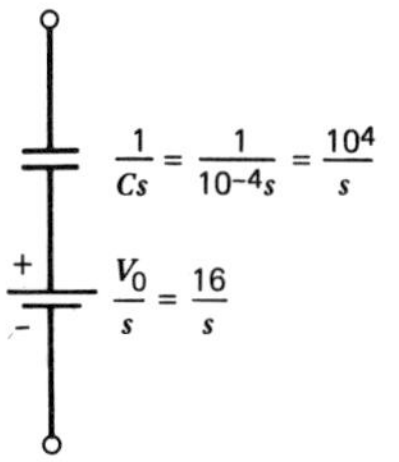

$$\frac{1}{Cs} = \frac{1}{10^{-4}s} = \frac{10^4}{s}$$

$$\frac{V_0}{s} = \frac{16}{s}$$

FIGURE 5.33 Charged capacitor for mesh analysis (Example 5.14).

And for the inductor, see Figure 5.34.

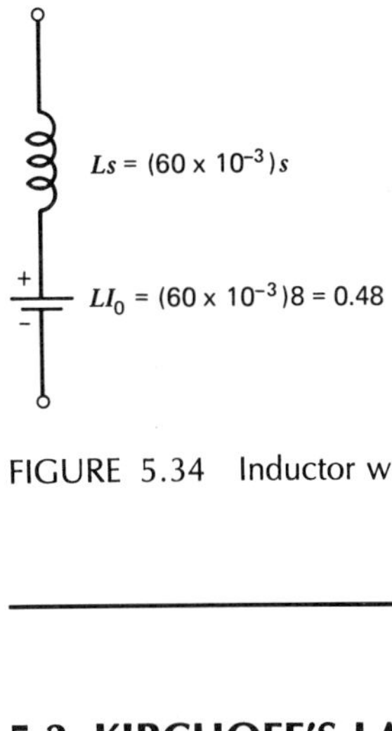

$Ls = (60 \times 10^{-3})s$

$LI_0 = (60 \times 10^{-3})8 = 0.48$

FIGURE 5.34 Inductor with initial current for mesh analysis (Example 5.14).

5.3 KIRCHOFF'S LAWS USING LAPLACE TRANSFORMS

The rules for applying Kirchoff's voltage law in the frequency domain are exactly the same as those reviewed in Section 5.1 for the time domain. We first transform each source and each impedance into the frequency domain. We then write Kirchoff's voltage law in terms of the variable s and solve the resulting equation for $I(s)$. Finally, we find the inverse transform $i(t)$ corresponding to $I(s)$. As indicated in Chapter 4, this is tantamount to solving a differential equation for $i(t)$.

EXAMPLE 5.15

Find $i(t)$ in the circuit shown in Figure 5.35 using LaPlace transforms. The applied voltage is a 10-V unit step occurring at $t = 0$ [$10u(t)$]. Assume zero initial current [$i(0) = 0$].

124 NETWORK ANALYSIS USING LAPLACE TRANSFORMS

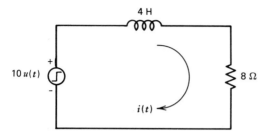

FIGURE 5.35   Find i(t) using LaPlace transforms (Example 5.15).

SOLUTION

The transformed version of this circuit is shown in Figure 5.36.

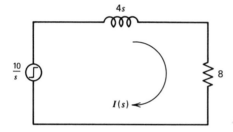

FIGURE 5.36   Transformed circuit of Figure 5.35 (Example 5.15).

Note that the applied 10-V step is the same as a 10-V dc source switched in at $t = 0$. *In future examples, we will simply depict all sources by their conventional symbols, with the understanding that they are applied at $t = 0$.* Writing Kirchoff's voltage law,

$$\frac{10}{s} = 4sI(s) + 8i(s) = I(s)\left(4s + 8\right)$$

$$I(s) = \frac{10/s}{4(s+2)} = \frac{2.5}{s(s+2)}$$

From Table 4-1, pair 7, we find: $i(t) = 1.25(1 - e^{-2t})$ A.

EXAMPLE 5.16

Find $i(t)$ in the circuit shown in Figure 5.37 using LaPlace transforms. The capacitor has an initial voltage of 10 V.

5.3 KIRCHOFF'S LAWS USING LAPLACE TRANSFORMS

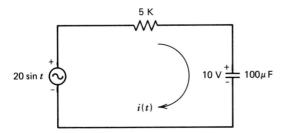

FIGURE 5.37 Find $i(t)$ using LaPlace transforms (Example 5.16).

SOLUTION

Draw the transform version of the circuit, including the Thevenin equivalent circuit for the charged capacitor as shown in Figure 5.38.

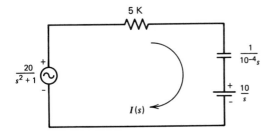

FIGURE 5.38 Transformed circuit of Figure 5.37 (Example 5.16).

Then

$$\frac{20}{s^2+1} = 5 \times 10^3 I(s) + \frac{10^4 I(s)}{s} + \frac{10}{s}$$

$$\frac{20}{s^2+1} - \frac{10}{s} = I(s)\left(5 \times 10^3 + \frac{10^4}{s}\right)$$

$$I(s) = \frac{[20s - 10(s^2+1)]/s(s^2+1)}{(5 \times 10^3 s + 10^4)/s} = \frac{20s - 10s^2 - 10}{5 \times 10^3(s+2)(s^2+1)}$$

$$= \frac{-2 \times 10^{-3}(s^2 - 2s + 1)}{(s+2)(s^2+1)}$$

$$= \frac{A}{(s+2)} + \frac{B}{s-j1} + \frac{C}{s+j1}$$

The portion of the inverse transform due to the pole at $s = -2$ may be found by evaluating A in the partial fraction expansion in the usual way

$$A = \left.\frac{-2 \times 10^{-3}(s^2 - 2s + 1)}{(s^2 + 1)}\right|_{s=-2}$$

$$= \frac{-2 \times 10^{-3}(4 + 4 + 1)}{4 + 1} = \frac{-18}{5} \times 10^{-3}$$

Thus, the portion of the inverse transform due to the pole at $s = -2$ is

$$i(t) = \frac{-18}{5} \times 10^{-3} e^{-2t}$$

To determine the portion of the inverse transform due to the quadratic $s^2 + 1$, that is, due to the pair of poles $s = \pm j$, we may use the formula discussed in Example 4.24. Here, $\alpha = 0$, $\omega = 1$, and

$$K(s) = \frac{-2 \times 10^{-3}(s^2 - 2s + 1)}{s + 2}$$

$$M\angle\theta = K(-\alpha + j\omega) = \left.\frac{-2 \times 10^{-3}(s^2 - 2s + 1)}{s + 2}\right|_{s=j1}$$

$$= \frac{-2 \times 10^{-3}(-1 - 2j + 1)}{2 + j1} = 1.79 \times 10^{-3} \angle 63.43°$$

So $M = 1.79 \times 10^{-3}$ and $\theta = 63.43°$. Substituting in Equation 23 of Chapter 4, the portion of the inverse transform due to the quadratic is

$$i(t) = 1.79 \times 10^{-3} \sin(t + 63.43°)$$

Therefore, the total current (due to all poles) is

$$i(t) = 1.79 \times 10^{-3} \sin(t + 63.43°) - \frac{18}{5} \times 10^{-3} e^{-2t}$$

$$= \left[1.79 \sin(t + 63.43°) - \frac{18}{5} e^{-2t}\right] \text{mA}$$

It is instructive to study this result and verify certain obvious facts about the original circuit. For example, at $t = 0$, we observe that the applied sinusoidal source $20 \sin t$ is zero and that the initial voltage on the capacitor is the only contributor to current in the circuit. The current at $t = 0$ should then be $i(0) = -10/5 \text{ K} = -2$ mA. (Note that the capacitor acts like a short circuit at $t = 0$.) To confirm this, we substitute $t = 0$ into our solution for $i(t)$ and find

$$i(0) = \left[1.79 \sin(63.43°) - \frac{18}{5}\right] \text{mA}$$

$$= (1.6 - 3.6) = -2 \text{ mA}$$

as required. Further, we note that in the original circuit, after a long period of time, the initial charge on the capacitor will have decayed and that the current will, therefore, eventually be

$$i(t) = \frac{e}{R - jX_c} = \frac{20 \sin t}{5 \times 10^{-3} - j10^4}$$

$$= \frac{20 \angle 0°}{11.18 \times 10^3 \angle -63.43°}$$

$$= 1.79 \times 10^{-3} \angle 63.43°$$

$$= 1.79 \sin(t + 63.43°) \text{ mA}$$

Once again, our solution for $i(t)$ is confirmed, since for large t, the term $-(18/5) \times 10^{-3}e^{-2t}$ is essentially zero, and we are left with $1.79 \sin(t + 63.43°)$. The transient solution is $18/5e^{-2t}$ mA, and the steady-state solution is $1.79 \sin(t + 63.43°)$ mA. Kirchoff's current law may also be applied in the frequency domain.

EXAMPLE 5.17

Find $v(t)$ in the circuit shown in Figure 5.39 using LaPlace transforms and Kirchoff's current law. Assume zero initial current in the inductor.

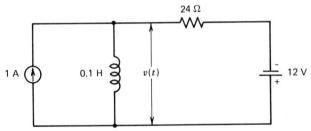

FIGURE 5.39 Find $v(t)$ using LaPlace transforms (Example 5.17).

SOLUTION

Converting the voltage source to an equivalent current source, transforming all quantities to the frequency domain, and rearranging the circuit slightly, we obtain the circuit shown in Figure 5.40.

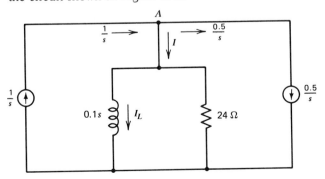

FIGURE 5.40 Transform of Figure 5.39 (Example 5.17).

Applying Kirchoff's current law at junction A,

$$I + 0.5/s = 1/s \quad \text{or} \quad I = 0.5/s$$

Then, by the current divider rule,

$$I_L = \frac{0.5}{s}\left(\frac{24}{0.1s + 24}\right) = \frac{0.5}{s}\left(\frac{240}{s + 240}\right) = \frac{120}{s(s + 240)}$$

Now, $v(t) = L\, di/dt$, or in LaPlace form $V(s) = LsI(s) - i(0) = LsI(s)$, since $i(0) = 0$. So

$$V(s) = (0.1s)\left(\frac{120}{s(s + 240)}\right) = \frac{12}{(s + 240)}$$

Therefore, $v(t) = 12e^{-240t}$. In this case, the steady-state solution is 0 V.

5.4 MESH ANALYSIS USING LAPLACE TRANSFORMS

Once again, the rules for successful mesh analysis in the frequency domain parallel those for mesh analysis in the time domain. We first transform the circuit and then write simultaneous equations for the loop currents $I_1(s)$, $I_2(s)$, and so forth, observing the same conventions as before. A slight complication arises when we attempt to solve the equations simultaneously for the loop currents. Since the coefficients of $I_1(s)$, $I_2(s)$, and so forth, generally involve the variable s, the conventional algebraic methods for solving equations simultaneously are not practical. Instead, we will use the method of determinants (Cramer's rule), which we'll review briefly as follows.

Suppose we wish to solve the following two equations simultaneously for unknowns X_1 and X_2:

$$C_1 = a_{11}X_1 + a_{12}X_2$$
$$C_2 = a_{21}X_1 + a_{22}X_2$$

where a_{11}, a_{12}, a_{21}, and a_{22} are coefficients and C_1 and C_2 are constants. Then

$$X_1 = \frac{\begin{vmatrix} C_1 & a_{12} \\ C_2 & a_{22} \end{vmatrix}}{\begin{vmatrix} a_{11} & a_{12} \\ a_{21} & a_{22} \end{vmatrix}} \quad \text{and} \quad X_2 = \frac{\begin{vmatrix} a_{11} & C_1 \\ a_{21} & C_2 \end{vmatrix}}{\begin{vmatrix} a_{11} & a_{12} \\ a_{21} & a_{22} \end{vmatrix}}$$

where the symbol $|\ |$ means the determinant of. The determinants are defined by

$$\begin{vmatrix} a_{11} & a_{12} \\ a_{21} & a_{22} \end{vmatrix} = a_{11}a_{22} - a_{21}a_{12}$$

$$\begin{vmatrix} C_1 & a_{12} \\ C_2 & a_{22} \end{vmatrix} = C_1 a_{22} - C_2 a_{12}$$

$$\begin{vmatrix} a_{11} & C_1 \\ a_{21} & C_2 \end{vmatrix} = C_2 a_{11} - C_1 a_{21}$$

5.4 MESH ANALYSIS USING LAPLACE TRANSFORMS

Note that in each case the determinant is found by subtracting the product of the elements on one diagonal from the product of the elements on the other diagonal.

EXAMPLE 5.18

Use determinants to solve the following two equations for X_1 and X_2:

$$5 = 2X_1 - 4X_2$$
$$-2 = -4X_1 + X_2$$

SOLUTION

$$X_1 = \frac{\begin{vmatrix} 5 & -4 \\ -2 & 1 \end{vmatrix}}{\begin{vmatrix} 2 & -4 \\ -4 & 1 \end{vmatrix}} = \frac{(5)(1) - (-2)(-4)}{(2)(1) - (-4)(-4)} = \frac{-3}{-14} = \frac{3}{14}$$

$$X_2 = \frac{\begin{vmatrix} 2 & 5 \\ -4 & -2 \end{vmatrix}}{\begin{vmatrix} 2 & -4 \\ -4 & 1 \end{vmatrix}} = \frac{(2)(-2) - (5)(-4)}{-14} = \frac{16}{-14} = \frac{-8}{7}$$

Note that the denominator determinant is the same (-14) in each case. This determinant, often called "delta" (Δ), is the determinant of the coefficients of the variables X_1 and X_2 when the equations are arranged so that the X_1-terms "line up" in a column and the X_2-terms do the same. For this reason, it is important to rearrange the equations to put them in this format. All constants should be on one side of each equation.

This method of solving equations simultaneously may be extended to three or more equations, but evaluating the determinants becomes somewhat more involved, and we will not take the time to review such cases. The reader may find details in any standard text in college algebra, linear algebra, or matrix analysis.

As indicated earlier, simultaneous equations in the frequency domain usually involve coefficients that are functions of the variable s. Nevertheless, we treat them in the same manner as constant coefficients when applying the method of determinants to solve the equations, as illustrated in Example 5.19.

EXAMPLE 5.19

Solve the following equations simultaneously for $I_1(s)$ and $I_2(s)$:

$$\frac{10}{s} = 8(I_1 - I_2) + 4sI_1$$

$$0 = 2sI_2 + 8(I_2 - I_1)$$

SOLUTION

We must first rearrange the equations so that they conform to the format in which the I_1-terms and I_2-terms "line up"

$$\frac{10}{s} = (8 + 4s)I_1 - 8I_2$$

$$0 = -8I_1 + (8 + 2s)I_2$$

Then

$$I_1(s) = \frac{\begin{vmatrix} 10/s & -8 \\ 0 & (8+2s) \end{vmatrix}}{\begin{vmatrix} (8+4s) & -8 \\ -8 & (8+2s) \end{vmatrix}} = \frac{(10/s)(8+2s)}{(8+4s)(8+2s)+64}$$

$$I_2(s) = \frac{\begin{vmatrix} (8+4s) & 10/s \\ -8 & 0 \end{vmatrix}}{\Delta} = \frac{80/s}{(8+4s)(8+2s)+64}$$

EXAMPLE 5.20

Find $i_1(t)$ and $i_2(t)$ using mesh analysis and LaPlace transforms (see Figure 5.41).

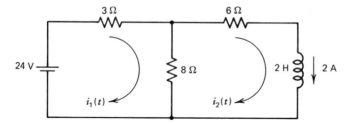

FIGURE 5.41 Find $i_1(t)$ and $i_2(t)$ using LaPlace transforms (Example 5.20).

SOLUTION

Transforming the circuit and recalling that the Thevenin equivalent circuit of the inductor with 2 A initial current is the impulse voltage source $(2\ \text{H})(2\ \text{A})\delta(t)$ whose transform is 4, we obtain Figure 5.42.

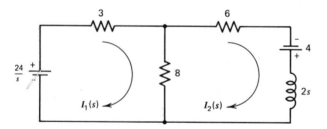

FIGURE 5.42 Transform of Figure 5.41 (Example 5.20).

Writing the loop equations and rearranging,

Loop 1: $24/s = 3I_1 + 8(I_1 - I_2)$ ⇒ $24/s = 11I_1 - 8I_2$

Loop 2: $4 = 6I_2 + 2sI_1 + 8(I_2 - I_1)$ ⇒ $4 = -8I_1 + (14 + 2s)I_2$

$$I_1 = \frac{\begin{vmatrix} 24/s & -8 \\ 4 & (14+2s) \end{vmatrix}}{\begin{vmatrix} 11 & -8 \\ -8 & (14+2s) \end{vmatrix}} = \frac{24/s\,(14+2s)+32}{11(14+2s)-64}$$

$$= \frac{336/s + 80}{22s + 90} = \frac{80s + 336}{s(22s + 90)} = \frac{3.64(s + 4.2)}{s(s + 4.09)}$$

$$I_2 = \frac{\begin{vmatrix} 11 & 24/s \\ -8 & 4 \end{vmatrix}}{\Delta} = \frac{44 + 192/s}{\Delta} = \frac{44s + 192}{s(22s + 90)} = \frac{2(s + 4.36)}{s(s + 4.09)}$$

To find the inverse transforms of $I_1(s)$ and $I_2(s)$, we may use pair 12 from Table 4-1, noting for our case that $a = 0$. Substituting $a = 0$ in pair 12, we see that

$$\mathcal{L}^{-1}\left[\frac{s + \alpha}{s(s + b)}\right] = \frac{1}{b}[\alpha - (\alpha - b)e^{-bt}]$$

Using this relation, we obtain

$$i_1(t) = 3.74 - .098e^{-4.09t} \text{ A}$$
$$i_2(t) = 2.13 - 0.132e^{-4.09t} \text{ A}$$

EXAMPLE 5.21

Find $i_1(t)$ and $i_2(t)$ using mesh analysis in the frequency domain (see Figure 5.43).

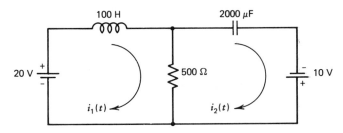

FIGURE 5.43 Find $i_1(t)$ and $i_2(t)$ (Example 5.21).

SOLUTION

Transforming the circuit, we obtain Figure 5.44.

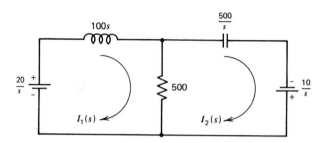

FIGURE 5.44 Transform of Figure 5.43 (Example 5.21).

Writing the loop equations and then rearranging, we have

Loop 1: $\dfrac{20}{s} = (100s)I_1 + 500(I_1 - I_2) = (100s + 500)I_1 - 500I_2$

Loop 2: $\dfrac{10}{s} = \dfrac{500}{s}I_2 + 500(I_2 - I_1) = -500I_1 + \left(\dfrac{500}{s} + 500\right)I_2$

Loop 1: $\dfrac{0.2}{s} = (s + 5)I_1 - 5I_2$

Loop 2: $\dfrac{.02}{s} = -I_1 + \left(\dfrac{1}{s} + 1\right)I_2$

$$I_1 = \dfrac{\begin{vmatrix} 0.2/s & -5 \\ 0.02/s & (1/s) + 1 \end{vmatrix}}{\begin{vmatrix} s + 5 & -5 \\ -1 & (1/s) + 1 \end{vmatrix}} = \dfrac{(0.2/s)[(1/s) + 1] + 5(0.02/s)}{(s + 5)[(1/s) + 1] - 5}$$

$$= \dfrac{(0.2/s^2) + (0.2/s) + (0.1/s)}{1 + s + (5/s) + 5 - 5}$$

$$= \dfrac{(1/s^2)(0.3s + 0.2)}{(1/s)(s^2 + s + 5)} = \dfrac{0.3[s + (2/3)]}{s(s^2 + s + 5)}$$

$$I_2 = \dfrac{\begin{vmatrix} s + 5 & 0.2/s \\ -1 & 0.02/s \end{vmatrix}}{\Delta} = \dfrac{(s + 5)(0.02/s) + 1(0.2/s)}{\Delta}$$

$$= \dfrac{0.02 + 0.1/s + 0.2/s}{\Delta}$$

$$= \dfrac{(1/s)(0.02s + 0.3)}{(1/s)(s^2 + s + 5)} = \dfrac{0.02(s + 15)}{s^2 + s + 5}$$

5.4 MESH ANALYSIS USING LAPLACE TRANSFORMS

To find the inverse transform of $I_1(s)$, we use the partial fraction expansion and Equation 23 in Example 4.24. Since

$$s^2 + s + 5 = [s + (0.5 + j2.18)][s + (0.5 - j2.18)]$$

we write,

$$I_1(s) = \frac{0.3[s + (2/3)]}{s(s^2 + s + 5)} = \frac{A}{s} + \frac{B}{s + (0.5 + j2.18)} + \frac{C}{s + (0.5 - j2.18)}$$

$$A = \frac{0.3[s + (2/3)]}{s^2 + s + 5}\bigg|_{s=0} = 0.04$$

Thus, the portion of $i_1(t)$ due to the pole at $s = 0$ is simply $i(t) = 0.04$ A. The response due to the complex poles is now calculated:

$$K(s) = \frac{0.3[s + (2/3)]}{s}$$

$$M\angle\theta = K(-0.5 + j2.18) = \frac{0.3[s + (2/3)]}{s}\bigg|_{s=-0.5+j2.18}$$

$$= \frac{0.3(0.167 + j2.18)}{-0.5 + j2.18} = \frac{(0.3)2.19\angle 85.62°}{2.24\angle 102.92°}$$

$$= 0.293\angle -17.3°$$

The contribution of the complex poles is then

$$i(t) = \frac{M}{\omega}e^{-\alpha t}\sin(\omega t + \theta)$$

$$= 0.134e^{-0.5t}\sin(2.18t - 17.3°) \text{ A}$$

and the total current is

$$i_1(t) = .04 + 0.134e^{-0.5t}\sin(2.18t - 17.3°) \text{ A}$$

We may find the inverse transform of $I_2(s)$ by completing the square in the denominator and noting that the result is in the form of pair 23 in Table 4-1

$$I_2(s) = \frac{0.02(s + 15)}{s^2 + s + 5} = \frac{0.02(s + 15)}{[s^2 + s + (1/4)] + 4\tfrac{3}{4}} = \frac{0.02(s + 15)}{[s + (1/2)]^2 + (2.18)^2}$$

From the table, we obtain directly

$$i_2(t) = .02\left[\frac{\sqrt{[15 - (1/2)]^2 + (2.18)^2}}{2.18}\right]e^{-t/2}\sin(2.18t + \theta)$$

$$\theta = \arctan\frac{2.18}{14.5} = 8.55°$$

$$i_2(t) = 0.134e^{-t/2}\sin(2.18t + 8.55°) \text{ A}$$

We see that both i_1 and i_2 have oscillatory transient solutions and that their steady-state values are 0.04 and 0 A, respectively.

5.5 NODAL ANALYSIS USING LAPLACE TRANSFORMS

When using LaPlace transforms to analyze a circuit by nodal analysis, the circuit should first be set up in the time domain following the steps outlined in Section 5.1. That is, all voltage sources should be converted to current sources, the nodes identified, a reference node chosen, an assumption made about the relative magnitudes of the node voltages, and all currents labeled. Then the circuit may be transformed and Kirchoff's current law applied at each node. Solve for the node voltages $V_1(s)$, $V_2(s)$, and so forth, in terms of s, and finally take inverse transforms to find $v_1(t)$, $v_2(t)$, and so forth.

EXAMPLE 5.22

Use nodal analysis and LaPlace transforms to find the current in the 1.5-K resistor shown in Figure 5.45.

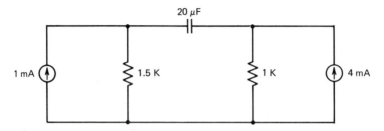

FIGURE 5.45   Find the current in 1.5-K resistor (Example 5.22).

SOLUTION

See Figure 5.46.

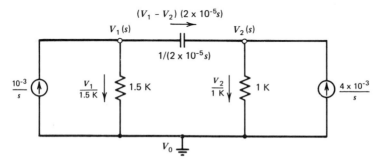

FIGURE 5.46   Transform of Figure 5.45 for nodal analysis (Example 5.22).

Node 1: $\dfrac{10^{-3}}{s} = \dfrac{V_1}{1.5\,K} + (V_1 - V_2)(2 \times 10^{-5}s)$

$$\frac{10^{-3}}{s} = \left(\frac{1}{1.5K} + 2 \times 10^{-5}s\right)V_1 - (2 \times 10^{-5}s)V_2$$

Node 2: $$\frac{4 \times 10^{-3}}{s} + (V_1 - V_2)(2 \times 10^{-5}s) = \frac{V_2}{1K}$$

$$\frac{4 \times 10^{-3}}{s} = -(2 \times 10^{-5}s)V_1 + \left(\frac{1}{1K} + 2 \times 10^{-5}s\right)V_2$$

$$V_1 = \frac{\begin{vmatrix} 10^{-3}/s & -(2 \times 10^{-5}s) \\ (4 \times 10^{-3})/s & (10^{-3} + 2 \times 10^{-5}s) \end{vmatrix}}{\begin{vmatrix} (2/3 \times 10^{-3} + 2 \times 10^{-5}s) & -(2 \times 10^{-5}s) \\ -(2 \times 10^{-5}s) & (10^{-3} + 2 \times 10^{-5}s) \end{vmatrix}}$$

$$= \frac{10^{-3}/s\,(10^{-3} + 2 \times 10^{-5}s) + (4 \times 10^{-3}/s)(2 \times 10^{-5}s)}{[(2/3) \times 10^{-3} + 2 \times 10^{-5}s](10^{-3} + 2 \times 10^{-5}s) - 4 \times 10^{-10}s^2}$$

$$= \frac{3(s + 10)}{s(s + 20)}$$

Using the same inverse transformation technique that was used in Example 5.20, we obtain

$$v_1(t) = 1.5 + 1.5e^{-20t} \text{ V}$$

Thus, the current through the 1.5-K resistor is

$$i(t) = \frac{v_1(t)}{1.5\,K} = (1 + e^{-20t}) \text{ mA}$$

5.6 SUPERPOSITION USING LAPLACE TRANSFORMS

The principle of superposition, as discussed in Section 5.1, is also valid in the frequency domain. That is, we may compute the effects of multiple sources one at a time, remembering to replace other current sources by open circuits and other voltage sources by short circuits. Once the individual effect of each source acting alone has been found, all of the effects may be added together to determine the total effect due to all sources acting simultaneously. It should be noted that the effect we are generally interested in finding is the current through, or voltage across, a component of the circuit. Superposition is *not* a valid technique for finding power, since power is not linearly related to voltage or current.

EXAMPLE 5.23

Use the principle of superposition and LaPlace transforms to find the voltage across the 1-H inductor shown in Figure 5.47.

136 NETWORK ANALYSIS USING LAPLACE TRANSFORMS

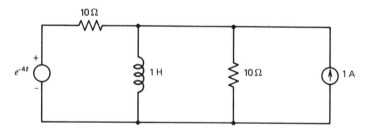

FIGURE 5.47 Find the voltage across the inductor (Example 5.23).

SOLUTION

Transforming the circuit and considering the effect of the voltage source first, we have Figure 5.48.

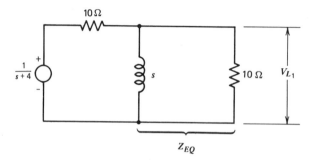

FIGURE 5.48 Transform of Figure 5.47 with current source removed (Example 5.23).

We will use the voltage-divider rule to find V_{L_1}, the voltage across the inductor due only to the voltage source. Since

$$V_{L_1} = \frac{Z_{EQ}}{10 + Z_{EQ}} \cdot \frac{1}{(s+4)} \quad \text{and} \quad Z_{EQ} = \frac{(10)(s)}{10 + s}$$

we find

$$V_{L_1} = \frac{(10)(s)/(10+s)}{10 + 10s/(10+s)} \cdot \frac{1}{(s+4)}$$

$$= \frac{10s}{(100 + 20s)(s+4)}$$

$$= \frac{(\frac{1}{2})s}{(s+5)(s+4)}$$

Using pair 11 from Table 4-1, we find $V_{L_1}(t)$ due to the voltage source

$$V_{L_1}(t) = 2.5e^{-5t} - 2e^{-4t}$$

5.7 THEVENIN'S AND NORTON'S THEOREMS

To find the contribution of the current source, we replace the voltage source by a short circuit and then use the current-divider rule to determine the current I_L through the inductor. Note that the two 10-Ω resistors have been replaced by their parallel equivalent in Figure 5.49.

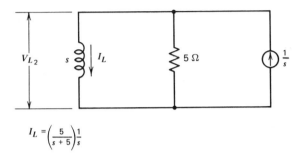

$$I_L = \left(\frac{5}{s+5}\right)\frac{1}{s}$$

FIGURE 5.49 Transform of Figure 5.47 with voltage source removed (Example 5.23).

Now the voltage across the inductor V_{L_2} is found using $V_{L_2} = LsI_L$

$$V_{L_2} = s\left(\frac{5}{s+5}\right)\frac{1}{s} = \frac{5}{s+5}$$

So

$$V_{L_2}(t) = 5e^{-5t}$$

Finally, the voltage across the inductor due to both sources simultaneously is

$$V_L(t) = V_{L_1}(t) + V_{L_2}(t)$$
$$= 7.5e^{-5t} - 2e^{-4t}$$

5.7 THEVENIN'S AND NORTON'S THEOREMS USING LAPLACE TRANSFORMS

As indicated in Section 5.1, network calculations can often be simplified by replacing a portion of the circuit under analysis with a Thevenin equivalent circuit. This is especially true when we wish to find how the voltage or current in one component changes as the component itself changes. A Thevenin equivalent circuit is constructed in the frequency domain following the same procedure reviewed in Section 5.1 for the time domain. The utility of Thevenin's theorem is illustrated in Example 5.24.

EXAMPLE 5.24

Use Thevenin's theorem and LaPlace transforms to find the current in the capacitor C when (a) C = .02 F, (b) C = .0625 F, and (c) C = .03125 F, after the switch is closed at $t = 0$ (see Figure 5.50). Assume all initial conditions are zero.

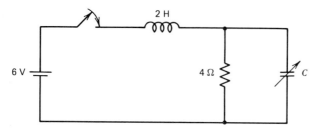

FIGURE 5.50   Find the current in C (Example 5.24).

SOLUTION

Since we are interested in the capacitor as a variable load, we remove it from the circuit temporarily and find the Thevenin equivalent circuit by looking into the terminals from which it is removed. Redrawing the circuit to reflect this fact and transforming it to the frequency domain, we obtain Figure 5.51.

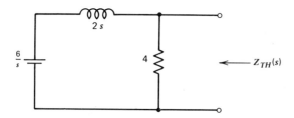

FIGURE 5.51   Transform of Figure 5.50 with C removed (Example 5.24).

Now

$$Z_{TH}(s) = \frac{(4)(2s)}{4 + 2s} = \frac{4s}{s + 2}$$

We find $E_{TH}(s)$ by applying the voltage-divider rule

$$E_{TH}(s) = \left(\frac{4}{4 + 2s}\right)\frac{6}{s} = \frac{12}{s(s + 2)}$$

The Thevenin equivalent circuit with the capacitor restored is then as shown in Figure 5.52.

5.7 THEVENIN'S AND NORTON'S THEOREMS

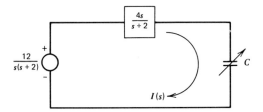

FIGURE 5.52 Thevenin equivalent circuit (Example 5.24).

Now

$$I(s) = \frac{E_{TH}(s)}{Z_{TH}(s) + 1/Cs} = \frac{12/s(s+2)}{4s/(s+2) + 1/Cs}$$

$$= \frac{3}{s^2 + (s/4C) + (1/2C)}$$

(a) For $C = .02$ F, we have

$$I(s) = \frac{3}{s^2 + 12.5s + 25}$$

The denominator may be factored, revealing real poles of $I(s)$ at $s = -10$ and $s = -2.5$.

$$I(s) = \frac{3}{(s+10)(s+2.5)}$$

Then

$$i(t) = 0.4(e^{-2.5t} - e^{-10t}) \text{ A}$$

(b) For $C = 0.0625$ F $= 1/16$ F, we find

$$I(s) = \frac{3}{s^2 + 4s + 8}$$

In this case, the roots of the denominator are complex; that is, $I(s)$ has complex conjugate poles. Completing the square in the denominator, we may write

$$I(s) = \frac{3}{s^2 + 4s + 4 + (8-4)} = \frac{3}{(s+2)^2 + (2)^2}$$

Using pair 18 of Table 4-1, we find $i(t) = 1.5e^{-2t} \sin 2t$

(c) When $C = .03125$ F $= 1/32$ F, we have

$$I(s) = \frac{3}{(s^2 + 8s + 16)} = \frac{3}{(s+4)^2}$$

From pair 6 of Table 4-1, $i(t) = 3te^{-4t}$.

140 NETWORK ANALYSIS USING LAPLACE TRANSFORMS

The solutions obtained in Example 5.24 could also be obtained by using a Norton equivalent circuit. Recall that a Norton equivalent circuit can be constructed from a Thevenin equivalent circuit simply by converting the Thevenin circuit to an equivalent current source. Alternatively, a Norton circuit can be found directly by computing the current that flows through the *shorted* terminals of that portion of the circuit to be replaced. This short-circuit current is the value used for the current source in the Norton equivalent. The Norton equivalent resistance is found in the same way as the Thevenin equivalent resistance.

EXAMPLE 5.25

Use LaPlace transforms and a Norton equivalent circuit to find the current in the capacitor of Example 5.24, when $C = .0625$ F.

SOLUTION

The Norton equivalent current source is found by computing the short-circuit current (see Figure 5.53).

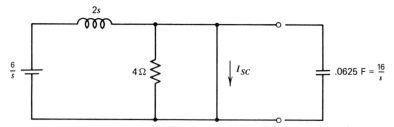

FIGURE 5.53 Find current in C using Norton equivalent circuit (Example 5.25).

Clearly,

$$I_{SC} = I_N = \frac{6/s}{2s} = \frac{3}{s^2}$$

Note also that

$$I_N = \frac{E_{TH}}{Z_{TH}} = \frac{[12/s(s+2)]}{4s/(s+2)} = \frac{3}{s^2}$$

The Norton equivalent impedance Z_N is the same as the Thevenin equivalent impedance Z_{TH}, so the Norton circuit is as shown in Figure 5.54.

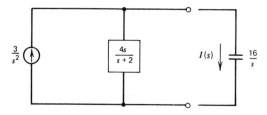

FIGURE 5.54 Norton equivalent circuit (Example 5.25).

5.8 THE EFFECTS OF DAMPING ON CIRCUIT RESPONSE

We now find $I(s)$ using the current-divider rule

$$I(s) = \frac{3}{s^2}\left[\frac{4s/(s+2)}{4s/(s+2) + 16/s}\right] = \frac{3}{s^2}\left(\frac{4s^2}{4s^2 + 16s + 32}\right)$$

$$= \frac{3}{s^2 + 4s + 8} = \frac{3}{(s+2)^2 + (2)^2}$$

This is the same transform obtained in Example 5.24 and has the same solution

$$i(t) = 1.5e^{-2t} \sin 2t$$

5.8 THE EFFECTS OF DAMPING ON CIRCUIT RESPONSE

Besides illustrating the ease with which we can find how one variable (current) changes when one component (capacitance) is changed in a circuit, Example 5.24 is instructive in another respect. The three values of capacitance used in the example were chosen to illustrate three distinct types of circuit response. With the smallest value of capacitance (.02 F) in the circuit, we saw that the current rose to an initial value [$i(t) = 0.4$ A at $t = 0$] and then decayed to zero after a long period of time. For the largest value of capacitance (.0625 F) that we considered, the response was a damped oscillatory current: $i(t) = 1.5e^{-2t} \sin 2t$, a sinusoidal current with frequency 2 rad/sec and with amplitude decaying exponentially to zero; this is often termed a "ringing" response. Finally, when C was set equal to .03125 F = 1/32 F, we obtained the response $i(t) = 3te^{-4t}$.

The reader is invited to solve this problem for arbitrary values of R, L, and C. If this is done, we find that

$$I(s) = \frac{6/L}{[s^2 + (1/RC)s + 1/LC]}$$

Using the quadratic formula, we find that the poles of this function (roots of the denominator) are

$$s = \frac{-(1/RC) \pm \sqrt{1/(RC)^2 - 4/LC}}{2} \qquad (1)$$

Thus, the poles are

1. Real if $1/(RC)^2 > 4/LC$
2. Real and equal if $1/(RC)^2 = 4/LC$
3. Complex if $1/(RC)^2 < 4/LC$

These three possibilities correspond to the three different values of C that we used in the previous example. In that example, the poles are equal (that is, we have a repeated pole or a pole of order two) when

$$\frac{1}{(RC)^2} = \frac{4}{LC} \quad \text{or} \quad \frac{1}{16C^2} = \frac{2}{C}$$

$$C = \frac{1}{32} F$$

We can conclude that for values of C greater than 1/32 F, the response will be a damped oscillation, and for values of C less than 1/32 F, the response will rise to a peak and then decay to zero. These are called *underdamped* and *overdamped* responses, respectively.

The damping of an *RLC* circuit such as the one just discussed is more frequently expressed in terms of the value of resistance present in relation to the inductance and capacitance. Referring back to Equation 1, we see that underdamping (complex poles) correspond to the case $1/(RC)^2 < 4/LC$, which, in terms of the resistance, may be expressed equivalently as

$$R > \frac{1}{2}\sqrt{\frac{L}{C}}$$

Overdamping occurs when the inequality is reversed, and when $R = \frac{1}{2}\sqrt{L/C}$, we say that the circuit is *critically* damped.

EXAMPLE 5.26

Find the conditions on the value of R that cause under, over, and critically damped responses in the series *RLC*-circuit shown in Figure 5.55.

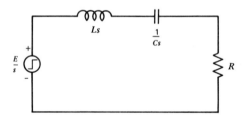

FIGURE 5.55 Find Values of R for under, over, and critically damped responses (Example 5.26).

SOLUTION

$$Z(s) = Ls + \frac{1}{Cs} + R = \frac{LCs^2 + RCs + 1}{Cs}$$

$$I(s) = \frac{E/s}{Z(s)} = \frac{EC}{LCs^2 + RCs + 1} = \frac{E/L}{s^2 + (R/L)s + (1/LC)}$$

In this case, the factors in the denominator are

5.8 THE EFFECTS OF DAMPING ON CIRCUIT RESPONSE

$$s = \frac{-R/L \pm \sqrt{(R/L)^2 - 4/LC}}{2} \qquad (2)$$

Consequently, the response is a damped oscillation (underdamped) due to the presence of complex poles when

$$\frac{R^2}{L} < \frac{4}{LC} \quad \text{or} \quad R < 2\sqrt{\frac{L}{C}} \qquad (3)$$

and overdamped when

$$R > 2\sqrt{\frac{L}{C}}$$

and critically damped when

$$R = 2\sqrt{\frac{L}{C}}$$

Contrast these results with those of Example 5.24.

When a quadratic with complex roots appears in the denominator of a transform, it is often written in the form:

$$s^2 + 2\zeta\omega_n s + \omega_n^2$$

For example, the current $I(s)$ in Example 5.25 was found to be

$$I(s) = \frac{E/L}{s^2 + (R/L)s + (1/LC)}$$

Comparing the coefficients of the quadratic in the denominator with the coefficients of the alternative form, we see that

$$2\zeta\omega_n = \frac{R}{L} \qquad (4)$$

$$\omega_n^2 = \frac{1}{LC} \qquad (5)$$

ζ is called the *damping ratio* and ω_n the *natural frequency*. From Equations 4 and 5, we obtain

$$\omega_n = \frac{1}{\sqrt{LC}} \qquad (6)$$

$$\zeta = \frac{R}{2}\sqrt{\frac{C}{L}} \qquad (7)$$

Referring to inequality 3, which is the condition for an underdamped response (complex roots) in the series *RLC* circuit, we see that

$$R < 2\sqrt{\frac{L}{C}} \quad \text{means} \quad \zeta = \frac{R}{2}\sqrt{\frac{C}{L}} < \frac{2\sqrt{L/C}}{2}\sqrt{\frac{C}{L}} = 1$$

We conclude that an underdamped response occurs in the series RLC-circuit for $\zeta < 1$. Indeed, this is the case for *any* circuit with a quadratic in the denominator of the form $s^2 + 2\zeta\omega_n s + \omega_n^2$ and for which $\zeta < 1$. This fact can be seen by using the quadratic formula to find the roots:

$$s = \frac{-2\zeta\omega_n \pm \sqrt{4\zeta^2 \cdot \omega_n^2 - 4\omega_n^2}}{2}$$

$$s = -\zeta\omega_n \pm \omega_n\sqrt{\zeta^2 - 1} \tag{8}$$

Clearly, Equation 8 yields complex roots when $\zeta^2 < 1$, that is, $\zeta < 1$. Critical damping occurs when $\zeta = 1$. As can be seen in Equation 8, critical damping results in repeated poles of order two at $s = -\zeta\omega_n$. Overdamping occurs for $\zeta > 1$, and since $\zeta > 1$ means $\zeta^2 - 1 > 0$, we see from Equation 8 that the overdamped response corresponds to the presence of two real, unequal poles.

EXAMPLE 5.27

Calculate and sketch $i(t)$ versus t in Figure 5.56 for (a) $R = 20\ \Omega$, (b) $R = 40\ \Omega$, and (c) $R = 50\ \Omega$.

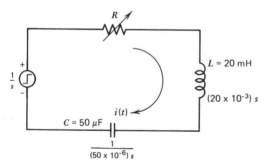

FIGURE 5.56 Find $i(t)$ (Example 5.27).

SOLUTION

(a) From Equation 7,

$$\zeta = \frac{R}{2}\sqrt{\frac{C}{L}} = 10\sqrt{\frac{50 \times 10^{-6}}{20 \times 10^{-3}}} = 0.5$$

Since $\zeta < 1$, we know the poles are complex and since

$$\omega_n = \frac{1}{\sqrt{LC}} = \frac{1}{\sqrt{(20 \times 10^{-3})(50 \times 10^{-6})}} = 10^3\ \text{rad/sec} \quad \text{and}$$

5.8 THE EFFECTS OF DAMPING ON CIRCUIT RESPONSE

$$2\zeta\omega_n = 2(0.5)\omega_n = 10^3 = \frac{R}{L}$$

we know

$$I(s) = \frac{50}{s^2 + 10^3 s + 10^6}$$

Now, there are two ways we can find $i(t)$. Since the poles are complex, we know we can complete the square in the denominator of $I(s)$ and, therefore, put $I(s)$ in the form of pair 18 in Table 4-1. Thus, we obtain

$$I(s) = \frac{50}{(s^2 + 10^3 s + 25 \times 10^4) + (10^6 - 25 \times 10^4)}$$
$$= \frac{50}{(s + 500)^2 + (\sqrt{0.75 \times 10^6})^2}$$

Noting that $a = 500$ and $b = \sqrt{0.75 \times 10^6} = 866$, we find from pair 18 that

$$i(t) = 50\left[\frac{1}{b}e^{-at} \sin bt\right] = .0578 e^{-500t} \sin 866t$$

Alternatively, pair 20 shows us that

$$i(t) = \frac{50}{\omega_n \sqrt{1 - \zeta^2}} e^{-\zeta \omega_n t} \sin \omega_n \sqrt{1 - \zeta^2}\, t$$

and substituting $\omega_n = 10^3$, $\zeta = 0.5$, leads to precisely the same result for $i(t)$. The current is sketched in Figure 5.57; note that the time constant of the envelope (e^{-500t}) is 1/500 sec or 2.0 msec and that the period of the oscillation is $T = 2\pi/866 = 7.26$ msec.

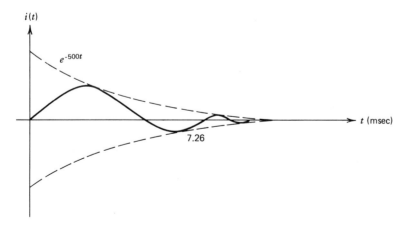

FIGURE 5.57 $i(t)$ for $R = 20\,\Omega$ (Example 5.27).

146 NETWORK ANALYSIS USING LAPLACE TRANSFORMS

In general, the smaller the value of ζ, that is, the less resistance there is in the circuit, the longer it will take for the oscillations to decay to zero. In the limit, if it were possible to reduce the resistance to zero, the oscillations would not decay at all, and we would have a perpetual oscillation. Of course, there is always some resistance associated with real inductors and capacitors. Conversely, as the resistance in the circuit is increased, we approach the condition of critical damping, as illustrated in (b).

(b) If $R = 40\ \Omega$, then substituting in Equation 7 shows $\zeta = 1.0$, the condition for critical damping. Referring to Equation 2, this condition also corresponds to $(R/L)^2 = 4/LC$, which results in a pole of order two at $s = -R/2L = -10^3$. Consequently,

$$I(s) = \frac{50}{(s + 10^3)^2}$$

and from pair 6,

$$i(t) = 50te^{-1000t}$$

This response is sketched in Figure 5.58.

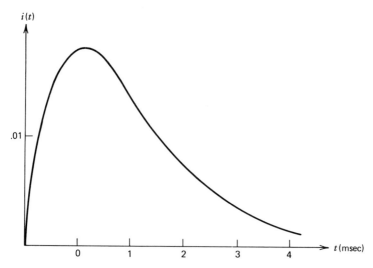

FIGURE 5.58 $i(t)$ for $R = 40\ \Omega$ (Example 5.27).

A critically damped response occurs when the value of resistance in the circuit is just sufficient to prevent an undershoot of current, that is, just sufficient to prevent the current from going negative. When ζ is just slightly less than unity, we obtain a response resembling Figure 5.59.

As discussed earlier, progressively smaller values of resistance and correspondingly smaller values of ζ result in an undershoot followed by an overshoot, possibly followed by another undershoot, and so on; in other words, a damped oscillation.

5.8 THE EFFECTS OF DAMPING ON CIRCUIT RESPONSE

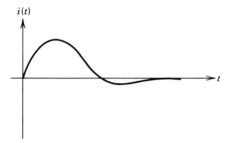

FIGURE 5.59 Response with ζ slightly less than unity.

(c) When $R = 50\ \Omega$, $R > 2\sqrt{L/C}$, and an overdamped response results. From Equation 2, the poles are

$$s = \frac{(-R/L) \pm \sqrt{(R/L)^2 - 4/LC}}{2} = \frac{-2.5 \times 10^3 \pm \sqrt{2.25 \times 10^6}}{2}$$

$$= -2 \times 10^3,\ -0.5 \times 10^3$$

Thus,

$$I(s) = \frac{50}{(s + 0.5 \times 10^3)(s + 2 \times 10^3)}$$

and from pair 10,

$$i(t) = 33.3 \times 10^{-3}(e^{-0.5 \times 10^3 t} - e^{-2 \times 10^3 t})$$

This response is sketched in Figure 5.60. Also, a sketch of $i(t)$ for the critically damped case is repeated on these same axes so that the two cases may be compared and contrasted.

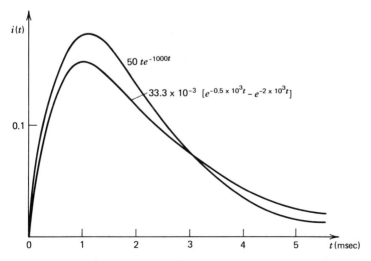

FIGURE 5.60 Critical and overdamped responses (Example 5.27).

5.9 ANALYSIS OF MECHANICAL NETWORKS

In Chapter 4, we discussed the relationships between the variable x (translational displacement) and the forces developed in a mechanical system. These relationships are summarized as follows:

For a spring with constant K, $F = Kx$
For a dashpot B, $F = Bv = B\,dx/dt$
For a mass M, $F = Ma = M\,d^2x/dt^2$

These relationships may be rewritten in terms of velocity, $v = (dx/dt)$, since

$$\frac{d^2x}{dt^2} = \frac{dv}{dt} \quad \text{and} \quad x = \int_0^t v\,dt$$

The force equations expressed in terms of velocity alongside the basic electrical equations that relate voltage and current in resistors, inductors, and capacitors are:

$$F = K\int_0^t v\,dt \qquad e = \frac{1}{C}\int_0^t i\,dt$$

$$F = Bv \qquad e = Ri$$

$$F = M\frac{dv}{dt} \qquad e = L\frac{di}{dt}$$

We cannot help but note the similarity in form of the mechanical and electrical equations. Indeed, the only distinction is in the letters chosen to represent the variables. This similarity in form is the basis for *analog simulation*, the use of electrical circuits to solve mechanical networks, about which we will have more to say in a later chapter. Note that force is analogous to voltage, velocity to current, mass to inductance, viscous damping to resistance, and spring constant to the reciprocal of capacitance. (Other analogs are also possible.)

It follows that any of the techniques that we have developed to solve electrical networks may also be used to find forces and velocities in a mechanical system. Furthermore, since the differential equation describing the behavior of any mechanical system has its direct counterpart (analog) in some electrical network, a solution to the equation may be found using LaPlace transform methods, that is, by analysis in the frequency domain just as we have done with electrical networks. The only additional skill that is required is the ability to recognize, understand, and be comfortable with the schematic symbols, notation, and variables when depicted in a mechanical network.

For the reader who through practice and repeated analysis of electrical networks has developed that certain feel for the abstract concepts and relations among variables that a symbolic diagram (schematic) conveys, it may be helpful to study analogous mechanical configurations to develop a similar facility. Some typical configurations are discussed as follows.

Spring-dashpot Combination

See Figure 5.61.

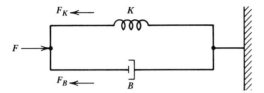

FIGURE 5.61 Spring-dashpot combination.

The applied force F is exactly counterbalanced by the sum of the spring and dashpot forces acting in the opposite direction.

$$F = F_K + F_B$$

$$F = Kx + B\frac{dx}{dt} \quad \text{or} \quad F = K\int_0^t v\,dt + Bv$$

This is analogous to $e = 1/C \int_0^t i\,dt + Ri$, which results from applying Kirchoff's voltage law around an RC-loop (see Figure 5.62).

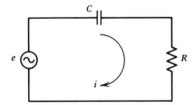

FIGURE 5.62 Electrical circuit analogous to Figure 5.61.

Note carefully that what appears to be a "parallel" mechanical connection corresponds, in fact, to a series electrical connection. But in the mechanical network the applied force must equal the sum of the resisting forces developed by the spring and dashpot, just as the applied voltage in the circuit must equal the sum of the (resisting) voltage drops. Remember that force is analogous to voltage. Also, the current is the same in each element, analogous to the fact that the velocity of the spring and dashpot are the same, since the right-hand side of each of these is fixed.

Consider now the configuration shown in Figure 5.63.

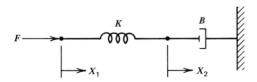

FIGURE 5.63 Another spring-dashpot combination.

At the point where the force is applied, we have

$$F = K(x_1 - x_2) = K\int_0^t (v_1 - v_2)\,dt$$

At the junction of the spring and dashpot, the force developed by the spring is applied to the dashpot, and we have

$$K\int_0^t (v_1 - v_2)\,dt = Bv_2 \quad \text{or} \quad 0 = K\int_0^t (v_2 - v_1)\,dt + Bv_2$$

These relations are analogous to the parallel RC-circuit shown in Figure 5.64.

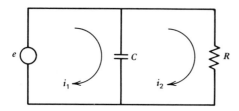

FIGURE 5.64 Electrical circuit analogous to Figure 5.63.

$$e = \frac{1}{C}\int_0^t (i_1 - i_2)\,dt$$

$$0 = \frac{1}{C}\int_0^t (i_2 - i_1)\,dt + Ri_2$$

Mass-spring Combination

Suppose a mass and spring are connected as shown in Figure 5.65.

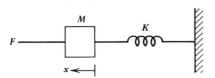

FIGURE 5.65 Spring-mass combination.

The mass is considered to be resting on a frictionless surface, so there are no vertical displacements involved. (Note that it is not possible to "ground" one end of a mass, that is, to fix it and still have a force developed due to its acceleration. For this reason, though every mechanical circuit has an electrical analog, the converse is not true.)

In the system in Figure 5.65,

$$F = M\frac{d^2x}{dt^2} + Kx \quad \text{or} \quad F = M\frac{dv}{dt} + K\int_0^t v\, dt$$

It is clear that the velocity of the mass and spring are equal, just as the current through the components of the analogous LC-circuit are equal (see Figure 5.66).

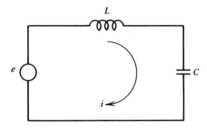

FIGURE 5.66 Electrical circuit analogous to Figure 5.65.

$$e = L\frac{di}{dt} + \frac{1}{C}\int_0^t i\, dt$$

In the absence of resistance or friction in the mechanical system, we know that sustained oscillations will result. This is an example of what appears to be a "series" mechanical configuration that is analogous to a series electrical circuit. The upshoot of all this is that we cannot think of "series" and "parallel" connections in a mechanical system in the same way we do in an electrical circuit. Instead, we must determine in each case where and how to sum forces correctly in order to obtain the correct differential equation for the mechanical network.

Spring-mass-dashpot Combinations

In the configuration shown in Figure 5.67, all components are seen to have the same velocity $v = dx/dt$.

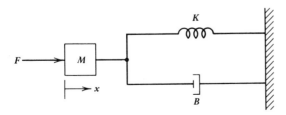

FIGURE 5.67 Spring-mass-dashpot combination.

Consequently,

$$F = M\frac{dv}{dt} + Bv + K\int_0^t v\, dt$$

This is analogous to the series RLC-circuit we analyzed previously. Consider now the combination of components shown in Figure 5.68.

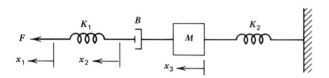

FIGURE 5.68 A combination of two springs, a mass, and a dashpot.

The appropriate differential equations are:

$$F = K_1(x_1 - x_2)$$

$$0 = K_1(x_2 - x_1) + B(v_2 - v_3)$$

$$0 = B(v_3 - v_2) + Ma_3 + K_2 x_3$$

Equivalently,

$$F = K_1 \int_0^t (v_1 - v_2)\, dt$$

$$0 = K_1 \int_0^t (v_2 - v_1)\, dt + B(v_2 - v_3)$$

$$0 = B(v_3 - v_2) + M\frac{dv_3}{dt} + K_2 \int_0^t v_3\, dt$$

Translating these equations to the electrical analogs, we have

$$e = \frac{1}{C_1}\int_0^t (i_1 - i_2)\, dt$$

$$0 = \frac{1}{C_1}\int_0^t (i_2 - i_1)\, dt + R(i_2 - i_3)$$

$$0 = R(i_3 - i_2) + L\frac{di_3}{dt} + \frac{1}{C_2}\int_0^t i_3\, dt$$

These are seen to be the loop equations for the network shown in Figure 5.69.

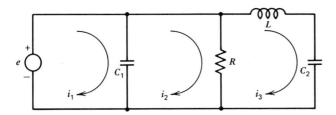

FIGURE 5.69 Elecrtrical circuit analogous to Figure 5.68.

EXAMPLE 5.28

Use LaPlace transforms to find the velocity and acceleration of the mass as functions of time (see Figure 5.70).

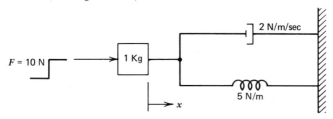

FIGURE 5.70 Find velocity and acceleration versus time (Example 5.28).

SOLUTION

$$\frac{10}{s} = sV(s) + 2V(s) + \frac{5}{s}V(s)$$

$$= V(s)\left(s + 2 + \frac{5}{s}\right)$$

$$V(s) = \frac{10}{s^2 + 2s + 5} = \frac{10}{(s^2 + 2s + 1) + 4} = \frac{10}{(s + 1)^2 + (2)^2}$$

Consequently, from pair 18,

$$v(t) = 5e^{-t}\sin 2t$$

Furthermore,

$$a(t) = \frac{dv}{dt} = 5e^{-t}(2\cos 2t - \sin 2t)$$

EXAMPLE 5.29

Use LaPlace transforms and the principle of superposition to find the velocity of the mass M as a function of time (see Figure 5.71).

154 NETWORK ANALYSIS USING LAPLACE TRANSFORMS

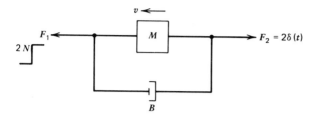

FIGURE 5.71 Find velocity v using superposition (Example 5.29).

SOLUTION

The velocity V_1 due to F_1 is found by "opening" the connection to F_2 (freeing the right side); see Figure 5.72.

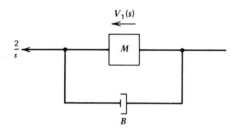

FIGURE 5.72 V_1 due to F_1 (Example 5.29).

$$\frac{2}{s} = MsV_1(s) + BV_1(s)$$

$$= V_1(s)(Ms + B)$$

$$V_1(s) = \frac{2/M}{s(s + B/M)}$$

We now find the velocity V_2 due to F_2; see Figure 5.73.

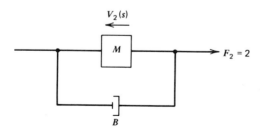

FIGURE 5.73 V_2 due to F_2 (Example 5.29).

$$-2 = MsV_2(s) + BV_2(s)$$

$$V_2(s) = -\frac{2/M}{s + B/M}$$

Consequently,

$$V(s) = V_1(s) + V_2(s) = \frac{2/M}{s(s + B/M)} - \frac{2/M}{(s + B/M)}$$

Therefore,

$$v(t) = \frac{2}{M}\left[\frac{M}{B}(1 - e^{(B/M)t})\right] - \frac{2}{M}(e^{-(B/M)t})$$

$$= \frac{2}{B} - 2e^{-(B/M)t}\left(\frac{1}{B} + \frac{1}{M}\right)$$

REFERENCES

1. **Aidala, Joseph B.**, and **Leon Katz**. *Transients in Electric Circuits*. Prentice-Hall, 1980.
2. **Alvarez, E. C.**, and **John Tontsch**. *Fundamental Circuit Analysis*. Science Research Associates, 1978.
3. **Angerbauer, George J.** *Principles of dc and ac Circuits*. Duxbury Press, 1978.
4. **Belove, Charles**, and **Melvyn M. Drossman**. *Systems and Circuits for Electrical Engineering Technology*. McGraw-Hill, 1976.
5. **Boylestad, Robert L.** *Introductory Circuit Analysis*. 3rd ed. Charles E. Merrill, 1977.
6. **Gardner, Murray F.**, and **John L. Barnes**. *Transients in Linear Systems*. John Wiley & Sons, 1956.
7. **Malvino, Albert P.** *Resistive and Reactive Circuits*. McGraw-Hill, 1974.
8. **Romanek, Richard J.** *Introduction to Electronic Technology*. Prentice-Hall, 1975.
9. **Stanley, William D.** *Transform Circuit Analysis for Engineering and Technology*. Prentice-Hall, 1968.

EXERCISES

5.1 A dc-voltage source produces an 8-V drop across a 1-Ω load and a 14-V drop across a 7-Ω load. Represent it as an ideal voltage source in series with an appropriate resistance.

5.2 How much current will be delivered to a 1-K load by a current source consisting of a 40-mA ideal current source with internal resistance 12.75 K?

5.3 (a) Convert the voltage source in Exercise 5.1 to an equivalent current source.

(b) Convert the current source in Exercise 5.2 to an equivalent voltage source.

5.4 Find the voltage drop across the 10-Ω resistor using Kirchoff's voltage law (see Figure 5.74).

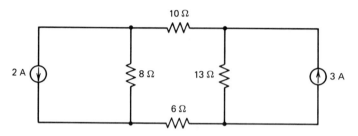

FIGURE 5.74

5.5 Use the voltage-divider rule to find e in the circuit shown in Figure 5.75.

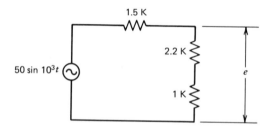

FIGURE 5.75

5.6 Use the current-divider rule to find the voltage across the 20-Ω resistor shown in Figure 5.76.

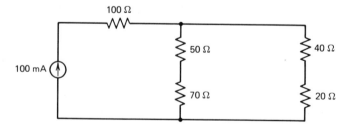

FIGURE 5.76

5.7 Use mesh analysis to find the current in the 8-Ω resistor shown in Figure 5.77.

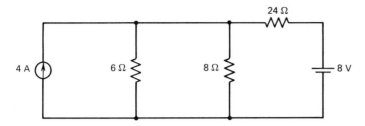

FIGURE 5.77

5.8 Use nodal analysis to find the current in the 5-Ω resistor shown in Figure 5.78.

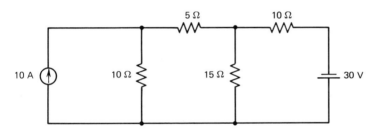

FIGURE 5.78

5.9 Use the principle of superposition to solve Exercise 5.7.

5.10 Use the principle of superposition to solve Exercise 5.8.

5.11 Find the Thevenin and Norton equivalent circuits with respect to the terminals AB (see Figure 5.79).

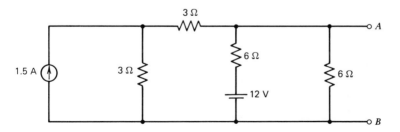

FIGURE 5.79

5.12 Transform the circuit shown in Figure 5.80, including initial conditions, so that it is suitable for analysis by Kirchoff's voltage law.

158 NETWORK ANALYSIS USING LAPLACE TRANSFORMS

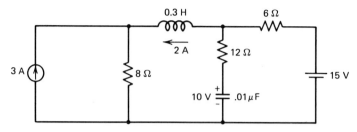

FIGURE 5.80

5.13 Use Kirchoff's voltage law and LaPlace transforms to find $i(t)$ in the circuit shown in Figure 5.81. Identify the transient and steady-state solutions.

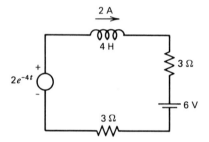

FIGURE 5.81

5.14 Use the current divider rule and LaPlace transforms to find the current in the inductor. Identify the transient and steady-state solutions. (See Figure 5.82.)

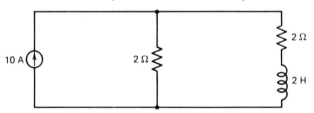

FIGURE 5.82

5.15 Use mesh analysis and LaPlace transforms to find the current in the 12-Ω resistor. Identify the transient and steady-state solutions. (See Figure 5.83.)

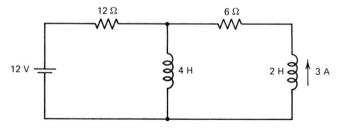

FIGURE 5.83

5.16 Write the transform-mesh equations for the circuit in Exercise 5.12. Set up the determinants for $I_1(s)$ and $I_2(s)$. It is not necessary to solve for $i_1(t)$ and $i_2(t)$.

5.17 Find $i_1(t)$ and $i_2(t)$ using mesh analysis and LaPlace transforms. Identify the transient and steady-state solutions. (See Figure 5.84.)

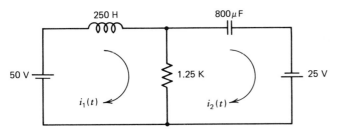

FIGURE 5.84

5.18 Use nodal analysis and LaPlace transforms to find the voltage across the 6-K resistor. Identify the transient and steady-state solutions. (See Figure 5.85.)

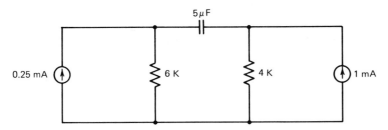

FIGURE 5.85

5.19 Use Thevenin's equivalent circuit in the frequency domain to find the current in the 4-Ω resistor. (See Figure 5.86.)

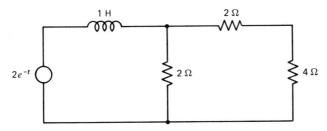

FIGURE 5.86

5.20 Repeat Exercise 5.19 using a Norton equivalent circuit.

5.21 Find and sketch $i(t)$ versus time. (See Figure 5.87.)

5.22 In Exercise 5.21, what should the value of R be in order for the response to be critically damped?

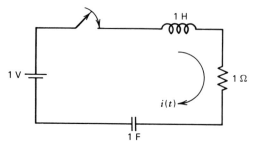

FIGURE 5.87

5.23 What is the damping ratio and natural frequency of the circuit in Exercise 5.21? How is the frequency of the oscillatory response found in Exercise 5.21 related to these two quantities?

5.24 For the mechanical system shown in Figure 5.88,

(a) Write a transform equation for the velocity of the mass after the step force is applied.
(b) Draw an equivalent electrical network.
(c) Find $v(t)$.
(d) Classify the response as over, under, or critically damped.

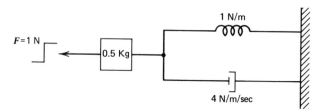

FIGURE 5.88

ANSWERS TO EXERCISES

5.1 $R = 1\,\Omega$; $V = 16$ V

5.2 $I_L = 37.09$ mA

5.3 (a) $I = 16$ A; $R = 1\,\Omega$
     (b) $E = 510$ V; $R = 12.75$ K

5.4 14.86 V

5.5 $e = 34 \sin 10^3 t$

5.6 1.33 V

ANSWERS TO EXERCISES **161**

5.7   1.625 A

5.8   5.6 A

5.11 $E_{TH} = 5.5$ V; $R_{TH} = 2\ \Omega$
       $I_N = 2.75$ A; $R_N = 2\ \Omega$

5.12   See Figure 5.89.

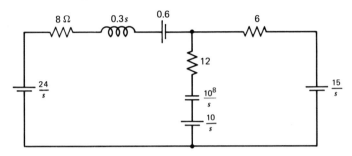

FIGURE 5.89

5.13 $i(t) = \dfrac{16}{5} e^{-3t/2} - 0.2 e^{-4t} - 1$

5.14 $i(t) = 5(1 - e^{-2t})$

5.15 $i(t) = 1 - 4.021\, e^{-11.2t} + .021 e^{-0.8t}$

5.16 $\dfrac{14}{s} - 0.6 = \left(20 + 0.3s + \dfrac{10^8}{s}\right) I_1 - \left(12 + \dfrac{10^8}{s}\right) I_2$

   $\dfrac{-5}{s} = -\left(12 + \dfrac{10^8}{s}\right) I_1 + \left(18 + \dfrac{10^8}{s}\right) I_2$

5.17 $i_1(t) = 0.04 + 0.134 e^{-0.5t} \sin(2.18t - 17.3°)$ A
       $i_2(t) = 0.134 e^{-t/2} \sin(2.18t + 8.55°)$ A
       (This is the same as Example 5.21 with circuit values scaled by a factor of 2.5.)

5.18 $v(t) = 1.5 + 1.5 e^{-20t}$ V
       (This is the same as Example 5.22 with circuit values scaled by a factor of 4.)

5.19 $i(t) = e^{-t} - e^{-1.5t}$

5.21 $i(t) = \dfrac{2}{\sqrt{3}} e^{-t/2} \sin \dfrac{\sqrt{3}}{2} t$

5.22 $R = 2\ \Omega$

5.23 $\xi = 0.5$

$\omega_n = 1$ rad/sec

$\omega = \sqrt{\omega_n(1 - \xi^2)} = \sqrt{\dfrac{3}{4}}$ rad/sec

5.24 (a) $\dfrac{1}{s} = 0.5sV(s) + 4V(s) + \dfrac{1}{s}V(s)$

(b) See Figure 5.90.

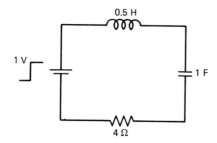

FIGURE 5.90

(c) $v(t) = \dfrac{1}{\sqrt{14}}\left[e^{-(4 - \sqrt{14})t} - e^{-(4 + \sqrt{14})t}\right]$

(d) Overdamped

CHAPTER

6

CONTROL SYSTEMS THEORY

6.1 OPEN AND CLOSED LOOP CONTROL SYSTEMS

A control system is a combination of devices connected together in such a way that the magnitude of some physical quantity (such as a voltage, displacement, angle, or velocity) is controlled by the magnitude of some other physical quantity. The latter is called input to the system, while the controlled quantity is called the output of the system. The input variable may or may not have the same units as the output variable. For example, when we turn the knob on an oven through a certain angular rotation, we are attempting to control the temperature of the oven; that is, maintain the oven temperature at a certain number of degrees proportional to the angular rotation of the knob. In this case, the output variable has units of temperature, while the input variable has units of angle. The transfer function of this system would then have units of, say, degrees F/rad.

An important concept in control-systems theory is the notion of *feedback*. Feedback takes place when the output variable being controlled is somehow measured, and this measured value is delivered too the input to the system, so that the input knows what the output is doing. A control system that incorporates feedback in some way to control the output variable is called a *closed-loop system*, while a system without feedback is said to be *open loop*.

An example of a closed-loop control system is the heating system in most modern homes. The input consists of an angular rotation of a thermostat dial. The temperature of the room is detected, or measured, by the thermostat and when the measured temperature reaches the preset value, the heater is shut off. This is an example of a bang-bang (on-off) closed-loop control system.

Another example of a closed-loop control system is a system used to position an antenna, that is, to rotate it through a prescribed angle. The input may again be an angular rotation of a potentiometer, whose output is amplified and used to drive a motor. The motor shaft rotates the antenna so long as the motor continues to turn. Another potentiometer could be used to feed back a voltage proportional to the angular rotation of the antenna. This second potentiometer might, for example, have its shaft geared to the motor shaft. At the input, some provision would be made to compare the voltage fed back to the controlling (input) voltage from the first potentiometer. When the two became equal, no further voltage would be applied to the motor, and the antenna would cease rotating. This system is illustrated in Figure 6.1.

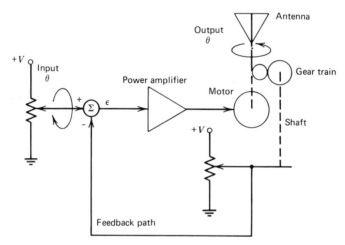

FIGURE 6.1 A position control system for an antenna.

In Figure 6.1, the symbol

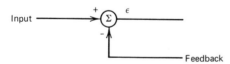

represents a device whose output (ε) is the *difference* between input and feedback: ε = (input) − (feedback). An example of such a device is a *differential amplifier*. Thus, when the voltage fed back is *equal* to the input voltage, signifying that the antenna has rotated through an angle equal to the angle specified by the input, we will have $\varepsilon = 0$, and there will be no further input to the power amplifier. That is, the motor will cease to turn, since the antenna is then in the desired position. This is an example of a *proportional* control system as opposed to a bang-bang control system.

An example of an open-loop control system is an unregulated dc power supply. A power supply consisting only of a transformer, a full or half-wave rectifier, and a filter capacitor is open-loop, since no provision is made to adjust the circuit to compensate for changes that may occur in the output dc voltage due, for example, to changes in the input ac voltage. A regulated power supply, on the other hand, continually monitors the output dc voltage and adjusts some circuit parameter to compensate for any tendencies of the output to change. The advantage of a closed-loop system over an open-loop system is that the closed-loop system will maintain the output at a desired level not only when external changes occur (such as a change in the input ac voltage) but also when certain internal changes take place. Suppose, for example, that the filter capacitor in the power-supply example begins to deteriorate. This (internal) change might cause the output voltage to drop off. In a closed-loop system, the decreased output would be sensed, and some other circuit parameter (such as voltage gain) would be adjusted automatically in order to raise the output back to the desired level. In an open-loop system (the unregulated power supply), the dc voltage would simply drop off.

The system used to position an antenna, previously described, is an example of a *servomechanism*. Although some authors adopt a very broad definition of the term servomechanism so that it might include any control system that incorporates feedback, we will be more restrictive. For our purposes, a servomechanism is a system of electrical, mechanical, and electromechanical devices that incorporate feedback to position a rotational load at some angular displacement prescribed by the input to the system. Position-control servomechanisms are designed and assembled using such commercially available components as servomotors, amplifiers, gear trains, and electronic-signal conditioning equipment. Completely assembled systems can also be purchased for specific applications. The Electro-craft Model E-586-BPC bidirectional position control system, whose specifications are shown in Appendix B, is an example.

6.2 BLOCK DIAGRAMS

Block diagrams are used to simplify the representation and analysis of control systems. Associated with each block in the diagram is the transfer function of a component of the system. The transfer function may be written as $G(j\omega)$ or $G(s)$, though the latter notation is preferred and will be used hereafter. When analyzing the system, it is generally the object to combine blocks in a way that will simplify the diagram. For example, two blocks in cascade (series) with transfer functions $G_1(s)$ and $G_2(s)$, respectively, may be combined in a single equivalent block with transfer function $G_1(s)G_2(s)$. We assume, of course, that each transfer function is *independent* of the other in the sense that neither transfer function changes when the components are actually connected in cascade, as discussed in Section 3.4. [Alternatively, we may assume that $G_1(s)$ and $G_2(s)$ are the actual transfer functions that result when the components *are* connected in cascade.]

EXAMPLE 6.1

An integrator with frequency independent gain 100 is connected in series with another integrator whose frequency independent gain is 0.5. Assume the integrators have infinite input impedance and zero output impedance. Find an equivalent transfer function for the series combination.

SOLUTION

Since integration is represented by 1/s in the frequency domain, we have

$$G_1(s) = \frac{100}{s} \quad \text{and} \quad G_2(s) = \frac{0.5}{s}$$

Since each integrator has infinite input impedance and zero output impedance, we know that connecting the two in series will not alter either transfer function (see Figure 6.2).

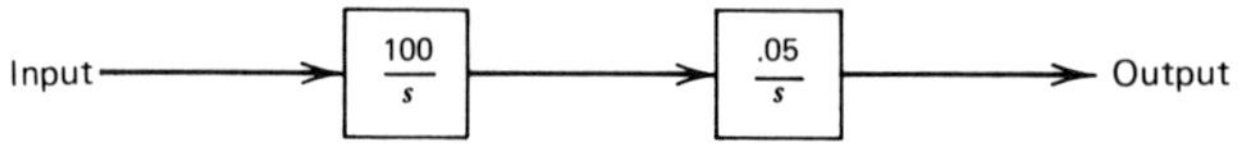

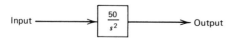

FIGURE 6.2 Block diagram of two integrators in series (Example 6.1).

Thus, the overall transfer function in Figure 6.2 may be found from

$$G(s) = G_1(s)G_2(s) = \frac{100}{s}\frac{0.5}{s} = \frac{50}{s^2}$$

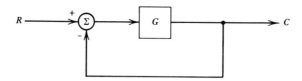

FIGURE 6.3 Block diagram equivalent to Figure 6.2 (Example 6.1).

It is conventional to use the letter C to represent the output of a control system, while the letter R is used to represent the input. The overall transfer function of the system would then be C/R. The simplest possible closed-loop control system may be shown in block diagram form as in Figure 6.4.

FIGURE 6.4 Block diagram of a fully fed-back control system.

Note that we have 100-percent feedback. That is, all of the output is fed back to the input, and the feedback loop contains no amplifiers, attenuators, or other devices that would alter the signal fed back. This system is often referred to as a fully fed-back or unity feedback system. The transfer function G is called the open-loop transfer function, since C/R would be exactly equal to G if the feedback loop were removed. Although we have written G instead of G(s), it will be understood hereafter that all transfer functions may have frequency dependence.

6.2 BLOCK DIAGRAMS

It is now our objective to find the closed-loop transfer function C/R for the control system. Note that if we can find the closed loop C/R, then we can replace the entire diagram by a single block with transfer function C/R and will, therefore, have achieved considerable simplification. Recall that ε = input − feedback. Therefore, $\varepsilon = R - C$. But ε is the input to the block with transfer function G. Thus, the output of that block is $G\varepsilon = G(R - C)$. The output of that block is also the output of the system C. Therefore, $C = G(R - C)$. (See Figure 6.5.)

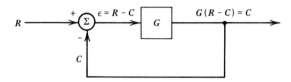

FIGURE 6.5 Input and output relations in the fully fed-back control system.

We now solve the equation $G(R - C) = C$ for C/R

$$GR - GC = C$$
$$-C - GC = -GR$$
$$C + GC = GR$$
$$C(1 + G) = GR$$
$$\frac{C}{R} = \frac{G}{1 + G} \tag{1}$$

Equation 1 is exceptionally important in control-systems theory; we will study the implications of this equation in considerable detail later. Let us first repeat that the entire fully fed-back control system can now be shown as a single block (see Figure 6.6).

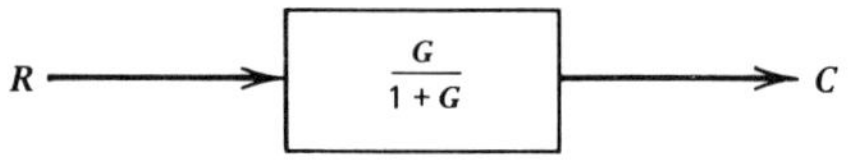

$$R \longrightarrow \boxed{\frac{G}{1+G}} \longrightarrow C$$

FIGURE 6.6 Simplified block diagram equivalent to Figure 6.4.

We will occasionally write G' to designate a closed-loop transfer function. Thus, for the fully fed-back system,

$$G' = \frac{C}{R} = \frac{G}{1 + G}$$

EXAMPLE 6.2

An integrator has a frequency independent gain of 50 and is enclosed in a fully fed-back loop. Find the closed-loop transfer function.

SOLUTION

$$G = \frac{50}{s}$$

Thus,

$$G' = \frac{G}{1 + G} = \frac{50/s}{1 + 50/s} = \frac{50}{s + 50}$$

(See Figure 6.7.)

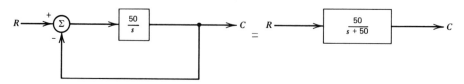

FIGURE 6.7 Block diagrams of an integrator in a fully fed-back system (Example 6.2).

There are two additional techniques for simplifying block diagrams. Suppose that the combination of blocks in Figure 6.8 appears somewhere in a block diagram.

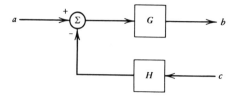

FIGURE 6.8 A typical combination of blocks.

Then Figure 6.8 is equivalent to Figure 6.9.

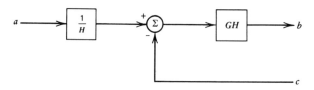

FIGURE 6.9 Block diagram equivalent to Figure 6.8.

Although this equivalent form does not necessarily represent a simplification, it is frequently useful when converting the diagram to some other form that can be simplified. To prove that the two forms are equivalent, we will show that the output b is the same in both cases. In the original configuration we have

$$\varepsilon = a - cH$$

and so
$$b = G(a - cH) = Ga - GcH$$
In the equivalent form, we have
$$\varepsilon = a\frac{1}{H} - c$$
$$b = GH\left(\frac{a}{H} - c\right) = Ga - GHc$$
We see that the two forms are equivalent.

EXAMPLE 6.3

Find C/R for the system in Figure 6.10.

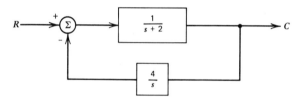

FIGURE 6.10 Find C/R (Example 6.3).

SOLUTION

The block diagram in Figure 6.10 is equivalent to the one shown in Figure 6.11.

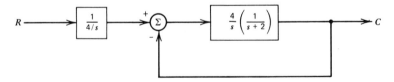

FIGURE 6.11 First simplification of Figure 6.10 (Example 6.3).

This can be simplified as shown in Figure 6.12.

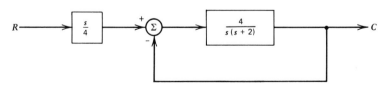

FIGURE 6.12 Second simplification of Figure 6.10 (Example 6.3).

Using Equation 1, the fully fed-back loop may be replaced by one equivalent block as in Figure 6.13.

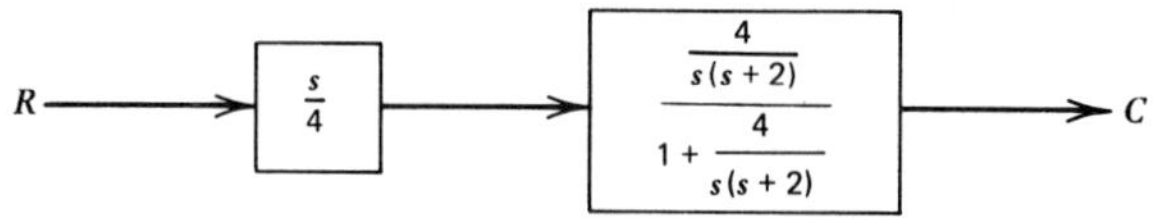

170 CONTROL SYSTEMS THEORY

R ─→ [$\frac{s}{4}$] ─→ [$\frac{\frac{4}{s(s+2)}}{1 + \frac{4}{s(s+2)}}$] ─→ C

FIGURE 6.13 Third simplification of Figure 6.10 (Example 6.3).

Now

$$\frac{4/(s+2)s}{1 + [4/(s+2)s]} = \frac{4/(s+2)(s)}{[(s+2)(s) + 4/(s+2)(s)]} = \frac{4}{(s+2)(s) + 4} = \frac{4}{s^2 + 2s + 4}$$

So we have the block diagram in Figure 6.14

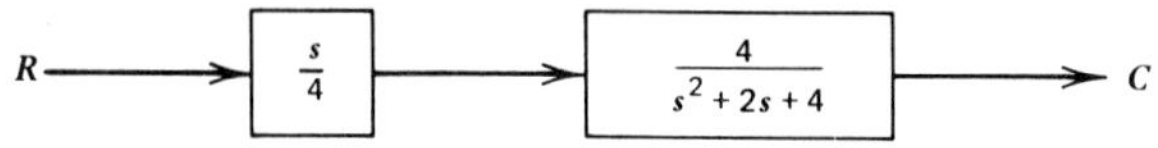

R ─→ [$\frac{s}{4}$] ─→ [$\frac{4}{s^2 + 2s + 4}$] ─→ C

FIGURE 6.14 Fourth simplification of Figure 6.10 (Example 6.3).

and, finally, the block diagram in Figure 6.15

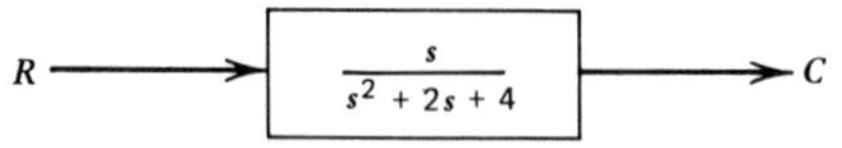

R ─→ [$\frac{s}{s^2 + 2s + 4}$] ─→ C

FIGURE 6.15 Final simplification of Figure 6.10 (Example 6.3).

so

$$\frac{C}{R} = \frac{s}{s^2 + 2s + 4}$$

We now describe another similar technique that may be used to reduce block diagrams. Suppose the combination of blocks shown in Figure 6.16 appears in a block diagram.

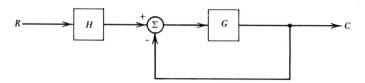

FIGURE 6.16 A typical combination of blocks.

This is equivalent to Figure 6.17. Proof that these are equivalent is left as an exercise (Exercise 6.6). The two methods of redrawing the block diagram as equivalent diagrams is summarized as follows: A block in the feedback loop may be placed in the forward loop provided it is divided into the input, and a block in the input may be placed in the forward loop provided it is divided into the feedback.

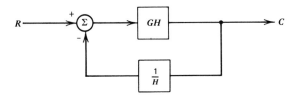

FIGURE 6.17   Block diagram equivalent to Figure 6.16.

EXAMPLE 6.4

Find C/R for the system shown in Figure 6.18.

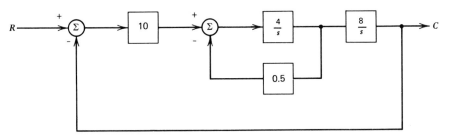

FIGURE 6.18   Find C/R (Example 6.4).

SOLUTION

The inner feedback loop may be transformed into a fully fed-back loop by dividing 0.5 into the input of that loop (see Figure 6.19).

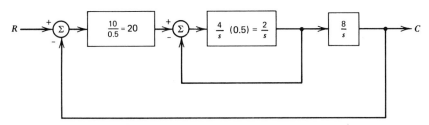

FIGURE 6.19   First simplification of Figure 6.18 (Example 6.4).

The fully fed-back loop may then be replaced by its equivalent

$$G' = \frac{G}{1 + G} = \frac{2/s}{1 + 2/s} = \frac{2}{s + 2}$$

(See Figure 6.20.)

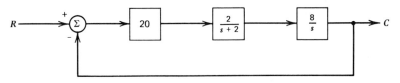

FIGURE 6.20   Second simplification of Figure 6.18 (Example 6.4).

172 CONTROL SYSTEMS THEORY

The transfer function of the forward loop may now be written as the product of the transfer functions in that loop

$$(20)\left(\frac{2}{s+2}\right)\left(\frac{8}{s}\right) = \frac{320}{s(s+2)}$$

(See Figure 6.21.)

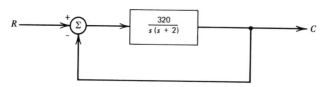

FIGURE 6.21 Third simplification of Figure 6.18 (Example 6.4).

Again, applying $G' = G/(1 + G)$, we have

$$G' = \frac{320/s(s+2)}{1 + [320/s(s+2)]} = \frac{320}{s(s+2) + 320} = \frac{320}{s^2 + 2s + 320}$$

The entire system is thus equivalent to the single block shown in Figure 6.22, and

$$\frac{C}{R} = \frac{320}{s^2 + 2s + 320}$$

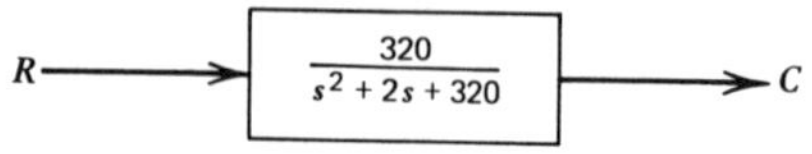

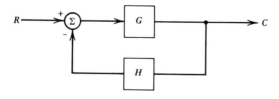

FIGURE 6.22 Final simplification of Figure 6.18 (Example 6.4).

A frequently encountered block diagram is the one shown in Figure 6.23. In contrast to the fully fed-back system, the feedback signal in this system is modified by the transfer function H before it is subtracted from the input R. H might, for example, be the transfer function of the potentiometer in the feedback of the antenna-positioning system described in Section 6.1.

FIGURE 6.23 Block diagram of a system with generalized-feedback transfer function H.

Applying one of the methods just described, we may redraw this diagram as shown in Figure 6.24. This is equivalent to Figure 6.25. Therefore,

$$\frac{C}{R} = \left(\frac{1}{H}\right)\left(\frac{GH}{1+GH}\right) = \frac{G}{1+GH} \tag{2}$$

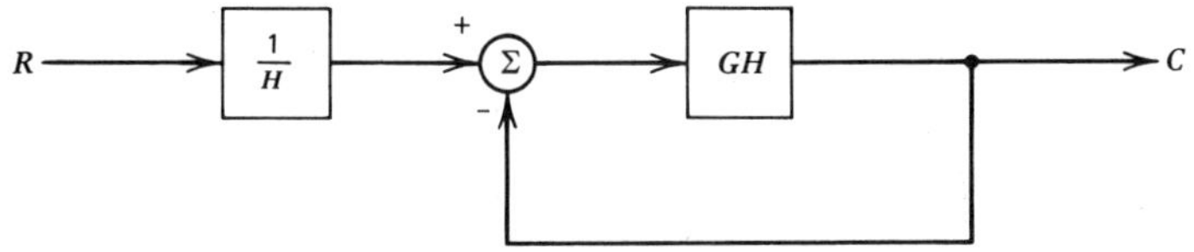

FIGURE 6.24 Block-diagram simplification of Figure 6.23.

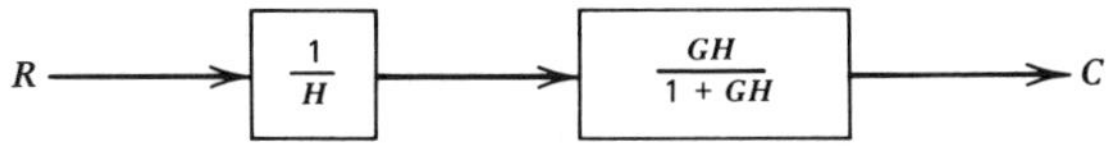

FIGURE 6.25 Further simplification of Figure 6.23.

Equation 2 is another very important relation in control-systems theory. Note that $H = 1$ corresponds to the fully fed-back case, and in this situation, Equation 2 reduces to Equation 1.

6.3 INTEGRATORS

The integrator is a particularly important component of a typical control system. For one thing, a position control system, or servomechanism, such as described in Section 6.1, incorporates a servomotor, which as a first approximation is an integrator. The ideal servomotor rotates at an angular velocity that is directly proportional to the voltage applied to it

$$\omega = KV \tag{3}$$

where

ω = angular velocity in rad/sec
V = applied voltage
K = constant of proportionality in rad/sec/V

(Note that K can be considered the "gain" of the servomotor.) In a position control system, the output may be the angular rotation θ of the shaft of a servomotor, while the input is a voltage V. Since $\omega = d\theta/dt$, we have from Equation 3:

$$\frac{d\theta}{dt} = KV$$

$$d\theta = KV\, dt$$

$$\int d\theta = \int KV\, dt$$

$$\theta = K\int V\, dt \tag{4}$$

Equation 3 shows that the output of a servomotor (angular position θ) is proportional to the integral of the input V. So the ideal servomotor is, indeed, an integrator in a position control system.

Integrators are of further importance in control-systems theory because electronic integration is frequently used to modify the amplitude and phase angle of electrical signals that are present in the system. As already discussed, integration of an electrical signal reduces the amplitude of that signal as frequency increases and causes a $-90°$ phase shift in the signal. Recall that the integrator has transfer function $G = 1/s$.

$$G = \frac{1}{s} = \frac{1}{j\omega}$$
$$|G| = \frac{1}{\omega}$$
$$\angle G = -90°$$

The low-pass filter consisting of a series resistor and shunt capacitor (see Figure 6.26) can be used to approximate the behavior of an integrator over a certain frequency range. As we know, the amplitude of the output from this filter decreases as frequency increases, just as it does for an integrator.

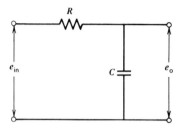

FIGURE 6.26 Low-pass filter.

6.4 THE SERVOMOTOR IN A CLOSED LOOP

Let us now consider a servomotor in a simple closed-loop control system. This system (see Figure 6.27) consists of a differential amplifier with gain K_1 whose output drives the servomotor. The servomotor is assumed to have frequency-independent gain K_2 rad/sec/V, and the shaft of the motor drives the shaft of a single-turn potentiometer. The terminals of the potentiometer are connected to $+V$ V and ground, while the output of the wiper arm is fed back to the inverting input of the amplifier. The input to the system is via another identical potentiometer.

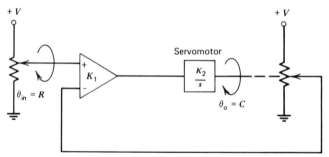

FIGURE 6.27 The servomotor in a closed-loop position control system.

6.4 THE SERVOMOTOR IN A CLOSED LOOP

We will determine the overall transfer function of this system by applying block-diagram algebra. Note that we are using the transfer function of an integrator K_2/s as a first approximation for the transfer function of the servomotor. Thus, the block diagram for this system may be drawn as in Figure 6.28.

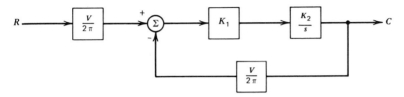

FIGURE 6.28 Block-diagram equivalent of Figure 6.27.

As discussed in Section 3.1, the transfer function of each potentiometer is $V/2\pi$, since a rotation of θ radians produces $\theta/2\pi \; V$ V at the wiper arm. By using Equation 2 with

$$G = \frac{K_1 K_2}{s} \qquad H = \frac{V}{2\pi} \qquad \text{and} \qquad \frac{G}{1 + GH} = \frac{K_1 K_2}{s + (K_1 K_2 V/2\pi)}$$

we may reduce the diagram to the equivalent one shown in Figure 6.29. Then

$$\frac{C}{R} = \frac{(V/2\pi)K_1 K_2}{s + (K_1 K_2 V/2\pi)}$$

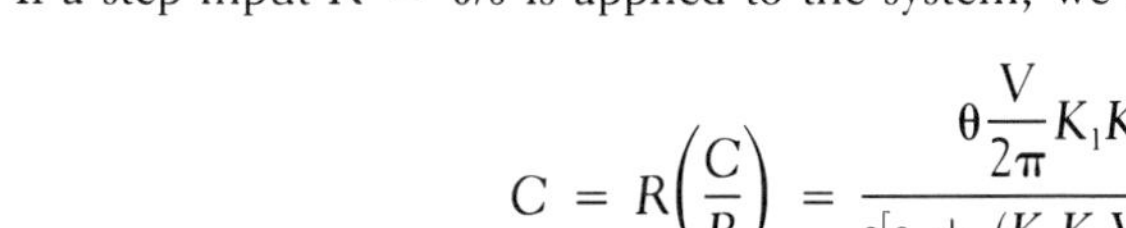

FIGURE 6.29 Block-diagram simplification of Figure 6.28.

If a step input $R = \theta/s$ is applied to the system, we see that

$$C = R\left(\frac{C}{R}\right) = \frac{\theta \dfrac{V}{2\pi} K_1 K_2}{s[s + (K_1 K_2 V/2\pi)]}$$

Then from pair 7 in Table 4-1, we find

$$C(t) = \theta\left(\frac{V}{2\pi}\right)K_1 K_2 \left\{ \frac{1}{K_1 K_2 V/2\pi}[1 - e^{-(K_1 K_2 V/2\pi)t}] \right\}$$

$$= \theta[1 - e^{-(K_1 K_2 V/2\pi)t}]$$

The input and response of the system are sketched in Figure 6.30. Note that the time constant of the exponential term is $\tau = 2\pi/K_1 K_2 V$.

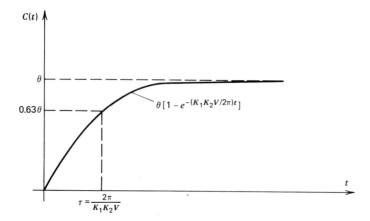

FIGURE 6.30 Response of the system in Figure 6.27 to a step input of θ rad.

In physical terms, this result tells us that if the system is suddenly told to rotate through a certain angle θ, then it will gradually begin to rotate, reaching 63% of the angle in $2\pi/K_1K_2V$ sec, eventually reach the correct angle, and then cease rotation. The steady-state response is this ultimate angular position θ, but we are more interested in the *transient* response, in particular the time constant τ, since this gives us an indication of how rapidly the system can respond to a command.

EXAMPLE 6.5

In the control system described in section 6.4, suppose that $K_2 = 4$, and $V = +10$ V. If the output must reach 63% of its final position in 0.1 sec in response to any step input, what must be the gain of the amplifier?

SOLUTION

We require that

$$0.1 = \frac{2\pi}{K_1K_2V} = \frac{2\pi}{K_1(40)}$$

Thus

$$K_1 = \frac{2\pi}{4} = 1.57$$

EXAMPLE 6.6

Suppose the control system in Example 6.5 is driven by a sinusoidal function $R = \theta \sin \omega t$ instead of a step. Using the same values for K_1 and K_2 as in Example 6.5, find the frequency at which $C = \theta/\sqrt{2} \angle -45°$ after a long period of time.

SOLUTION

$$\frac{C}{R} = \frac{(V/2\pi)K_1K_2}{[s + (K_1K_2V/2\pi)]} = \frac{10}{s + 10}$$

6.4 THE SERVOMOTOR IN A CLOSED LOOP

For an input $R = \theta \sin \omega t$, we have

$$R(s) = \frac{\theta \omega}{s^2 + \omega^2}$$

and therefore

$$C = \frac{10\theta\omega}{(s^2 + \omega^2)(s + 10)} = \frac{A}{s + 10} + \frac{B}{s + j\omega} + \frac{C}{s - j\omega}$$

Examining the partial fraction expansion for C reveals that $C(t)$ will consist of an exponential term (Ae^{-10t}) and a sinusoidal term due to the poles at $\pm j\omega$. Since we are only interested in $C(t)$ after a long period of time when the exponential can be assumed to be zero, we need only find the sinusoidal portion of the response. We will use Equation 23 from Chapter 4, with $-\alpha + j\omega = 0 + j\omega$.

$$K(-\alpha + j\omega) = \frac{10\theta\omega}{s + 10}\bigg]_{s = j\omega} = \frac{10\theta\omega}{10 + j\omega} = \frac{10\theta\omega}{\sqrt{100 + \omega^2}} \angle - \arctan\frac{\omega}{10} = M\angle\theta$$

Then

$$C(t) = \frac{M}{\omega} e^{-\alpha t} \sin(\omega t + \theta)$$

$$= \frac{10\theta\omega/\sqrt{100 + \omega^2}}{\omega} \sin(\omega t + \theta) = \frac{10\theta}{\sqrt{100 + \omega^2}} \sin(\omega t + \theta)$$

In order for $\theta = -45°$, we must have $-\arctan \omega/10 = -45°$ or $\omega/10 = 1$; that is, $\omega = 10$ rad/sec.

With $\omega = 10$ rad/sec, we see that

$$\frac{10\theta}{\sqrt{100 + \omega^2}} = \frac{10\theta}{\sqrt{200}} = \frac{\theta}{\sqrt{2}}$$

Thus, when $\omega = 10$, we have

$$C(t) = \frac{\theta}{\sqrt{2}} \angle -45° \quad \text{(as required)}$$

Example 6.6 illustrates an exceptionally important fact that we can now state as a generalization. First, we note that the frequency ($\omega = 10$ rad/sec) that causes the output to be $1/\sqrt{2}$ times the input magnitude and the output to lag the input by 45°, is, by definition, the cutoff frequency of the system. If we think of this position control apparatus as being driven by a sinusoidal input (unlikely in a practical control system, for we are telling the output to continually reverse itself), then when we reach a frequency of 10 rad/sec, the servomotor shaft will only be able to rotate back and forth with amplitude $1/\sqrt{2}$ times that of the input amplitude, and, furthermore, these oscillations will lag the input oscillations by 45°. In short, the system is not fast enough (responsive enough) to keep up with the input any better than this. We see that the system acts like a low-pass filter with cutoff at $\omega = 10$ rad/sec.

178 CONTROL SYSTEMS THEORY

We also note that the time constant of the system's response to a step input was $\tau = 0.1$ sec. That is, the output reached 63% of its final value in 0.1 sec. Not by coincidence is the cutoff frequency ($\omega = 10$) equal to the reciprocal of the time constant ($\tau = 0.1$)! Indeed, for any system with transfer function of the form

$$\frac{C}{R} = \frac{a}{(s + a)}$$

it will always be true that

$$a = \omega_{co} = \frac{1}{\tau} \tag{5}$$

where ω_{co} = the cutoff frequency. In other words, the greater the bandwidth of the system, the faster it responds to a step input. This is an important example of how information about a system's characteristics in the frequency domain allow us to predict its behavior in the time domain.

EXAMPLE 6.7

How much frequency-independent gain should be put in series with an integrator whose frequency-independent gain is 25 in order for it to have a bandwidth of 200 Hz when enclosed by a fully fed-back loop?

SOLUTION

Let K_1 = the required additional gain. (See Figure 6.31.)

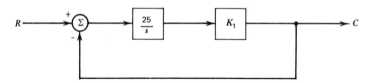

FIGURE 6.31 Integrator with cascade gain (Example 6.7).

$$\frac{C}{R} = \frac{(25K_1/s)}{1 + (25K_1/s)} = \frac{25K_1}{s + 25K_1}$$

From the form of this transfer function, we know that $\omega_{co} = 25K_1$. Since we want ω_{co} to be $2\pi(200 \text{ Hz}) = 400\pi$ rad/sec, we must have

$$25K_1 = 400\pi \quad \text{or} \quad K_1 = 16\pi$$

6.5 BODE PLOT FOR THE SIMPLE POSITION CONTROL SYSTEM

Bode plots were described in Chapter 3 as an effective way of depicting frequency-response information of a system or network. Recall that the log of gain magnitude

6.5 BODE PLOT FOR THE SIMPLE POSITION CONTROL SYSTEM

is plotted versus log frequency. We wish now to investigate the Bode plot for the closed-loop position control system described in Section 6.4.

We have already shown that this simple-position control system has a closed-loop transfer function in the form

$$\frac{C}{R} = \frac{a}{s + a}$$

Now

$$\left|\frac{C}{R}\right| = \left|\frac{a}{j\omega + a}\right| = \frac{a}{\sqrt{a^2 + \omega^2}}$$

At $\omega = 0$, $\left|\frac{C}{R}\right| = 1$

At $\omega = a$, $\left|\frac{C}{R}\right| = \frac{1}{\sqrt{2}} \simeq 0.707$

And,

$$\lim_{\omega \to \infty} \left|\frac{C}{R}\right| = 0$$

Figure 6.32 shows the Bode plot for the closed-loop system along with its asymptotic approximation. As we might expect from our previous discussion, the plot

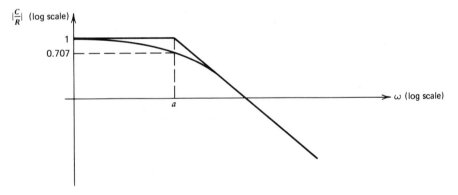

FIGURE 6.32 Bode plot for the closed-loop transfer function $C/R = s/(s + a)$.

resembles that of a low-pass filter with cutoff frequency at $\omega = a$. We note that $\log 1 = 0$, so if Figure 6.32 were truly plotted on log-log paper, the asymptotic approximation for $\log |C/R|$ would be shown negative for all frequencies above $\omega = a$. If, however, we add frequency-independent gain $K > 1$ outside the closed loop, so that the transfer function is then of the form

$$\frac{C}{R} = \frac{Ka}{s + a}$$

the effect is to shift the entire Bode plot upwards. The unity-gain frequency ω_1 would then be found as follows:

$$\left|\frac{C}{R}\right| = \frac{Ka}{\sqrt{a^2 + \omega_1^2}} = 1$$

$$\omega_1 = a\sqrt{K^2 - 1} \tag{6}$$

It is important to realize that adding gain outside the closed loop does not affect the phase-angle response. The phase angle of C/R is the same in both cases:

$$\angle \frac{a}{s + a} = -\arctan\frac{\omega}{a} = \angle \frac{Ka}{s + a}$$

If, on the other hand, frequency-independent gain K is added *inside* the closed loop (as, for example, by increasing the gain of the differential amplifier), then we obtain

$$\frac{C}{R} = \frac{Ka}{s + Ka}$$

In this case $|C/R|$ is again unity at $\omega = 0$, and we have increased the cutoff frequency to $\omega_{co} = Ka$. In this case, the frequency at which the output phase angle is $-45°$ is also $\omega_{co} = Ka$. When we consider system *stability* in Chapter 7, we will see that knowledge of the unity-gain frequency and the phase angle at that frequency is of critical importance.

EXAMPLE 6.8

How much gain should be placed outside the closed loop of the position control system in Example 6.5 in order for the unity-gain frequency to be 20 Hz? What is the phase angle of C/R at that frequency?

SOLUTION

From Equation 6, $\omega_1 = 2\pi \times 20 = 10\sqrt{K^2 - 1}$

Then

$$K^2 = \frac{(2\pi \times 20)^2}{10^2} + 1$$

$$K = 12.61$$

$$\theta = -\arctan\left[(2\pi \times 20)/10\right] = -85.45°$$

EXAMPLE 6.9

Suppose the gain K calculated in Example 6.8 were placed *inside* the closed loop instead of outside. What then would be:

(a) The unity-gain frequency?
(b) The cutoff frequency?

(c) The phase angle at cutoff?
(d) The time required for the system to reach 63% of its final value in response to a step input?

SOLUTION

(a) The transfer function becomes
$$\frac{C}{R} = \frac{126.1}{s + 126.1}$$
The unity-gain frequency is thus $\omega = 0$.

(b) The cutoff frequency now equals 126.1 rad/sec = 20.07 Hz.

(c) $\theta = -45°$

(d) $\tau = \dfrac{1}{\omega_{co}} = 49.8$ msec

6.6 THE DYNAMICS OF ROTATING SYSTEMS

As we have seen, a position control system typically involves the use of rotating components: motors, gears, shafts, and so on. We must, therefore, investigate the dynamics of rotating components in order to be able to predict and analyze their effects on the performance of a control system. Fortunately, the equations that describe the behavior of rotating systems are completely analogous to those describing electrical networks and translational mechanical systems, which we have already covered in reasonable depth.

The analog of force in a translational mechanical system or of voltage in an electrical circuit is *torque* in the rotational system. As we know, torque is the product of a force and the perpendicular distance from the line of action of that force to an axis of rotation (see Figure 6.33). The units of torque are force × distance or newton-meters (Nm).

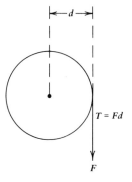

FIGURE 6.33 Torque = force × distance.

Displacement in a rotational system is *angular* displacement Θ in radians, and it is analogous to linear displacement x in a translational system. Angular velocity is $\omega = d\Theta/dt = \dot{\Theta}$ rad/sec and is analogous to $v = dx/dt$ or to current in an electrical ciruit. Angular acceleration is, of course, equal to

$$\dot{\omega} = \frac{d^2\Theta}{dt^2} = \ddot{\Theta} \text{ rad/sec}^2$$

Inertia J is the analog of mass. Just as $F = Ma = M\ddot{x}$, we have

$$T = J\ddot{\Theta} \tag{7}$$

Thus, the units of J are those of torque divided by angular acceleration, namely, N-m per rad/sec^2, or (N-m-sec^2)/rad.

Rotational damping B produces a torque that is proportional to angular velocity, just as translational damping produces a force proportional to linear velocity.

$$T = B\dot{\Theta} \tag{8}$$

The units of rotational damping are therefore (N-m-sec)/rad. An example of rotational damping exists in an automobile transmission in which gears and shafts must rotate in a body of oil.

Finally, rotational spring constant K is a measure of the torque produced by a body in response to an angular displacement. This can be visualized as the tendency of a fixed shaft to restore itself when it is twisted. The shaft exerts a torque that opposes the torque twisting it through an angle Θ, and the magnitude of this torque is directly proportional to the total twist Θ. In equation form,

$$T = K\Theta \tag{9}$$

Clearly, the units of K are N-m/rad.

Power is the rate of doing work and may be calculated in a rotational system from the relation

$$P = T\dot{\Theta} \tag{10}$$

If T is in N-m and Θ in rad/sec, then the units of P are watts (W).

EXAMPLE 6.10

A servomotor with armature inertia 0.04 (N-m-sec^2)/rad has a load inertia of 0.26 (N-m-sec^2)/rad.

(a) How much torque must it develop to accelerate at 50 rad/sec^2?
(b) What power must it develop to reach an angular velocity of 50 rad/sec?

SOLUTION

(a) $T = J\ddot{\Theta} = (0.04 + 0.26)50 = 15$ N-m
(b) $P = T\dot{\Theta} = (15)(50) = 750$ W

6.6 THE DYNAMICS OF ROTATING SYSTEMS

Equations 7, 8, and 9 may be transformed to the frequency domain in the usual way:

$$T(s) = Js^2\Theta(s) \tag{11}$$
$$T(s) = Bs\Theta(s) \tag{12}$$
$$T(s) = K\Theta(s) \tag{13}$$

The dynamic equations of a rotational system can now be solved using LaPlace transforms in the same way we have solved circuit equations.

EXAMPLE 6.11

Find $\Theta(t)$ and $\omega(t) = \dot{\Theta}(t)$ in the rotational system in Figure 6.34 using LaPlace transforms. A step torque of 8 N-m is applied to the system at $t = 0$.

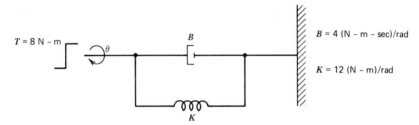

FIGURE 6.34 Find $\omega(t)$ (Example 6.11).

SOLUTION

$$T(s) = Bs\theta(s) + K\theta(s)$$
$$\frac{8}{s} = 4s\theta(s) + 12\theta(s)$$
$$\theta(s) = \frac{8/s}{4s + 12} = \frac{2}{s(s + 3)}$$

Then

$$\theta(t) = \frac{2}{3}(1 - e^{-3t}) \text{ rad}$$

and

$$\omega(t) = 2e^{-3t} \text{ rad/sec}$$

These results tell us that the shaft will rotate until it reaches a steady-state angle of 2/3 rad. The angular velocity will be initially 2 rad/sec and then gradually decay to zero. Note that when the rest position is reached, the torque developed by the rotational spring

$$\left(\frac{2}{3} \text{ rad}\right)[12(\text{N-m})/\text{rad}] = 8 \text{ N-m}$$

is exactly equal to the applied torque. The angle of rotation θ will reach 63% of its final value in $\tau = 1/3$ sec.

6.7 GEAR TRAINS

When a gear with N_1 teeth rotating at angular velocity ω_1 drives a second gear with N_2 teeth, the angular velocity of the second gear is

$$\omega_2 = \frac{N_1}{N_2} \omega_1 \tag{14}$$

Clearly, the driven gear may rotate faster or slower than the driving gear, depending upon the gear ratio N_1/N_2.

EXAMPLE 6.12

At what angular velocity is gear G_4 in Figure 6.35 rotating?

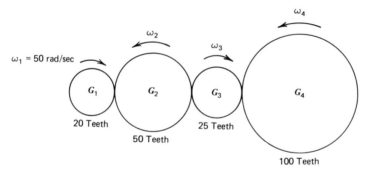

FIGURE 6.35 Find ω_4 (Example 6.12).

SOLUTION

$$\omega_2 = \frac{20}{50} \omega_1 = 20 \text{ rad/sec}$$

$$\omega_3 = \frac{50}{25} \omega_2 = 40 \text{ rad/sec}$$

$$\omega_4 = \frac{25}{100} \omega_3 = 10 \text{ rad/sec}$$

One effect of a set of gears is to transform the value of torque applied through it to a load. Suppose the torque applied to gear G_1 is T_1. Then, as illustrated in Figure 6.36, the force F on gear G_2 is $F = T_1/r_1$. Therefore, the torque on G_2 is

$T_2 = Fr_2 = T_1 r_2/r_1$. Assuming equally spaced teeth on each gear, the ratio of the radii is the same as the gear ratio. Hence,

$$T_2 = T_1 \frac{N_2}{N_1} \tag{15}$$

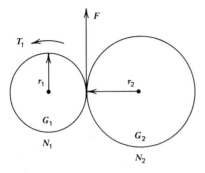

FIGURE 6.36 Force and torque relations in gears.

We see that when G_2 is the larger gear, its angular velocity is decreased by the factor N_1/N_2, but the torque it develops is increased by the factor N_2/N_1. A familiar example is shifting into low gear in an automobile.

Another effect that gears have on a rotational system is transforming the load inertia (connected to the driven gear) into a larger or smaller inertia at the driving gear. Consider the gears shown in Figure 6.37, where a load inertia J_2 is attached to the shaft on which G_2 is mounted.

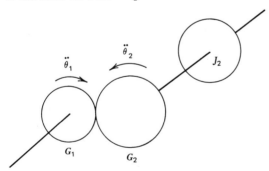

FIGURE 6.37 Gear system with an inertia load.

By Equation 15, the torque T_1 at G_1 is

$$T_1 = T_2 \left(\frac{N_1}{N_2}\right)$$

But

$$T_2 = J_2 \ddot{\theta}_2 = J_2 \left[\frac{d\dot{\theta}_2}{dt}\right] = J_2 \left[\frac{d(N_1/N_2)\dot{\theta}_1}{dt}\right] = J_2 \left(\frac{N_1}{N_2}\right) \ddot{\theta}_1$$

186 CONTROL SYSTEMS THEORY

Therefore,

$$T_1 = J_2\left(\frac{N_1}{N_2}\right)^2 \ddot{\theta}_1 \quad \text{or} \quad \frac{T_1}{\ddot{\theta}_1} = J_1 = J_2\left(\frac{N_1}{N_2}\right)^2 \tag{16}$$

We see that a load inertia on a driven gear is reflected to the driving gear by a factor equal to the *square* of the gear ratio. As an exercise, show that B and K are also reflected through a set of gears by the square of the gear ratio.

For most practical purposes, gears themselves can be considered to have negligible inertia. In subsequent examples, we will assume that all gears are inertialess.

EXAMPLE 6.13

Four Vernitech Model 15010 ten-turn potentiometers are geared through 1:1 gear ratios to the output shaft of a Vernitech Model 08s-016 speed reducer. Use the specification sheets provided in Appendix B to determine (a) the torque required at the input shaft of the reducer in order to accelerate the potentiometers at 50 rad/sec² and (b) the total inertia reflected to the input shaft of the speed reducer.

SOLUTION

(a) Examining the specification sheet for the Model 15010 potentiometer, we find that the inertia of each potentiometer is given as 0.6 gm-cm². Note that the units we have used heretofore for inertia J, namely, N-m-sec²/rad, can be expressed equivalently as kg-m²/rad, since 1 N is the same as 1 kg-m/sec². Since radians are considered dimensionless, they are often omitted from specifications. Thus, the specified potentiometer inertia of 0.6 gm-cm² is equivalent to

$$J = (0.6 \text{ g-cm}^2)\left(\frac{1 \text{ kg}}{10^3 \text{ g}}\right)\left(\frac{1 \text{ m}^2}{10^4 \text{ cm}^2}\right)$$
$$= 0.6 \times 10^{-7} \text{ kg-m}^2$$
$$= 0.6 \times 10^{-7} \text{ N-m-sec}^2/\text{rad}$$

The total inertia of the four potentiometers is therefore,

$$J_2 = 2.4 \times 10^{-7} \text{ (N-m-sec}^2)/\text{rad}$$

The 08S-016 speed reducer has a negligibly small inertia (0.0012 g/cm²) and a gear ratio of $N_1/N_2 = 50/142$, where N_1 refers to the input (high speed) side of the reducer. Therefore, the required torque T_1 is found as follows:

$$T_2 = J_2\ddot{\theta}_2 = (2.4 \times 10^{-7})(50) = 1.2 \times 10^{-6} \text{ N-m}$$

$$T_1 = \left(\frac{N_1}{N_2}\right)T_2 = \frac{50}{142}(1.2 \times 10^{-6}) = 0.423 \times 10^{-6} \text{ N-m}$$

(b) The inertia reflected to the input shaft is
$$J_1 = J_2(N_1/N_2)^2 = 2.4 \times 10^{-7} (50/142)^2$$
$$= 0.298 \times 10^{-7} \text{ (N-m-sec}^2\text{)/rad}$$

EXAMPLE 6.14

Use LaPlace transforms to find the angular rotation ω_1 of the shaft to which G_1 is connected when G_1 is subjected to a step input torque of T N-m (see Figure 6.38).

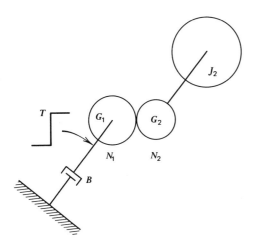

FIGURE 6.38 Find ω_1 (Example 6.14).

SOLUTION

$$T = B\left(\frac{d\theta_1}{dt}\right) + J_1\left(\frac{d^2\theta_1}{dt^2}\right)$$

$$\frac{T}{s} = B\omega_1(s) + \left(\frac{N_1}{N_2}\right)^2 J_2 s\, \omega_1(s)$$

$$\omega_1(s) = \frac{T/s}{(N_1/N_2)^2 J_2 s + B} = \frac{T/(N_1/N_2)^2 J_2}{s\{s + [B/(N_1/N_2)^2 J_2]\}}$$

Then
$$\omega_1(t) = \frac{T}{B}\left[1 - e^{-B/(N_1/N_2)^2 J_2\, t}\right]$$

We note that the angular velocity begins at zero and eventually reaches a constant value of T/B, a velocity at which the dashpot produces a constant torque T equal and opposite to the applied torque T.

188 CONTROL SYSTEMS THEORY

If a rotational dashpot or spring is connected to the driven side of a gear train, we can find angular displacement, velocity, or acceleration on either side by remembering and applying the fact that

$$\omega_2 = \frac{N_1}{N_2}\omega_1$$

implies

$$\theta_2 = \frac{N_1}{N_2}\theta_1 \quad \text{and} \quad \ddot{\theta}_2 = \frac{N_1}{N_2}\ddot{\theta}_1$$

EXAMPLE 6.15

Find $\theta_1(t)$, $\theta_2(t)$, $\omega_1(t)$, and $\omega_2(t)$ when the system shown in Figure 6.39 is subjected to a step torque of 10 N-m.

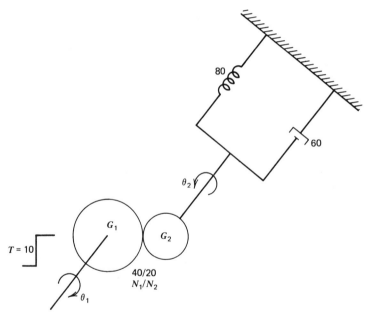

FIGURE 6.39 Find angular displacements and velocities (Example 6.15).

SOLUTION

From Equation 15,

$$T_2 = 10u(t)\left(\frac{20}{40}\right) = 5u(t)$$

Then

$$5u(t) = 80\,\theta_2(t) + 60\,\dot{\theta}_2(t)$$

Transforming,

$$\frac{5}{s} = 80\,\theta_2(s) + 60\,s\theta_2(s)$$

$$\theta_2(s) = \frac{5}{s(60s + 80)} = \frac{1/12}{s(s + 4/3)}$$

$$\theta_2(t) = \frac{1}{16}(1 - e^{-4t/3})\ \text{rad}$$

$$\omega_2(t) = \dot{\theta}_2(t) = \frac{1}{12}e^{-4t/3}\ \text{rad/sec}$$

$$\theta_1(t) = \frac{N_2}{N_1}\theta_2(t) = 0.5\,\theta_2(t) = \frac{1}{32}(1 - e^{-4t/3})$$

$$\omega_1(t) = \dot{\theta}_1(t) = \frac{N_2}{N_1}\dot{\theta}_2(t) = 0.5\,\dot{\theta}_2(t) = \frac{1}{24}e^{-4t/3}$$

6.8 THE SECOND APPROXIMATION OF A SERVOMOTOR

In the electrical-mechanical analog, inertia J is analogous to inductance (with angular velocity analogous to current). Just as inductance resists a change in current through it, an inertia resists any attempt to change its angular velocity. It is not possible to change instantaneously the current through an inductor nor the angular velocity of an inertia. In other words, a certain *time* (delay) must elapse before the angular velocity can undergo a change in response to a sudden stimulus. We indicated earlier that response time is inversely related to bandwidth. The inertia of the servomotor (and that of any load connected or reflected to it) is, therefore, responsible for a reduction in bandwidth. Equivalently, this inertia is responsible for introducing another pole in the servomotor's open-loop transfer function. Let us therefore write the transfer function as

$$G = \frac{\theta_o}{V_{in}} = \frac{K}{s(s + \omega_J)} \tag{17}$$

where ω_J is the break frequency caused by the presence of inertia. The asymptotic Bode plot for this transfer function is sketched in Figure 6.40.

190 CONTROL SYSTEMS THEORY

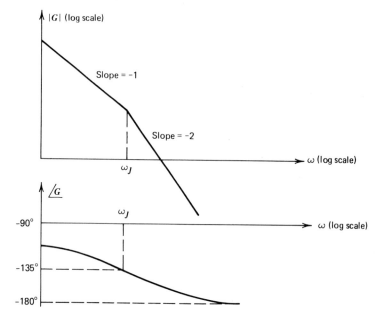

FIGURE 6.40 Asymptotic Bode plot for the second approximation of a servomotor.

Let us now consider the servomotor with transfer function given by Equation 17 in a closed loop (see Figure 6.41).

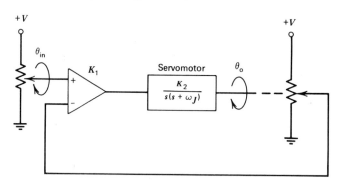

FIGURE 6.41 The servomotor in a closed loop.

$$\frac{C}{R} = \frac{V}{2\pi}\left\{\frac{K_1K_2/s(s+\omega_J)}{1 + [K_1K_2V/s(s+\omega_J)2\pi]}\right\}$$

$$= \left[\frac{VK_1K_2}{s(s+\omega_J)2\pi + K_1K_2V}\right]$$

$$= \left[\frac{VK_1K_2/2\pi}{s^2 + \omega_J s + (K_1K_2V/2\pi)}\right]$$

6.8 THE SECOND APPROXIMATION OF A SERVOMOTOR

Let $K = VK_1K_2/2\pi$ and suppose $R(S) = 1/s$. Then

$$C(s) = \frac{K}{s(s^2 + \omega_f s + K)}$$

The poles of $C(s)$ are

$$s = 0 \quad \text{and} \quad s = \frac{-\omega_f \pm \sqrt{\omega_f^2 - 4K}}{2}$$

EXAMPLE 6.16

Find $C(t)$ for the unit-step input to the closed-loop system just discussed, given $\omega_f = 10$ rad/sec when:

(a) $K = 16$
(b) $K = 25$
(c) $K = 50$

SOLUTION

(a) $\dfrac{-\omega_f \pm \sqrt{\omega_f^2 - 4K}}{2} = \dfrac{-10 \pm 6}{2} = -2, \; -8$

Thus,

$$C(s) = \frac{16}{s(s + 2)(s + 8)}$$

From pair 8 of Table 4-1,

$$C(t) = 1 - \frac{4}{3}e^{-2t} + \frac{1}{3}e^{-8t}$$

This is the *overdamped* case where both factors of the quadratic are real. The response is sketched in Figure 6.42.

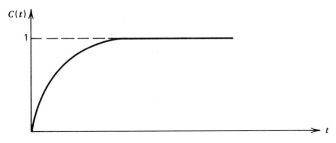

FIGURE 6.42 Overdamped response (Example 6.16).

(b) $\dfrac{-\omega_f \pm \sqrt{\omega_f^2 - 4K}}{2} = -5 \pm 0$

192 CONTROL SYSTEMS THEORY

Thus,

$$C(s) = \frac{25}{s(s+5)^2} = \frac{A}{s} + \frac{B}{(s+5)^2} + \frac{C}{s+5}$$

$$A = \left.\frac{25}{(s+5)^2}\right|_{s=0} = 1$$

$$B = \left.\frac{25}{s}\right|_{s=-5} = -5$$

$$C = \left.\frac{d(25/s)}{ds}\right|_{s=-5} = \left.\frac{-25}{s^2}\right|_{s=-5} = -1$$

$$C(s) = \frac{1}{s} - \frac{5}{(s+5)^2} - \frac{1}{(s+5)}$$

$$C(t) = 1 - 5te^{-5t} - e^{-5t}$$

This is the *critically damped case* due to a repeated pole of order 2; it is sketched in Figure 6.43.

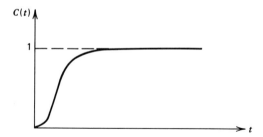

FIGURE 6.43 Critically damped response (Example 6.16).

(c) $$\frac{-\omega_j \pm \sqrt{\omega_j^2 - 4K}}{2} = -5 \pm j5$$

Thus,

$$C(s) = \frac{50}{s[s + (5 + j5)][s + (5 - j5)]}$$

$$= \frac{A}{s} + \frac{B}{s + (5 + j5)} + \frac{C}{s + (5 - j5)}$$

$$A = \left.\frac{50}{(s^2 + 10s + 50)}\right|_{s=0} = 1$$

6.8 THE SECOND APPROXIMATION OF A SERVOMOTOR

The portion of the response due to the pair of complex conjugate poles is found using Equation 23 from Chapter 4

$$C(t) = \frac{M}{\omega} e^{-\alpha t} \sin(\omega t + \theta)$$

where $\alpha = 5$, $\omega = 5$, and

$$M\angle\theta = K(s)\Big|_{-5+j5} = \frac{50}{s}\Big|_{-5+j5}$$

$$= \frac{50}{-5+j5} = \frac{50}{5\sqrt{2}} \angle -135° = 7.07\angle -135°$$

so the response due to these poles is

$$C(t) = 1.414 e^{-5t} \sin(5t - 135°)$$

The total response is then

$$C(t) = 1 + 1.414 e^{-5t} \sin(5t - 135°)$$

This is the *underdamped* case, which occurs when the poles are complex conjugates; it is sketched in Figure 6.44.

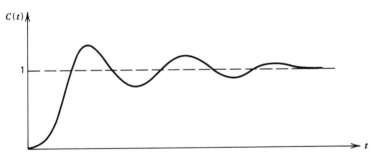

FIGURE 6.44 Underdamped response (Example 6.16).

EXAMPLE 6.17

A servomotor with transfer function $2.36/s(s + \omega_1)$ is to be used to position a rotary antenna. The antenna has inertia $J_A = 80(\text{N-m-sec}^2)/\text{rad}$ and is geared to the servomotor by a set of gears with ratio $N_1/N_2 = 10/20$. From the servomotor specifications provided by the manufacturer, it is known that

$$\omega_1 = \frac{125}{J_L} + 1.75$$

where J_L is the load inertia on the servomotor shaft. The feedback signal is obtained from a ten-turn potentiometer that is geared through a 40/10 gear ratio to the motor shaft. The input signal is obtained through another ten-turn potentiometer. The gain of the differential amplifier is one. Find the angular rotation $\theta_A(t)$ of the antenna when a unit-step input is applied to the system. See Figure 6.45.

194 CONTROL SYSTEMS THEORY

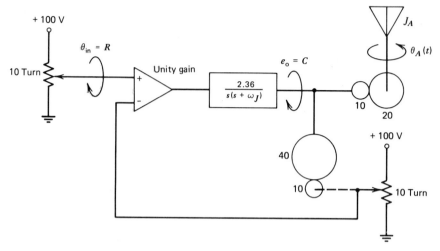

FIGURE 6.45 A servomotor used to position a rotary antenna (Example 6.17).

SOLUTION

The inertia reflected to the servomotor is

$$J_L = \left(\frac{N_1}{N_2}\right)^2 J_A = \left(\frac{10}{20}\right)^2 80 = 20(\text{N-m-sec}^2)/\text{rad}$$

Therefore,

$$\omega_1 = \frac{125}{20} + 1.75 = 8 \text{ rad/sec}$$

The combined gain of the 40/10 gears and the feedback potentiometer may be calculated as follows.

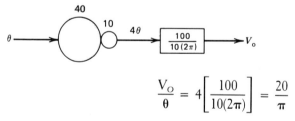

$$\frac{V_O}{\theta} = 4\left[\frac{100}{10(2\pi)}\right] = \frac{20}{\pi}$$

The block diagram of the system, excluding the antenna, is then as shown in Figure 6.46.

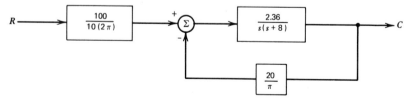

FIGURE 6.46 Block diagram of the system in Figure 6.45 (Example 6.17).

This diagram is equivalent to the one in Figure 6.47.

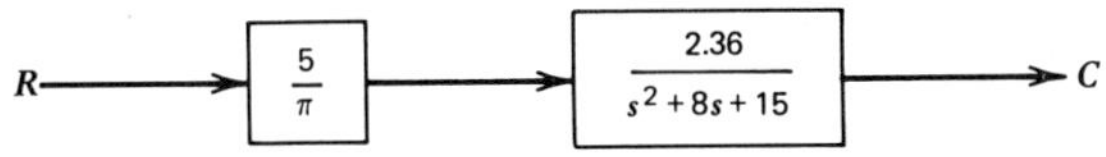

FIGURE 6.47 Block-diagram equivalent of Figure 6.46 (Example 6.17).

Thus,
$$\frac{C}{R} = \frac{3.76}{s^2 + 8s + 15} = \frac{3.76}{(s+5)(s+3)}$$

When $R(s) = 1/s$, we have
$$C(s) = \frac{3.76}{s(s+5)(s+3)}$$

Then, from pair 8 of Table 4-1,
$$C(t) = 0.251(1 - 2.5e^{-3t} + 1.5e^{-5t}) \text{ rad}$$

Since $\theta_A(t) = 0.5C(t)$ due to the gearing, we have
$$\theta_A(t) = 0.126(1 - 2.5e^{-3t} + 1.5e^{-5t}) \text{ rad}$$

We see that the antenna response falls into the overdamped category.

6.9 A VELOCITY CONTROL SYSTEM

Suppose we wish to control the angular *velocity* of a load, rather than its position. The speed-control apparatus on some automobiles is an example of such a system. Clearly, we will need some device to detect velocity, that is, a device whose output is proportional to angular velocity. A *tachometer* is such a device; it is simply a dc generator that produces an output voltage proportional to its speed of rotation. Whatever the physical nature of the device used, it will have a transfer function of the form Ks, since

$$\omega(s) = \frac{d\theta(s)}{dt} = s\theta(s)$$

As an example, the Electro-craft Model M-100-A tachometer, whose data sheet may be found in Appendix B, is specified to generate 3 V/KRPM or 28.65 (mV/rad)/sec. Thus, it has transfer function

$$G(s) = 28.65 \times 10^{-3}s$$

Let us use the servomotor with transfer function $\frac{K_1}{s(s+\omega_j)}$ to derive a load whose velocity is detected by a tachometer. The tachometer output will be fed back to the input as shown in Figure 6.48. Then

$$\frac{C}{R} = \frac{G}{1 + GH} = \frac{K_1/s(s + \omega_j)}{1 + [K_1K_2/(s + \omega_j)]} = \frac{K_1}{s[s + (\omega_j + K_1K_2)]}$$

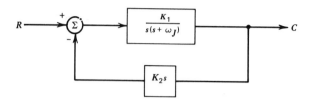

FIGURE 6.48 A velocity-control system with tachometer feedback.

For a unit-step input, $R(s) = 1/s$

$$C(s) = \frac{K_1}{s^2[s + (\omega_J + K_1K_2)]}$$

From pair 21 of Table 4-1,

$$C(t) = \frac{K_1}{(\omega_J + K_1K_2)^2}[(\omega_J + K_1K_2)t - 1 + e^{-(\omega_J + K_1K_2)t}]$$

Since $C(t)$ is the angular position of the output, we see that after a long period of time (after the exponential term has decayed to zero), the angle increases linearly with t. This is to be expected when angular velocity remains constant. To find an expression for the angular velocity, we may differentiate $C(t)$ above with respect to time or simply use the fact that

$$\omega(s) = \dot{\theta}(s) = s\theta(s) - \theta(0) = \frac{K_1}{s[s + (\omega_J + K_1K_2)]}$$

Then from pair 7,

$$\omega(t) = \frac{K_1}{(\omega_J + K_1K_2)}[1 - e^{-(\omega_J + K_1K_2)t}]$$

This shows us that when the exponential decays to zero, the velocity will have constant value

$$\omega = \frac{K_1}{(\omega_J + K_1K_2)}$$

6.10 A COMPREHENSIVE ANALYSIS OF THE SERVOMOTOR IN A CONTROL SYSTEM WITH MECHANICAL COMPONENTS

In order to study the interaction of a servomotor and the various mechanical components that it drives, our analysis must include some provision for connecting a device (the motor) whose input is voltage and whose output is angular displacement to components whose inputs are torques and whose outputs are angles. Thus, we must develop a third approximation for the servomotor, a more detailed model that

6.10 A COMPREHENSIVE ANALYSIS OF THE SERVOMOTOR

includes its capability to develop output torque. The servomotor typically develops torque in direct proportion to armature *current*, so we may write

$$T_m = K_{TI}I_A \tag{18}$$

where

T_m = torque developed by the motor
I_A = armature current
K_{TI} = proportionality constant for a given motor, relating output torque to input current

To relate armature current to input voltage, we must take into consideration two other factors: the armature impedance and the back *emf* of the motor.

The armature impedance may be considered to consist principally of armature resistance R_A and armature inductance L_A. Thus,

$$Z_A = R_A + sL_A \tag{19}$$

The back *emf* is proportional to the angular velocity of the motor. Within the motor itself, back *emf* is effectively fed-back and subtracted from the applied armature voltage. The greater the speed of rotation, the greater the back *emf*, and, hence, the smaller the armature current. Therefore, we surmise that greater torques (due to larger armature currents) are developed at lower speeds. Let us define

$$emf_B = K_{V\omega}\omega \tag{20}$$

where

emf_B = back *emf*
ω = angular velocity
$K_{V\omega}$ = proportionality constant relating back *emf* to angular velocity.

The angular velocity of the motor ω is related to the components of mechanical "impedance" reflected to the motor, as we know, by the equations:

$$T = Js\omega(s) \tag{21}$$

$$T = B\omega(s) \tag{22}$$

$$T = \frac{K}{s}\omega(s) \tag{23}$$

If we let Z_M = the total equivalent mechanical impedance reflected to the driving motor, then we can define a transfer function

$$G_M = \frac{\omega}{T} = \frac{1}{Z_M} \frac{\text{rad/sec}}{\text{N-m}} \tag{24}$$

where G_M relates the angular velocity of the mechanical components to the torque applied to them.

We can tie all these diverse considerations together in a block diagram. Figure 6.49 treats the motor itself as a control system and allows us to relate all of the factors we have discussed to each other. It is important to realize, however, that while the diagram applies only to the motor, it includes the effects of all mechanical components in the system by including the total equivalent mechanical impedance reflected to the motor. We may now reduce this diagram to obtain our third,

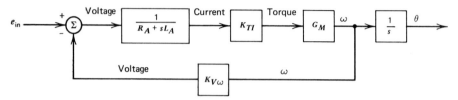

FIGURE 6.49 Comprehensive block diagram of a servomotor.

most accurate approximation of the servomotor, including the effects of mechanical impedance. Letting

$$G = \frac{G_M K_{TI}}{R_A + sL_A} \qquad H = K_{V\omega}$$

we have the diagram shown in Figure 6.50.

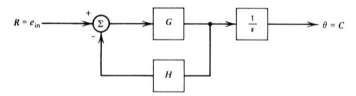

FIGURE 6.50 Block-diagram equivalent of Figure 6.49.

$$\frac{C}{R} = \left(\frac{G}{1 + GH}\right)\frac{1}{s} \text{ rad/V} \qquad (25)$$

Suppose the mechanical impedance Z_M consists of an inertia J that is the sum of armature inertia and any load inertia reflected to the armature. We will find the transfer function in a form that allows us to identify the resulting poles.

Since $G_M = 1/Js$ in this case, we have

$$\frac{C}{R} = \frac{[(K_{TI}/Js)/(R_A + sL_A)]\,1/s}{1 + \left[\dfrac{(K_{TI}K_{V\omega}/Js)}{R_A + sL_A}\right]} = \frac{(K_{TI}/Js^2)}{R_A + sL_A + (K_{TI}K_{V\omega}/Js)}$$

$$= \frac{K_{TI}/J}{s^2(R_A + sL_A) + s(K_{TI}K_{V\omega}/J)}$$

$$= \frac{K_{TI}/J}{s[L_A s^2 + R_A s + (K_{TI}K_{V\omega}/J)]}$$

$$\frac{C}{R} = \frac{K_{TI}/JL_A}{s[s^2 + (R_A/L_A)s + (K_{TI}K_{V\omega}/JL_A)]} \text{ rad/V} \qquad (26)$$

6.10 A COMPREHENSIVE ANALYSIS OF THE SERVOMOTOR

We see that the transfer function has a pole at $s = 0$ and a pair of poles due to the quadratic expression.

EXAMPLE 6.18

Find the transfer function of the Electro-craft servomotor Model E-510-A whose specifications are given in Appendix B. Also find the electrical and mechanical (unloaded) break frequencies.

SOLUTION

We note from the units used by the manufacturer to specify the motor characteristics that the manufacturer uses symbols which are somewhat different from those we have used to represent the same quantities. We must, therefore, convert the English system units that are given to the units that we have been using in order to obtain the transfer function in Equation 26 in V/rad.

Thus, we find

$$K_T = 5.88 \text{ oz-in./A}$$

from which we deduce

$$K_{TI} = [5.88 \text{ (oz-in.)/A}] [7.06 \times 10^{-3} \text{ (N-m/oz-in.)}]$$

$$= 42 \times 10^{-3} \text{ (N-m/A)}$$

and

$$K_E = 4.35 \text{ V/KRPM}$$

from which we deduce

$$K_{V\omega} = (4.35 \text{ V/KRPM}) \frac{\text{KRPM}}{10^3 \text{RPM}} \frac{1 \text{ rev}}{2\pi \text{ rad}} (60 \text{ sec/min})$$

$$= .042 \text{ V/rad/sec}$$

We also find from the specifications that

$$R_A = 1.24 \text{ }\Omega$$

$$L_A = 3.39 \times 10^{-3} \text{ H}$$

Assuming no load inertia, the value we use for J will be the armature inertia, which is given as 0.0038 (oz-in.)sec² or

$$J = [3.8 \times 10^{-3} \text{(oz-in-sec}^2)](7.06 \times 10^{-3} \text{ N-m/oz-in.})$$

$$= 26.83 \times 10^{-6} \text{(N-m)sec}^2/\text{rad}$$

(Note that radians are often omitted in compound units, since they are considered dimensionless.)

Using the quantities in Equation 26, we find

$$\frac{C}{R} = \frac{\dfrac{42 \times 10^{-3}}{(26.83 \times 10^{-6})(3.39 \times 10^{-3})}}{s\left[s^2 + \left(\dfrac{1.24}{3.39 \times 10^{-3}}\right)s + \dfrac{(42 \times 10^{-3})(.042)}{(26.83 \times 10^{-6})(3.39 \times 10^{-3})}\right]}$$

$$= \frac{4.62 \times 10^5}{s(s^2 + 366s + 1.94 \times 10^4)} \text{ V/rad}$$

From the specification sheet, we find that the electrical time constant is 2.06 ms, and that the mechanical time constant is 2.47 msec. Therefore, the electrical break frequency is

$$\omega = \frac{1}{2.06 \times 10^{-3}} = 485.4 \text{ rad/sec} = 77.3 \text{ Hz}$$

and the mechanical break frequency is

$$\omega = \frac{1}{24.7 \times 10^{-3}} = 40.49 \text{ rad/sec} = 6.44 \text{ Hz}$$

EXAMPLE 6.19

Determine the frequency-domain response $C(s)$ of the servomotor with transfer function given by Equation 26 when it is enclosed in a fully fed-back loop and is subjected to a unit step-function input. Assume the following system parameters:

$K_{TI} = 2 \times 10^{-3}$ (N-m)/A
$K_{V\omega} = .05$ (V/rad)/sec
$R_A = 10 \, \Omega$
$L_A = 5$ mH
$J = 10^{-6}$ (N-m-sec^2)/rad

SOLUTION

Using the parameters specified, the transfer function of the motor may be written

$$\left(\frac{C}{R}\right)_{Motor} = \frac{(2 \times 10^{-3})/(10^{-6})(5 \times 10^{-3})}{s\{s^2 + (10/5 \times 10^{-3})s + [2 \times 10^{-3})(0.05)/(10^{-6})(5 \times 10^{-3})]\}} \text{ rad/V}$$

$$= \frac{4 \times 10^5}{s[s^2 + (2 \times 10^3)s + 2 \times 10^4]}$$

$$\left(\frac{C}{R}\right)_{Motor} = \frac{4 \times 10^5}{s(s + 1.99 \times 10^3)(s + 10)} \text{ rad/V}$$

Before continuing, we note that the two poles due to the quadratic term are quite far apart; this is typical. The pole at $s = -1.99 \times 10^3$ is very nearly equal to $-(R_A/L_A) = -2 \times 10^3$, which is the break frequency due to the time constant

6.10 A COMPREHENSIVE ANALYSIS OF THE SERVOMOTOR

of the armature inductance and armature resistance. The portion of the response due to this pole will then have time constant $\tau = L_A/R_A = 0.5$ msec. In other words, its contribution will decay very rapidly. This is an important point: *The pole at $s = -10$ (with time constant 0.1 sec) is much more significant in determining overall response than is the pole at $s = -1.99 \times 10^3$.* The pole at $s = -10$ is, in fact, very nearly equal to the break frequency we called ω_I in our second approximation of the motor. The electrical break frequency (R_A/L_A) only appeared when we refined our model considerably to achieve a more accurate approximation. A Bode plot for the refined model will reflect this fact; see Figure 6.51. Continu-

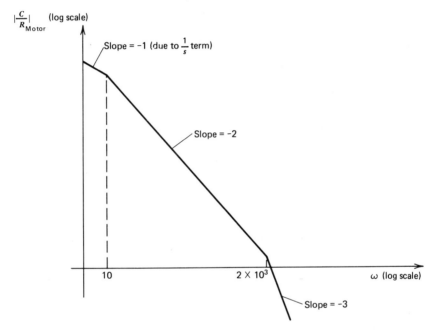

FIGURE 6.51 Bode plot for $|C/R|_{\text{Motor}}$ (Example 6.19).

ing with our analysis, we enclose the motor in a fully fed-back loop and obtain Figure 6.52.

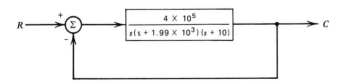

FIGURE 6.52 Block diagram for Example 6.19.

Let $K = 4 \times 10^5$, $a = 1.99 \times 10^3$, and $b = 10$

$$\frac{C}{R} = \frac{G}{1+G} = \frac{K/s(s+a)(s+b)}{1+[K/s(s+a)(s+b)]} = \frac{K}{s(s+a)(s+b)+K}$$

For $R = 1/s$, we obtain

$$C(s) = \frac{4 \times 10^5}{s[s(s + 1.99 \times 10^3)(s + 10) + 2 \times 10^5]}$$

Inverting this expression to find $C(t)$ involves considerable algebraic effort. However, for frequency domain analysis, we do not need to invert $C(s)$. As we will see, the information we need to evaluate the performance of the system is contained in the system's open-loop transfer function (the motor transfer function, as given by Equation 26). We have already provided some insight into this fact by sketching the Bode plot and by relating cutoff frequencies to time-domain response times. When we investigate system stability, we will see that a knowledge of the frequency response of the open-loop transfer function is invaluable. Suffice it to say for now that we have shown how to determine the effects that mechanical components have on this frequency response. In Exercise 25, the reader has an opportunity to investigate the effect that introducing mechanical damping B in combination with inertia J has on the pole locations. Successfully completing this exercise will reveal the analogy between B/J and R_A/L_A.

6.11 SPEED-TORQUE CURVES

Before concluding our discussion of servomotors, we should mention a practical technique that is frequently used to display motor characteristics graphically. Besides permitting the user to read certain important motor parameters directly, this graphical display helps provide insight into, and understanding of, basic motor operation. We will use previously described relationships between motor variables to obtain a graph of motor speed ω versus motor torque T_m.

In this analysis, we are interested only in the steady-state behavior of the motor. Thus, we will not be concerned with inductance and inertia, which contribute only to the transient response of the motor when the input is suddenly changed. Suppose, then, that the motor input voltage (or terminal voltage) e_{in} has been applied for a long period of time and that a steady-state motor speed ω has been reached. Then the back emf generated at this speed is from Equation 20

$$emf_B = (K_{V\omega})\omega$$

and since the resultant current that flows through the armature resistance R_A must then be the net voltage across it divided by R_A

$$I_A = \frac{e_{in} - emf_B}{R_A}$$

we have

$$I_A R_A = e_{in} - emf_B \quad \text{or} \quad (K_{V\omega})\omega = -I_A R_A + e_{in} \tag{27}$$

6.11 SPEED-TORQUE CURVES

From Equation 18, we know that motor torque is directly proportional to armature current

$$T_m = K_{TI} I_A \quad \text{or} \quad I_A = \frac{T_m}{K_{TI}} \tag{28}$$

Substituting Equation 27 into Equation 26, we obtain

$$(K_{V\omega})\omega = -\left(\frac{R_A}{K_{TI}}\right) T_m + e_{in}$$

or

$$\omega = -\left(\frac{R_A}{K_{V\omega} K_{TI}}\right) T_m + \left(\frac{e_{in}}{K_{V\omega}}\right) \tag{29}$$

Regarding ω and T_m as the variables in Equation 29 and all other quantities as constants, we see that the graph of speed ω versus torque T_m is a straight line with slope $-(R_A/K_{V\omega}K_{TI})$ and ω-intercept equal to $e_{in}/K_{V\omega}$. See Figure 6.53.

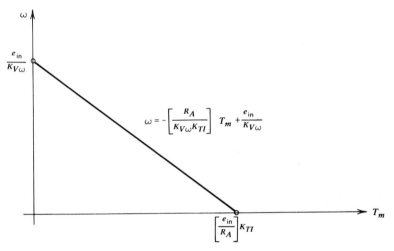

FIGURE 6.53 Motor speed ω versus motor torque T_m.

Note that $\omega = e_{in}/K_{V\omega}$ (at $T_m = 0$) is the *no-load speed*. Also, at $\omega = 0$,

$$0 = -\frac{R_A}{K_{V\omega} K_{TI}} T_m + \frac{e_{in}}{K_{V\omega}}$$

$$T_m = \frac{e_{in}}{R_A} K_{TI} = \text{the } stall \ torque$$

If the motor input voltage is changed, the no-load speed and stall torque values (i.e., the intercepts on the graph) are changed, but the slope remains the same.

Thus, choosing different values for e_{in} (e_1, e_2, etc.) leads to a family of *speed-torque curves*, as shown in Figure 6.54.

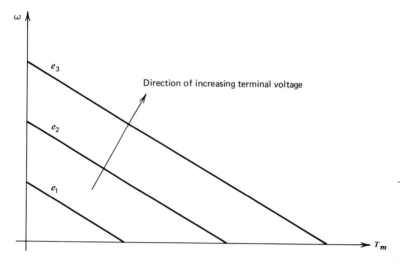

FIGURE 6.54 A family of speed-torque curves for a servomotor.

If torque is fixed (i.e., if a vertical line is drawn on the graph) and if equal increments of e_{in} along this line result in equal increments in ω, we conclude that motor speed is directly proportional to terminal voltage. This is generally the case and was the assumption we made in Equation 3 to show that the motor is an integrator. In practice, motor operation is limited to a restricted portion of the speed-torque curves, since excessive power dissipation occurs at high torque loads; Example 6.20 illustrates this fact.

EXAMPLE 6.20

Using the specifications provided in Appendix B for the Electro-Craft Model E-531-A motor tachometer, determine:

(a) The value of $K_{V\omega}$ (specified as K_E) in V/KRPM
(b) The no-load speed at e_{in} = 34.8 V
(c) The slopes of the speed-torque curves
(d) Whether or not motor speed is directly proportional to e_{in} for a constant motor torque of 20 oz-in.
(e) The maximum permissible torque for continuous operation at e_{in} = 34.8 V

SOLUTION

(a) From the specification sheet, $K_E = K_{V\omega}$ = 8.7 V/KRPM.
(b) Since $4K_E$ = 34.8, the intersection of the $4K_E$ line with the vertical axis on the speed-torque curves for the model E-531 is equal to the no-load speed. This value is seen to be 4000 RPM.

(c) The slopes of the speed-torque curves are equal and have value
$$\frac{\Delta\omega}{\Delta T_m} = \frac{-1000 \text{ RPM}}{50 \text{ oz-in.}} = -20 \text{ (RPM/oz-in.)}.$$

(d) Motor speed is directly proportional to e_{in} at $T_m = 20$ oz-in., since voltage increments of $2K_E$ result in speed increments of 2000 RPM.

(e) The $4K_E = 34.8$-V line intersects the S.O.A.C. (safe operating area, continuous) line at a torque of approximately 38 oz-in.

6.12 ERROR COEFFICIENTS

Having examined a variety of control-systems configurations, we are now in a position to seek criteria by which we can judge the performance of the various configurations. In other words, we require some means for determining whether one system is better than another, and if so, in what respects. We need to be able to determine, for example, what the desirable or undesirable effects are on changes in gain, a parameter over which we usually have some control. We have already noticed that response time and bandwidth are inversely related, and it is easy to show that increasing the gain of an integrator in a closed loop causes the bandwidth to be increased proportionally. But is there a limit on the extent to which we can increase gain? We answer this question in Chapter 7 when we consider *stability*, which is certainly an important performance criterion. An unstable system is of no use whatsoever.

For the time being, we will assume that we are dealing with stable systems and we will define certain *error coefficients* that provide us with measures of how accurately a control system responds to various types of inputs. An error coefficient is, in fact, a measure of the difference between what the control system is doing and what we want it to do; that is, a measure of its error. These coefficients are sometimes called static-error coefficients, because they are defined and computed a long period of time after an input has been applied to the system. Thus, we assume that all *transient* responses such as exponential decays have died out, and we examine only the steady-state response. Error coefficients thus do not yield any information on response time or how quickly the system can adjust itself to a new input.

In order to facilitate calculating error coefficients, we will need a new mathematical tool, called the final-value theorem. The final-value theorem allows us to determine the steady-state value of a time-domain function from its LaPlace transform. Mathematically, the final-value theorem states

If
$$\mathcal{L}[f(t)] = F(s)$$
then
$$\lim_{t \to \infty} f(t) = \lim_{s \to 0} sF(s) \tag{30}$$

In other words, we can find the steady-state value of a function by finding the limit as s approaches 0 of s times the transform of the function.

EXAMPLE 6.21

Find the steady-state value of $f(t) = 1 - e^{-at}$ using the final-value theorem.

SOLUTION

$$F(s) = \frac{1}{s} - \frac{1}{s+a}$$

$$\lim_{t \to \infty} f(t) = \lim_{s \to 0} s\left(\frac{1}{s} - \frac{1}{s+a}\right) = \lim_{s \to 0}\left(1 - \frac{s}{(s+a)}\right) = 1$$

This result is obviously confirmed by the fact that $\lim_{t \to \infty} (1 - e^{-at}) = 1 - 0 = 1$.

The three types of error coefficients are called the position-, velocity-, and acceleration-error coefficients. These names do *not* imply that they are applicable only to position, velocity, or acceleration control systems, respectively; all three are applicable to any type of control system.

In order for any measure of error to be useful, it should be normalized with respect to the magnitude of the input. For example, it does us no good from a comparison standpoint to state that one control system has half the error of another, unless we know that both have the same input. Thus, we define normalized error coefficients as follows:

$$\varepsilon = \frac{\text{steady-state error}}{\text{coefficient of input function}} \quad (31)$$

For the position-error coefficient ε_P, the input is a step function $Ku(t)$; for the velocity-error coefficient ε_V, the input is a ramp function Kt; and for the acceleration-error coefficient ε_A, the input is parabolic, Kt^2. In each of these cases, the "coefficient of the input function" is K, an arbitrary constant.

Consider the general block diagram of the fully fed-back control system with open-loop transfer function g shown in Figure 6.55. Since $C = GE$ where E is

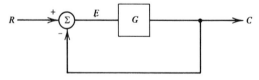

FIGURE 6.55 The fully fed-back control system with error term E.

the error, we have $E = C/G$. But since $C/R = G/(1 + G)$, that is, $C = RG/(1 + G)$, we may write

$$E = \frac{1}{G}\left(\frac{RG}{1+G}\right) = \frac{R}{1+G}$$

6.12 ERROR COEFFICIENTS

By the final-value theorem,

$$e_{ss} = \lim_{s \to 0} s\left(\frac{R}{1+G}\right)$$

where e_{ss} = steady-state error. Thus, Equation 31 may be written in the form

$$\varepsilon = \frac{\lim_{s \to 0} s[R/(1+G)]}{K} \qquad (32)$$

Since $R(s) = K/s$ for ε_P, K/s^2 for ε_V, and K/s^3 for ε_A, we have, on substituting in Equation 32,

$$\varepsilon_P = \frac{\lim_{s \to 0} s\left[\dfrac{K/s}{1+G}\right]}{K} = \lim_{s \to 0}\left[\frac{1}{1+G}\right] \qquad (33)$$

$$\varepsilon_V = \frac{\lim_{s \to 0} s\left[\dfrac{K/s^2}{1+G}\right]}{K} = \lim_{s \to 0} 1/s\left[\frac{1}{1+G}\right] \qquad (34)$$

$$\varepsilon_A = \frac{\lim_{s \to 0} s\left[\dfrac{K/s^3}{1+G}\right]}{K} = \lim_{s \to 0} 1/s^2\left[\frac{1}{1+G}\right] \qquad (35)$$

EXAMPLE 6.22

Find the position-, velocity-, and acceleration-error coefficients for each of the following position control systems (see Figures 6.56–6.58):

(a)

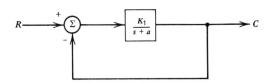

FIGURE 6.56 Find the error coefficients (Example 6.22a).

(b)

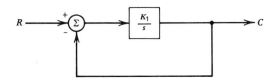

FIGURE 6.57 Find the error coefficients (Example 6.22b).

(c)

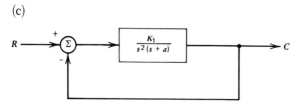

FIGURE 6.58 Find the error coefficients (Example 6.22c).

SOLUTION

(a) $\varepsilon_P = \lim\limits_{s \to 0} \dfrac{1}{1 + G} = \lim\limits_{s \to 0} \left[\dfrac{1}{1 + K_1/(s + a)} \right] = \dfrac{1}{1 + (K_1/a)}$

$\varepsilon_V = \lim\limits_{s \to 0} \dfrac{1}{s}\left[\dfrac{1}{1 + G} \right] = \lim\limits_{s \to 0} \left\{ \dfrac{1}{s + [K_1 s/(s + a)]} \right\} = \infty$

$\varepsilon_A = \lim\limits_{s \to 0} \dfrac{1}{s^2}\left[\dfrac{1}{1 + G} \right] = \infty$

(b) $\varepsilon_P = \lim\limits_{s \to 0} \left[\dfrac{1}{1 + (K_1/s)} \right] = 0$

$\varepsilon_V = \lim\limits_{s \to 0} \left[\dfrac{1}{s + K_1} \right] = \dfrac{1}{K_1}$

$\varepsilon_A = \lim\limits_{s \to 0} \left[\dfrac{1}{s^2 + K_1 s} \right] = \infty$

(c) $\varepsilon_P = \lim\limits_{s \to 0} \left\{ \dfrac{1}{1 + [K_1/s^2(s + a)]} \right\} = 0$

$\varepsilon_V = \lim\limits_{s \to 0} \left\{ \dfrac{1}{s + [K_1/s(s + a)]} \right\} = 0$

$\varepsilon_A = \lim\limits_{s \to 0} \left\{ \dfrac{1}{s^2 + [K_1/(s + a)]} \right\} = \dfrac{a}{K_1}$

Let us interpret some of the results in Example 6.22. In the control system in (a), suppose the input is the step function $10u(t)$. Then after a long period of time, the error in angular position would be $e = (\varepsilon_P)(\text{input coefficient}) = 10/[(1 + K_1)/a]$ rad. Clearly, increasing the gain K_1 will have the effect of reducing the positional error. For the control systems in (b) and (c), on the other hand, there will eventually be zero positional error, irrespective of gain K_1.

The velocity-error coefficient for the system in (a) is infinite. This means that when the system is subjected to a continually increasing input ($R = Kt$), the output will lag further and further behind, so that the error in angular position accumulates without bound. The system is unable to keep up with this type of input. For the control system in (b), $\varepsilon_V = 1/K_1$. Thus, if the input is changing at a constant rate of, say $10t$, the output will eventually keep up to the extent that there will be a *constant* error in position of $10/K_1$ rad. Visualize the situation in these terms: We are continually commanding the system to rotate to newer, ever larger angular displacements; the system eventually settles down to the point that it is always lagging behind our command by a constant angle. If our input is telling the system to assume an angular rotation of, say 30°, and it has rotated 25°, then when the input command has increased to 35°, the output reaches 30°, and so forth, for a constant error of 5°. Again, it can be seen that increasing the gain K_1 reduces the error. The acceleration-error coefficient, on the other hand, is infinite. Thus, if the input command is parabolic, $R(t) = Kt^2$, the position error will increase indefinitely.

Finally, the control system in (c) has zero position and velocity-error coefficients. Given sufficient time, it is able to occupy an angular position commanded by a step input or by a constantly increasing input with zero error. The output eventually catches up to the input in these cases. However, if it is subjected to an input Kt^2, it will eventually maintain a constant positional error of $K(a/K_1)$ rad.

We note in Example 6.22 that the error coefficients generally improved (became smaller) as the degree of s in the denominator of G increased; in other words, as the number of integrations increased. In case (a), there are zero integrations, in case (b), one integration, and in case (c), two integrations. The trade-off here is that we cannot continually improve error by increasing both gain and the number of integrations without threatening stability. Again, this fact will become evident from material to be covered in the next chapter.

EXAMPLE 6.23

Find the steady-state error in angular position if the input to the system in Figure 6.59 is $R(t) = 40t$ deg/sec.

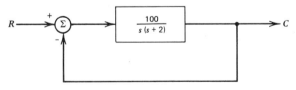

FIGURE 6.59 Find the steady-state error (Example 6.23).

SOLUTION

$$\varepsilon_V = \lim_{s \to 0} \frac{1}{s}\left[\frac{1}{1 + G}\right]$$

$$= \lim_{s \to 0} \frac{1}{s}\left\{\frac{1}{1 + [100/s(s + 2)]}\right\} = \lim_{s \to 0} \left[\frac{1}{s + [100/(s + 2)]}\right] = .02$$

Then $e = (.02)40 = 0.8°$.

210 CONTROL SYSTEMS THEORY

Suppose we wish to determine the steady-state error in a control system that has a transfer function other than unity in the feedback loop (see Figure 6.60). In

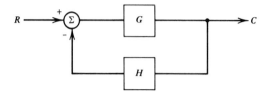

FIGURE 6.60 Control system with nonunity feedback H.

this case, we can no longer define error as the output of the junction where the feedback is subtracted from the input. Instead, error is defined as the difference between input and output: $E = R - C$. This error quantity no longer appears at any given point in the system. Since

$$\frac{C}{R} = \frac{G}{1 + GH}$$

we have

$$E = R - R\left(\frac{G}{1 + GH}\right) = R\left(1 - \frac{G}{1 + GH}\right) \tag{36}$$

By the final-value theorem,

$$e_{ss} = \lim_{s \to 0} sR(s)\left[1 - \frac{G(s)}{1 + G(s)H(s)}\right]$$

$$= \lim_{s \to 0} sR(s)\left[\frac{1 + G(s)H(s) - G(s)}{1 + G(s)H(s)}\right] \tag{37}$$

For a step input $R(s) = K/s$, we have

$$e_{ss} = \lim_{s \to 0} K\left[\frac{1 + G(s)H(s) - G(s)}{1 + G(s)H(s)}\right] \tag{38}$$

To obtain zero position error, we see it is necessary for

$$\lim_{s \to 0} [1 + G(s)H(s)] = \lim_{s \to 0} G(s) \tag{39}$$

This is equivalent to the condition

$$\lim_{s \to 0} \frac{1}{G(s)} = \lim_{s \to 0} [1 - H(s)] \tag{40}$$

or

$$\lim_{s \to 0} H = \lim_{s \to 0}\left[1 - \frac{1}{G}\right] \tag{41}$$

EXAMPLE 6.24

Find H required for zero steady-state position error in the system shown in Figure 6.61.

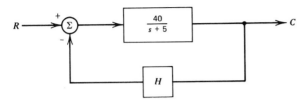

FIGURE 6.61 Find H for zero steady-state position error (Example 6.24).

SOLUTION

From Equation 41, we require

$$\lim_{s \to 0} H(s) = \lim_{s \to 0} \left[1 - \frac{1}{G(s)} \right] = \lim_{s \to 0} \left[1 - \frac{s+5}{40} \right] = \frac{7}{8}$$

There are many transfer functions H that satisfy $\lim_{s \to 0} H(s) = 7/8$, for example, $7/(s + 8)$, $(s + 7)/(s^2 + 8)$, and so on. We could simply choose a resistive voltage divider with ratio 7/8. Recall from Example 6.19 that a transfer function of the form $K_1/(s + a)$ with unity feedback has nonzero position error. We have shown that it is possible to eliminate the error in this type of system by modifying the feedback in an appropriate way.

REFERENCES

1. **Byrd, Roy D.** *Automatic Controls.* Delmar Publishers, 1972.
2. **Charkey, Edward D.** *Electromechanical System Components.* Wiley, 1972.
3. **Davis, Sidney A.**, and **Byron K. Ledgerwood.** *Electromechanical Components for Servomechanisms.* McGraw-Hill, 1961.
4. **D'Azzo, John J.**, and **Constantine Houpis.** *Feedback Control-System Analysis and Synthesis.* McGraw-Hill, 1960.
5. **Dorf, Richard C.** *Modern Control Systems.* Addison-Wesley, 1967.
6. **Fortmann, Thomas E.**, and **Konrad L. Hitz.** *An Introduction to Linear Control Systems.* Marcel Dekker, 1977.
7. **Hale, Francis J.** *Introduction to Control-System Analysis and Design.* Prentice-Hall, 1973.
8. **McNeil, P. A.**, and **Dewey A. Yeager.** *Servomechanisms.* Delmar Publishers, 1972.
9. **Miller, Richard W.** *Servomechanisms: Devices and Fundamentals.* Reston, 1977.
10. **Yeager, D. A.**, and **Robert L. Gourley.** *Introduction to Electron and Electromechanical Devices.* Prentice-Hall, 1976.
11. **Zeines, Ben.** *Automatic Control Systems.* Prentice-Hall, 1972.

EXERCISES

6.1 What is the open-loop transfer function of the control system in Figure 6.62?

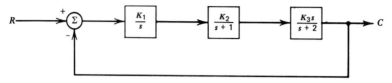

FIGURE 6.62

6.2 If $C(s) = \dfrac{K_2}{s + 2}$ in Exercise 6.1, find an expression for the error $\varepsilon(s)$.

6.3 Find the closed-loop transfer function of Exercise 6.1.

6.4 An integrator has frequency-independent gain equal to 0.1 and is enclosed in a fully fed-back loop. Find the closed-loop transfer function.

6.5 Find C/R using block-diagram manipulations for each of the systems in Figure 6.63.

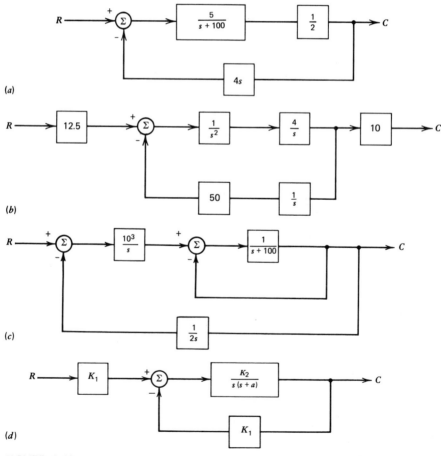

FIGURE 6.63

6.6 Prove that the two systems shown in Figure 6.64 are equivalent.

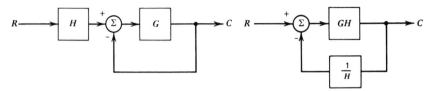

FIGURE 6.64

6.7 Without using block-diagram manipulations, prove that $\dfrac{C}{R} = \dfrac{G}{1 + GH}$ (see Figure 6.65).

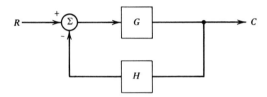

FIGURE 6.65

6.8 Design an RC-network to approximate an integrator at a frequency of $\omega = 10^3$ rad/sec. The output amplitude should not differ from that of an ideal integrator by more than 5% at $\omega = 10^3$ rad/sec.

6.9 In the control system in Example 6.5, suppose $K_2 = 10$ and $V = +5$ V.

(a) What must be the gain of the differential amplifier in order for the output to reach 63% of its final value in response to a step input in .01 sec.?
(b) With the amplifier gain determined in (a), at what frequency will the output of the system lag the input by 45° when driven by a sinusoidal input?
(c) With the same gain, at what frequency will the output magnitude be 50% of the input magnitude when driven by a sinusoidal input?
(d) With the same gain, how long will it take the output to reach 90% of its final value in response to a step input?

6.10 What voltage gain in decibels should be added to the integrator in Figure 6.66 so that the system

(a) Has a bandwidth of 500 Hz?
(b) Has a bandwidth of 10 rad/sec?

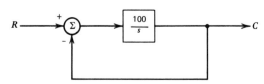

FIGURE 6.66

214 CONTROL SYSTEM THEORY

6.11 Suppose the system shown in Figure 6.67 is driven by a *square* wave that alternates between $+5$ V and zero V with a period of 0.1 sec. Sketch $C(t)$ versus time.

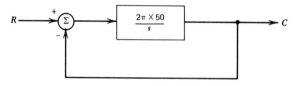

FIGURE 6.67

6.12 Each of the control systems shown in Figure 6.68 is subjected to a step input $R = 10u(t)$. Use LaPlace transform analysis to find $C(t)$ in each case. Sketch $C(t)$ versus time, and identify the transient and steady-state responses in each case.

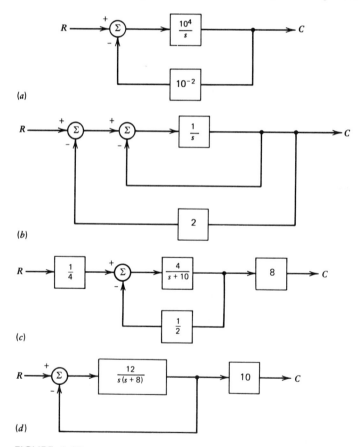

FIGURE 6.68

6.13 A servomotor develops a torque of 15 N-m. Its armature inertia is 0.25 (N-m-sec^2)/rad. How much load inertia can it accelerate at 10 rad/sec^2?

6.14 It requires 4.2 N-m of torque to rotate a rotary dashpot at a constant angular velocity of 18.8 rad/sec. How much torque is requried to rotate it at 30 rad/sec?

6.15 A servomotor can develop a maximum torque of 0.8 N-m. Through what angle is it capable of rotating a flexible shaft that has a rotary spring constant of 76 N/rad?

6.16 Find ω(t) after a step torque of 24 N-m is applied to the system shown in Figure 6.69.

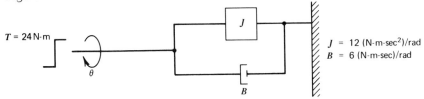

FIGURE 6.69

6.17 In the gear train shown in Figure 6.70, G_3 is driving G_2, and G_2 is driving G_1. G_3 is rotating at 64 rad/sec and has 30 teeth, G_2 is rotating at 16 rad/sec, and G_1 is rotating at 40 rad/sec. Find the number of teeth on G_1 and G_2.

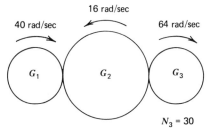

FIGURE 6.70

6.18 In the diagram in Figure 6.71, the shaft connected to G_2 has angular velocity given by $\dot{\theta}_2(t) = 45 + 8e^{-t}$ rad/sec.

  (a) How much torque is the driving shaft connected to G_1 developing at $t = 1$ sec?
  (b) Write an expression for $\dot{\theta}_1(t)$ and $\ddot{\theta}_1(t)$.

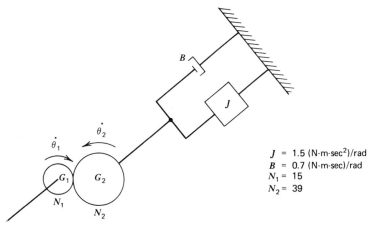

FIGURE 6.71

6.19 If the system in Exercise 6.18 starts at rest and is then subjected to a step input torque of 2 N-m on the driving shaft, find expressions for $\dot\theta_1(t)$ and $\dot\theta_2(t)$. (The expression for $\dot\theta_1(t)$ in Exercise 6.18 is no longer applicable.)

6.20 In Example 6.16, if $\omega_I = 15$ rad/sec, find values of the gain K that yield under, over, and critically damped responses to a step input.

6.21 In Example 6.17, assuming the same amplifier gains, gears, and potentiometers, what is the maximum antenna inertia that could be used in the system without having an underdamped response to a step input?

6.22 In Example 6.17, suppose the differential amplifier gain is increased to 89/60. Find and sketch $C(t)$ in response to a unit-step input.

6.23 In the velocity control system in Figure 6.72, it is desired that the output velocity increase to 63% of its final value in response to a step input 5 msec after the step is applied. What should be the value of K_1?

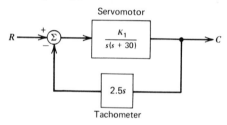

FIGURE 6.72

6.24 In Exercise 6.23, with the value of K_1 calculated, what will be the final output velocity if the system starts from rest and a unit step is applied to the input?

6.25 Find the transfer function of the Electro-craft servomotor Model E-512-B whose specifications are given in Appendix B.

6.26 Using the specifications provided in Appendix B for the Electro-craft Model E-532-C motor tachometer, determine
  (a) The transfer function of the tachometer.
  (b) The value of $K_{V\omega}$ for the motor (specified as K_E) in V/KRPM.
  (c) The no-load speed of the motor at $e_{in} = 34.76$ V.
  (d) The slopes of the motor's speed-torque curves.
  (e) Whether or not the motor can operate continuously within specifications at 1200 RPM with 55 oz-in. of torque.

6.27 In deriving the transfer function of a servomotor in Section 6.10, it was assumed that the only significant mechanical component was inertia J; this led to Equation 25. Derive a similar transfer function when a damping term B is included ($Z_M = Js + B$). Express any quadratics obtained in a form such that the coefficient of s^2 is 1. In this form, you should be able to identify a frequency term analogous to the electrical break frequency R_A/L_A. What is it?

6.28 Use the final-value theorem to find the limit as t approaches infinity of each of the following:

(a) $f(t) = 4te^{-2t}$
(b) $f(t) = 12[1 - e^{-6t} + e^{-t}]$
(c) $f(t) = 1/4(1 - \cos 2t)$

6.29 Find the position, velocity, and acceleration-error coefficients for each of the following control systems (Figure 6.73):

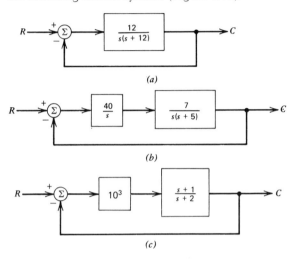

FIGURE 6.73

6.30 How much gain should be placed in series with G in part (a) of Exercise 6.29 in order to obtain a velocity-error coefficient of 0.25?

6.31 Find the steady-state position error if the system in Figure 6.74 is subjected to an input $R(t) = 2.5t$

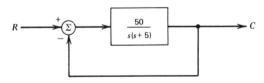

FIGURE 6.74

6.32 Determine a feedback transfer function H necessary to provide zero position error ($e = R - C$) in the following position-control system (Figure 6.75).

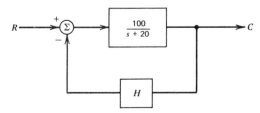

FIGURE 6.75

ANSWERS TO EXERCISES

6.1 $G(s) = \dfrac{K_1 K_2 K_3}{(s+1)(s+2)}$

6.2 $\varepsilon(s) = (s+1)/K_1 K_3$

6.3 $C/R = \dfrac{K_1 K_2 K_3}{(s+1)(s+2) + K_1 K_2 K_3}$

6.4 $C/R = 0.1/(s+0.1)$

6.5 (a) $C/R = \dfrac{2.5}{11s + 100} = \dfrac{0.227}{s + 9.09}$

(b) $C/R = \dfrac{500s}{s^4 + 200}$

(c) $C/R = \dfrac{10^3 s}{s^3 + 101 s^2 + 500}$

(d) $C/R = \dfrac{K_1 K_2}{s^2 + as + K_1 K_2}$

6.8 $.952 \le RC \le 1.05$

6.9 (a) $K_1 = 12.57$
(b) $\omega = 100$ rad/sec; $f = 15.91$ Hz
(c) $\omega = 173$ rad/sec; $f = 27.57$ Hz
(d) $t = .023$ sec

6.10 (a) 29.94 dB
(b) -20 dB

6.11 See Figure 6.76.

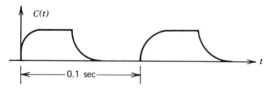

FIGURE 6.76

6.12 (a) $C(t) = 10^3(1 - e^{-100t})$
(b) $C(t) = (10/3)(1 - e^{-3t})$
(c) $C(t) = 6.67(1 - e^{-12t})$
(d) $C(t) = 100 - 150 e^{-2t} + 50 e^{-6t}$

6.13 $J_L = 1.25$ N-m-sec²/rad

ANSWERS TO EXERCISES 219

6.14 $T = 6.7$ N-m

6.15 $\theta = .0105$ rad

6.16 $\omega = 4(1 - e^{-0.5t})$ rad/sec

6.17 $N_1 = 48;\quad N_2 = 120$

6.18 (a) $T = 11.21$ N-m
(b) $\ddot{\theta}_1(t) = 117 + 20.8e^{-t}$
$\ddot{\theta}_2(t) = -20.8e^{-t}$

6.19 $\dot{\theta}_1(t) = 19.18(1 - e^{-0.47t})$ rad/sec
$\dot{\theta}_2(t) = 7.38(1 - e^{-0.47t})$ rad/sec

6.20 Critical: $K = 56.25$
Under: $K > 56.25$
Over: $K < 56.25$

6.21 $J_A \leq 83.32$ N-m-sec²/rad

6.22 $C(t) = 0.25 + 0.471e^{-4t} \sin(2.5t - 148°)$

6.23 $K_1 = 68$

6.24 $\omega_{ss} = 0.34$ rad/sec

6.25 $\dfrac{C}{R} = \dfrac{3.12 \times 10^5}{s(s^2 + 195s + 1.62 \times 10^4)}$

6.26 (a) $G(s) = 0.136s$ V/rad/sec
(b) 17.38 V/KRPM
(c) 1900 RPM
(d) -11.67 RPM/oz-in.
(e) No

6.27 $\left(\dfrac{C}{R}\right)_{Motor} = \dfrac{K_{TI}/JL_A}{s\{s^2 + [(R_A/L_A) + (B/J)]s + [(BR_A + K_{TI}K_{V\omega})/JL_A]\}}$
$\dfrac{B}{J}$ is analogous to $\dfrac{R_A}{L_A}$.

6.28 (a) 0
(b) 12
(c) ¼

220 CONTROL SYSTEM THEORY

6.29 (a) $\varepsilon_P = 0$, $\varepsilon_V = 1$, $\varepsilon_A = \infty$
(b) $\varepsilon_P = 0$, $\varepsilon_V = 0$, $\varepsilon_A = 1/56$
(c) $\varepsilon_P = 1.996 \times 10^{-3}$, $\varepsilon_V = \infty$, $\varepsilon_A = \infty$

6.30 $K = 4$

6.31 $\varepsilon_{ss} = 0.25$ rad

6.32 $H = 0.8$

CHAPTER

7

STABILITY AND COMPENSATION

7.1 STABLE AND UNSTABLE CONTROL SYSTEMS

As the name implies, a stable control system is one that responds in a predictable and desirable manner to an input. While the response of an unstable system may often be predictable, it is certainly not desirable. Instability is usually characterized by uncontrolled oscillations. The output of the unstable-position control system will typically alternate between a position that is greater than the one desired and one that is smaller, with a frequency determined by the mechanical and electrical characteristics of the system. In some systems, instability is characterized by the output increasing or decreasing without restraint, irrespective of any input command, until it reaches the limit of its travel (if one exists).

As the reader may have guessed, a system may become unstable in practice when *positive feedback* occurs. An electronic oscillator is a device into which positive feedback has been intentionally introduced. By altering the frequency-dependent components in such a device (e.g., inductors and capacitors), we are able to control the frequency of oscillation. A control system, on the other hand, is like a high gain amplifier: We wish to design it so that it has as high a gain as possible without becoming an oscillator.

We have emphasized on several occasions that gain is a frequency-dependent characteristic of any system. The implications of this fact are quite significant when we investigate stability. In general, the gain of a system will be somewhat different at every frequency, and if there exists some frequency—any frequency—at which the gain is sufficient to cause positive feedback, the system will oscillate at that frequency. We have seen that response time in a control system is generally im-

222 STABILITY AND COMPENSATION

proved by increasing the bandwidth of the system. But given the fact that increased bandwidth means a greater gain over a wider range of frequencies, we may very well encounter positive feedback at some frequency in the band and thus experience instability. This then is the tradeoff: improved stability for a more sluggish response. The situation is further complicated by the fact that real physical components tend to change their characteristics with age, temperature, and other unpredictable factors. We must, therefore, have a good *margin of stability* to ensure that a stable system will not suddenly become unstable. The principal task of the control-systems designer is to optimize the response of the system in the frequency range for which it is to be used and at the same time, *compensate* the system by various techniques of gain-frequency control to ensure a good margin of stability.

Positive feedback occurs when the magnitude of the feedback signal *adds* to the input rather than subtracts from it. It is instructive to think of the feedback junction where the feedback signal is subtracted from the input signal as causing a 180° phase shift in the feedback. Thus, the feedback junction effectively inverts the feedback signal and adds it to the input signal. Symbolically, the diagrams in Figure 7.1 are equivalent. Therefore, if the feedback signal undergoes a 180° phase in-

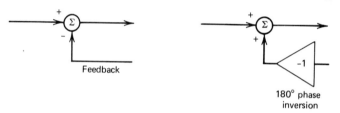

FIGURE 7.1 Equivalent representations of the feedback junction.

version *before* it reaches the junction, the net phase shift will be 360° when it is combined with the input. In other words, addition rather than subtraction will occur. As we will see in Section 7.2, we can predict whether this situation will occur by examining the open-loop transfer function.

7.2 STABILITY AND THE OPEN-LOOP TRANSFER FUNCTION

As we have seen, the fully fed-back control system with open-loop transfer function G has closed-loop transfer function

$$\frac{C}{R} = \frac{G}{1 + G}$$

This closed-loop transfer function is clearly undefined if $G = -1$, for then the denominator is zero. We must remember that G is a transfer function and, as such, has both magnitude and angle, both of which are dependent on frequency. Thus, the statement $G = -1$ means that there exists some frequency at which $|G| =$

7.2 STABILITY AND THE OPEN-LOOP TRANSFER FUNCTION

1 and $\angle G = -180°$. This then is the mathematical equivalent of a condition for instability. In practical terms, we say that *if the gain of the open-loop system at a frequency where the phase shift is 180° is sufficiently large (greater than or equal to 1), then instability will result.* Note that it is theoretically impossible for a pure integrator with transfer function $G = K/s$ and enclosed in a fully fed-back loop to become unstable. The phase shift caused by a pure integrator is a constant $-90°$ regardless of frequency, so no matter what the gain K is, we will never achieve the conditions for instability. Of course, from a practical standpoint, we cannot construct a perfect integrator. The servomotor that we represented by an integrator as a first approximation does, in fact, have break points (poles) at higher frequencies, as we saw in Chapter 6, and these contribute additional phase shift to the overall transfer characteristic.

Consider the open-loop transfer function with poles at $s = 0$, $s = -a$, and $s = -b$

$$G = \frac{K}{s(s + a)(s + b)}$$

Substituting $s = j\omega$, we have

$$G = \frac{-j(K/\omega)}{(j\omega + a)(j\omega + b)} = \frac{-j(K/\omega)}{(ab - \omega^2) + j\omega(a + b)}$$

$$|G| = \frac{(K/\omega)}{\sqrt{(ab - \omega^2)^2 + \omega^2(a + b)^2}} \qquad (1)$$

$$\angle G = -90 - \arctan\left[\frac{\omega(a + b)}{ab - \omega^2}\right] \qquad (2)$$

From Equation 2, we see that in order for $\angle G$ to be $-180°$, we must have $\omega^2 = ab$ [which gives us $\arctan(\infty) = 90°$]. Then if $\omega^2 = ab$, Equation 1 becomes

$$|G| = \frac{K/\omega}{\omega(a + b)} = \frac{K}{\omega^2(a + b)} = \frac{K}{ab(a + b)} \qquad (3)$$

Thus, in order to have $|G| < 1$ at the frequency where $\angle G = -180°$, we must have from Equation 3,

$$K < ab(a + b) \qquad (4)$$

EXAMPLE 7.1

Suppose

$$G = \frac{K}{s(s + 5)(s + 20)}$$

Find $|G|$ and $\angle G$ at $\omega = 5$, 10, and 20 rad/sec.

SOLUTION

(a) $G = \dfrac{K}{j5(5 + j5)(20 + j5)} = \dfrac{K}{j5(75 + j125)}$

$|G| = \dfrac{K}{5\sqrt{(75)^2 + (125)^2}} = \dfrac{K}{728.9}$

$\angle G = -90° - \arctan \dfrac{125}{75} = -149.04°$

(b) $G = \dfrac{K}{j10(5 + j10)(20 + j10)} = \dfrac{K}{j10(0 + j250)}$

$|G| = \dfrac{K}{2500}$

$\angle G = -90° - \arctan \infty = -180°$

(c) $G = \dfrac{K}{j20(5 + j20)(20 + j20)} = \dfrac{K}{j20(-300 + j500)}$

$|G| = \dfrac{K}{20\sqrt{(300)^2 + (500)^2}} = \dfrac{K}{11662}$

$\angle G = -90° - \arctan \dfrac{500}{-300} = -211°$

This example shows that as frequency increases, the magnitude of G decreases and that the total phase shift becomes more and more negative. The critical frequency is found in part (b) where $\omega = 10$, for at this frequency, the total phase shift is $-180°$. Since $|G| = K/2500$ at this frequency, we must be certain that $K < 2500$ to ensure that $|G| < 1$ and thus ensure stability. Note also that this result agrees with the relation in Equation 4

$$K < ab(a + b) = (5)(20)(5 + 20) = 2500$$

The theory we have developed for the fully fed-back control system is equally applicable to a system with an arbitrary transfer function H in the feedback path. We are concerned, after all, with any amplitude and phase modifications that the signal undergoes as it travels the complete path from point A to point B in Figure 7.2. Since $G' = G/(1 + GH)$ in this case, we must be certain that the situation $GH = -1$ does not occur. The equation $1 + GH = 0$, which is the condition for instability, is called the *characteristic equation*.

7.2 STABILITY AND THE OPEN-LOOP TRANSFER FUNCTION

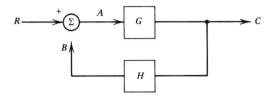

FIGURE 7.2 Signal path through which signal modifications affect stability.

Suppose that

$$G = \frac{K_1}{(s + a)(s + b)} \quad \text{and} \quad H = \frac{K_2}{(s + a)}$$

We will find conditions on the gain $K_1 K_2$ in terms of a and b to ensure a stable system

$$GH = \frac{K_1 K_2}{(s + a)^2(s + b)}$$

Substituting $s = j\omega$,

$$GH = \frac{K_1 K_2}{(a + j\omega)^2(b + j\omega)} = \frac{K_1 K_2}{(a^2 + 2aj\omega - \omega^2)(b + j\omega)}$$

$$= \frac{K_1 K_2}{[(a^2 - \omega^2) + 2aj\omega](b + j\omega)}$$

$$= \frac{K_1 K_2}{[b(a^2 - \omega^2) - 2a\omega^2] + j\omega(2ab + a^2 - \omega^2)}$$

Now

$$\angle GH = -\arctan \frac{\omega(2ab + a^2 - \omega^2)}{[b(a^2 - \omega^2) - 2a\omega^2]}$$

so in order for this angle to equal $-180°$, it is necessary that

$$\omega(2ab + a^2 - \omega^2) = 0 \tag{5}$$

provided that

$$b(a^2 - \omega^2) - 2a\omega^2 < 0 \tag{6}$$

Solving Equation 5, we find

$$\omega^2 = (a^2 + 2ab) \tag{7}$$

Substituting Equation 7 into Equation 6, we obtain

$$b[a^2 - (a^2 + 2ab)] - 2a(a^2 + 2ab) = a^2 b - a^2 b - 2ab^2 - 2a^3 - 4a^2 b$$
$$= -2a(b^2 + a^2 + 2ab) < 0 \tag{8}$$

226 STABILITY AND COMPENSATION

Inequality (8) is true for any positive values of a and b, and so Equation 6 is satisfied when $\omega^2 = (a^2 + 2ab)$. We conclude that the phase angle of GH will be $-180°$ whenever $\omega^2 = (a^2 + 2ab)$. At this frequency, we find that

$$|GH| = \frac{K_1 K_2}{|b(a^2 - \omega^2) - 2\omega^2|}$$

$$= \frac{K_1 K_2}{2a(b^2 + a^2 + 2ab)} \tag{9}$$

Therefore, to ensure stability, we require

$$K_1 K_2 < 2a(b^2 + a^2 + 2ab) \tag{10}$$

This result is, of course, applicable to any combination of G and H that yields

$$GH = \frac{K_1 K_2}{(s + a)^2 (s + b)}$$

for example,

$$G = \frac{K_1}{(s + a)^2} \quad \text{and} \quad H = \frac{K_2}{s + b}$$

EXAMPLE 7.2

Suppose $G = K_1/(s + 1)(s + 12)$, $H = K_2/(s + 1)$. Find $|GH|$ and $\angle GH$ at frequencies $\omega = 1$ rad/sec, 5 rad/sec, and 12 rad/sec.

SOLUTION

(a) At $j\omega = j1$,

$$GH = \frac{K_1 K_2}{[12(1 - 1) - 2] + j1(24 + 1 - 1)} = \frac{K_1 K_2}{-2 + j24}$$

$$|GH| = \frac{K_1 K_2}{24.08} \quad \angle GH = -85.24°$$

(b) At $j\omega = j5$,

$$GH = \frac{K_1 K_2}{[12(1 - 25) - 50] + j5(24 + 1 - 25)}$$

$$|GH| = \frac{K_1 K_2}{338} \quad \angle GH = -180°$$

(c) At $j\omega = j12$,

$$GH = \frac{K_1 K_2}{[12(1 - 144) - 288] + j12(24 + 1 - 144)}$$

$$= \frac{K_1 K_2}{-2004 - j1428}$$

$$|GH| = \frac{K_1 K_2}{2461} \qquad \angle GH = -215.5°$$

In Example 7.2, we note that the frequency at which the phase angle is $-180°$ is $\omega = 5$ rad/sec $= (a^2 + 2ab)^{1/2}$, in agreement with Equation 7. We further note that we must have $K_1 K_2 < 338$ to ensure stability. This is in agreement with Equation 10, since $2a(b + a + 2ab) = 338$ in this example.

7.3 STABILITY ANALYSIS USING BODE PLOTS

Recall that a Bode plot is a graph of the log magnitude of a transfer function versus log frequency. If we include a plot of the phase angle of the transfer function, then we have a convenient way of determining the magnitude of the transfer function at the frequency where the phase angle is $-180°$. If the magnitude is greater than 1 (log 1 = 0 on the Bode plot) at that frequency, we know that we have an unstable system. The plot affords the further convenience of allowing us to determine immediately how much gain must be removed to achieve stability (or how much can be added without introducing instability), since increases in gain simply translate the entire frequency-response plot upwards on the graph.

It is rare that we would design a system with gain so large that the magnitude of G (or GH) was just barely less than unity at the $-180°$ phase-shift frequency. For reasons discussed earlier, we want to include a good safety margin of stability. Therefore, for most practical purposes, the asymptotic Bode plot, as discussed in Chapter 3, is sufficiently accurate for stability investigations. Recall that the asymptotic approximation of a term $a/(s + a)$ is flat (horizontal) out to $\omega = a$, at which point the plot changes to maintain a constant slope of -1. If we draw a composite Bode plot representing the behavior of a system with several poles, then the slope changes by an additional -1 each time we reach a new pole. For example, the asymptotic Bode plot for $ab/(s + a)(s + b)$ with $a < b$, would have slope -2 at frequencies above $\omega = b$. However, the accuracy of the asymptotic approximation becomes poor in a region where two or more poles (break frequencies) are close together. If the frequency at which the phase shift is $-180°$ is in the vicinity of break frequencies that are close together, we should not rely on the asymptotic approximation of the magnitudes, but calculate and plot them precisely.

EXAMPLE 7.3

Find the actual and asymptotic values of $|G|$ for each of the following open-loop transfer functions at $\omega = 2$ rad/sec.

(a) $G = \dfrac{(1)(3)}{(s + 1)(s + 3)}$

(b) $G = \dfrac{(1)(20)}{(s + 1)(s + 20)}$

SOLUTION

(a) $G(j2) = \dfrac{3}{(1 + j2)(3 + j2)} = \dfrac{3}{-1 + j8}$

$|G(j2)| = 0.372$

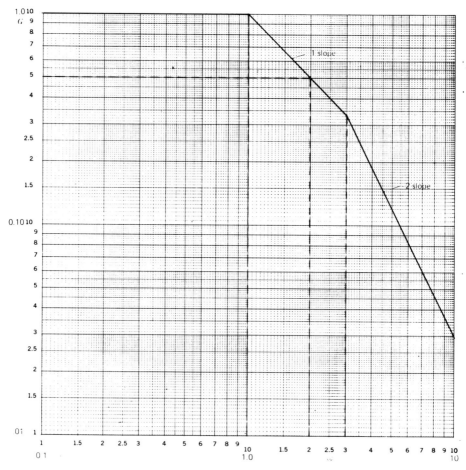

FIGURE 7.3 Asymptotic plot for $|G|$ with $G = 3/(s + 1)(s + 3)$ (Example 7.3).

At $\omega = 0$, $|G| = 1$, so the asymptotic approximation has log $|G| = \log 1$ (0 dB) out to $\omega = 1$ rad/sec. It then has slope -1, as shown in Figure 7.3. A slope of -1 corresponds to -6 dB/octave. From $\omega = 1$ to $\omega = 2$ is one octave of frequency change, and so the magnitude changes by 6 dB from 1.0 to 0.5 [6 dB $= 20 \log (1/0.5)$]. Thus, the asymptotic plot would tell us that $|G| = 0.5$ at $\omega = 2$ rad/sec. The discrepancy between actual and asymptotic values is 2.6 dB ($= 20$ [log 0.5 $- \log 0.372$]).

(b) $G(j2) = \dfrac{20}{(1 + j2)(20 + j2)} = \dfrac{20}{16 + j42}$

$|G(j2)| = 0.445$

The asymptotic approximation would be the same in this case as in part (a), $|G| = 0.5$, which we see is much closer to the actual value of $|G|$ at $\omega = 2$ (1.0-dB error). The improved accuracy is due to the fact that the break frequency $\omega = 20$ rad/sec in part (b) is much further removed from $\omega = 1$ than is the break frequency $\omega = 2$ in part (a).

Let us illustrate stability analysis using Bode plots by considering the servomotor, whose transfer function is given in Equation 25 of Chapter 6, in a control system with feedback transfer function $H = K$.

EXAMPLE 7.4

Suppose the servomotor with transfer function

$$\frac{C}{R} = \frac{K_{TI}/JL_A}{s[s^2 + (R_A/L_A)s + K_{TI}K_{V\omega}/JL_A]}$$

is used in a position control system with feedback transfer function $H = H$ (a constant frequency-independent gain).
Let

$$\frac{K_{TI}}{JL_A} = 10^3$$

$$\frac{R_A}{L_A} = 200$$

$$\frac{K_{TI}K_{V\omega}}{JL_A} = 900$$

Determine if the system is stable when $H = K = 200$.

SOLUTION

$$GH = \frac{(200 \times 10^3)}{s(s^2 + 200s + 900)} = \frac{2 \times 10^5}{s(s + 195.4)(s + 4.61)}$$

230 STABILITY AND COMPENSATION

The asymptotic Bode plot is sketched in Figure 7.4. Since the phase angle is $-180°$ at $\omega = 30$ rad/sec and since $|GH| = 1.11$ at that frequency, we conclude that the system is unstable. The gain K in the feedback must, therefore, be reduced by a factor of *at least* $1/1.11 = 0.9$. In practice, we would reduce the gain K even further in order to have some *gain margin*, a concept that will be discussed later.

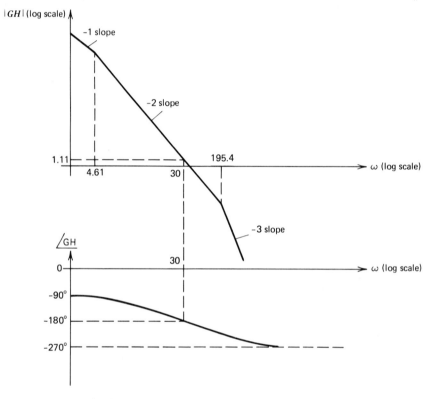

FIGURE 7.4 Bode plot of the open-loop transfer function for a servomotor in a Loop with $H = 200$ (Example 7.4).

7.4 STABILITY ANALYSIS OF THE SYSTEM WITH COMPLEX CONJUGATE POLES

In Example 7.4 of Section 7.3, the quadratic expression in the denominator was factored and we obtained two real poles. If a quadratic expression can *not* be expressed as the product of two real factors, then we know that the poles are a complex conjugate pair. Recall that in these cases it is convenient to express the quadratic in the form

$$G = \frac{\omega_n^2}{s^2 + 2\zeta\omega_n s + \omega_n^2}$$

7.4 STABILITY ANALYSIS OF THE SYSTEM WITH CONJUGATE POLES

(We write ω_n^2 in the numerator in order to normalize the gain; i.e., so that $|G| = 1$ at $j\omega = 0$.) The asymptotic Bode plot for a quadratic expression in this form is a horizontal line from $\log |G| = \log 1 = 0$ dB out to ω_n. The asymptote for frequencies greater than ω_n is a line with slope -2 intersecting the horizontal line at ω_n, as shown in Figure 7.5.

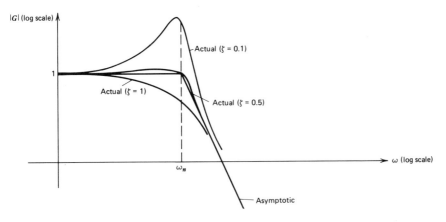

FIGURE 7.5 Log $|G|$ plots versus damping for a transfer function with quadratic denominator.

The actual gain magnitude may be found from

$$|G| = \{\sqrt{[1 - (\omega^2/\omega_n^2)]^2 + (2\zeta\omega/\omega_n)^2}\}^{-1}$$

and has maximum value

$$M_m = \frac{1}{2\zeta\sqrt{1 - \zeta^2}}$$

The approximation is reasonably accurate provided the damping ratio ζ is in the range $0.5 < \zeta < 0.7$ (recall that $\zeta < 1$ is required for complex poles). For ζ in the range indicated, the error between the actual magnitude and the asymptotic approximation does not exceed 3 dB. For ζ close to 1, the error can be as much as 6 dB at $\omega = \omega_n$, and for $\zeta \doteq 0.1$, the error is 14 dB at $\omega = \omega_n$. However, at frequencies one octave above or below ω_n, the error does not exceed 3 dB regardless of ζ, and one decade away from ω_n the error is less than 1 dB for any ζ. For $\zeta = 0.5$, the error is very small in the vicinity of ω_n and is, in fact zero, at ω_n. Once again, if the frequency where we have an overall phase shift of $-180°$ is in the vicinity of the ω_n corresponding to some quadratic expression, it would be good practice to calculate $|G|$ directly rather than rely on the asymptotes.

When plotted versus log frequency, the asymptotic phase response is a constant $0°$ through frequencies up to ω_n, where it abruptly drops to $-180°$. (The actual phase at $\omega = \omega_n$ is $-90°$.) See Figure 7.6.

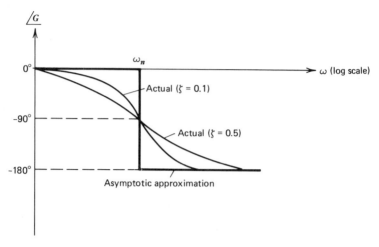

FIGURE 7.6 Phase plots versus damping for a transfer function with quadratic denominator.

As can be seen in Figure 7.6, the approximation is better for small values of ζ. Again, it is good practice to calculate the phase angle directly if the $-180°$ phase-shift frequency for the overall transfer function is in the vicinity of ω_n. The actual phase angle due to a quadratic expression may be found from

$$\theta = -\arctan\left[\frac{2\zeta\omega/\omega_n}{1 - (\omega^2/\omega_n^2)}\right] \qquad (11)$$

When terms such as s, $s + a$, and $(s^2 + 2\zeta\omega_n s + \omega_n^2)$ appear in the *numerator* of the open-loop transfer function, the asymptotic approximations are the mirror images of their counterparts in the denominator. Thus, the slope of the Bode plot becomes less negative when numerator break frequencies are reached. Example 7.5 illustrates such a case.

EXAMPLE 7.5

Determine whether the system shown in Figure 7.7 is stable using a Bode plot analysis.

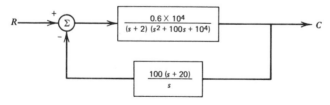

FIGURE 7.7 Control system for stability analysis of Example 7.5.

SOLUTION

$$GH = \frac{0.6 \times 10^6 (s + 20)}{s(s + 2)(s^2 + 100s + 10^4)}$$

7.4 STABILITY ANALYSIS OF THE SYSTEM WITH CONJUGATE POLES

In the quadratic term in the denominator, $\omega_n = 100$ and $2\zeta\omega_n = 100$, so $\zeta = 0.5$, and we can use the asymptotic approximation with reasonable confidence. We also note that all break frequencies (2, 20, and 100 rad/sec) are far enough apart that our asymptotic approximation will be reasonably valid. The $1/s$ term in the denominator will cause the plot to begin with a slope of -1. At $\omega = 2$, the slope will break to -2; at $\omega = 20$, the slope will break back upwards to -1; and at $\omega = 100$, it will become -3 and remain -3 thereafter. At $\omega = 1$, $|G| = 537$. The asymptotic gain magnitude and phase responses are sketched in Figure 7.8.

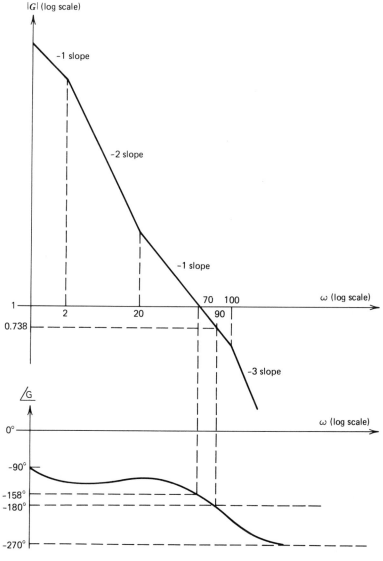

FIGURE 7.8 Bode plot for Example 7.5.

234 STABILITY AND COMPENSATION

Note that the frequency at which the total phase shift is $-180°$ is approximately $\omega = 90$ rad/sec. At this frequency, the gain magnitude is 0.738. Therefore, the system is stable. Also note that the frequency at which the gain magnitude is unity is approximately 70 rad/sec, and the phase shift at this frequency is $-158°$.

Example 7.5 illustrates a result that is generally true in control-systems theory. The inclusion of a zero (at $s = -20$) in the open-loop transfer function widened the bandwidth without compromising stability. This is the case because the zero increases the overall gain yet adds *positive* phase shift in the frequency region where it has significant influence. Thus, the overall phase shift does not reach $-180°$ quite so rapidly as frequency increases, and we can tolerate higher gains at lower frequencies. The judicious selection of a zero location can be of significant benefit, and it is one of the tools that the control-systems designer can use to improve performance without seriously affecting stability.

7.5 GAIN AND PHASE MARGINS

We have indicated that it is desirable to have a margin of stability incorporated into a control system's design, since we would not wish a small change in gain or phase shift to nudge a stable system into instability. There are two circumstances that can produce that result: (1) an increase in gain at the frequency where the open-loop phase shift is $-180°$; or (2) an increase in open-loop phase shift at the frequency of unity gain. We thus define gain and phase margins as follows:

Gain margin = M_G = the additional gain that when added to the system will cause the gain at the frequency of $-180°$ phase shift to equal unity.
Phase margin = M_P = the additional phase shift that will result in a $-180°$ phase shift at the frequency where the gain is unity.

These definitions, of course, refer to the gain and phase of the open-loop transfer function. Gain margin is often expressed in decibels, while phase margin is expressed in degrees. The Bode plot of the open-loop transfer function provides a convenient way of reading gain and phase margins directly. In Example 7.5, the gain margin is seen to be

$$M_G = 20 \log \left(\frac{1}{0.738}\right) = 2.64 \text{ dB}$$

while the phase margin is

$$M_P = (180° - 158°) = 22°$$

Thus, it would require an unexpected increase in gain of 2.64 dB (approximately 35%) or an unexpected increase of 22° in phase shift to cause the system in Example 7.5 to become unstable. Though it is difficult to generalize, these margins would be considered rather small in most practical control systems.

7.6 POLAR PLOTS

Stability criteria have also been developed in terms of polar (phasor) plots of the open-loop transfer function. The Nyquist stability criterion, discussed in the next section, is probably the best known example. In this section, we will develop some methods and general rules that aid in constructing polar plots.

A polar plot of any transfer function is simply the locus of the tip of the phasor representing the function as it changes in magnitude and angle through the frequency range from zero to infinity. Consider, for example, the transfer function

$$G = \frac{10}{s+2} = \frac{10}{j\omega + 2}$$

We know that

$$|G| = \frac{10}{\sqrt{\omega^2 + 4}} \quad \text{and} \quad \angle G = -\arctan \omega/2$$

The following table lists values for $|G|$ and $\angle G$ at selected frequencies:

ω	$|G|$	$\angle G$
0	5.00	0
1	4.47	$-26.6°$
2	3.54	$-45°$
4	2.24	$-63.4°$
8	1.21	$-76°$
16	0.62	$-86.4°$
∞	0	$-90°$

Sketching the phasors corresponding to each frequency and joining the tips of these phasors, we obtain a plot resembling the one shown in Figure 7.9. The specially designed graph paper shown in Figure 7.9 is available to facilitate the construction of polar plots. The coordinates on this paper are polar coordinates: magnitude and angle. The student should acquire this paper and use it for the examples and exercises of this and the following section.

For any transfer function of the form $G = K/(s+a)$,

since $\quad \lim_{s \to 0}|G| = K/a \quad \lim_{s \to 0}\angle G = 0$

and $\quad \lim_{s \to \infty}|G| = 0 \quad \lim_{s \to \infty}\angle G = -90°$

we can conclude that its polar plot will have the general shape shown in Figure 7.10. Using a similar analysis, we can deduce the shapes of polar plots for various other forms.

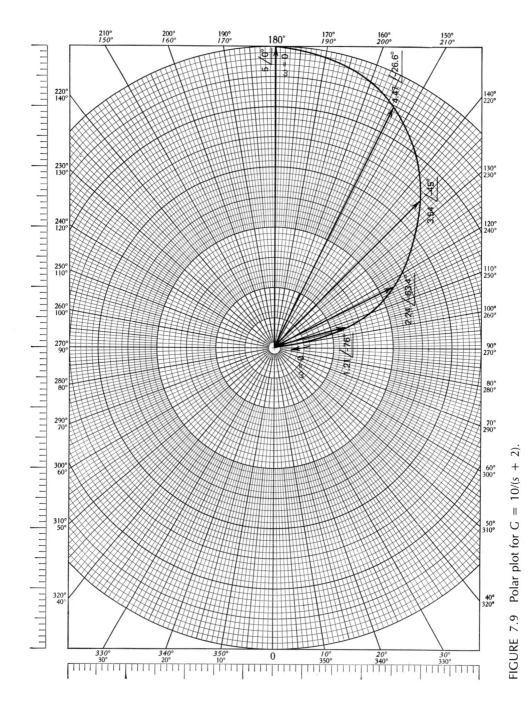

FIGURE 7.9 Polar plot for $G = 10/(s + 2)$.

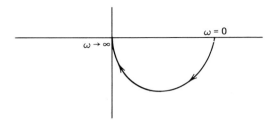

FIGURE 7.10 Polar plot for $G = K/(s + a)$.

EXAMPLE 7.6
Sketch the polar plots for the following transfer functions:

(a) $G = \dfrac{K}{(s + a)(s + b)}$

(b) $G = \dfrac{K}{(s + a)(s + b)(s + c)}$

SOLUTION

(a) We may assume without loss of generality that $b > a$.

$$G(j\omega) = \frac{K}{(j\omega + a)(j\omega + b)} = \frac{K}{(ab - \omega^2) + j\omega(a + b)}$$

$$|G| = \frac{K}{\sqrt{(ab - \omega^2)^2 + \omega^2(a + b)^2}} \qquad \angle G = -\arctan\frac{\omega(a + b)}{(ab - \omega^2)}$$

1. $\lim_{\omega \to 0} |G| = K/ab \qquad \lim_{\omega \to 0} \angle G = 0°$

2. $|G(ja)| = \dfrac{K}{a\sqrt{(b - a)^2 + (b + a)^2}} < K/ab$

   $\angle G(ja) = -\arctan\left(\dfrac{b + a}{b - a}\right)$

   $-90° < \angle G(ja) < -45° \qquad \left(\text{since } \dfrac{b + a}{b - a} > 1\right)$

3. $|G(jb)| = \dfrac{K}{b\sqrt{(a - b)^2 + (a + b)^2}} = \dfrac{a}{b}|G(ja)| < |G(ja)|$

   $\angle G(jb) = -\arctan\left(\dfrac{a + b}{a - b}\right)$

   $-180° < \angle G(jb) < -135°$

4. At $\omega = \sqrt{ab} \qquad |G| = \dfrac{K}{\sqrt{ab}(a + b)} \qquad \angle G = -90°$

5. $\lim_{\omega\to\infty}|G| = 0$ $\lim_{\omega\to\infty}\angle G = -180°$

We see that the polar plot begins at $K/ab \angle 0°$, gradually decreases in magnitude and in negative angle, and eventually approaches $0\angle -180°$. See Figure 7.11.

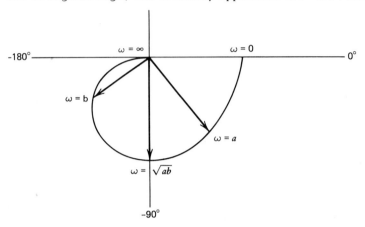

FIGURE 7.11 Polar plot for $G = K/(s + a)(s + b)$ (Example 7.6a).

It is usually not necessary to calculate all the intermediate values of $|G|\angle\theta$ that we have shown in order to obtain a sketch of the general shape of the polar plot. However, when the polar-plot technique is used in conjunction with stability analysis, it may be necessary to calculate a few critical values precisely, so we have reviewed how to go about finding angles and magnitudes at different frequencies in Example 7.6a.

(b) $G = \dfrac{K}{(a + j\omega)(b + j\omega)(c + j\omega)}$

$\lim_{\omega\to 0}|G| = K/abc$ $\lim_{\omega\to 0}\angle G = 0°$

$\lim_{\omega\to\infty}|G| = 0$ $\lim_{\omega\to\infty}\angle G = -270°$

The polar plot has the general shape shown in Figure 7.12.

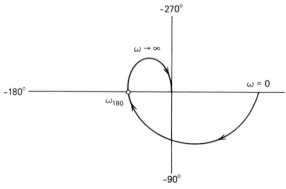

FIGURE 7.12 Polar plot for $G = K/(s + a)(s + b)(s + c)$ (Example 7.6b).

7.6 POLAR PLOTS

In Example 7.6, the frequency at which the plot crosses the negative real axis is of particular importance. This frequency is labeled ω_{180} in the plot. (ω_{180} is the frequency at which the total phase is 180°.) ω_{180} may be found by setting the imaginary part of $G(j\omega)$ equal to zero and solving for ω. It is left as an exercise (see Exercise 7.14) to show that in the present case

$$\omega_{180} = \sqrt{ab + ac + bc} \qquad (12)$$

When the transfer function contains a term of the form $(s + a)$ in the *numerator*, the effect is to increase magnitude and shift phase in a positive direction. In the frequency range where the magnitude and phase contributions of this term are changing significantly (a decade or so above and below $\omega = a$), the resulting polar plot may have indentations in it. This is illustrated in Example 7.7.

EXAMPLE 7.7

Sketch the polar plot for $G(s) = 100(s + 3)/(s + 1)(s + 20)$.

SOLUTION

$$G(j\omega) = \frac{100(3 + j\omega)}{(1 + j\omega)(20 + j\omega)}$$

$$|G| = 100 \sqrt{\frac{9 + \omega^2}{(20 - \omega^2)^2 + 441\omega^2}}$$

$$\angle G = \arctan \frac{\omega}{3} - \arctan \omega - \arctan \frac{\omega}{20}$$

$$\lim_{\omega \to 0} |G| = 15 \qquad \lim_{\omega \to 0} \angle G = 0°$$

$$\lim_{\omega \to \infty} |G| = 0 \qquad \lim_{\omega \to \infty} \angle G = -90°$$

In Figure 7.13, we have indicated the phasor magnitudes and angles at several frequencies in the vicinity of the indentation.

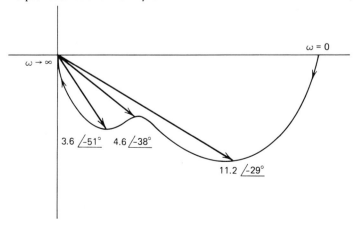

FIGURE 7.13 Polar plot for $G = 100(s + 3)/(s + 1)(s + 20)$ (Example 7.7).

A transfer function of the form $K/(s^2 + 2\zeta\omega_n s + \omega_n^2)$, with $\zeta < 1$, has a polar plot whose shape is very much dependent on the value of ζ. Since

$$|G| = \frac{K/\omega_n^2}{\sqrt{[1 - (\omega^2/\omega_n^2)]^2 + (2\zeta\omega/\omega_n)^2}} \quad \text{and} \quad \angle G = -\arctan\frac{2\zeta\omega/\omega_n}{1 - (\omega^2/\omega_n^2)}$$

then regardless of the value of ζ, we have

$$\lim_{\omega \to 0} |G| = \frac{K}{\omega_n^2} \qquad \lim_{\omega \to 0} \angle G = 0°$$

$$\lim_{\omega \to \infty} |G| = 0 \qquad \lim_{\omega \to \infty} \angle G = -180°$$

However, at $\omega = \omega_n$,

$$|G(j\omega_n)| = \frac{K}{2\zeta\omega_n^2} \quad \text{and} \quad \angle G(j\omega_n) = -90° \qquad (13)$$

In addition, the maximum value of the magnitude $|G|_{MAX}$ as well as the frequency at which it occurs ω_{GMAX} are functions of ζ

$$|G|_{MAX} = \frac{K/\omega_n^2}{2\zeta\sqrt{1 - \zeta^2}} \qquad (14)$$

$$\omega_{GMAX} = \omega_n\sqrt{1 - 2\zeta^2} \qquad (15)$$

Figure 7.14 shows the polar plots for the cases $\zeta = 0.1$ and $\zeta = 0.5$, with $K = 100$ and $\omega_n = 10$ rad/sec in each case. Phasor values are noted at several particular frequencies of interest.

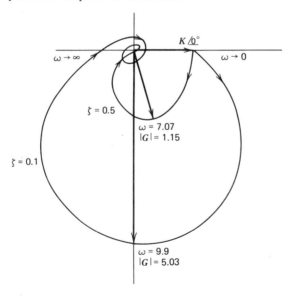

FIGURE 7.14 Polar plots versus damping for $G = K/(s^2 + 2\zeta\omega_n s + \omega_n^2)$.

7.6 POLAR PLOTS

When the transfer function includes a pure integrator (corresponding to an s in the denominator), the effect is to add a constant $-90°$ phase shift at all frequencies in addition to the usual effect of decreasing magnitude in proportion to frequency. One complication arises in finding the asymptotic magnitude when $\omega \to 0$, as illustrated in Example 7.8.

EXAMPLE 7.8

Sketch the form of the polar plot for $G = K/s(s + a)(s + b)$.

SOLUTION

$$G(j\omega) = \frac{K}{j\omega(a + j\omega)(b + j\omega)} = \frac{K}{j\omega[(ab - \omega^2) + j\omega(a + b)]}$$

$$|G| = \frac{K/\omega}{\sqrt{(ab - \omega^2)^2 + \omega^2(a + b)^2}}$$

$$\angle G = -90° - \arctan\left[\frac{\omega(a + b)}{ab - \omega^2}\right]$$

$$\lim_{\omega \to 0} |G| = \infty \qquad \lim_{\omega \to 0} \angle G = -90°$$

$$\lim_{\omega \to \infty} |G| = 0 \qquad \lim_{\omega \to \infty} \angle G = -270°$$

Even though these equations indicate that $|G|$ approaches infinity and $\angle G$ approaches $-90°$ as ω approaches zero, it is not necessarily true that the $-90°$ axis is an asymptote for the plot. Seemingly peculiar results can, nonetheless, be valid when we move an *infinite* distance in any direction. For example, if we were to travel an infinite distance down some line *parallel* to the $-90°$ axis, then our polar position relative to the origin would also be $\infty \angle -90°$. This is, in fact, the case in Example 7.8. The true asymptote, which is a line parallel to the $-90°$ axis, can be found from Equation 16

$$A = \lim_{\omega \to 0} \text{Re}[G(j\omega)] \qquad (16)$$

where Re stands for "the real part of." In Example 7.8, the real part of $G(j\omega)$ is found by rationalizing the transfer function

$$G(j\omega) = \frac{-jK/\omega}{(ab - \omega^2) + j\omega(a + b)} \cdot \left[\frac{(ab - \omega^2) - j\omega(a + b)}{(ab - \omega^2) - j\omega(a + b)}\right]$$

$$= \frac{-K}{\omega}\left[\frac{\omega(a + b) + j(ab - \omega^2)}{(ab - \omega^2)^2 + \omega^2(a + b)^2}\right]$$

Thus,

$$\text{Re}[G(j\omega)] = \frac{-K(a + b)}{(ab - \omega^2)^2 + \omega^2(a + b)^2} \qquad (17)$$

$$\text{Im}[G(j\omega)] = \frac{-K(ab - \omega^2)}{\omega[(ab - \omega^2)^2 + \omega^2(a + b)^2]} \qquad (18)$$

where Im stands for "the imaginary part of." Then, from Equation 17,

$$A = \lim_{\omega \to 0} \left[\frac{-K(a + b)}{(ab - \omega^2)^2 + \omega^2(a + b)^2} \right] = \frac{-K(a + b)}{(ab)^2} \qquad (19)$$

We note that the phase angle changes from $-90°$ to $-270°$ as frequency changes from 0 to ∞. The plot must, therefore, cross the $-180°$ axis at some frequency. As we have indicated before, this is an important frequency and can be found by setting the imaginary part of G equal to zero. From Equation 18, we have

$$ab - \omega_{180}^2 = 0 \Rightarrow \omega_{180} = \sqrt{ab} \qquad (20)$$

The polar plot for Example 7.8 is sketched in Figure 7.15.

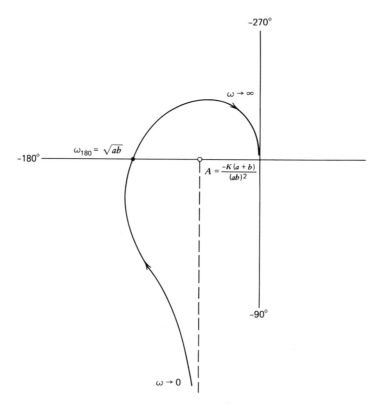

FIGURE 7.15 Polar plot for $G = K/s(s + a)(s + b)$ (Example 7.8).

The results of the examples we have studied above permit us to arrive at some rather broad conclusions about the shape and asymptotes of polar plots. We now state them as general rules for constructing polar plots. Let the general form of the open-loop transfer function be

$$G = \frac{K(s + a_1)(s + a_2) + \ldots + (s + a_m)}{s^p(s + b_1)(s + b_2) + \ldots (s + b_n)}$$

1. The power p of s determines the angle of G as $\omega \to 0$

$$p = 0 \Rightarrow \lim_{\omega \to 0} \angle G = 0°$$

$$p = 1 \Rightarrow \lim_{\omega \to 0} \angle G = -90°$$

$$p = 2 \Rightarrow \lim_{\omega \to 0} \angle G = -180°$$

In short, $\lim_{\omega \to 0} \angle G = p(-90°)$ (21)

2. $\lim_{\omega \to \infty} G = 0 \angle (m - p - n)\, 90°$ (22)

3. For $p = 1$, the asymptote as $\omega \to 0$ is found from

$$A = \lim_{\omega \to 0} [\text{Re}(G)] \quad (23)$$

4. The frequency at which the plot crosses the $-180°$ axis is found by solving for ω in the equation

$$\text{Im}[(G)] = 0 \quad (24)$$

From the standpoint of stability analysis, the most important point on the polar plot is the one where it crosses the $-180°$ axis. Therefore, after determining the frequency at which this crossing occurs, it is necessary to determine accurately the magnitude of $|G|$ at that frequency.

7.7 THE NYQUIST STABILITY CRITERION

Like Bode plots, the procedure based upon the Nyquist criterion is a graphical technique for investigating the stability of a closed-loop control system. Application of this technique requires using a polar plot of the open-loop transfer function GH, as discussed in Section 7.6. We will not develop the mathematics necessary to demonstrate the validity of the Nyquist procedure, as it involves complex-variable theory that is beyond the scope of this book. Suffice it to say that the Nyquist criterion is theoretically applicable to a wider range of control-systems types than is the Bode plot procedure. In particular, the Nyquist criterion may be applied to a control system whose open-loop transfer function has poles or zeroes in the right half of the complex plane (poles or zeroes with positive real parts). The Bode procedure is not valid for such systems.

For all of the relatively straightforward control-systems types we have considered thus far, the Nyquist stability criterion reduces to a simple observation of whether or not the polar plot of the open-loop transfer function encloses the point $-1 + j0 = -1 \angle 180°$. If it does, the system is unstable. In Example 7.9, we will reexamine a control system that has been shown to be unstable in Example 7.4.

EXAMPLE 7.9

Apply the Nyquist criterion to the control system whose open-loop transfer function was given in Example 7.4

$$GH = \frac{2 \times 10^5}{s(s + 195.4)(s + 4.61)}$$

SOLUTION

We use the rules in Section 7.7 to sketch the polar plot of GH. We note that GH is in the form of the general transfer function considered in Example 7.8, and we may, therefore, apply the results from that example.

$$\lim_{\omega \to 0} |GH| = \infty \qquad \lim_{\omega \to 0} \angle GH = -90°$$

$$\lim_{\omega \to \infty} |GH| = 0 \qquad \lim_{\omega \to \infty} \angle GH = -270°$$

From Equation 19,

$$A = \frac{-K(a + b)}{(ab)^2} = -49.3$$

From Equation 20,

$$\omega_{180} = \sqrt{ab} = \sqrt{(195.4)(4.61)} = 30 \text{ rad/sec}$$

At $\omega = \omega_{180} = 30$, we find by using the formula for $|G|$ that we developed in Example 7.8 that $|G(j30)| = 1.11$.

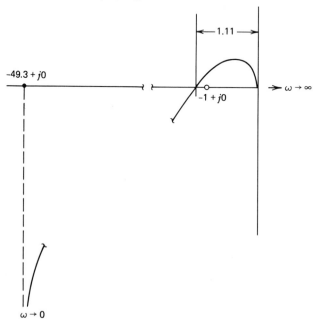

FIGURE 7.16 Application of the Nyquist criterion (Example 7.9).

The polar plot is sketched in Figure 7.16. Since the polar plot encloses −1 + j0, we conclude again that we have an unstable system.

EXAMPLE 7.10

Apply the Nyquist criterion to a control system with open-loop transfer function

$$G(s) = \frac{200}{(s + 1)(s + 5)(s + 10)}$$

SOLUTION

$$\lim_{\omega \to 0} |G| = 4 \qquad \lim_{\omega \to 0} \angle G = 0°$$

$$\lim_{\omega \to \infty} |G| = 0 \qquad \lim_{\omega \to \infty} \angle G = -270°$$

From Equation 12,

$$\omega_{180} = \sqrt{ab + ac + bc} = 8.06 \text{ rad/sec}$$

$$G(j8.06) = \frac{200}{(1 + j8.06)(5 + j8.06)(10 + j8.06)}$$

$$= \frac{200}{(8.12)(9.49)(12.84)} = 0.202$$

The polar plot is sketched in Figure 7.17.

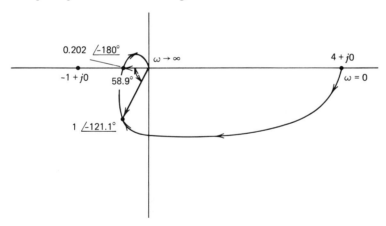

FIGURE 7.17 Application of the Nyquist criterion (Example 7.10).

We see the system is stable. The gain margin is

$$M_G = \frac{1}{0.202} = 4.95 = 13.89 \text{ dB}$$

A careful, accurately scaled plot would allow us to measure the angle of the G phasor where its measured length equals one. In Figure 7.17, this phasor is seen to have angle $-121.1°$. Thus, the phase margin is $58.9°$.

246 STABILITY AND COMPENSATION

Some systems with more complex transfer functions have the property that they become unstable if the gain is either increased *or* decreased. Such a system is said to be *conditionally stable*. The Nyquist procedure provides a very vivid means for identifying such a system. Consider, for example, the open-loop transfer function

$$GH = \frac{K(s + a)}{(s + b)(s + c)(s + d)(s + e)}$$

For certain values of the open-loop poles and zeroes and a sufficiently large gain K, a careful sketch of the polar plot will reveal a situation resembling that in Figure 7.18.

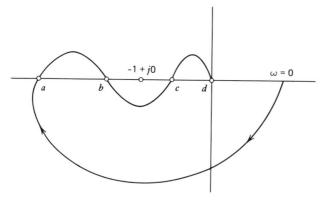

FIGURE 7.18 Application of the Nyquist criterion to a conditionally stable system.

Note that the $(-1 + j0)$ point is not enclosed by the polar plot, so the system is stable. However, an increase in gain will cause the $(-1 + j0)$ point to fall between points c and d on the sketch and thus be enclosed. Also, a decrease in gain will cause the $(-1 + j0)$ point to fall between points a and b and again be enclosed. Thus, we have a conditionally stable system.

7.8 THE NICHOLS CHART

Another graphical method that is widely used for investigating control systems performance is based upon the *Nichols chart*. This is a specially designed chart that allows the user to determine the values of closed-loop gain and phase that correspond to plotted values of open-loop gain and phase. Study the Nichols chart shown in Figure 7.19. Note that the vertical axis is labeled "20 log|G|", so this axis represents values of the open-loop gain magnitude in decibels. Note also that these dB values are plotted on a linear scale. The horizontal scale is labeled "phase G" and represents the phase angle of the open-loop transfer function plotted on a linear scale. The Nichols chart is symmetrical about the vertical axis,, so the lines corresponding to angles from $-60°$ to $-180°$ may also be used for angles from $-180°$ to $-300°$, respectively.

The curves on the Nichols chart are lines of constant closed-loop gain magnitude in decibels and closed-loop phase angle. These curves have been constructed using the fact that any value of open-loop gain G uniquely defines a corresponding closed-loop value: $G' = G/(1 + G)$. The Nichols chart, therefore, has the very useful and convenient feature that a closed-loop gain can be read directly from the chart

7.8 THE NICHOLS CHART

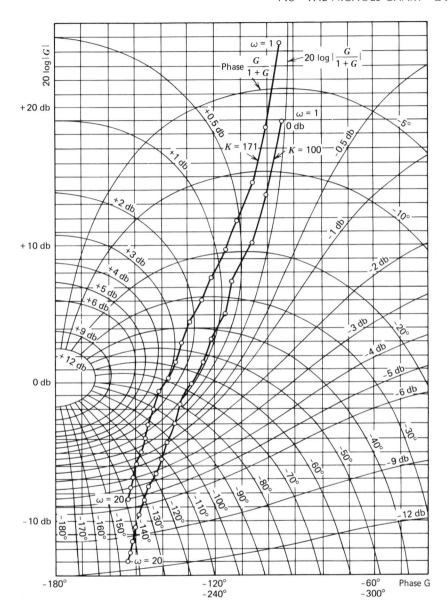

FIGURE 7.19 Nichols chart $G = K/s(s + 10)$.

simply by locating the coordinate corresponding to open-loop gain magnitude (in dB) and open-loop phase angle. For example, suppose $G = 2.24 \angle -130°$. Then, $20 \log|G| = 7$ dB. Locate the intersection of the 7-dB open-loop phase line on the Nichols chart. Verify that the closed-loop gain magnitude is 2 dB and that its phase angle is approximately $-26°$. Thus, $G' = 1.26 \angle -26°$.

When the values of $20 \log|G|$ and $\angle G$ are calculated for different values of frequency and then plotted on the Nichols chart, we see that the result may also be read as a plot of closed-loop gain $20 \log|G'|$ and $\angle G'$ as frequency varies. We,

248 STABILITY AND COMPENSATION

therefore, obtain a useful graphical display of the closed-loop behavior of the system. It is easy to determine whether the system is stable simply by determining whether or not the open-loop gain magnitude is greater than 0 dB at $-180°$. If the system is stable, the gain and phase margins can be read directly from the chart. The gain margin in dB is the value of open-loop gain where the plot intersects the $-180°$ open-loop phase line, and the phase margin is the difference between $-180°$ and the open-loop phase angle where the plot intersects the 0-dB open-loop gain line.

One criterion that is often used to optimize the performance of a control system is the degree to which the output *overshoots* its ultimate (steady-state) value in response to a step input. We know that a heavily damped system tends to have a sluggish response and no overshoot at all. On the other hand, a very lightly damped system responds quickly and has considerable overshoot. The magnitude of the overshoot in such a system may be unacceptably large, and the system may, in fact, overshoot several times (exhibit a damped oscillatory response) before settling into its steady-state value. Many practical systems are optimized by reaching a compromise between quickness of response and degree of overshoot. As a rule of thumb, it has been determined that an overshoot of approximately 30 to 40% (between 2 and 3 dB) is a good compromise. Since the Nichols chart plot displays closed-loop gain values, it provides a convenient means of determining overshoot magnitude and for adjusting the gain of a system to achieve a desired overshoot.

To illustrate the use of the Nichols chart for system optimization, we will first consider the control system with open-loop transfer function of the general form

$$G = \frac{K}{s(s + a)}$$

This choice of G leads to the closed-loop gain function

$$G' = \frac{K}{s^2 + as + K}$$

The denominator is a quadratic, which, as before, we can identify with the general quadratic form

$$s^2 + 2\zeta\omega_n s + \omega_n^2$$

There will be *no* overshoot for values of ζ greater than $\sqrt{2}/2 \simeq 0.707$. When ζ is less than 0.707, an overshoot occurs, and its maximum value is conventionally termed M_m. By plotting gain values on the Nichols chart as frequency is allowed to vary, we can determine the value of M_m simply by observing the maximum closed-loop value of gain that the plot reaches. (For $\zeta > 0.707$, the plot will never cross the line of 0-dB closed-loop gain.) Since changing the open-loop gain simply amounts to moving the Nichols chart plot up or down, we can conveniently determine the additional gain required to achieve a desired M_m (between 2 and 3 dB). Simply measure the total vertical translation necessary to make the plot intersect the required M_m. This translation, measured along the vertical axis, is the total dB gain that should be added or removed from the system.

Suppose we want a 3-dB overshoot in a system whose open-loop transfer function is

$$G = \frac{100}{s(s + 10)}$$

To use the Nichols chart, we must first calculate values of $20 \log|G|$ and $\angle G$ at different frequencies. (These computations are facilitated using a simple computer program, as will be demonstrated in Chapter 10.) Table 7-1 shows the results of the required computations for frequencies between 1 and 20 rad/sec.

TABLE 7-1 VALUES OF $\angle G$ AND $20 \log|G|$ FOR NICHOLS CHART PLOT $G = 100/s(s + 10)$

ω(rad/sec)	$|G|$	$\angle G$ (degrees)	$20 \log|G|$
1	9.95	−95.7	19.96
2	4.90	−101.3	13.81
3	3.19	−106.7	10.08
4	2.32	−111.8	7.31
5	1.79	−116.6	5.05
6	1.43	−121.0	3.10
7	1.17	−125.0	1.37
8	0.976	−128.7	−0.210
9	0.825	−132.0	−1.66
10	0.707	−135.0	−3.01
11	0.711	−137.7	−4.27
12	0.533	−140.2	−5.46
13	0.469	−142.4	−6.58
14	0.415	−144.5	−7.64
15	0.369	−146.3	−8.64
16	0.331	−148.0	−9.60
17	0.298	−149.5	−10.51
18	0.270	−150.9	−11.38
19	0.245	−152.2	−12.21
20	0.224	−153.4	−13.01

The data from Table 7-1 is plotted on the Nichols chart in Figure 7.19 (labeled $K = 100$). The peak-overshoot M_m is seen to be slightly over +1 dB and occurs for ω between 8 and 9 rad/sec. (The exact values are +1.25 dB at 8.66 rad/sec.) The open-loop phase where the plot crosses the 0-dB open-loop gain line is approximately −128°, so the phase margin is 180 − 128 = 52°. To achieve a value of M_m = +3 dB, approximately 4.6 dB of gain must be added to the system. This can be seen from the second plot in Figure 7.19, which represents the original plot translated upward by 4.6 dB. (Translation is best accomplished using a pair of dividers.) Note that the second line yields the desired value of M_m (3 dB). Note also that this 3-dB value occurs on the translated plot at ω equals approximately 12 rad/sec. It is *not* correct to add an amount of gain equal to 3 dB minus the old

250 STABILITY AND COMPENSATION

value of M_m. If we had done that, we would have added only $3 - 1.25 = 1.75$ dB of gain to the system. Note that adding gain to the open-loop system does not increase the closed-loop gain (which is what we are interested in changing) by the same amount.

In this example, we saw that it was necessary to increase the open-loop by 4.6 dB to achieve 3 dB of overshoot. This means that K must be increased from 100 to approximately 171. Note that on the Nichols chart the new phase margin resulting from the increased gain is approximately $180 - 138 = 42°$. This reduction in phase margin (from 52°) is to be expected, since we have once again nudged the system slightly in the direction of instability for the sake of improved response.

As another illustration of the use of the Nichols chart, we will investigate the system with open-loop transfer function

$$G = \frac{K(s + 100)}{s(s + 10)(s + 40)}$$

Table 7-2 shows calculated values of $|G|$, $\angle G$, and $20 \log|G|$, when $K = 500$.

TABLE 7-2 VALUES OF G AND 20 log|G| FOR NICHOLS CHART PLOT
$G = 500(s + 100)/s(s + 10)(s + 40)$

ω(rad/sec)	|G|	∠G (degrees)	20 log|G|
6	17.70	−126.1	25.00
7	14.45	−130.9	23.19
8	12.00	−135.4	21.59
9	10.11	−139.5	20.10
10	8.62	−143.3	18.71
11	7.41	−146.8	17.40
12	6.43	−150.1	16.17
13	5.62	−153.0	15.00
14	4.95	−155.8	13.89
15	4.38	−158.3	12.82
16	3.89	−160.7	11.81
17	3.48	−162.9	10.83
18	3.13	−165.0	9.90
19	2.82	−166.9	9.00
20	2.55	−168.7	8.13
21	2.31	−170.4	7.29
22	2.11	−172.0	6.48
23	1.93	−173.4	5.70
24	1.77	−174.8	4.94
25	1.62	−176.2	4.21
26	1.50	−177.4	3.49
27	1.38	−178.6	2.80
28	1.28	−179.7	2.13
29	1.18	−180.7	1.47

7.8 THE NICHOLS CHART

The Nichols chart plot of values taken from Table 7-2 is shown in Figure 7.20 (labeled $K = 500$). The plot is seen to intersect the $-180°$ phase line with 20 log $|G|$ equal to approximately 2 dB, and the system is clearly unstable. Shown in the same figure is the plot that results when the open-loop gain is reduced by 20 dB (to $K = 50$). We note that this gain reduction produces a stable system with phase margin approximately 40°. The M_m value is between 3 and 4 dB. In practice, it would probably be desirable to reduce the open-loop gain somewhat more to improve the phase margin and reduce the overshoot.

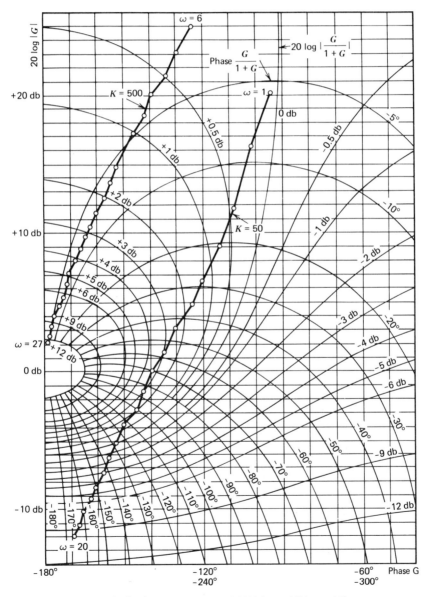

FIGURE 7.20 Nichols chart $G = K(s + 100)/s(s + 10)(s + 40)$.

7.9 ROOT LOCUS

Still another widely used technique for investigating stability is the root-locus method. This method requires plotting (in the complex plane) all the possible solutions of the characteristic equation $1 + GH = 0$ as gain K varies from zero to infinity. That is, we draw a curve through all of the possible roots of $1 + GH$ for all of the possible values of gain. All of these possible roots correspond to possible poles of the closed-loop transfer function $C/R = G/(1 + GH)$. The criterion for stability is simply that no value of gain should be used that results in any root of $1 + GH$ lying on the imaginary axis (except possibly at the origin) or in the right half of the complex plane. The principal advantage of the root-locus method is that it not only tells whether or not a system is stable, but also allows us to determine quite readily the effect that gain changes have on overall system performance. Before elaborating on this point, we will consider a simple example to demonstrate how the roots of a characteristic equation change with changes in gain.

EXAMPLE 7.11

Plot the locus of all roots of the characteristic equation for the system in Figure 7.21.

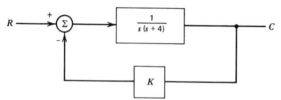

FIGURE 7.21 Control system for root-locus investigation (Example 7.11).

SOLUTION

In this case, $G = 1/s(s + 4)$ and $H = K$. So the characteristic equation is

$$1 + \frac{K}{s(s + 4)} = 0$$

or

$$s^2 + 4s + K = 0 \tag{25}$$

For $K = 0$, $s^2 + 4s = 0 \Rightarrow s(s + 4) = 0$, and we see that the roots are at $s = 0$ and $s = -4$. [More precisely, in the complex plane, these roots would be plotted at the origin $(0 + j0)$ and at the point $(-4 + j0)$ on the negative real axis.] In general, the roots may be found for any K by using the quadratic formula

$$s = \frac{-4 \pm \sqrt{4^2 - 4K}}{2} = -2 \pm \sqrt{4 - K}$$

We see that all roots will be *real* provided $K \leq 4$. For example, with

$K = 1$ $s = -2 \pm \sqrt{3} = -0.27, -3.73$
$K = 2$ $s = -2 \pm \sqrt{2} = -0.59, -3.41$
$K = 3$ $s = -2 \pm 1 = -1, -3$
$K = 4$ $s = -2$ (repeated twice)

We notice that these roots move closer together along the negative real axis when K is increased, as illustrated in Figure 7.22.

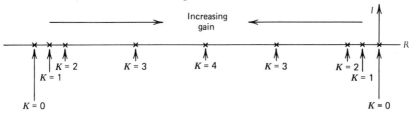

FIGURE 7.22 Root locus for $K \leq 4$ (Example 7.11).

We use the symbol X in Figure 7.22 to represent the root locations. It is conventional to use this symbol for a pole, and, of course, each root *is* a pole of the closed-loop transfer function. Note that we have only plotted a few of the possible pole locations for a few possible values of gain between 0 and 4. The root locus up to this point is a continuous line between -4 and 0 on the real axis, representing the infinitely many possible pole locations for all values of K, with $4 < K < 0$.

Now if $K > 4$, we will obtain complex conjugate roots. For example, with

$K = 5 \quad s = -2 \pm j$
$K = 8 \quad s = -2 \pm 2j$
$K = 13 \quad s = -2 \pm 3j$
$\equiv \quad\quad \equiv$
$K = K_1 > 4 \quad s = -2 \pm j\sqrt{K_1 - 4}$

We observe that as the gain K is increased, the real part of the complex roots remains fixed at -2, while the imaginary parts become larger and larger in opposite directions. These roots are, in effect, moving further apart from each other along a vertical line through $-2 + j0$. Combining this part of the locus with the part for $K < 4$, we obtain the complete root locus as shown in Figure 7.23. The roots are seen to come together at -2, then separate, and move in opposite directions indefinitely along a line parallel to the imaginary axis as the gain K is increased toward infinity. Since there is no value of K that results in a root lying on the imaginary axis (other than the origin) nor in the right-half plane, we conclude that the system is stable for all values of gain.

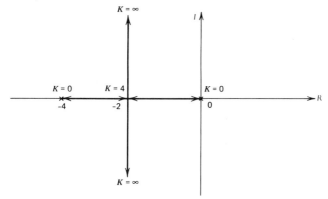

FIGURE 7.23 Complete root locus (Example 7.11).

Let us evaluate system performance in terms of the location of the roots in the complex plane. Toward this end, we will generalize the open-loop transfer function discussed in Example 7.12.

Consider the system shown in Figure 7.24. This leads to the closed-loop trans-

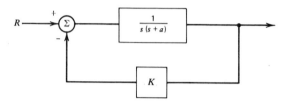

FIGURE 7.24 Generalized control system for root-locus investigation.

fer function $C/R = 1/(s^2 + as + K)$, with corresponding characteristic equation

$$s^2 + as + K = 0 \tag{26}$$

Expressing Equation 26 in the more general form,

$$s^2 + 2\zeta\omega_n s + \omega_n^2 = 0 \tag{27}$$

we see that

$$\omega_n^2 = K \quad \text{or} \quad \omega_n = \sqrt{K} \tag{28}$$

and

$$2\zeta\omega_n = a \quad \text{or} \quad \zeta = \frac{a}{2\sqrt{K}} \tag{29}$$

Recall that ω_n and ζ are related to the (underdamped) response to a step input when the damping factor ζ is less than one. The response includes a time-domain expression of the form

$$e^{-\zeta\omega_n t} \sin \sqrt{1 - \zeta^2}\, \omega_n t \tag{30}$$

which is a damped sine wave. Also recall that the solutions of Equation 26 are found from the quadratic formula to be

$$s = \frac{-2\zeta\omega_n \pm \sqrt{4\zeta^2\omega_n^2 - 4\omega_n^2}}{2}$$

or when $\zeta < 1$,

$$s = -\zeta\omega_n \pm j\omega_n\sqrt{1 - \zeta^2} \tag{31}$$

These roots are shown in the complex plane in Figure 7.25. By comparing this figure with Equation 30, we see that the further the root is to the left of the imaginary axis (i.e., the greater the value of $\zeta\omega_n$), the faster the sinusoidal response dampens out or decays to zero. Also, the further the zero is above the real axis (i.e., the greater the value of $\omega_n\sqrt{1 - \zeta^2}$), the higher the frequency of oscillation of the

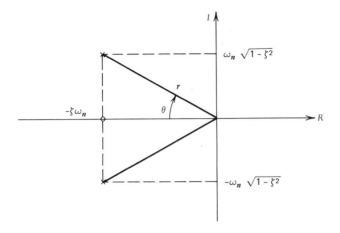

FIGURE 7.25   Location of roots for the system in Figure 7.22.

sinusoidal response. We further note that the length r of a line drawn from the origin to a pole is, by the Pythagorean theorem, equal to

$$r = \sqrt{(-\zeta\omega_n)^2 + (\omega_n\sqrt{1-\zeta^2})^2} = \omega_n$$

regardless of the value of ζ. Now referring once again to Figure 7.25, we see that

$$\cos\theta = \frac{\zeta\omega_n}{r} = \frac{\zeta\omega_n}{\omega_n} = \zeta$$

Thus, the closer the root is to the imaginary axis, the greater the angle θ, and, therefore, the smaller the value of ζ (since $\cos\theta$ decreases as θ increases). If it were possible for ζ to become zero (no damping), then the roots would fall directly on the imaginary axis. With zero damping, we would have a sustained oscillation at frequency ω_n rad/sec and no damping out. We would, thus, have an unstable system.

For the cases where $\zeta > 1$, the rooots lie on the real axis, and we have the overdamped case. The time-domain response is then the sum of two exponentials whose time constants are inversely related to the distance each root is from the origin. There are no oscillations in this case. The further the roots are from the origin, the faster the transient portion of the response will decay to zero. Finally, if the roots occupy the same position on the real axis, we have a repeated root and a critically damped response.

Summarizing these interpretations of root locations, we can state that it is generally desirable for roots of the open-loop transfer function to be as far to the left of the imaginary axis as possible. This results in faster response times and ensures that the system is well removed from the region where instability occurs. In Figure 7.26, we indicate the nature of time-domain responses to step inputs that correspond to various root locations.

256 STABILITY AND COMPENSATION

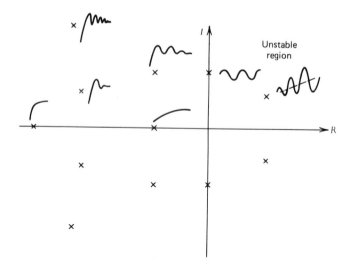

FIGURE 7.26 Effect of root locations on time-domain response.

Let us now examine the root locus corresponding to some other typical open-loop transfer functions. We first consider the case where a zero has been added to the transfer function we just studied.

Let

$$GH = \frac{K(s + b)}{s(s + a)}$$

Then

$$1 + GH = 0 \Rightarrow s^2 + (a + K)s + Kb = 0 \tag{32}$$

For $K = 0$,

$$s^2 + as = 0 \Rightarrow s = 0, -a$$

From the quadratic formula,

$$s = \frac{-(a + K) \pm \sqrt{(a + K)^2 - 4Kb}}{2}$$

and we see that both roots lie on the real axis provided $(a + K)^2 > 4Kb$.

EXAMPLE 7.12

Sketch the root locus for the characteristic equation of the system in Figure 7.27.

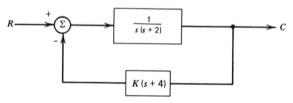

FIGURE 7.27 Control system for root-locus investigation (Example 7.12).

SOLUTION

From Equation 32,

$$s^2 + (2 + K)s + 4K = 0$$

$$K = 0 \Rightarrow s = 0, \quad -2$$

$$s = \frac{-(2 + K) \pm \sqrt{(2 + K)^2 - 16K}}{2} = \frac{-(2 + K) \pm \sqrt{K^2 - 12K + 4}}{2}$$

$$K = \frac{1}{3} \Rightarrow s = \frac{-2\frac{1}{3} \pm \frac{1}{3}}{2} = -1, \quad \frac{-4}{3}$$

$$K = 4 \Rightarrow s = -3 \pm j\sqrt{7}$$

$$K = 12 \Rightarrow s = \frac{-7 \pm 2}{2} = -2.5, \quad -4.5$$

We notice that as the gain is increased, the roots go from real to complex and back to real. A careful analysis (Exercise 7.21) would show that the roots are real for $0 \leq K \leq (6 - 4\sqrt{2})$ and for $K \geq (6 + 4\sqrt{2})$. In the limit as $K \to \infty$, one real root approaches $-\infty$ and the other approaches -4. The locus is sketched in Figure 7.28. [The conventional symbol for a zero (0) is shown on the plot at $s = -4$.]

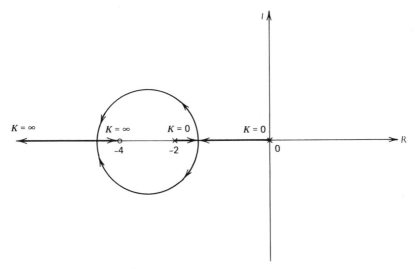

FIGURE 7.28 Root locus for Example 7.12.

Clearly, no value of gain will cause instability.

A significant result is illustrated in Example 7.12, where we see that the addition of a zero to the open-loop transfer function has the effect of generally shifting the entire root locus to the left in comparison to the locus of the previous example. In accordance with our interpretation of root locations, this means that we have a more stable, faster responding system. Recall that we arrived at a similar conclusion when studying the effect of a zero on the Bode plot.

258 STABILITY AND COMPENSATION

Consider now an open-loop transfer function of the form

$$GH = \frac{K}{s(s + a)(s + b)}$$

The characteristic equation is

$$s(s + a)(s + b) + K = 0 \tag{33}$$

For $K = 0$, the roots are $s = 0$, $s = -a$, and $s = -b$. As we might expect from our previous analyses of this type of transfer function, there are values of gain that will cause this system to become unstable. We will not involve ourselves with the algebra required to find the roots of Equation 33 for various values of K. In a subsequent section, we will discuss the availability of techniques other than algebraic for sketching a root locus for the more complex transfer functions. Suffice it for our purposes now to display the general shape of the root locus of Equation 33 (see Figure 7.29). We see that there are, indeed, values of gain that force the root locus into the right-half plane and, therefore, result in unstable systems. We also note that the overall effect of adding an additional pole (at $s = -b$) has shifted the root locus to the right in comparison with Example 7.11, resulting in a generally slower responding and less stable system.

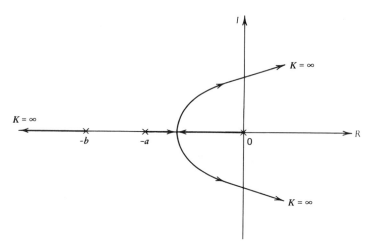

FIGURE 7.29 Root locus for Equation 33.

Suppose now that the open-loop transfer function is of the form

$$GH = \frac{K(s + a)}{s^2 + 2\zeta\omega_n s + \omega_n^2} \quad \text{with } \zeta < 1$$

Then

$$1 + GH = 0 \Rightarrow s^2 + (2\zeta\omega_n + K)s + \omega_n^2 + Ka = 0 \tag{34}$$

Since $\zeta < 1$, $K = 0$ implies that Equation 34 has a pair of complex conjugate roots

$$s = -\zeta\omega_n \pm j\omega_n\sqrt{1 - \zeta^2}$$

As K is increased, the roots remain complex until K is sufficiently large, so that

$$(2\zeta\omega_n + K)^2 > 4(\omega_n^2 + Ka)$$

As $K \to \infty$, the roots approach $-\infty$ and $-a$. The locus is sketched in Figure 7.30.

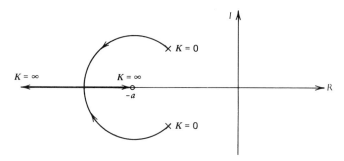

FIGURE 7.30 Root locus for Equation 34.

The system is seen to be stable for all K. Again, the removal of the zero in the transfer function of this example and the addition of a pole will cause the system to be less responsive and less stable, as evidenced by the locus plotted in Figure 7.31 for the transfer function

$$GH = \frac{K}{(s + a)(s^2 + 2\zeta\omega_n s + \omega_n^2)} \quad \text{with } \zeta < 1$$

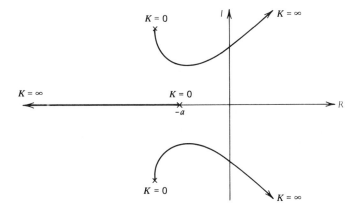

FIGURE 7.31 Root locus modified by removal of a zero and addition of a pole.

260 STABILITY AND COMPENSATION

There is a great deal of literature describing techniques that have been developed to simplify sketching and interpreting root loci. The method was originated by W. R. Evans, who invented a device called the spirule for purposes of simplifying root-locus plots. This device resembles a transparent, plastic combination of a ruler and a protractor and has been widely used by control-system engineers and designers. Although we are not able in this book to devote the space required for a thorough discussion of techniques for plotting root loci or the details of the spirule, it would behoove the serious student of control-system theory to consult the literature* on these topics. Insofar as graphical techniques for investigating system stability and control-systems performance are concerned, the root-locus method is probably the most useful and most widely used.

7.10 COMPENSATION

When the term compensation is used in connection with control-systems theory, it means the improvement of system performance through the addition of carefully selected frequency-sensitive components in appropriate locations. The reader who has experience in the design of high-gain amplifiers will recognize compensation as a term applied to the procedure of rolling off the amplifier by the use of reactive networks in order to reduce its gain at high frequencies and thus prevent it from oscillating. In an analogous manner, compensation of a control system would certainly include whatever procedure is required to improve its performance by eliminating instability. In a broader context, compensation of a control system may also include improving other characteristics such as response time and error coefficients. In practice, the design and selection of compensators usually takes place after the basic control system has been constructed. One usually has little or no control over the specification of many of the basic system parameters: mechanical properties of the load, motor size, amplifier gain, and so forth. It then becomes the designer's job to *compensate* after the fact for any undesirable performance characteristics that are inherent in the basic system. We cannot begin to cover this topic in the depth it deserves, for herein lies the *art* of control-systems design, and much has been written about a great variety of compensating techniques. We will content ourselves with an introduction to two popular and straightforward approaches to the compensation problem: lead compensators and lag compensators placed in the forward loop of the control system.

Lead compensation in the forward loop is used to improve the responsiveness of a control system. As we will presently see, the addition of a lead compensator to a system results in the addition of a zero to the open-loop transfer function, and we have observed on several occasions the benefits that such a zero produces. The term lead is used because this type of compensator has the basic property that its output leads its input; that is, the output phase angle for sinusoidal inputs is more positive than the phase angle of the input, at least over some frequency range. In a more general sense, the output of a lead compensator is proportional to the time rate of change, or derivative, of the input. We know that the output of a differentiator

---

*See, for example, Reference 2, or Appendix D of Reference 1.

leads its input, and, indeed, a differentiator with transfer function Ks is an example of a lead compensator.

Lead compensation may be accomplished by placing a lead network, or differentiator, around an amplifier whose input is the error signal developed at the feedback junction. This is illustrated in Figure 7.32. When the output C of the

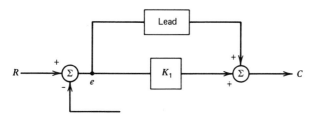

FIGURE 7.32 Lead compensation.

control system has the correct value for a given input R, then the error is zero, and the lead compensator contributes nothing. Also, for slowly varying changes in input and output, the lead compensator has little effect. However, if a sudden and rapid change in position is required corresponding to a sudden and rapid change in the input R or if a sudden and rapid disturbance at the output causes the output to deviate from its prescribed value, then the error e will undergo a sudden and rapid change. The lead compensator, whose output is proportional to the rate of change of the input, will then produce a large signal that is added to the forward control signal that ultimately drives the output. The effect, then, is a sudden increase in gain, which causes the output to respond rapidly to the sudden change required.

Having explained the purpose of the lead compensator in a qualitative manner, we will now refine our mathematical model of the system so we can study the effects of the compensator quantitatively. Consider the fully fed-back control system with lead compensator $K_1 s$ connected as shown in Figure 7.33. The open-loop transfer function is $G(K + K_1 s)$. Therefore,

$$\frac{C}{R} = \frac{G(K + K_1 s)}{1 + G(K + K_1 s)} \tag{35}$$

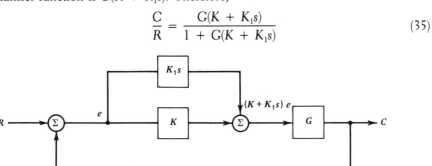

FIGURE 7.33 Lead compensation using $K_1 s$ compensator.

If, for example, $G = 1/s(s + a)$, then the characteristic equation is

$$1 + \frac{K + K_1 s}{s(s + a)} = 0 \Rightarrow s^2 + (a + K_1)s + K = 0 \tag{36}$$

262 STABILITY AND COMPENSATION

Note the addition of the zero due to the term $(K + K_1 s)$.

Suppose we have an open-loop transfer function of the form $G = K/s(s + a)(s + b)$. We have already seen that this type of system may be unstable for sufficiently large values of K. The root locus is repeated in Figure 7.34. If a zero is added to this system between $s = -a$ and $s = -b$ by using a lead compensator, the root locus is improved considerably, as seen in Figure 7.35. We see that the root locus has been shifted to the left and that we have eliminated any possibility of instability.

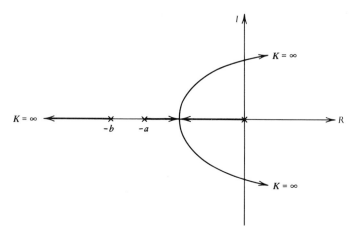

FIGURE 7.34 Root locus for uncompensated system with $G = K/s(s + a)(s + b)$.

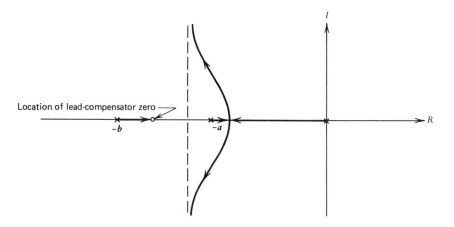

FIGURE 7.35 Root locus for lead-compensated system.

EXAMPLE 7.13

Design a lead compensator for the control system shown in Figure 7.36.

7.10 COMPENSATION

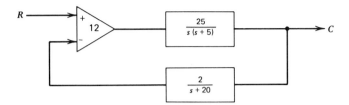

FIGURE 7.36 Control system to be lead-compensated (Example 7.13).

SOLUTION

A common choice for the location of the zero contributed by the lead compensator is the geometric mean of a and b, namely, $\sqrt{ab}$. In Example 7.13,

$$\sqrt{ab} = \sqrt{(5)(20)} = 10$$

so we will choose the gain of our lead compensator in such a way that a zero is added at $s = -10$. The compensator will be placed around the differential amplifier, so that we obtain the modified system shown in Figure 7.37. The open-

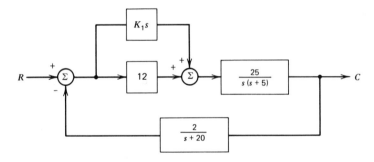

FIGURE 7.37 Lead-compensated system (Example 7.13).

loop transfer-function of this system is

$$GH = (K_1 s + 12)\left[\frac{50}{s(s+5)(s+20)}\right]$$
$$= \frac{50K_1(s + 12/K_1)}{s(s+5)(s+20)}$$

In order for the zero to occur at $s = -10$, we require

$$\frac{12}{K_1} = 10 \quad \text{or} \quad K_1 = 1.2$$

Hence, the gain of our differentiator should be set to 1.2. Operational-amplifier circuits can be designed to perform differentiation with a specified gain.

264 STABILITY AND COMPENSATION

One of the disadvantages of using a pure differentiator for a lead compensator is that it effectively amplifies high frequency noise inputs. Since broad-band noise is present in every electronic system and since the differentiator produces an output whose amplitude is directly proportional to the frequency of the input, we might experience unacceptable noise amplification at some higher frequencies. In addition to this consideration, the pure differentiator must be constructed from *active* circuitry, with its attendant increase in power consumption, cost, and complexity. Consequently, instead of a true differentiator, we can frequently use a passive RC-network to perform approximate differentiation over a limited frequency range. This strategy eliminates both the high frequency noise problem and the need for complex circuitry.

An RC-network configuration that is often used for this purpose is shown in Figure 7.38. The transfer function of this network was derived in Example 3.20 (Equation 16 in Chapter 3). We found that

$$G(j\omega) = \frac{(1/T) + j\omega}{(1/\alpha T) + j\omega}$$

where $\alpha = R_1/(R_1 + R_2)$ and $T = R_2C$.

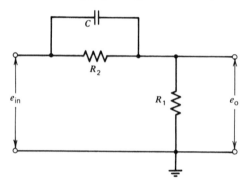

FIGURE 7.38 An RC-lead-network compensator.

The asymptotic Bode plot, also shown in Chapter 3, was found to produce a differentiating characteristic (output amplitude proportional to input frequency) over the frequency range from $\omega_1 = 1/R_2C$ to $\omega_2 = \omega_1/\alpha$. Writing the transfer function in terms of s, ω_1, and ω_2, we obtain

$$G(s) = \frac{\omega_1 + s}{\omega_1/\alpha + s} = \frac{s + \omega_1}{s + \omega_2} \tag{37}$$

The correct application of this network requires it to be placed in series with the forward loop, as, for example, at the input of a high input-impedance amplifier (see Figure 7.39). Therefore, the transfer function of the lead network simply multiplies the transfer function of the remainder of the open loop. The net effect of the lead network then by Equation 37 is to add both a pole and a zero to the open-loop transfer function. In order to prevent the additional pole from having detrimental effects on the transient response, it is necessary for it to occur far to

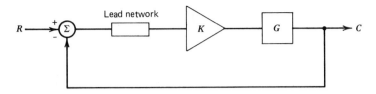

FIGURE 7.39 Location of the RC-lead-network compensator.

the left of the dominant poles. That is, it is necessary for ω_2 to be large so that its corresponding time constant is small. Since $\omega_2 = \omega_1/\alpha$, this may be accomplished by choosing a small α; in practice, a value of $\alpha = 0.1$ is frequently used. Unfortunately, there are no hard or fast rules for selecting ω_1. But it is usually the case that a significant improvement in transient response will result if the zero at ω_1 is placed so that it cancels one of the poles of the original open-loop transfer function.* As is the case in many compensator design situations, a trial-and-error procedure is often the most practical approach. Trial and error is a technique that is greatly facilitated by computer simulation, which will be discussed in Chapter 8.

EXAMPLE 7.14

Design an RC lead compensator for the control system in Figure 7.40. The low frequency gain of the compensated system should be the same as the low frequency gain of the original system.

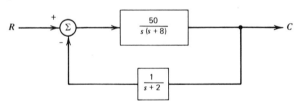

FIGURE 7.40 Control system to be compensated using RC lead network (Example 7.14).

SOLUTION

The open-loop transfer function including compensator is

$$GH = \left(\frac{s + \omega_1}{s + \omega_2}\right)\left[\frac{50}{s(s + 2)(s + 8)}\right]$$

We will design the lead compensator so that it produces a zero at $s = -2$ in order to cancel* the pole at $s = -2$. Thus, $\omega_1 = 2$, and choosing $\alpha = 0.1$, we have $\omega_2 = (\omega_1/\alpha) = 20$. Let $R_1 = 10$ K. Then

$$\alpha = \frac{R_1}{R_1 + R_2} \Rightarrow 0.1 = \frac{10\text{ K}}{R_2 + 10\text{ K}} \Rightarrow R_2 = 90\text{ K}$$

---

*In practice, this may be difficult due to component tolerances, parameter variations with age, and so forth. Place the zero as close to the pole as possible.

Since $\omega_1 = 2 = 1/(R_2 C)$, we find $1/(90 \times 10^3)C = 2 \Rightarrow C = 5.55 \ \mu F$. The required lead network is thus as shown in Figure 7.41. At low frequencies, where the capacitor has negligible effect, the gain is

$$\frac{R_1}{R_1 + R_2} = \alpha = 0.1$$

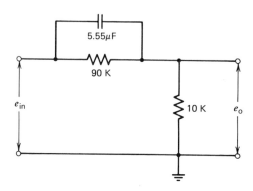

FIGURE 7.41 RC lead network for Example 7.14.

We must, therefore, increase the forward gain of the open-loop transfer function by a factor of 10 to compensate for this attenuation. The resulting open-loop transfer with compensator is

$$GH = \frac{500}{s(s + 8)(s + 20)}$$

We observe that the poles are now at $s = 0, -8,$ and -20, in comparison with uncompensated poles at $s = 0, -2,$ and -8. Thus, we expect a significant improvement in transient response.

We now turn our attention to lag compensators in the forward loop. The principal purpose of a lag compensator is to improve the steady-state error in a system. As might be expected from our discussion of lead compensators, the lag compensator has precisely the opposite characteristics: Its output amplitude is inversely proportional to the rate of change of the input. Thus, a pure *integrator* is an example of a lag compensator, and the effect of such a device is to produce a (K_1/s) term in the open-loop transfer function. Recall from Section 6.11 that the steady-state error coefficients of a system are improved as the number of pure integrations is increased. A pure integrator compensator may be placed in a control loop as shown in Figure 7.42. From a qualitative standpoint, any steady-state error resulting from a discrepancy between C and R will be integrated. Since the integral of a constant voltage is a linearly increasing voltage, the output of the integrator will increase until the error is corrected. From another viewpoint, the integrator's output magnitude increases as its input frequency decreases, so slowly varying error signals are effectively amplified in comparison to rapidly varying signals.

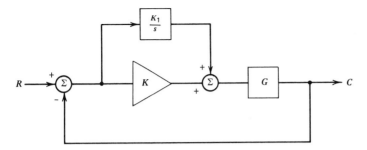

FIGURE 7.42   Integrator (lag) compensation in a control system.

The open-loop transfer function corresponding to the lag-compensated system shown in Figure 7.42 becomes

$$\left(K + \frac{K_1}{s}\right)G = \left[\frac{s + K_1/K}{s}\right]KG \qquad (38)$$

We see that a zero has been introduced at $s = -K_1/K$. In order to compensate for the detrimental effects that the new pole at $s = 0$ has upon the transient response and upon stability, the new zero should be placed as close to the origin as possible. Thus, K_1, the frequency-independent gain of the integrator, should be as small as possible to make (K_1/K) as small as possible.

EXAMPLE 7.15

Compensate the control system in Figure 7.43 so that it has zero-position error coefficient ε_p.

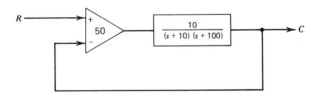

FIGURE 7.43   Control system to be lag-compensated (Example 7.15).

SOLUTION

In order to achieve zero position error, it is necessary to introduce a pole at the origin in the open-loop transfer function. This can be accomplished by using an integrator as a lag compensator. Assuming that we want the zero that results from using the integrator to be at $s = -.01$, we have from Equation 38,

$$.01 = \frac{K_1}{K} = \frac{K_1}{50} \Rightarrow K_1 = 0.5$$

Our compensated system thus appears as shown in Figure 7.44. The open-loop transfer function is now $G = 500(s + 0.01)/s(s + 10)(s + 100)$.

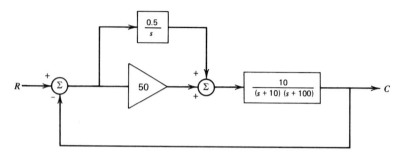

FIGURE 7.44 Lag-compensated system for Example 7.15.

One advantage that the pure integrator enjoys in contrast with the pure differentiator is that it attenuates high frequency noise. Nonetheless, it is frequently desirable to lag compensate a system using a passive RC-network to approximate integral characteristics. One such network is shown in Figure 7.45. It is left as

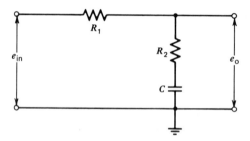

FIGURE 7.45 An RC lag compensator.

an exercise (Exercise 7.29) to show that when this network is placed in series with the forward loop of a control system whose open loop transfer function is G, the new transfer function becomes

$$\left(\frac{R_2}{R_1 + R_2}\right)\left[\frac{s + 1/(R_2C)}{s + 1/(R_1 + R_2)C}\right]G \tag{39}$$

From Equation 39, we see again that both a new pole and a new zero have been introduced to the open-loop transfer function. We also note that the compensated system will attenuate high frequencies by a factor of $R_2/(R_1 + R_2)$, so it may be necessary to increase system gain by a corresponding amount.

REFERENCES

1. **D'Azzo, John J.**, and **Constantine Houpis.** *Feedback Control-System Analysis and Synthesis.* McGraw-Hill, 1960.
2. **Dorf, Richard C.** *Modern Control Systems.* Addison-Wesley, 1967.
3. **Dransfield, Peter,** and **Donald F. Haber.** *Introducing Root Locus.* Cambridge University Press, 1973.

4. Fortmann, Thomas E., and Donrad L. Hitz. *An Introduction to Linear Control Systems*. Marcel Dekker, 1977.
5. Hale, Francis J. *Introduction to Control-System Analysis and Design*. Prentice-Hall, 1973.
6. Miller, Richard W. *Servomechanisms: Devices and Fundamentals*. Reston, 1977.
7. Zeines, Ben. *Automatic Control Systems*. Prentice-Hall, 1972.

EXERCISES

7.1 Find $\dfrac{C}{R} = \dfrac{G}{1 + G}$ when:

  (a) $G = 1\angle 45°$
  (b) $G = 1\angle -90°$
  (c) $G = 1\angle 180°$
  (d) $G = 1\angle -175°$
  (e) $G = -1\angle 0°$

For which value (s) of G does C/R represent the transfer function of an unstable system?

7.2 If the open-loop transfer function of a system is $G = \dfrac{K}{s(s + 2)(s + 50)}$,

  (a) at what frequency does $\angle G = -180°$?
  (b) what value of gain K will cause the closed-loop system to be unstable?

7.3 Given the control system shown in Figure 7.46, the servomotor has transfer function $G = \dfrac{2 \times 10^5}{s(s^2 + 1700s + 16 \times 10^4)}$.

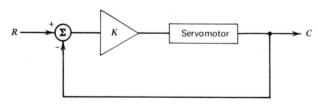

FIGURE 7.46

What value of amplifier gain K will cause the closed-loop system to become unstable?

7.4 Suppose $G = \dfrac{K_1}{(s + 10)^2}$ and $H = \dfrac{K_2}{s + 50}$.

(a) Find $|GH|$ and $\angle GH$ in terms of K_1 and K_2 for:
1. $\omega = 10$ rad/sec
2. $\omega = \sqrt{1100}$ rad/sec
3. $\omega = 40$ rad/sec

(b) If $K_1 = 150$, what value of K_2 will cause the closed-loop system to be unstable?

7.5 (a) Derive an equation for the frequency ω at which $\angle G = -180°$ when
$$G = \frac{K}{(s + a)^3}.$$

(b) Using this result, derive an expression for the value of K that will cause the closed-loop system to be unstable.

7.6 (a) Sketch the asymptotic Bode plot (magnitude and phase) for each of the following open-loop transfer functions:

1. $G = \dfrac{200}{s(s + 50)}$

2. $G = \dfrac{2500(s + 500)}{s(s + 50)(s + 100)}$

3. $G = \dfrac{5 \times 10^3(s + 2000)}{s(s + 100)(s + 400)}$

4. $G = \dfrac{5 \times 10^3}{s(s + 5)(s + 25)}$

5. $G = \dfrac{10^5(s + 1)}{s(s + 10)(s + 500)}$

(b) By analyzing the Bode plot, determine which transfer functions represent stable closed-loop systems.

7.7 A servomotor has $\left(\dfrac{R_A}{L_A}\right) = 260$, $\dfrac{K_{TI}K_{V\omega}}{J_{LA}} = 2500$, and $\left(\dfrac{K_{TI}}{J_{LA}}\right) = 10{,}000$. An amplifier with gain 70 is used to drive the motor. Sketch the Bode plot of the open-loop transfer function and determine whether this system would be stable if enclosed in a fully fed-back loop.

7.8 Sketch an asymptotic Bode plot for the gain magnitude of $G = \dfrac{400}{s^2 + 24s + 400}$. Is the asymptotic approximation reasonably accurate? Why?

7.9 Use a Bode analysis to determine whether the system shown in Figure 7.47 is stable.

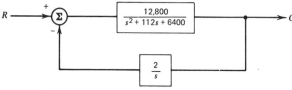

FIGURE 7.47

7.10 Insert a zero in the system in Exercise 7.9 so that it will become a stable system.

7.11 Using Bode plots, determine the approximate gain and phase margin (where possible) for each stable system in Exercise 7.6.

7.12 How much attenuation must be placed in series with the amplifier in Exercise 7.7 in order to achieve a gain margin of 12 dB?

7.13 Sketch the polar plots for the following transfer functions:

(a) $G = \dfrac{20}{(s + 20)}$

(b) $G = \dfrac{100}{(s + 1)(s + 10)}$

(c) $G = \dfrac{5 \times 10^5}{(s + 20)(s + 100)(s + 1000)}$

7.14 Derive Equation 7.12.

7.15 Sketch polar plots for the following transfer functions:

(a) $G = \dfrac{10^4(s + 50)}{(s + 10)(s + 1000)}$

(b) $G = \dfrac{800}{s(s + 100)(s + 200)}$

7.16 Sketch polar plots for the following transfer functions:

(a) $G(s) = \dfrac{500}{s^2 + 20s + 2500}$

(b) $G(s) = \dfrac{800}{s^2 + 112s + 6400}$

7.17 Apply the Nyquist criterion to determine which of the following open-loop transfer functions correspond to stable closed-loop systems:

(a) $\dfrac{3000}{s(s + 8)(s + 20)}$

(b) $\dfrac{2 \times 10^4}{s(s^2 + 4s + 400)}$

7.18 Determine graphically the gain and phase margins for those transfer functions that represent stable systems in Exercise 7.17.

7.19 The polar plot for the open-loop transfer function of a certain system is shown in Figure 7.48.

(a) What increase in gain (in dB) will cause this system to be unstable?
(b) What decrease in gain (in dB) will cause this system to be unstable?

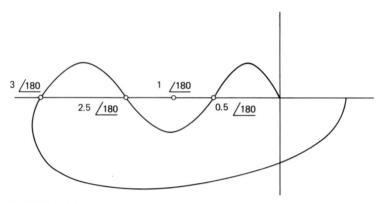

FIGURE 7.48

7.20 Calculate the values of ζ that correspond to each of the plots shown in the Nichols chart in Figure 7.19 (i.e., for $M_m = 1.25$ dB and for $M_m = 3.0$ dB).

7.21 Construct Nichols chart plots for each of the following open-loop transfer functions, and in each case, determine M_m and the phase margin (provided the system is stable). Determine what open-loop gain adjustments are necessary in each case to produce $M_m = 3$ dB.

(a) $G = \dfrac{10{,}000}{s(s + 50)}$

(b) $G = \dfrac{210}{s(s + 20)}$

(c) $G = \dfrac{8000(s + 0.2)}{s(s + 50)^2}$

7.22 (a) Sketch the root locus for the case
$$G(s) = \frac{1}{s(s + 2)} \qquad H(s) = K$$
(b) Using the root locus, graphically determine ζ and ω_n when:
1. $K = 4$
2. $K = 8$
(c) Are there any values of K that cause this system to be unstable?
(d) For each value of K in part (b), describe the closed-loop response to a step input.

7.23 For $G(s) = \dfrac{1}{s(s + 2)}$ and $H(s) = \dfrac{K}{(s + 4)}$, prove that the root locus is on the real axis when
$$0 \leq K \leq (6 - 4\sqrt{2}) \quad \text{or when} \quad K \geq (6 + 4\sqrt{2})$$

7.24 (a) Sketch the root locus for $GH = \dfrac{K(s+8)}{s(s+4)}$.

(b) Are there any values of K that will cause this system to be unstable?

7.25 Sketch the root locus for the following open-loop transfer functions:

(a) $GH = \dfrac{K}{s(s+4)(s+10)}$

(b) $GH = \dfrac{K}{s(s+8)(s+100)}$

(c) $GH = \dfrac{K}{(s+1)(s^2+s+100)}$

For what values of K do these transfer functions represent unstable closed-loop systems?

7.26 Design a lead compensator for the control system shown in Figure 7.49, using $K = 500$. Sketch the root locus before and after compensation.

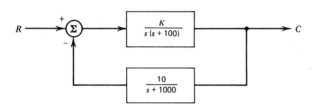

FIGURE 7.49

7.27 Design an RC lead compensator for the system in Exercise 7.26. The dc gain of the system should be the same before and after compensation. Sketch the root locus before and after compensation.

7.28 Compensate the control system in Figure 7.50 so that it has zero position-error coefficient ε_p.

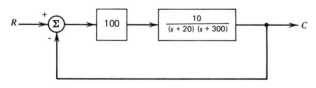

FIGURE 7.50

7.29 Derive Equation 39.

ANSWERS TO EXERCISES

7.1 (a) $0.541 \angle 22.5°$
(b) $0.707 \angle -45°$
(c) ∞
(d) $11.46 \angle -87.5°$
(e) ∞
Unstable for (c) and (e)

7.2 (a) $\omega = 10$ rad/sec
(b) $K = 5200$

7.3 $K = 1360$

7.4 (a) 1. $|GH| = \dfrac{K_1 K_2}{10{,}198}$; $\angle GH = -101.31°$

2. $|GH| = \dfrac{K_1 K_2}{72{,}000}$; $\angle GH = -180°$

3. $|GH| = \dfrac{K_1 K_2}{108{,}853}$; $\angle GH = -190.59°$

(b) $K_2 = 480$

7.5 (a) $\omega = \sqrt{3}\, a$
(b) $K = 8a^3$

7.6 1. Stable   4. Unstable
 2. Unstable   5. Stable
 3. Stable

7.7 Unstable

7.8 The asymptotic approximation is within 3 dB of the actual gain magnitude, since $\zeta = 0.6$.

7.9 Unstable. Gain must be less than 1.0 at $\omega = \omega_n = 80$.

7.12 0.233

7.17 (a) Stable
(b) Unstable

7.19 (a) 6 dB
(b) -7.95 dB

7.20 $\zeta = 0.5$ for $M_m = 1.25$ dB
$\zeta = 0.383$ for $M_m = 3$ dB

ANSWERS TO EXERCISES **275**

7.22 (b) 1. $\omega_n = 2$ rad/sec; $\zeta = 0.5$
2. $\omega_n = 2.83$ rad/sec; $\zeta = 0.354$

7.24 (b) No

7.25 (a) $K = 560$
(b) $K = 86{,}400$
(c) $K = 2$

CHAPTER

8

ANALOG COMPUTATION AND SIMULATION

8.1 ANALOG COMPUTERS

An electronic analog computer contains components that are capable of performing mathematical operations such as addition, subtraction, integration, and the like, on continuous time-varying voltage inputs. It simultaneously produces a continuous time-varying output voltage representing the solution to a mathematical problem. The essence of an analog computer, then, is that it produces a time-domain solution to a time-varying problem and it produces this solution in *real time*. For example, if we wished to find the integral of $5\cos 100t$, then an analog computer, which has been "programmed" to do integration, could be used to produce the solution .05 $\sin 100t$ as a time-varying voltage in response to the time-varying input voltage $5\cos 100t$. Programming an analog computer consists of patching the various electronic components together in such a way that the desired mathematical operations are performed on any time-varying inputs. In contrast, a digital computer would be programmed to solve a mathematical problem one number at a time, so to speak. The digital computer requires a certain time to perform the required calculation on each new number, that is, for each new value of the input as it changes with time, and, therefore, does not truly produce a solution in real time.

Analog computers are used principally to solve differential equations. Since the behavior of many physical systems can be described by differential equations, an analog computer may be used to *simulate* these systems. For example, the computer can be configured to solve the differential equation relating the angular displacement of a spring-dashpot-inertia combination to the torque applied to it. Then the solution, angular displacement as a function of time, can be obtained for a variety of input torques: step functions, sinusoids, and so forth, simply by generating input signals that vary in a corresponding manner. The analog computer is particularly useful when there is no practical way to find a solution by purely mathematical analysis. This is the case for many differential equations, particularly *nonlinear* differential equations.

In recent years, advances in digital-computer technology and programming techniques have made it possible to obtain digital solutions to differential equations very rapidly and to a high degree of accuracy. The digital computer has, thus, largely supplanted the analog computer in many of the latter's traditional roles. However, the ability of an analog computer to produce a solution in real time still makes it indispensible in some situations, as, for example, when it is necessary to solve a large number of differential equations simultaneously. The analog computer is also a valuable tool for providing real-time insight into the physical properties of complex systems, since it permits the investigator to monitor the variation of a number of different variables simultaneously. Hybrid computers, which will be discussed later, combine some of the advantages of both analog and digital computation.

8.2 OPERATIONAL AMPLIFIERS

The most important active component of an analog computer is the operational amplifier. Depending on the complexity of the problems that the computer is designed to solve, an analog computer may have six or a hundred such amplifiers. Apart from their role in the construction of an analog computer, our study of these devices is further motivated by the variety of their other important applications, among which are active filters, instrumentation systems, buffering, and signal conditioning of all kinds. The term operational describing the amplifier stems from its ability to perform mathematical operations on the voltages applied to it. The particular mathematical operations it performs is determined and may be changed by the passive components connected to it.

From the standpoint of performing a mathematical operation *accurately*, the two most important properties of the amplifier are its input impedance and its voltage gain; both must be very high. We will use the symbol in Figure 8.1 to represent an operational amplifier. Note that the typical operational amplifier has two

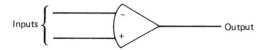

FIGURE 8.1 Symbolic representation of an operational amplifier.

inputs, a negative, or inverting, input and a positive, or noninverting, input. When the inverting input is used, the output is shifted 180° in phase from the input. For analog computer applications, the inverting input is most often used, while the noninverting input is grounded (or connected through a resistor to ground). In our subsequent treatment of this subject, we will assume that only the inverting input is used, and we will not bother to show the noninverting input on circuit diagrams.

278 ANALOG COMPUTATION AND SIMULATION

We will, thus, refer to the gain of the amplifier as being $-A$. We will denote the input impedance of the amplifier by Z_g. (The subscript g is a carryover from earlier days when the input was the grid of a vacuum tube; now, it is more likely to be the gate of a field effect transistor.)

Because of the extremely high voltage gain of the operational amplifier, we never connect an input voltage directly to the input terminal. Instead, the signal is coupled to the input terminal through a passive component such as a resistor. Also, a passive component must be connected between the output of the amplifier and its input. Consider the configuration shown in Figure 8.2. This circuit is used for *scaling*,

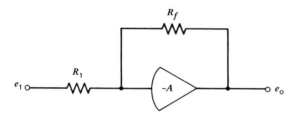

FIGURE 8.2 An operational amplifier with input and feedback resistors.

that is, for multiplying an input voltage by a fixed constant. We will demonstrate that provided A and Z_g are very large (ideally infinite), then

$$e_o = -\frac{R_f}{R_1} e_{in} \qquad (1)$$

The voltage and current relations in the amplifier are shown in Figure 8.3. As-

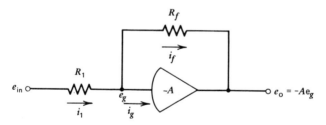

FIGURE 8.3 Voltage and current relations in an operational-amplifier circuit.

suming a positive input e_{in}, the output voltage e_o is negative, and the directions of current flow are as shown. Applying Kirchoff's current law at the junction of R_1 and R_f, we have

$$i_1 = i_g + i_f \qquad (2)$$

Assuming an extremely large amplifier input impedance Z_g, the current i_g into the amplifier will be negligible (ideally zero for infinite Z_g). Thus, we can neglect i_g and write Equation 2 as

$$i_1 = i_f \qquad (3)$$

But $i_1 = (e_{in} - e_g)/R_1$ and $i_f = (e_g - e_o)R_f$, so Equation 3 becomes

$$\frac{e_{in}}{R_1} - \frac{e_g}{R_1} = \frac{e_g}{R_f} - \frac{e_o}{R_f} \tag{4}$$

Since $e_o = -Ae_g$ or $e_g = -e_o/A$,

$$\frac{e_{in}}{R_1} + \frac{e_o}{AR_1} = \frac{-e_o}{AR_f} - \frac{e_o}{R_f} \tag{5}$$

If the amplifier has extremely high gain A, ideally infinite, then the terms in Equation 5 that have A in the denominator are negligibly small. Consequently, we obtain

$$\frac{e_{in}}{R_1} = -\frac{e_o}{R_f} \quad \text{or} \quad \frac{e_o}{e_{in}} = -\frac{R_f}{R_1} \tag{6}$$

which is equivalent to Equation 1. The significance of this result is that the scale factor R_f/R_1 depends only on the resistor values and not on the amplifier characteristics, provided Z_g and A are both very large.

EXAMPLE 8.1

Given an input voltage $e_1 = 5 + 10 \sin 20t$, design operational-amplifier circuits that produce outputs:

(a) $10 + 20 \sin 20t$
(b) $-2.5 - 5 \sin 20t$

SOLUTION

(a) We require a gain of 2 and an output that is in phase with the input; thus, two amplifiers (two phase inversions) will be required. The gain may be concentrated in one amplifier, while the second amplifier has unity gain, or the gain may be distributed over both amplifiers in such a way that their product is 2. Choosing the former option, let R_1 for the first amplifier be 100 K. Then

$$\frac{R_f}{R_1} = 2 \Rightarrow R_f = 200 \text{ K}$$

Figure 8.4 shows a possible solution.

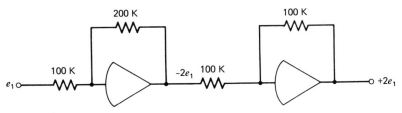

FIGURE 8.4 An operational-amplifier circuit that produces $2e_1$ (Example 8.1).

(b) We require $e_o = -0.5e_{in}$. Let $R_f = 0.5 \text{ M}\Omega$.
Then, $R_f/R_1 = 0.5 \Rightarrow R_1 = 1 \text{ M}\Omega$.

280 ANALOG COMPUTATION AND SIMULATION

The reader might wonder why it is necessary to use an amplifier to obtain an attenuation of an input signal by 0.5, for instance, in Example 8.1. Why wouldn't a potentiometer suffice? The answer lies in the *accuracy* of the attenuation. Remember that operational amplifiers are used to perform mathematical operations, and in order for these operations to lead to valid, accurate solutions, it is imperative for all operations to be as accurate as possible. A potentiometer cannot generally be set accurately enough nor remain accurate enough to satisfy these demands. Since the scaling operation that we obtain from an operational amplifier is virtually independent of changes in amplifier characteristics provided gain and impedance remain high, we have a consistently accurate attenuation. Large scale, laboratory-grade analog computers use very high precision components and even temperature-controlled ovens to eliminate the effects of temperature variations on component values. A further advantage of using operational amplifiers is that they generally have low output impedances and are not, therefore, loaded by the circuits they drive as a potentiometer or voltage divider might be. And, of course, the amplifier produces power gain for driving loads, which passive components cannot do.

Since $e_o = -Ae_g$ or $e_g = -e_o/A$ and A is assumed to be very large, the voltage e_g will always be quite small. In fact, it is so close to 0 V that the junction where it is measured is said to be *virtual ground*. Thus, the impedance seen by the source that is driving the input through R_1 is essentially R_1 Ω. (See Figure 8.5.) This

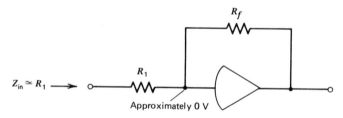

FIGURE 8.5 Virtual ground at the summing junction of an operational amplifier.

result is important from the standpoint of determining how many operational-amplifier circuits may be driven by the output of a single operational amplifier. If, for example, the amplifier specifications limit output current to 2 mA and the maximum output voltage is 10 V, then the resistive load connected to the amplifier cannot be less than 10 V/2 mA = 5 K. With these limitations, we could *not* make the connections shown in Figure 8.6. In this case, the load is equivalent to 10 K in parallel with 1 K, which is less than 5 K. In general, we require

$$R_T \geq \frac{e_{max}}{i_{max}} \tag{7}$$

where

R_T = total parallel equivalent resistance of all loads
e_{max} = maximum rated output voltage of amplifier
i_{max} = maximum rated output current of amplifier

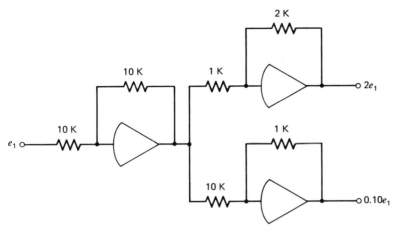

FIGURE 8.6 Excessive loading of an operational amplifier.

There are several other operational amplifier characteristics that are of practical importance when patching these circuits on an analog computer or when selecting amplifiers for a particular application. A knowledge of the *frequency response* of the amplifier is often very important. Because of the very high gain of the amplifier, it is usually compensated (rolled off) severely to prevent high frequency oscillations. In other words, the high gain we rely on to obtain accuracy may not be there at frequencies perhaps as low as 100 Hz. Related to the frequency response is the *slew rate* of the amplifier, which specifies its maximum rate of change of voltage. A low slew rate will significantly reduce the permissible frequency of operation when signal magnitudes are large. If, for example, the rated slew rate were 0.1 V/μsec, then the amplifier could just handle a 1-V peak sine wave at 15.9 kHz, but it could not handle a 10-V peak sine wave at greater than 1.59 kHz. For sinusoidal signals,

$$f_{max} = \frac{S}{2\pi E_p} \text{ MHz} \qquad (8)$$

where

f_{max} = maximum sinusoidal frequency as limited by slew rate alone
S = rated slew rate in V/μsec
E_p = peak sine-wave voltage, volts

Of course, the gain at either frequency, 1.59 kHz or 15.9 kHz, might be unacceptably low anyway due to the frequency response of the amplifier.

Another important characteristic of the amplifier is its dc *offset*, that is, the dc voltage appearing at the amplifier input when the signal is zero. Since operational amplifiers are generally direct-coupled so they may be used at dc, the output should be exactly 0 V when the input is 0 V. This is particularly important when the amplifier is used as an integrator, for integration of a constant will generate a ramp voltage output and eventually saturate the amplifier. Some amplifiers have an offset adjustment, which permits zeroing the output when the input is grounded. Some are *chopper-stabilized*, a means of converting the offset to ac, amplifying it, and using this signal for automatic zeroing (a feedback control system). Offset can also be compensated for by judicious selection of a resistor connected between the nonverting input and ground. Related to offset is the *drift* specification: the amount that the offset will vary with time, usually a function of temperature.

Finally, the noise characteristics of the amplifier are important, especially when we have to perform accurate mathematical operations on low-level signals. When using an analog computer to solve a differential equation, we frequently do not know how small one of the variables is liable to become in the course of the computation. Consequently, a noisy amplifier may mask out or distort signals used to generate the solution. It should be noted that using large values of input resistance (coupling the input signal to the amplifier input) contributes to the noise problem, since every resistor generates a thermal noise voltage proportional to the square root of its resistance value. The presence of noise is the reason that analog computers used to solve differential equations do *not* perform differentiation! Recall that a differentiator acts like a high pass filter and effectively amplifies noise voltages in direct proportion to their frequency.

EXAMPLE 8.2

Use the specifications provided in Appendix B for the Analog Devices AD 517J operational amplifier to answer the following questions:

(a) What is the open-loop gain magnitude at 100 Hz? At 1 kHz? What is the dB change in open-loop gain magnitude over this decade?

(b) Should this amplifier be used in an application requiring unity gain for a 10-V peak, 100-kHz sine-wave signal?

(c) What is the smallest resistive load that should be connected to the output of this amplifier if it must be driven to its maximum output voltage?

(d) In a unity gain application when the input is 0 V, what is the maximum output voltage after a temperature change from 20°C to 50°C?

SOLUTIONS

(a) By interpolating the open-loop magnitude frequency-response plot, we find that the gain magnitude is approximately 70 dB at 100 Hz and 50 dB at 1 kHz. These correspond to gain magnitudes of antilog(3.5) = 3162 and antilog(2.5) = 316, respectively. The open-loop gain-magnitude falls 20 dB over the decade.

(b) The frequency response at unity gain is flat out to approximately 250 kHz, so the 100-kHz signal is within this specification. The unity-gain slew rate is specified to be 0.1 V/μsec. From Equation 8,

$$f_{max} = \frac{0.1}{(2\pi \times 10)} \text{ MHz} = 1.59 \text{ kHz}$$

Therefore, the amplifier should not be used for this application.

(c) The specified output current is 10 mA and maximum output voltage is 10 V. Therefore, the resistive load should not be less than 10 V/10 mA = 1 K.

(d) The input-offset-voltage specifications state that the maximum initial offset is 150 μV and that the drift is 3 μV/°C. Thus, at unity gain, the maximum output (offset) voltage is 150 μV + 30(3 μV) = 240 μV.

8.3 OPERATIONAL AMPLIFIER CIRCUITS

Operational amplifiers can be used to *sum* two voltages accurately by using the configuration in Figure 8.7. Using an analysis similar to that used to demon-

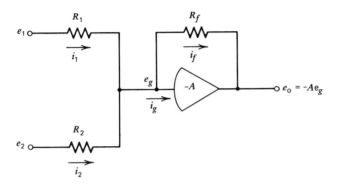

FIGURE 8.7 Summing two voltages in an operational-amplifier circuit.

strate scaling in Section 8.2, we first apply Kirchoff's current law at the junction of the three resistors

$$i_1 + i_2 = i_g + i_f$$

Assuming i_g is negligible due to a very large Z_g, we have

$$i_1 + i_2 = i_f$$

or

$$\frac{e_1}{R_1} - \frac{e_g}{R_1} + \frac{e_2}{R_2} - \frac{e_g}{R_2} = \frac{e_g}{R_f} - \frac{e_o}{R_f}$$

Since $e_g = -e_o/A$, we have

$$\frac{e_1}{R_1} + \frac{e_o}{AR_1} + \frac{e_2}{R_2} + \frac{e_o}{AR_2} = -\frac{e_o}{AR_f} - \frac{e_o}{R_f}$$

For very large A, this reduces to

$$\frac{e_1}{R_1} + \frac{e_2}{R_2} = -\frac{e_o}{R_f} \quad \text{or} \quad e_o = -\left(\frac{R_f}{R_1}e_1 + \frac{R_f}{R_2}e_2\right) \quad (9)$$

Equation 9 implies that we can sum and independently scale two voltages simultaneously. Of course, if no scaling is required, we simply set $R_1 = R_2 = R_f$.

EXAMPLE 8.3

Design operational amplifier circuits to generate the following signals:

(a) $-(e_1 + 0.5e_2)$
(b) $2e_1 - 4e_2$

SOLUTION

(a) Let $R_f = 100$ K. Then $R_f/R_1 = 1 \Rightarrow R_1 = 100$ K, and $R_f/R_2 = 0.5 \Rightarrow R_2 = 200$ K. See Figure 8.8.

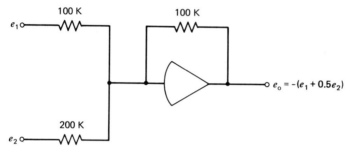

FIGURE 8.8 An operational-amplifier circuit designed to produce $-(e_1 + 0.5e_2)$ (Example 8.3a).

(b) We will need *two* amplifiers to perform subtraction. Note that $2e_1$ should be generated first, since it must ultimately have a positive sign. The circuit in Figure 8.9 will generate the required signal.

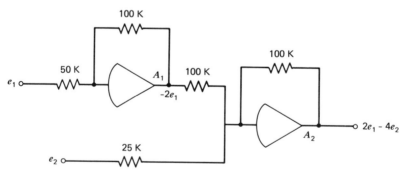

FIGURE 8.9 An operational-amplifier circuit designed to produce $2e_1 - 4e_2$ (Example 8.3b).

There is usually more than one correct way to generate a specified signal. In Figure 8.9, we could have set the gain of amplifier A_1 equal to unity and then scaled e_1 by a factor of 2 in amplifier A_2. The choice of circuit is sometimes dictated by other considerations, such as resistor values available, amplifier loading, and so on. In any case, the least possible number of amplifiers should always be used.

Adding voltage signals can be extended to three or more inputs simply by adding the additional input resistors required. In general,

$$e_o = -\left(\frac{R_f}{R_1}e_1 + \frac{R_f}{R_2}e_2 + \cdots + \frac{R_f}{R_n}e_n\right)$$

EXAMPLE 8.4

Design an operational amplifier circuit to generate $8e_1 - .05e_2 + 1.5e_3 - 0.2e_4$.

SOLUTION

The positive terms can be combined and scaled in one 2-input amplifier circuit, and the result added and scaled with the negative terms in a 3-input amplifier. See Figure 8.10.

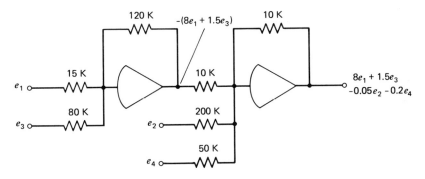

FIGURE 8.10 Adding and scaling voltages (Example 8.4).

It is frequently necessary to add or subtract a *constant* from a time-varying signal input. This can be accomplished simply by applying the desired dc voltage (of opposite polarity) to an input resistor of the amplifier in the same way signal voltages are applied. Scaling a dc input may also be performed in the same way that a signal is scaled. In an analog computer, constant (dc) inputs are usually obtained from a precision multiturn potentiometer with a grounded center tap and with well-regulated positive and negative reference voltages connected to either end. Care must be taken to use an input resistance to the amplifier large enough to prevent loading the potentiometer.

EXAMPLE 8.5

Design an operational amplifier circuit to generate $0.4 - 2e_1$.

SOLUTION

See Figure 8.11.

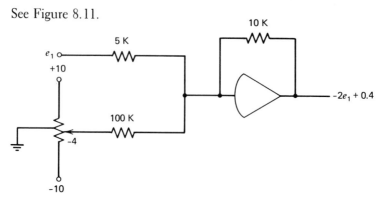

FIGURE 8.11 An operational-amplifier circuit designed to produce $-2e_1 + 0.4$ (Example 8.5).

In Figure 8.11, the precision potentiometer is set to produce -4 V, which is inverted and scaled by $\frac{1}{10}$ in the amplifier. The source impedance of a potentiometer depends upon the wiper-arm setting. This can be shown by finding the Thevenin equivalent circuit of a potentiometer, as illustrated in Figure 8.12. The source

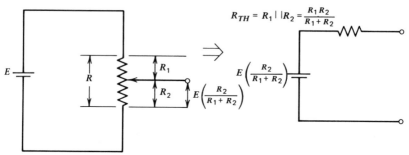

FIGURE 8.12 Thevenin equivalent circuit of a potentiometer.

impedance is a maximum when $R_1 = R_2 = R/2$. In that case, $R_{TH} = (R/2) \| (R/2) = R/4$. In Example 8.4, if the total potentiometer resistance between center tap and one end is 10 K, then the maximum source impedance it presents to the amplifier is 10 K/4 = 2.5 K. The 100-K input resistance of the amplifier would not seriously load 2.5 K. In any event, loading effects can and should always be eliminated by measuring and setting the potentiometer voltage with the wiper arm *connected* to the input resistor of the amplifier.

An operational amplifier can be used to perform integration by using a capacitor as the feedback path from output to input instead of a resistor. We will use a LaPlace transform analysis of the circuit to demonstrate this fact. Consider the circuit shown in Figure 8.13. Under the usual assumption that I_g is negligible,

$$I_1(s) = I_f(s)$$

$$\frac{E_1(s)}{R_1} - \frac{E_g(s)}{R_1} = \frac{E_g(s) - E_o(s)}{1/Cs}$$

8.3 OPERATIONAL AMPLIFIER CIRCUITS

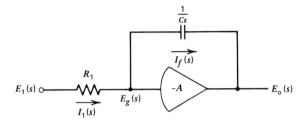

FIGURE 8.13 An operational amplifier with capacitor feedback used to perform integration.

Since $E_g(s) = E_o(s)/-A$ is negligibly small due to large A, we have

$$\frac{E_1(s)}{R_1} = -\frac{E_o(s)}{1/Cs} \quad \text{or} \quad E_o(s) = -\frac{1}{R_1 Cs} E_1(s) \quad (10)$$

Equation 10 shows that the output of the amplifier is $(1/R_1C)$ times the integral of the input, since division of the voltage transform by s corresponds to integration of that voltage in the time domain.

EXAMPLE 8.6

Design an operational amplifier circuit to generate the signal $\int_0^t 50 e_1 \, dt$.

SOLUTION

$$\int_0^t 50 e_1 \, dt = 50 \int_0^t e_1 \, dt = \frac{1}{R_1 C} \int_0^t e_1 \, dt$$

Thus, we require that $1/R_1C = 50$. Choosing $C = 0.1\mu F$, we have

$$\frac{1}{10^{-7} R_1} = 50 \qquad R_1 = \frac{10^7}{50} = 200 \text{ K}$$

A second amplifier is required to obtain a positively signed integral (see Figure 8.14).

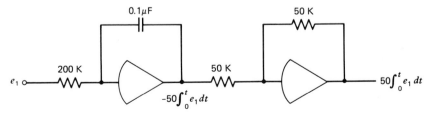

FIGURE 8.14 Operational-amplifier circuit used to produce $50 \int e \, dt$.

By connecting additional input resistors to the integrator, we can simultaneously scale and integrate several inputs. See Example 8.7.

288 ANALOG COMPUTATION AND SIMULATION

EXAMPLE 8.7

Design operational amplifier circuits to generate the following:

(a) $-\int_0^t (2e_1 + 4e_2)\,dt$

(b) $6\int_0^t (e_1 - 5)\,dt + 2e_1$

SOLUTION

The solutions shown in Figures 8.15 and 8.16 are not unique, but each does utilize the minimum number of amplifiers possible.

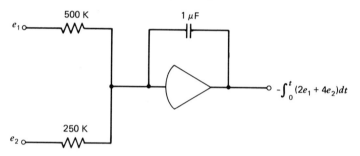

FIGURE 8.15 Integrating and scaling two variables (Example 8.7a).

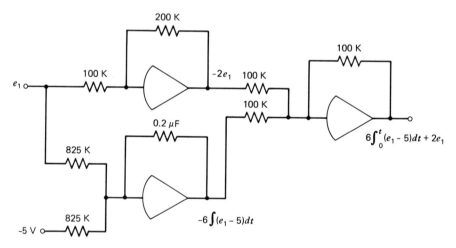

FIGURE 8.16 Integration and summation of variables (Example 8.7b).

We have seen that the transfer function of an operational amplifier with resistive input and resistive feedback is $(e_o/e_{in}) = -(R_f/R_1)$, while resistive input and capacitive feedback gives us transfer function $(e_o/e_{in}) = -1/R_1Cs$. These are both special cases of a general result that we now derive for an arbitrary impedance $Z_1(s)$ in the input path and arbitrary impedance $Z_f(s)$ in the feedback. See Figure 8.17.

8.3 OPERATIONAL AMPLIFIER CIRCUITS

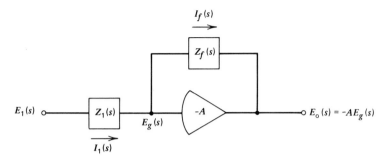

FIGURE 8.17 Operational amplifier with generalized input and feedback impedances.

Using the same approach we have followed on several previous occasions, we apply Kirchoff's current law, neglect I_g and use the fact that terms involving E_g may be dropped when A is very large.

$$I_1(s) = I_f(s)$$

$$\frac{E_1(s)}{Z_1(s)} - \frac{E_g(s)}{Z_1(s)} = \frac{E_g(s)}{Z_f(s)} - \frac{E_o(s)}{Z_f(s)}$$

$$\frac{E_1(s)}{Z_1(s)} = -\frac{E_o(s)}{Z_f(s)}$$

$$\frac{E_o(s)}{E_1(s)} = -\frac{Z_f(s)}{Z_1(s)} \tag{11}$$

Equation 11 tells us that the transfer function of an operational amplifier circuit is simply the (negative) ratio of feedback impedance to input impedance, a very useful fact. This result is often applied in the design of *active filters* and in the *simulation* of components with frequency dependent gains.

EXAMPLE 8.8

Find the transfer function of the circuit in Figure 8.18.

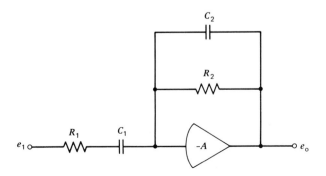

FIGURE 8.18 Find the transfer function (Example 8.8).

SOLUTION

$$Z_f(s) = \frac{R_2(1/C_2s)}{R_2 + (1/C_2s)} = \frac{R_2}{R_2C_2s + 1} = \frac{1/C_2}{s + (1/R_2C_2)}$$

$$Z_1(s) = R_1 + \frac{1}{C_1s} = \frac{R_1C_1s + 1}{C_1s} = \frac{R_1}{s}\left(s + \frac{1}{R_1C_1}\right)$$

Then from Equation 11,

$$\frac{E_o(s)}{E_1(s)} = -\frac{\dfrac{1/C_2}{s + (1/R_2C_2)}}{R_1/s[s + (1/R_1C_1)]}$$

$$= -\frac{(1/R_1C_2)s}{[s + (1/R_1C_1)][s + (1/R_2C_2)]} \tag{12}$$

8.4 ANALOG SOLUTION OF DIFFERENTIAL EQUATIONS

As we have already noted, analog computers are used principally to find solutions to differential equations. Recall that a differential equation is simply a relationship between a variable and one or more of its derivatives, while a solution of the differential equation is a function that satisfies it. One of the simplest examples of a differential equation is $y + (dy/dt) = 0$. $y = e^{-t}$ is a solution of this equation, since it is a function $y(t)$ that satisfies the equation.

We use an analog computer to solve a differential equation by connecting operational amplifiers together in such a way that they perform mathematical operations on the variable and its derivatives as specified by the equation. Using the computer, it is our responsibility to generate a voltage that represents the variable as well as a voltage for each of its derivatives and then to add, subtract, scale, or otherwise combine these voltages in accordance with the equation. We noted earlier that we do not ordinarily use differentiators in analog computation, so we must have some other way of generating voltages that represent the derivatives of the variable. The method that is used can be summarized as follows:

1. Rearrange the differential equation to obtain an expression for the *highest order* derivative present in the equation.
2. *Assume* that a voltage proportional to this derivative is available.
3. Integrate this voltage as many times as necessary to generate voltages proportional to all the lower order derivatives that appear in the equation. (When a derivative of order n is integrated, we obtain a derivative of order $n - 1$.)
4. Combine these lower order derivatives as specified by the equation so that their combination is equal to the highest order derivative. The highest order derivative that we assumed we had in step 2 has now been generated.

If this all sounds like magic, be assured that it is a legitimate procedure. A careful study of Example 8.9 will convince the reader of this fact as well as clarify the steps in the procedure.

EXAMPLE 8.9

Draw an operational amplifier circuit that can be used to solve the differential equation

$$2\ddot{y} + 8\dot{y} + 4y = 12 \tag{12}$$

SOLUTION

We first obtain an expression for the highest order derivative in the equation. In this example, the second derivative $\ddot{y}$ is the highest order.

$$\ddot{y} = -4\dot{y} - 2y + 6 \tag{13}$$

Now assume that a voltage proportional to $\ddot{y}$ is available. If the second derivative is the input to an operational amplifier connected as an integrator, then the output will be (minus) the first derivative, that is, $-\dot{y}$. See Figure 8.19. The minus sign

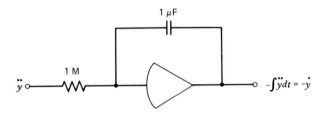

FIGURE 8.19 Integration produces a lower order derivative (Example 8.9).

is due to the phase inversion of the amplifier. According to Equation 13, we will need $-4\dot{y}$. We can obtain $+4\dot{y}$ by multiplying $-\dot{y}$ by 4 in an operational amplifier with gain 4. See Figure 8.20.

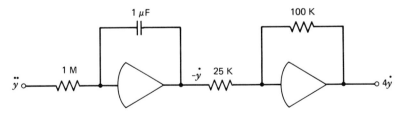

FIGURE 8.20 Scaling the first derivative (Example 8.9).

Referring again to Equation 13, we see that we also need $-2y$. We can obtain $2y$ by integrating $-\dot{y}$ and simultaneously multiplying by 2, as shown in Figure 8.21.

292 ANALOG COMPUTATION AND SIMULATION

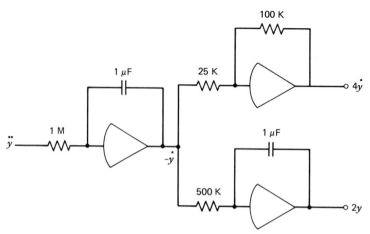

FIGURE 8.21 Generating 2y (Example 8.9).

We can now combine $4\dot{y}$ and $2y$ along with the constant 6 in a single summing amplifier. Note that 6 has the opposite sign of $4\dot{y}$ and $2y$ in Equation 13, so we add -6 to $+(4\dot{y} + 2y)$, as shown in Figure 8.22.

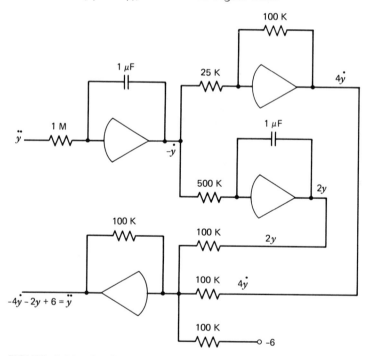

FIGURE 8.22 Combining terms to generate a voltage proportional to $\ddot{y}$ (Example 8.9).

Note that the final summing amplifier inverts the sign of all terms, and we have, therefore, produced a voltage that by Equation 13 is exactly equal to $\ddot{y}$. Since we have generated ÿ, we connect it back to the input of the first amplifier where we assumed we had ÿ to begin with. The final circuit thus appears as shown in Figure 8.23.

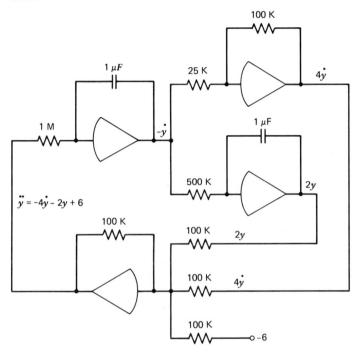

FIGURE 8.23 Closing the loop to complete the circuit required to solve the differential equation (Example 8.9).

In order to obtain the solution $y(t)$, it is only necessary to connect the circuit in Figure 8.23, apply power (or use a switch to close the loop), and let the circuit take over. The instant the switch is closed corresponds to $t = 0$. Although y itself is not generated explicitly in the circuit, we can take the voltage proportional to $2y$, multiply by ½, and thus obtain a voltage that varies with time exactly as the solution y does. This voltage can be observed on an oscilloscope or recorded on a plotter. In practice, it would usually not be necessary to scale down the voltage $2y$ to obtain y, since this can be conveniently accomplished by using the sensitivity controls on the oscilloscope or plotter. Generally, if the circuit produces any fixed multiple of the solution y, this is adequate for recording purposes. We can also use this circuit to obtain, observe, and record $\dot{y}$ and $\ddot{y}$ if we wish. Observing these derivatives often provides insight into the problem or valuable additional information about the behavior of the system represented by the differential equation.

As usual, we should use the minimum number of amplifiers possible to solve an equation. If we are not interested in observing $\ddot{y}$ explicitly, then it is not necessary to generate a voltage proportional to it. In these cases, it is often possible to eliminate the extra amplifiers required to produce $\ddot{y}$. In Example 8.9, the amplifier that sums

294 ANALOG COMPUTATION AND SIMULATION

$4\dot{y}$, $2y$, and -6 can be eliminated, since the integrator that produces $\dot{y}$ can also be used to perform summation and scaling. Of course, this integrator inverts the sign as well, so the terms to be summed must be of polarities opposite to those we have developed in the example. This requires a slight modification of the circuit, as illustrated in Figure 8.24.

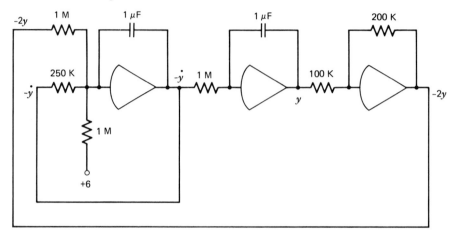

FIGURE 8.24 An analog-computer circuit that uses one less amplifier than Figure 8.23.

Note that $-\dot{y}$ is scaled by a factor of 4 in the first integrator $[1/(10^{-6} \times 250 \text{ K}) = 4]$ in order to obtain $-4\dot{y}$, as required by Equation 13. The first integrator provides unity gain to the $-2y$ input and to the $+6$ input as well as summing and integrating.

To simplify the diagram of an operational amplifier circuit used in an analog computer, it is conventional to omit the individual resistor and capacitor symbols. Instead, certain special symbols are used to represent integrators and summing amplifiers. The scaling factor, or gain, provided by the amplifier to each of its inputs is indicated by writing the appropriate number at the input. A summing amplifier that multiplies one input by 5 and the other by 0.2 would be depicted as shown in Figure 8.25. An integrator that also multiplies both inputs by 2 would be depicted as shown in Figure 8.26.

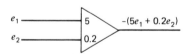

FIGURE 8.25 Abbreviated notation for representing closed-loop amplifier gains.

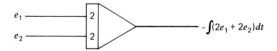

FIGURE 8.26 Abbreviated notation for representing integrator gains.

8.4 ANALOG SOLUTION OF DIFFERENTIAL EQUATIONS

Using these symbols, the three-amplifier solution to Example 8.9 would be shown as in Figure 8.27. We will use this abbreviated scheme in future circuit diagrams.

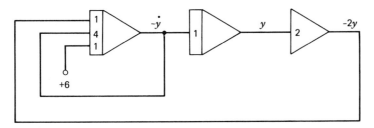

FIGURE 8.27 Solution of Figure 8.24 using abbreviated notations.

The integrators in an analog computer are equipped with *initial-condition* inputs. These inputs are used to set the initial values of the dependent variable and its derivatives prior to problem solution. The initial conditions for the differential equation that is to be solved on the computer must always be specified and set in the integrators in order to obtain the particular solution required. We will discuss the initial-condition circuitry associated with each integrator and the mechanics for switching initial conditions into the problem in a later section. For now, we will simply state that initial-condition inputs are shown symbolically in the computer diagram by using vertical lines drawn to each integrator, as illustrated in Example 8.10.

EXAMPLE 8.10

Draw an operational amplifier circuit that can be used to solve the differential equation

$$\ddot{y} + a^2 y = 0$$

with initial conditions

$$y(0) = 0 \quad \text{and} \quad \dot{y}(0) = a$$

SOLUTION

$y'' = -a^2 y$. See Figure 8.28. The solution to this differential equation happens to be $y = \sin at$. (Verify this.) Thus, the circuit in Figure 8.28 can be used as a precision oscillator whose frequency can be varied by varying the gain of the third amplifier. It is often necessary to produce driving functions—functions of time such as $\sin at$, $u(t)$, Kt^2, and so forth—in order to implement the circuit required to solve a differential equation. Example 8.10 shows how we might generate a sinusoidal driving function.

296 ANALOG COMPUTATION AND SIMULATION

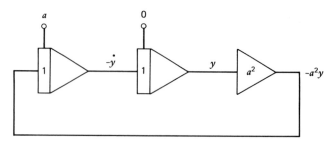

FIGURE 8.28 Analog-computer circuit used to solve $\ddot{y} + a^2 y = 0$ (Example 8.10).

EXAMPLE 8.11

Draw an operational amplifier circuit that could be used to solve the differential equation

$$2\ddot{y} + 18\dot{y} + 12y = 3t - 6 \sin 5t$$

Assume all initial conditions are zero.

SOLUTION

$$\ddot{y} = -9\dot{y} - 6y + \frac{3t}{2} - 3 \sin 5t$$

In Example 8.11, we have to produce the driving function $3t/2 - 3 \sin 5t$. We will concentrate on that task first. The sine wave can be generated by the method in Example 8.9. The ramp $3t/2$ can be generated by integrating a constant $3/2$, since

$$\int_0^t \frac{3}{2} dt = \frac{3}{2} t$$

See Figure 8.29.

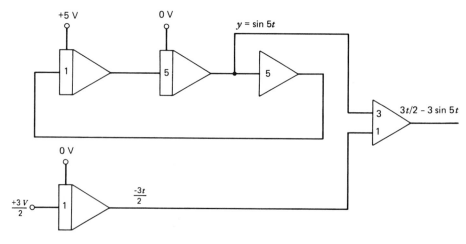

FIGURE 8.29 Circuitry for implementing a driving function (Example 8.11).

8.4 ANALOG SOLUTION OF DIFFERENTIAL EQUATIONS

Note that the gain $a^2 = 25$ required for the oscillator has been distributed over two amplifiers, since concentrating the entire gain in one amplifier may lead to unreasonable or impractical component values. We now implement the rest of the circuitry required to solve the equation as shown in Figure 8.30.

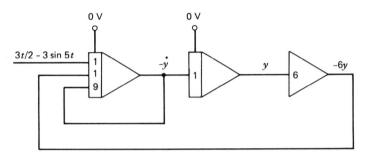

FIGURE 8.30 Analog-computer implementation with driving-function input (Example 8.11).

EXAMPLE 8.12

Draw an analog computer circuit that will generate the response $C(t)$ of the control system shown in Figure 8.31 when it is subjected to a unit-step input. Assume zero initial conditions.

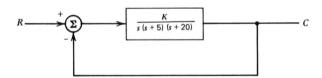

FIGURE 8.31 Draw an analog-computer circuit for simulating the control system (Example 8.12).

SOLUTION

We first express C as a differential equation by writing

$$\frac{C}{R} = \frac{K/s(s+5)(s+20)}{1 + [K/s(s+5)(s+20)]} = \frac{K}{s(s+5)(s+20) + K}$$

$Cs(s+5)(s+20) + CK = RK$

$Cs^3 + 25Cs^2 + 100Cs + CK = RK$

Since multiplying the variable C by s^n corresponds to taking the n-derivative of C, we have

$$\dddot{C} + 25\ddot{C} + 100\dot{C} + CK = RK$$

We may now proceed in the usual way to implement an analog computer circuit.

Solving for the highest order derivative $\dddot{C}$ yields

$$\dddot{C} = -25\ddot{C} - 100\dot{C} - CK + RK$$

Figure 8.32 shows the analog computer simulation of the control system.

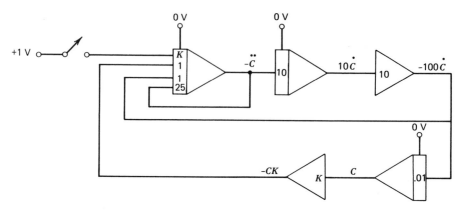

FIGURE 8.32 Analog simulation of the control system in Figure 8.31 (Example 8.12).

By closing the switch shown in Figure 8.32, we simulate the application of a unit step to the system. The response $C(t)$ can then be observed or plotted and recorded. We can easily investigate the effect that the magnitude of the gain K has on system response by changing the input resistance (using a precision potentiometer) at the two locations shown in Figure 8.32. Example 8.12 illustrates the great utility of analog simulation: It is a great deal easier and cheaper to construct an analog circuit and vary the circuit parameters to determine system characteristics than to construct the actual control system and change *its* parameters.

EXAMPLE 8.13

Draw an analog-computer circuit that will generate a voltage proportional to the angular displacement $\theta(t)$ of the system shown in Figure 8.33 after it is subjected to a step-unit torque of T N-m. Assume zero initial conditions.

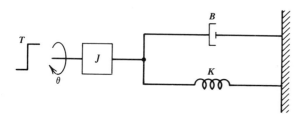

FIGURE 8.33 Draw an analog simulation for $\theta(t)$ (Example 8.13).

SOLUTION

The differential equation of motion is $T = J\ddot{\theta} + B\dot{\theta} + K\theta$. Solving for $\ddot{\theta}$, we obtain

$$\ddot{\theta} = \frac{T}{J} - \frac{B}{J}\dot{\theta} - \frac{K}{J}\theta$$

An analog-computer circuit that will solve this equation is shown in Figure 8.34. Figure 8.34 shows one of several possible ways of distributing gains among the amplifiers. Another possibility would be to make the gain of the second integrator unity and the gain of the inverting amplifier equal to K/J. The choice of individual amplifier gains ultimately depends on the limitations imposed by amplifier characteristics and practical component values, which in turn depend on the actual values of T, K, B, and J. In practice, general-purpose analog computers (those that can be programmed by patch-panel connections to solve a number of different problems) are equipped with *coefficient potentiometers*. These potentiometers may be inserted anywhere in the circuit to multiply a variable by any fixed scale factor between 0 and 1.0. Their use will be discussed later.

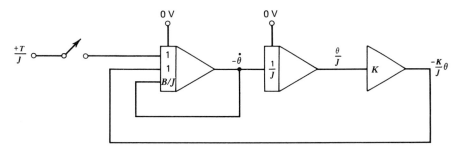

FIGURE 8.34 Analog simulation of Figure 8.33 (Example 8.13).

8.5 ANALOG COMPUTER MODES

A general-purpose analog computer whose components can be patched together in different ways to solve a variety of differential equations will usually be capable of operating in one of several different *modes*. Every such computer will have a switch by which the operator can cause the machine to begin problem solution. When the computer is actively engaged in problem solution, it is said to be in the COMPUTE (or RUN or OPERATE) mode. When the operator wishes to terminate problem solution, he or she switches the machine into the RESET mode. The RESET mode is also called the INITIAL CONDITION mode, for this is the mode that is used to set initial-condition voltages (corresponding to initial values of the variables and their derivatives) in the computer. Recall that n initial conditions must be specified in order to obtain a particular solution to an n-order differential equation. Having been set in the INITIAL CONDITION mode, the initial-condition voltages will appear at the proper locations in the analog circuit at the instant the computer is switched to the COMPUTE mode, that is, at $t = 0$.

Since the output of an integrator used in an analog computer represents a derivative (the $n - 1$ derivative when the input represents the n-derivative), each integrator is equipped with a means for setting an initial-condition voltage. This may be accomplished by automatically switching an adjustable, floating-voltage supply directly across the feedback capacitor when the computer is switched to RESET. The capacitor charges to the value of the floating supply, and this value is then present at the integrator output when the computer is switched to OPERATE. The independent, floating-voltage supplies are automatically switched out of all integrators at $t = 0$ when the computer is switched to the OPERATE mode. Another means of setting initial conditions eliminates the need for floating supplies and is illustrated in Figure 8.35.

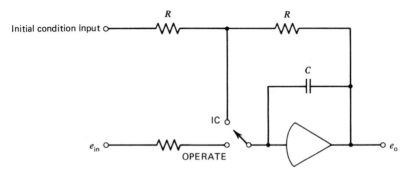

FIGURE 8.35 The initial-condition input to an integrator.

When the switch is in the IC (initial-condition) position, the output voltage e_o is simply the negative of the voltage applied to the initial-condition input. Normal integrator operation of e_{in} occurs when the switch is put in the OPERATE position. Note that the switch must remain in the IC position for several RC time constants in order to allow the capacitor to charge fully before switching back to OPERATE. The initial-condition input to an integrator is shown symbolically in Figure 8.36.

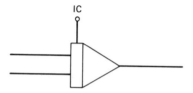

FIGURE 8.36 Symbolic representation of initial-condition input.

Many computers are capable of operation in a REPETITIVE mode. In this mode, the computer automatically switches back and forth between OPERATE and RESET at a relatively rapid rate. The solution to a differential equation up to a certain time T [that is, the voltage $y(t)$ for $0 < t < T$] is thus obtained over and over again. This mode is referred to as REP-OP, short for "repetitive operation." The purpose of REP-OP is to permit the operator to display the solution on an oscilloscope. By synchronizing the sweep rate of the oscilloscope to the period of the REP-OP switching, a stable display of the voltage representing the solution may be viewed. The investigator may then observe instantaneously the effects that changes in problem parameters have on the solution. For example, if we are investigating the effect of damping on the response of a spring-mass-dashpot system, the change in response may be observed while manually rotating a potentiometer connected to the circuit in such a way that it causes the damping to increase or decrease.

Many computers equipped with a REP-OP capability are also equipped with the capability to vary the rate of repetition. Two additional factors that must be considered in designing and implementing an analog computer to solve a differential equation in the REP-OP mode deserve discussion. First, the switching function that causes the computer to alternate between OPERATE and RESET must permit the computer to remain in the RESET mode each time it is switched to RESET long enough to reestablish initial conditions on the integrators. Example 8.14 illustrates this point.

EXAMPLE 8.14

An analog computer that uses integrators whose initial conditions are set using the scheme shown in Figure 8.35 is to be used in the REP-OP mode. If $R = 1$ K and $C = .01$ µF, how long should the computer remain in the RESET mode each time it is switched to RESET?

SOLUTION

We may assume that the feedback capacitor will be essentially fully charged five time-constants after applying an initial-condition voltage. Thus, the time T_R during which the computer must remain reset is

$$T_R = 5RC = 5 \times 10^3 \times 10^{-8} = 50 \text{ µsec}$$

A second consideration in using repetitive operation is the fact that a relatively long period of time may be required to develop the real-time solution to some equations. If, for example, the solution to a differential equation turned out to be $y = e^{-0.1t} \sin 0.3t$, it would take at least 10 seconds to display a representative portion of the voltage $y(t)$. The repetition rate in these cases would have to be so slow that an effective oscilloscope display would not be possible. For this reason, repetitive operation is usually accompanied by *time scaling*. Time scaling permits the computer to develop the solution to a differential equation in a much faster time than real time. This scaling is accomplished by increasing the gain $1/RC$ of all integrators by the same proportion. If it is desired to speed up solution time by a factor k, then all integrator gains must be multiplied by k. This may be accomplished by reducing the feedback capacitance in each integrator by a factor of k, the method used by computers that automatically speeds up solution time when switched into the REP-OP mode.

EXAMPLE 8.15

Figure 8.37 shows an integrator with a switch that is automatically placed in position 1 for real-time solution, in position 2 to speed the operation by a factor of 10, and in position 3 to speed the operation by a factor of 50. If $C_1 = 1.5$ μF, what should be the values for C_2 and C_3?

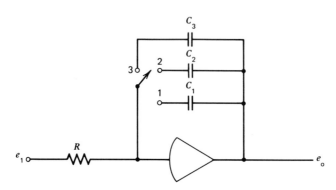

FIGURE 8.37 Find capacitor values for time scaling (Example 8.15).

SOLUTION

When the switch is in position 2, $k = 10$, and we require

$$\frac{1}{RC_2} = \frac{10}{RC_1}$$

That is,

$$C_2 = \frac{C_1}{10} = 0.15 \text{ μF}$$

Similarly,

$$C_3 = \frac{C_1}{50} = 0.03 \text{ μF}$$

When the computer is sped up in this fashion, it is said to be operating in "fast time," while real time is often referred to as "slow time." Fast-time solutions present no problem in interpretation, since it is only necessary to multiply each time division on the time axis of the display by the factor k.

EXAMPLE 8.16

The oscilloscope display shown in Figure 8.38 was obtained from a computer using the integrators in Example 8.14 switched into position 2. The horizontal sweep sensitivity was 0.1 sec/cm. What is the (real-time) frequency of the damped sinusoid?

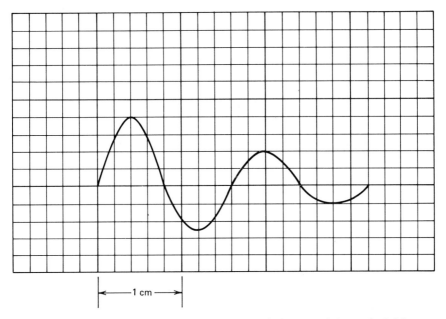

FIGURE 8.38 Find the frequency of the time-scaled sinusoid (Example 8.16).

SOLUTION

The period of the sinusoid is seen to be 1.6 cm. In terms of the oscilloscope sweep, this is

$$(1.6 \text{ cm})(0.1 \text{ sec/cm}) = 0.16 \text{ sec}$$

Since the computer is operating in fast time with $k = 10$, the period is $10 \times 0.16 = 1.6$ sec real time. Thus the frequency is $f = 1/1.6 = 0.625$ Hz.

Another useful mode of operation available in some analog computers is the HOLD mode. When in HOLD, the computer suspends operation and maintains, or holds, the voltages present at each point in the circuit. The HOLD capability permits the operator to interrupt problem solution at any desired time and observe or record values of the variables at that instant of time. Computation is resumed at the same point where it is interrupted by switching back to OPERATE. The computer is put into a HOLD mode by automatically disconnecting all inputs to each integrator. In this configuration, the feedback capacitors are acting as storage elements, and since the capacitors must ultimately discharge, the computer cannot be left in a HOLD mode for a long period of time.

Figure 8.39 shows a switching arrangement that could be used to place an integrator in any one of the modes OPERATE, HOLD, or RESET. In older computers, these switching functions were performed by relays. Modern analog and hybrid computers employ electronic switches, which are faster, generate less noise, and consume less power than relays. Mode control by electronic switching is discussed further in Section 8.8, Hybrid Computers.

304 ANALOG COMPUTATION AND SIMULATION

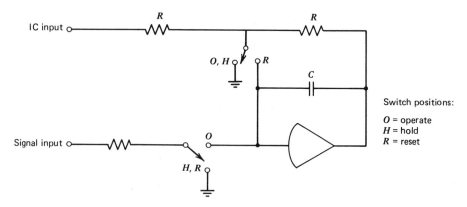

FIGURE 8.39 Mode switching arrangement for an integrator.

8.6 MAGNITUDE SCALING

Our discussion of programming techniques for the analog computer has thus far omitted one important consideration, How do we translate voltages measured at the outputs of the operational amplifiers into the units of the variables of the actual problem we are solving? If we wish to simulate a control system whose output is angular rotation θ and if θ in the actual system varies between 0° and 300°, must we have an amplifier capable of producing 300 V? Clearly, if that were the case, it would be a most impractical situation. Most modern general-purpose analog computers have amplifiers whose outputs are limited to plus or minus 10 V. Obviously, we must devise some way to scale the magnitudes of the *computer variables* (voltages) in such a way that these voltages have practical values and yet are able to represent the magnitudes we expect the *problem variables* to have.

We must perform this magnitude scaling in such a way that the resulting voltages are neither too large nor too small. A voltage becomes too large if in order to represent the total variation of a problem variable, it must exceed the rated maximum output voltage of an amplifier. This maximum voltage is usually called the computer's *reference* voltage. Magnitude scaling produces voltages that are too small if the maximum values attained are in or near the noise region of the amplifiers. We would not want the maximum voltage appearing at an amplifier during the course of a problem solution to be one millivolt (mV), for example. As a general rule, magnitude scaling should result in output voltages at each amplifier that are as near as possible to the maximum rated voltage of the amplifier without exceeding it.

As mentioned in an earlier section, the general-purpose analog computer is equipped with coefficient potentiometers that can be used to scale voltages by a fixed factor between 0 and 1.0. Figure 8.40 illustrates the symbol for a coefficient potentiometer that has been set to 0.25 as well as the actual circuit schematic and its effect on the signal. Coefficient potentiometers are very useful in magnitude scaling and are, in fact, a necessary part of the general-purpose computer whose amplifiers can only be patched to produce certain fixed gains, as 1 and 10.

A number of different methods have been developed for the actual procedure of magnitude scaling. These methods lead to a variety of ways in which variable

8.6 MAGNITUDE SCALING

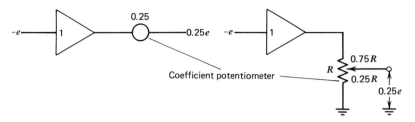

FIGURE 8.40 Equivalent representations of a coefficient potentiometer.

values are represented on the computer diagrams and to a corresponding variety of interpretations. We will discuss only a method that seems to have gained the most popularity in recent years, a method that does *not* depend on a knowledge of the computer's reference voltage. It is called *normalized*, or *dimensionless* scaling, and has the advantage that diagrams produced by this method can be used on any machine. Whatever method is used, it must be emphasized that it is necessary to have some knowledge of the maximum value that each problem variable will attain during the course of a computer run. This knowledge may be based on estimates, insight, previous experience, or trial and error. We will describe one method for estimating such values for one type of problem in a later section.

The normalized method of scaling requires each problem variable to be multiplied and divided in the original differential equation by the maximum value it might attain. Thus, for example, if it is known that $X < 4$ m, $\dot{X} < 20$ m/sec, and $\ddot{X} < 80$ m/sec^2 in the differential equation $\ddot{X} + 5\dot{X} + 25X = 50$, we would write

$$80(\ddot{X}/80) + 5 \cdot 20(\dot{X}/20) + 25 \cdot 4(X/4) = 50$$

We then proceed to implement the computer solutions in the way we have previously described, except that the variables we develop and show on the diagram are $\ddot{X}/80$, $\dot{X}/20$, and $X/4$, instead of $\ddot{X}$, $\dot{X}$, and X. Since the maximum of each of the implemented quantities is *one*, we interpret actual voltages obtained on the computer as percentages of the reference voltage. Thus, for example, if we were using a computer with a 10-V reference to solve our problem and we happened to read 2 V for X, we would interpret this as $(2/10)(4) = 0.8$ m. Similarly, an 8-V reading at the output of the amplifier producing $\dot{X}$ would be interpreted as $(8/10)(20) = 16$ m/sec. Coefficient potentiometers are used as necessary to develop the correct multiplying factors for each of the quantities $(X/4)$, $(\dot{X}/20)$, and so forth. Example 8.17 illustrates the method.

EXAMPLE 8.17

Assuming that $\theta \leq 2.5$ rad, $\dot{\theta} \leq 20$ rad/sec, and $\ddot{\theta} \leq 160$ rad/sec^2, draw a magnitude-scaled analog computer diagram to solve the equation

$$\ddot{\theta} + 2\dot{\theta} + 64\theta = -80$$

for $\ddot{\theta}$, $\dot{\theta}$, and θ. Assume zero initial conditions.

SOLUTION

$$160\left(\frac{\ddot{\theta}}{160}\right) + 2 \times 20\left(\frac{\dot{\theta}}{20}\right) + 64 \times 2.5\left(\frac{\theta}{2.5}\right) = -80$$

Solving for $\ddot{\theta}/160$ in the usual way,

$$\frac{\ddot{\theta}}{160} = \frac{-2 \times 20}{160}\left(\frac{\dot{\theta}}{20}\right) - \frac{64 \times 2.5}{160}\left(\frac{\theta}{2.5}\right) - \frac{80}{160}$$

$$\frac{\ddot{\theta}}{160} = -0.25\left(\frac{\dot{\theta}}{20}\right) - 0.4\left(\frac{\theta}{2.5}\right) - 0.5$$

The computer diagram is shown in Figure 8.41.

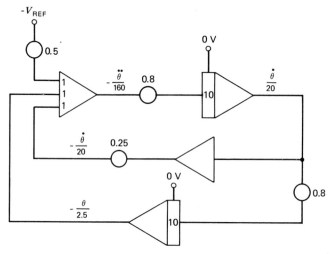

FIGURE 8.41 A magnitude-scaled circuit using coefficient potentiometers (Example 8.17).

Note that in Figure 8.41 the output of each integrator will be a voltage that is near the maximum computer-reference voltage. If, for example, $\dot{\theta}$ were actually to reach 20 rad/sec and if $V_{REF} = 10$ V, then the output of the integrator producing $\dot{\theta}/20$ would be 10 V, since

$$\frac{10}{10}(20) = 20 \text{ rad/sec}$$

Thus, an integrator should never be followed by an amplifier with a gain greater than one unless that gain is in another integrator. An integrator following an integrator will generate another variable (derivative) that will in general have a different scale factor. Note also how the coefficient potentiometers are used in Figure 8.41. For example, a potentiometer set to 0.8 is used to multiply $\dot{\theta}/20$ by

0.8, and this result is then simultaneously integrated and multiplied by 10 to produce

$$-10 \int 0.8(\dot{\theta}/20)\,dt = -0.8\theta = -\theta/2.5$$

A plot of $-\theta$ versus time as produced by this computer simulation on a 10-V machine would resemble that shown in Figure 8.42. We see that the absolute value of the maximum value of θ is

$$\left(\frac{10}{10}\right)(2.5) = 2.5 \text{ rad}$$

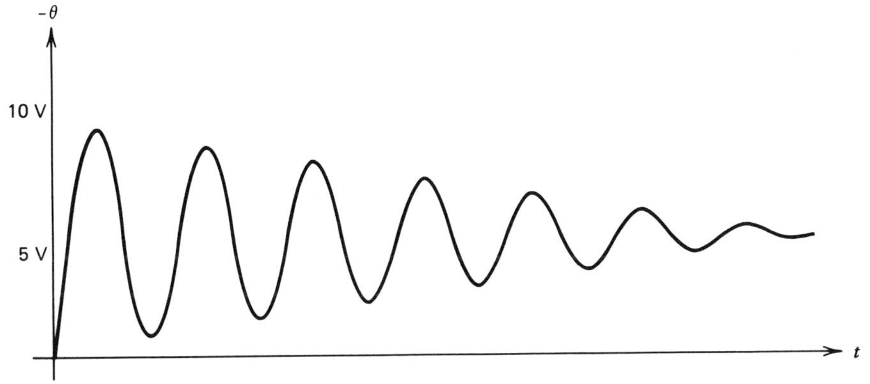

FIGURE 8.42 $-\theta$ versus time (Example 8.17).

and that after a long period of time

$$\theta = \left(\frac{5}{10}\right)(2.5) = 1.25 \text{ rad}$$

As indicated earlier, an estimate of the maximum values of the problem variables must be made in order to perform magnitude scaling. In complicated problems, this estimate may be nothing more than an educated guess, which may then be further refined by trial and error programming. In simpler problems, analyzing the mathematics of the problem may pay off in time saved or at least provide the programmer with some insight as a basis for estimated magnitudes. Before proceeding with an example, let us point out that what we seek are actually *upper bounds* on the variables and not necessarily maximum values. Note that a maximum value is a value that is actually reached by a variable, and though it would be nice if we knew the maximum of each problem variable, in practice we seldom have enough information about the problem to know actual maximum values. (Most books are very misleading in this respect: The term maximum is used when upper bound is meant.) An upper bound is simply a value that we know a variable will never exceed. Infinity is an upper bound for every variable, though not a very useful one. Obviously, we would like our upper bound to be as close to the actual maximum as possible, since scaling on the basis of upper bounds that are too large will result in computer voltages that are unnecessarily small.

Consider the second-order differential equation

$$\ddot{X} + 2\xi\omega_n \dot{X} + \omega_n^2 X = k$$

where k is a constant. Recall that this equation could represent the behavior of a spring-mass-damper system subjected to a step input of k units. We know that the underdamped system ($\zeta < 1$) will have an oscillatory response whose maximum value increases as ζ decreases. In the zero-damped case ($\zeta = 0$), the response is a sustained oscillation that rises k/ω_n^2 units above and below k/ω_n^2. See Figure 8.43.

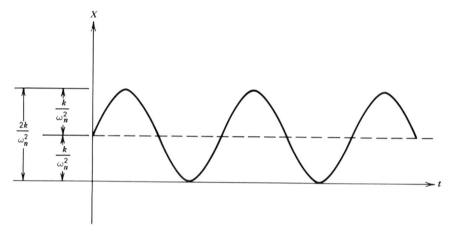

FIGURE 8.43 Underdamped response of a second-order system showing maximum excursions.

Thus, an upper bound for this differential equation is $2k/\omega_n^2$. This is a good choice for scaling X in second-order systems with small damping. In Example 8.16, $\zeta = 0.125$, we used $2k/\omega_n^2 = 2 \times 80/64 = 2.5$ to scale θ. Note that in Figure 8.42 θ reached (actually very nearly reached) the maximum of 2.5. In these oscillatory cases, a good choice of upper bounds for $\dot{X}$ and $\ddot{X}$ is

$$\dot{X} \leq \omega_n X_{ub}$$

$$\ddot{X} \leq \omega_n^2 X_{ub}$$

where X_{ub} is the upper bound chosen for X. These results follow from the fact that the derivative of a sinusoidal function of the form $\sin \omega_n t$ is another sinusoidal function of the form $\omega_n \sin \omega_n t$.

In the overdamped case ($\zeta > 1$), there is no overshoot in the response and $X \leq k/\omega_n^2$. In the spring-mass-damper system, this is the steady-state value of displacement X that results when the spring force equals the applied force. Of course, ω_n^2 has no frequency interpretation in this case.

8.7 NONLINEAR DIFFERENTIAL EQUATIONS

A differential equation is *nonlinear* if (1) a variable or one of its derivatives is raised to a power other than unity or (2) a variable or one of its derivatives is multiplied by a variable or a derivative. Example 8.18 illustrates several nonlinear differential equations.

EXAMPLE 8.18

The following are nonlinear differential equations:

(a) $\ddot{y}^2 + 4\dot{y} - 2y = \sin 10t$

(b) $\ddot{y} + 3y(\dot{y} + 1) = 2t$

(c) $\ddot{y} - 8(\dot{y})^{1/2} + yt = 0$

SOLUTION

(a) is nonlinear because the second derivative is squared; (b) is nonlinear because of the product term $3y\dot{y}$; and (c) is nonlinear because $\dot{y}$ is raised to the 0.5 power.

In order to construct the analog-computer circuitry necessary to solve a nonlinear differential equation, it is frequently necessary to have an electronic device that is capable of multiplying two *time-varying* signals together. In part (b) of Example 8.16, we would have to be able to multiply $\dot{y}$ and y, both of which are varying with time. Contrast this multiplication requirement with that of multiplying by a constant, namely, scaling.

Modern analog computers are equipped with electronic multipliers that can be patched into operational-amplifier circuits to solve nonlinear differential equations. Electronic multipliers can also be obtained in separate modules for incorporation into many analog systems, including control systems. Many such multipliers are based upon the *quarter-square* principle. Quarter-square multiplication requires using two squaring circuits, that is, two devices each of whose outputs is proportional to the square of its input. Given two such devices, the product of variables x and y can be obtained by implementing the quantity

$$\frac{1}{4}[(x + y)^2 - (x - y)^2] \qquad (14)$$

Algebraic expansion of Equation 14 will show that it is equal to the product xy. The circuit in Figure 8.44 illustrates how this multiplication could be implemented.

Multipliers are classified as being two-quadrant or four-quadrant. A four-quadrant multiplier can accept inputs that are either positive or negative and produces a product with the correct algebraic sign according to the signs of the inputs. In a two-quadrant multiplier, however, only one input may be either positive or negative,

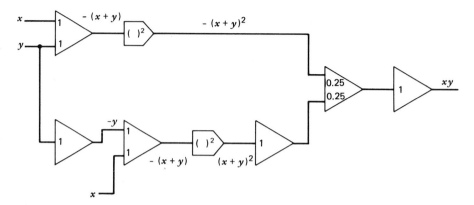

FIGURE 8.44 Implementation of a multiplier using the quarter-square principle.

while the other input must always be positive. The symbol for a multiplier is shown in Figure 8.45. In practice, the output of the multiplier is frequently an attenuated version of the product, for example, $xy/10$. Example 8.19 illustrates this case.

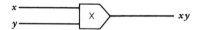

FIGURE 8.45 Symbolic representation of a multiplier.

EXAMPLE 8.19

Use the specifications provided in Appendix B for the Analog Devices Model 429 Multiplier, Divider, Square Rooter, to write a mathematical expression for the output of this device when it is connected as a multiplier, and its inputs are as follows:

(a) $X = +5$ V dc, $Y = +8$ V dc
(b) $X = +5$ V dc, $Y = 10 \sin 40t$
(c) $X = -4 \sin 100t$, $Y = 4 \sin 100t$

SOLUTION

The product literature and specifications for the Model 429 reveal that is is a four-quadrant multiplier with transfer function $XY/10$, has maximum output voltage ± 11 V, and a 300-kHz bandwidth for 1% maximum error. Therefore, the X and Y inputs in (a), (b), and (c), above all, fall within the specified ranges and produce outputs of

(a) $\dfrac{XY}{10} = +4$ V dc

(b $\dfrac{XY}{10} = 5 \sin 40t$

(c) $\dfrac{XY}{10} = -1.6 \sin^2 100t$

EXAMPLE 8.20

Draw an operational amplifier circuit capable of solving the differential equation

$$2\ddot{y} - 4y\dot{y} = 8yt$$

Initial conditions are $\dot{y}(0) = y(0) = 0$.

SOLUTION

We first solve the the highest order derivative in the usual way

$$\ddot{y} = 2y\dot{y} + 4yt$$

The function $4yt$ will be developed using a multiplier to form the product of $4y$ and t. To obtain a voltage proportional to t, we integrate the constant voltage -1 V, as shown in Figure 8.46. The complete circuit is shown in Figure 8.47.

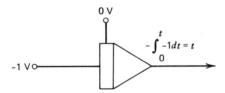

FIGURE 8.46 Generating t-term required for solving a nonlinear differential equation (Example 8.20).

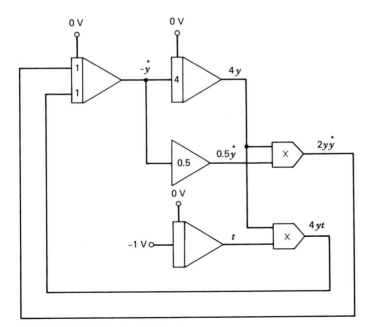

FIGURE 8.47 Analog-computer solution of a nonlinear differential equation (Example 8.20).

A nonlinear differential equation may require that either or both the dependent and independent variables (y and t) be subjected to a variety of other nonlinear operations. Programmable *function generators* are often used to simulate these operations. With a function generator, we can approximate a nonlinear operation by programming *break points* corresponding to the ends of straight line segments that approximate the desired function. These straight line segments determine different values of gain, which the function generator switches in at appropriate levels of the input. Figure 8.48 illustrates a straight-line-segment approximation of a particular nonlinear function.

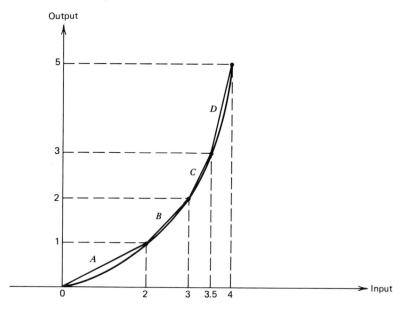

FIGURE 8.48 Straight-line-segment approximation of a nonlinear function.

In Figure 8.48, we see that the function is approximated by four straight-line segments labeled A through D. Since the slope of segment A is 0.5, the function generator must have gain 0.5 for all input voltages with levels between zero and 2 V. The slope of segment B is $[(2 - 1)/(3 - 2)] = 1$, and so the function generator has unity gain for input levels between 2 and 3 V. Similarly, it has gain 2 for inputs in the range 3 to 3.5 V and gain 4 for inputs greater than 3.5 V. Notice that it has zero gain for all negative inputs. The value of the output when the input is zero (also zero in this example) is called the *parallax*.

Nonlinear function generators are often constructed by using biased diodes in the input and feedback paths of an operational amplifier. The bias voltage on each diode is set for the operator in such a way that no current flows through the path containing the diode until the desired break-point voltage is reached. Function generators that are constructed using this technique are called diode function generators (DFGs). Figure 8.49 illustrates how the nonlinear function of Figure 8.48 could be implemented by a DFG.

For voltages in the range 0 to 2 V, diode D1 conducts the signal through an input resistance of $2R$, and so the amplifier gain is $R/2R = 0.5$. For inputs between

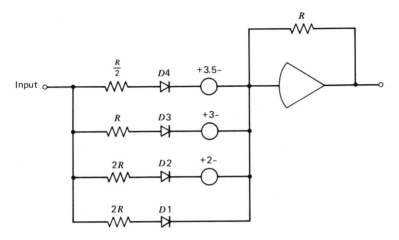

FIGURE 8.49 A diode-function generator for the function in Figure 8.48.

2 and 3 V, diode D2 also conducts, making the total resistance in the input path equal to $2R \| 2R = R$. Thus, the gain becomes $R/R = 1$ in this region. Each time the input reaches a breakpoint, another diode conducts, and the amplifier gain increases due to reduced overall input resistance. Diode D1 ensures zero gain for negative inputs. Due to inversion by the amplifier, the output is, of course, negative and would have to be inverted again with a unity-gain amplifier to achieve the characteristic of Figure 8.48.

Figure 8.49 illustrates the biased-diode principle of function generation but is not necessarily the best implementation. The voltage sources shown in this circuit are floating, which is undesirable from a practical standpoint. Also, the nonlinear characteristics of real diodes will cause the gain values to be in error until there is sufficient current flowing through each diode to bias it well into its conduction region. The actual gain, for example, will be much smaller than the desired value of 0.5 when the input voltage is 0.1 V due to the fact that D1 will not yet be fully conducting. There are more complex circuits available to overcome these difficulties when greater precision is required, but we cannot pursue this subject in any further depth here.

A small general-purpose analog computer, the Comdyna Model GP-6, is illustrated in Appendix B. This computer can be patched to multiply and divide variables as well as to perform the square and square-root functions. The computer contains eight amplifiers, four of which may be patched as integrators or as summers, two more of which may be patched only as summers, and two of which are used only as unity-gain inverters. Also clearly visible are the eight coefficient potentiometers. The computer may be operated conventionally or in fast-time repetitive operation.

8.8 HYBRID COMPUTERS

Most modern analog computers incorporate digital-logic circuitry to control, monitor, record, or aid in simulating analog variables and are, therefore, said to be hybrid computers. The degree of hybridization, that is, the extent to which digital-

314 ANALOG COMPUTATION AND SIMULATION

logic circuitry is employed, varies from computer to computer. In some versions, the digital equipment is limited to digital voltmeters and switching circuits that are used to monitor amplifier outputs and as an aid in setting initial conditions and coefficient potentiometers. These computers cannot truly be characterized as hybrid. At the other extreme, the analog computer may share computational responsibilities with a digital computer, and communicate with the digital computer through analog-to-digital (A/D) and digital-to-analog (D/A) interface circuitry. This arrangement would more properly be termed a hybrid computer *system*. A majority of the truly hybrid computers utilize digital-logic circuits to control the automatic iteration of analog computations. That is, the function of the logic circuitry is to monitor certain analog variables during a computer run, automatically adjust systems parameters, initiate another run, and repeat until the variables meet some predetermined criterion. In a sense, this kind of operation is similar to REP-OP, where the human function of monitoring the response and making adjustments until the desired response is observed has been replaced by digital-logic circuitry. In this section, we will discuss the basic hardware complement of a typical hybrid computer, the terminology and conventions, but avoid an in-depth treatment of applications. Suffice it to say that hybrid computers are capable of solving some very sophisticated mathematical problems and of simulating complex sampled-data systems, topics that are beyond the scope of this book.

One fundamental component of the hybrid computer is the voltage *comparator*. The comparator, as the name implies, compares two input voltages and produces a binary output (0 or 1) depending on the result of this comparison. The standard symbol for a comparator is shown in Figure 8.50. The output of the com-

FIGURE 8.50 The voltage comparator.

parator is labeled U, while its complement is $\overline{U}$. The operation of the comparator is defined by

$$\left.\begin{array}{l} U = 0 \\ \overline{U} = 1 \end{array}\right\} \text{ if } E_1 + E_2 \leq 0$$

$$\left.\begin{array}{l} U = 1 \\ \overline{U} = 0 \end{array}\right\} \text{ if } E_1 + E_2 > 0$$

Note that these relations are equivalent to stating that $U = 0$ if $E_1 \leq -E_2$ and $U = 1$ if $E_1 > -E_2$.

EXAMPLE 8.21

Assuming the switch S in Figure 8.51 closes at $t = 0$, at what time t will U change from 1 to 0?

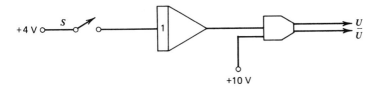

FIGURE 8.51 Find the time t when the comparator switches (Example 8.21).

SOLUTION

The output of the integrator is $-\int 4\,dt = -4t$, that is, a ramp voltage that decreases in magnitude at the rate of 4 V/sec. When $-4t + 10 = 0$, the comparator switches state. Thus, U becomes 0 after $t = 2.5$ sec. Example 8.21 illustrates one technique by which a comparator can be used to cause the computer to switch modes (as from COMPUTE to RESET) after a specified length of time.

Another fundamental component of a hybrid computer is the *digital-analog switch* (DAS), more familiarly known as an electronic switch. The DAS is simply a single-pole switch that is controlled by a digital voltage (binary 1 closes the switch and binary 0 opens it). The switch is used to connect or disconnect an analog signal in a circuit and is shown schematically in Figure 8.52.

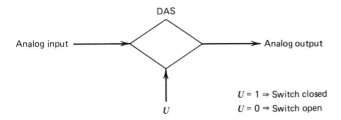

FIGURE 8.52 The digital-analog switch.

Figure 8.53 illustrates how two such switches could be used with a comparator to generate the absolute value of an analog signal x. In reference to Figure 8.53, when the analog voltage x exceeds zero, the output U of the comparator closes DAS1, which connects x to the output, while $\overline{U} = 0$ keeps DAS2 open. If x goes negative, then $U = 0$ and $\overline{U} = 1$, so DAS2 connects $-x$ to the output. In either case, the output is always positive and, therefore, equal to $|x|$.

We have already discussed the fact that the circuit configuration of each integrator in an analog computer will be changed when the computer undergoes changes in mode (OPERATE, RESET, and HOLD). (See, for example, Figure 8.39.) These changes in circuit configuration are usually accomplished using digital-analog switches, which are in turn driven by comparators used to monitor variables being generated during a computer run. This is the essence of iterative computation using repetitive operation. When an integrator is switched to the configuration corre-

316 ANALOG COMPUTATION AND SIMULATION

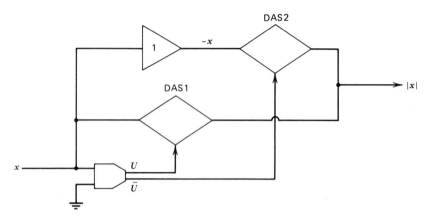

FIGURE 8.53 Using a voltage comparator and digital-analog switches to generate absolute value.

sponding to RESET, or IC, then whatever voltage is applied to its IC input appears (inverted) at its output. If the voltage applied to its IC terminal is varying with time, then so will the output of the integrator. In this situation, the integrator is said to TRACK the (IC) input. If the integrator is then switched to its HOLD configuration, its output will hold whatever value it was tracking (except for sign inversion) at the instant when it was switched. When an integrator is used in this fashion, it is said to be a TRACK-HOLD circuit and is given the special symbol shown in Figure 8.54. [Note that some authors refer to this as a track and *store* (TS) circuit.] Specially designed *sample/hold* amplifier modules, which perform the same functions as TRACK-HOLD circuits, are also available commercially. The Analog Devices Model AD582, for which product literature may be found in Appendix B, is an example. Note the oscilloscope display of a sine wave that is sampled (tracked) and held at alternately positive and negative voltage levels.

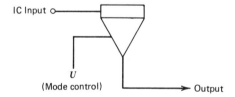

FIGURE 8.54 The TRACK-HOLD circuit.

When two T-H circuits are cascaded and their mode control inputs are driven by complementary logic signals, we obtain a new device known variously as a *memory pair, ratchet,* or *zero-order hold.* Figure 8.55 illustrates this device.

Let us assume that the mode-control input on each T-H circuit is such that a logical 1 applied there causes it to track, while a logical 0 causes it to hold. Suppose then that a time-varying signal $y(t)$ is applied to the input of the memory pair, while a square wave (alternately 0 and 1) is applied to the control input. When the square

8.8 HYBRID COMPUTERS 317

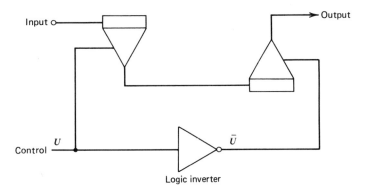

FIGURE 8.55 A memory-pair, ratchet, or zero-order hold.

wave is logical 1, the first T-H is tracking while the second T-H is holding. When the square wave switches to zero, the second T-H tracks the output of the first T-H. But since the first T-H is now holding, the output of the second T-H is constant. Thus, the output of the memory pair (that is, of the second T-H) is a series of *steps* that represent the value of $y(t)$ at intervals of time corresponding to the duty cycle of the square wave. This result is illustrated in Figure 8.56, where we have assumed a triangular-pulse input to the memory pair.

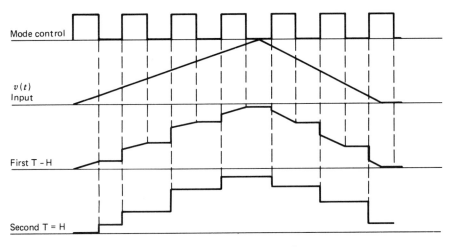

FIGURE 8.56 Output of a memory pair with a triangular input.

Figure 8.56 shows that the output of a memory pair is a discrete-level approximation of the input. The signal is said to have been *quantized* and is typical of so-called *sampled-data* control systems, which respond to periodic signal samples rather than to continuous, analog-type signals. Hybrid computers can be used effectively to simulate such systems and to investigate the effects of quantizing error, the error inherent in approximating a continuous signal by a sequence of fixed voltage levels. The memory pair is also used in hybrid computation to save computed variable values at the end of a computer run and to provide these as starting values in a successive computer run, that is, during iterative computations.

Figure 8.57 shows how a memory pair can be used with two integrators and a comparator to generate a sequence of discrete voltage levels proportional to the product of two time-varying functions x and t. When this configuration is operated in REP-OP, the output is the quantized product of the inputs. This technique is called *time-division multiplication*. Integrators I_1 and I_2 are both operated in

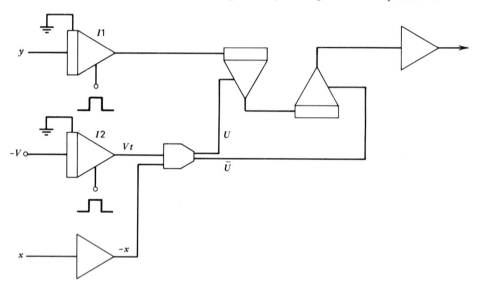

FIGURE 8.57 Time-division multiplication.

REP-OP, that is, they are alternately switched between RESET and OPERATE. Let t_0 be the total time they are in the OPERATE mode during each REP-OP cycle. $-V$ is a fixed reference voltage, so during OPERATE time, the output of integrator $I2$ is the ramp voltage Vt (a sawtooth waveform when viewed over several REP-OP cycles). When $Vt = x$, that is, when $t = x/V$, the comparator switches and causes the memory pair to hold the output of $I1$ at that instant. Let t_1 be the time at which this switching occurs, $t_1 = x/V$. The output of integrator $I1$ at this instant is

$$-\int_0^{t_1} y\, dt \qquad (15)$$

If the REP-OP frequency is much greater than the frequency content of y, that is if t_0 is very small compared to time variations in y, then y is essentially constant from $t = 0$ to $t = t_0$. Since $t_1 < t_0$, the same can be said for y in the interval from 0 to t_1. In this case, Equation 15 is a good approximation of the quantity $-yt_1$. But since $t_1 = x/V$, we have

$$-yt_1 = -y\left(\frac{x}{V}\right) = \frac{-xy}{V}$$

Thus, the memory pair holds a voltage value proportional to the (negative) product of x and y. The inverting amplifier produces a positive output. This sequence is repeated during each COMPUTE cycle of the REP-OP, so the output of the multiplier over a period of time is a stairstep (quantized) approximation of the product xy/V. We note that it would also be possible to hold x constant and allow V to represent some time-varying signal, in which case the system would perform mathematical *division*.

REFERENCES

1. **Bekey, George A.**, and **Walter J. Karplus.** *Hybrid Computation.* Wiley, 1968.
2. **Burak, Aram.** *Passive and Active Network Analysis and Synthesis.* Houghton-Mifflin, 1974.
3. **Coughlin, Robert F.**, and **Frederick F. Driscoll.** *Operational Amplifiers and Linear Integrated Circuits.* Prentice-Hall, 1977.
4. **Faulkenberry, Luces M.** *An Introduction to Operational Amplifiers.* Wiley, 1977.
5. **Graeme, Jerald D.** *Applications of Operational Amplifiers, Third-Generation Techniques.* McGraw-Hill, 1973.
6. **Graeme, Jerald G.** *Designing with Operational Amplifiers, Applications, and Alternatives.* McGraw-Hill, 1977.
7. **Jackson, Albert S.** *Analog Computation.* McGraw-Hill, 1960.
8. **Johnson, David E.**, and **John L. Hilburn.** *Rapid Practical Design of Active Filters.* Wiley, 1975.
9. **Jung, Walter G.** *IC Op-Amp Cookbook.* Howard W. Sams, 1977.
10. **Kochenburger, Ralph J.** *Computer Simulation of Dynamic Systems.* Prentice-Hall, 1972.
11. **Korn, Granino A.**, and **Theresa M. Korn.** *Electronic Analog and Hybrid Computers.* McGraw-Hill, 1964.
12. **Lancaster, Don.** *Active Filter Cookbook.* Howard E. Sams & Co., 1975.
13. **Stephenson, Robert E.** *Computer Simulation for Engineers.* Harcourt Brace Jovanovich, 1971.
14. **Stout, D. F.**, and **Milton Kaufman.** *Handbook of Operational Amplifier Circuit Design.* McGraw-Hill, 1976.
15. **Wong, Y. J.**, and **William E. Ott.** *Function Circuits, Design and Applications.* McGraw-Hill, 1976.

EXERCISES

8.1 A certain operational amplifier has the following specifications:

1. Maximum output voltage: ± 10 V
2. Maximum output current: ± 4 mA
3. Frequency response: ± 1 DB, 0 to 1 kHz

4. Maximum slew rate: 1 V/μsec

   (a) How many operational amplifiers with 6-K input resistors can it drive simultaneously?

   (b) Will the slew-rate specification be exceeded if the amplifier is driven by a 10-V peak sine wave at 1 kHz?

   (c) What is the minimum permissible rise time of a 5-V pulse for the rated slew rate of the amplifier not to be exceeded?

8.2 Use the specifications provided in Appendix B for the Analog Devices operational-amplifier model AD517J to answer the following questions:

(a) Over what frequency range is the amplifier response flat when its closed-loop gain is 100?

(b) With the specified slew rate, can the amplifier be used in a unity-gain application for a triangular wave that varies between 0 and 10 V and has frequency 100 kHz? Explain.

(c) What is the maximum permissible frequency of a triangular wave that varies between 0 and 10 V?

8.3 (a) An operational amplifier has an open-loop voltage gain of 10,000. It is used in a circuit designed to amplify a signal voltage by 2 by connecting a feedback resistor of 100 K and an input resistor of 50K to it. If the amplifier input impedance is so large that input current to the amplifier is negligible, calculate the *exact* voltage gain of the configuration (taking into account the fact that the open-loop gain is not infinite).

(b) Repeat, if the open-loop gain is 1000.

8.4 Design operational-amplifier circuits that produce the following outputs:

(a) $2e_1 - 0.5e_2$

(b) $40(e_1 + 0.1e_2) - 5e_3$

(c) $\dfrac{2e_1}{3} - e_2 + .04e_3 - 12$

(d) $5 - 18(e_1 - e_2)$

8.5 A 10-K, ten-turn precision potentiometer with $+10$ V connected to one terminal and the opposite terminal grounded has a calibrated dial to indicate the precise setting of its wiper arm. If the calibrated setting 0.00 corresponds to zero turns (0 V out) and 10.00 corresponds to a full ten turns ($+10$ V out), what should the setting be in order to provide $+6.75$ V to a 20-K input resistor of an operational amplifier?

8.6 Use LaPlace transform analysis to prove that

$$e_o = -\frac{1}{R_1 C}\int e_1\, dt - \frac{1}{R_2 C}\int e_2\, dt$$

in the circuit shown in Figure 8.58.

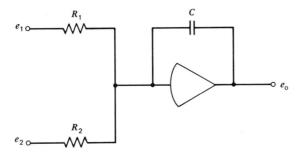

FIGURE 8.58

8.7 Design operational-amplifier circuits to produce the following:

(a) $2\int(e_2 - e_2)dt$
(b) $\int 0.5 e_1 dt + 10\int e_2 dt - 20$
(c) $e_1 + 8\int(e_1 - 2e_2)dt + 5(e_2 - 2)$
(d) $25 e_1 + 12.5 \int e_2 dt - 1.5 \int(e_3 + 4)dt$

8.8 Find the transfer functions of the circuits in Figure 8.59:

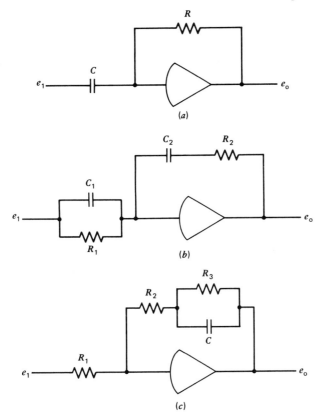

FIGURE 8.59

322 ANALOG COMPUTATION AND SIMULATION

8.9 Draw operational-amplifier circuits that can be used to solve the following differential equations (it is not necessary to show actual component values used):

(a) $3\ddot{y} - 6\dot{y} + 18y = 9$
(b) $\ddot{y} + 4\dot{y} + 2y = 12t$
(c) $8.5\ddot{y} - 17\dot{y} - 21.25y = 38.25 \sin t$
(d) $2\ddot{y} + 8\dot{y} = y + 6t - 10$

8.10 Draw operational-amplifier circuits that will generate the response $C(t)$ of the control systems shown in Figure 8.60 when each is subjected to a unit-step input (it is not necessary to show actual component values used).

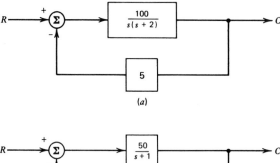

(a)

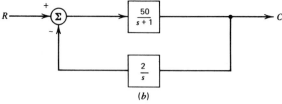

(b)

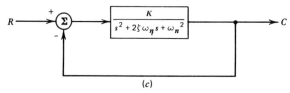

(c)

FIGURE 8.60

8.11 Draw an analog-computer circuit that could be used to solve for the displacement $x(t)$ in the system shown in Figure 8.61 after being subjected to a step-force input of F N at $t = 0$.

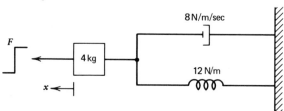

FIGURE 8.61

EXERCISES 323

8.12 The circuit in Figure 8.35 is to be used to switch the integrators of an analog computer between OPERATE and RESET during repetitive operation. If C = 0.1 μF and the time allowed for RESET is 0.1 msec, what should be the value of R?

8.13 The solution to a particular differential equation is $10 + 8 \sin 0.4\pi t$. It is necessary to observe one full cycle of the sinusoidal portion of this solution in a REP-OP mode that generates ten solutions per second. If the integrating capacitors of the analog computer are 1.0 μF in slow time, what should their value be in REP-OP?

8.14 Draw analog-computer-circuit diagrams, including amplitude scaling, to solve the following equations:

(a) $y + 49y = 100$
(b) $2y + 20y + 128y = -80$
(c) $y + 2y + 16y = -64$

8.15 Draw operational-amplifier circuits capable of solving the following differential equations:

(a) $4\ddot{y} - 2\dot{y}y = 8t$
(b) $\ddot{y} + 8y = \dot{y} \sin 10t$
(c) $9.3\ddot{y} + 37.2(\dot{y})^2 = 23.25 - (y)(13.95t)$

8.16 Prove that implementing Equation 14 yields the product xy.

8.17 Draw a straight-line-segment approximation of the function $y = 1 - e^{-t}$ for $0 < t < 3$, using 5 segments. Determine the slope of each segment and draw a DFG circuit, using an operational amplifier that will generate the approximation. What is the maximum percent deviation between your straight-line-segment approximation and the actual function?

8.18 What parallax is required for a DFG that generates a straight-line-segment approximation of the function $y = 12e^{-10t} + 2$?

8.19 the circuit in Figure 8.62 is designed to swtich the mode of integrator I1 20 msec after switch S1 is closed. What should be the value of V?

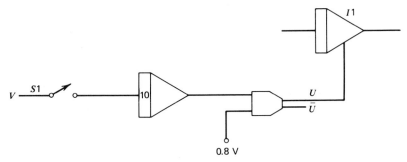

FIGURE 8.62

324 ANALOG COMPUTATION AND SIMULATION

8.20 Use an integrator and a voltage comparator to design a circuit that will produce a sawtooth waveform with peak value 5 V and period 0.1 sec.

8.21 The control input in Figure 8.54 is a square wave that alternates between logic 1 and logic 0 with a period of .05 sec. The IC input to the first integrator is $10 \sin 4\pi t$. Plot the output of the memory pair from $t = 0$ to $t = 0.5$ sec.

8.22 The mode control signal on integrators $I1$ and $I2$ in the multiplier in Figure 8.56 is a square wave that alternates between logic 0 and logic 1 with a period of 10 msec. If $x = 5 \sin 20\pi t$, y is a sawtooth with peak value 1 V and period 0.1 sec, and $V = 1$ V, plot the output of the multiplier from $t = 0$ to $t = 0.1$ sec.

ANSWERS TO EXERCISES

8.1 (a) 2
    (b) No
    (c) 5 μsec

8.2 (a) Approximately 4 kHz.
    (b) No. The waveform rises 10 V in 5 μsec, or 2 V/μsec, which exceeds the slew rate.
    (c) 5 kHz

8.3 (a) -1.9994
    (b) -1.994

8.5   7.40

8.8 (a) $-RCs$
    (b) $-(R_1C_1s + 1)(R_2C_2s + 1)/R_1C_2s$
    (c) $-(R_2R_3Cs + R_2 + R_3)/R_1(R_3Cs + 1)$

8.12   200 Ω

8.13   0.02 μf

8.18   14 V

8.19   4 V

CHAPTER 9
DIGITAL COMPUTER SIMULATION

9.1 ROLE OF THE DIGITAL COMPUTER

The advantages of using a digital computer to analyze and simulate a control system, as opposed to an analog computer, include the following:

(1) Accuracy

The accuracy of digital-computer simulation is limited only by the number of significant digits the investigator is willing to use, the accuracy of any input data that may have been gathered experimentally, and by the ability of the programmer to model the system faithfully (see Section 1.4). Unlike the analog computer, accuracy is not affected by measurement error, instrument error, or changes in hardware characteristics due to environmental factors.

(2) Ease of Programming

The background and training required to program a digital computer is generally less extensive than that required to program an analog computer. Knowledge of a user-oriented digital-computer language and basic mathematical skills are usually sufficient. The programmer need not be concerned with the electronics of the computer he or she is using. This is not always the case with an analog computer, since the user must often consider such factors as circuit loading, limits on voltage swings, amplifier overloading, noise, frequency response and bandwidth, environmental effects, and so forth. Debugging an analog-computer setup is also complicated by these factors.

(3) Convenience and Accessibility

Nearly every educational and industrial facility of any size has access to a digital computer these days. The advent and rapid profusion of microcomputers provides even the very small organization with an inexpensive means of simulating complex systems. Analog computers are not so widespread, nor can they be accessed by telephone lines or time shared by a large number of users.

A digital-computer simulation of a control system may range in complexity from a simple program to generate the gain and phase response of a transfer function to a program that simulates the response of an entire system and performs iterative computations to optimize its parameters. The simpler programs are often written in one of the popular general-purpose languages such as FORTRAN or BASIC. For more complex problems, specialized software (programming packages with instructions or subroutines oriented to linear-systems simulation) have been developed. We will later demonstrate how interactive programs may be written in BASIC to study a variety of typical problems in control-systems analysis. BASIC was designed specifically for such on-line interactive programming and is, therefore, useful when the investigator wishes to make on-the-spot parameter changes and study their effects on overall performance. Finally, we will discuss briefly one of the specialized software packages designed for large-scale system simulation.

9.2 COMPUTER ANALYSIS OF TRANSFER FUNCTIONS

In this section, we will illustrate how the BASIC programming language can be used to write simple on-line interactive programs for analyzing transfer functions. It is assumed that the reader is familiar with basic BASIC. The programs displayed here were written in BASIC, version D02, for running on a Xerox Sigma 9 computer. Minor modifications may be necessary before running on other versions of BASIC. Flowcharts are provided to assist the reader in understanding the logic and sequence of operations performed by each program.

Beginning with a simple example, consider the transfer function

$$G(s) = \frac{K}{s(s + a)} \quad (1)$$

Suppose we wish to investigate the gain magnitude and phase angle of Equation 1 as a function of frequency ω. We might, for example, want to construct a Bode plot or sketch a polar plot using the results of such an investigation. The computations that must be performed by the computer are

$$|G| = \frac{K}{\omega\sqrt{\omega^2 + a^2}} \quad (2)$$

and

$$\angle G = -90° - \arctan\frac{\omega}{a} \quad (3)$$

Equations 2 and 3 are relatively easy to program. But we wish to take full advantage of the interactive capabilities of the BASIC language and will, therefore, provide

9.2 COMPUTER ANALYSIS OF TRANSFER FUNCTIONS

the user with opportunities to change the value of K as well as to select the frequency range and the total number of frequencies at which computations are to be performed. This flexibility will allow the investigator to see immediately the effects of changes in gain K as well as to "home in" on prescribed response values by repeatedly narrowing the frequency interval. The program can be thought of as requiring three parts: (1) initialization when the user inputs values for K, the frequency range and the number of frequencies n of computation; (2) the repeated computation and printout of values determined from Equations 2 and 3; and (3) interrogation by the computer to determine whether or not the user wishes to change any of the initial values in (1) and repeat the computations. The flowchart is shown in Figure 9.1.

In the flowchart in Figure 9.1, ω_1 is the lowest frequency and ω_2 the highest frequency at which the computations will be performed. These will be designated W1 and W2 in the actual program. n is the total number of frequency intervals and will be designated by N in the program. The program will be required to

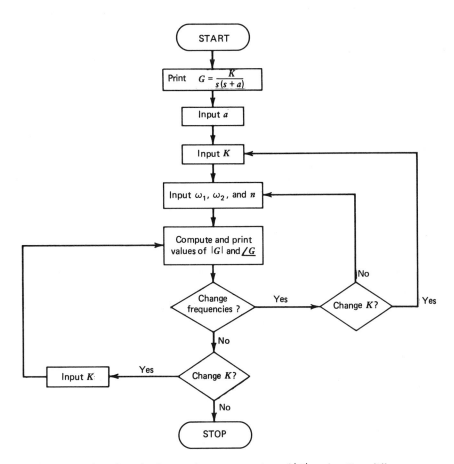

FIGURE 9.1 Flowchart for interactive computation of $|G|$ and $\angle G$ at different frequencies.

compute the size of each frequency interval, namely, $\Delta\omega = (\omega_1 - \omega_2)/n$. To discourage any attempt at division by zero, we will include remarks at the beginning of the program to the effect that neither a nor n nor any value of ω may be zero. We could have included loops within the program that would automatically send the programmer back to the appropriate input steps if zero values were entered, but this would have added unnecessary complication to a relatively simple program. It may be desirable to incorporate such safeguards in more complex simulations where it may not be obvious that certain inputs are not allowed as well as to print a message advising the user of an invalid entry whenever one is detected. However, in the present case, the user is assumed to be sufficiently acquainted with the theory to realize that certain inputs would not be permissible. In any event, the only consequence of an attempt to divide by zero will be the printout of an error message from the computer. The program listing is shown in Figure 9.2.

In the program listing in Figure 9.2, note that we have used the define function capability of the BASIC language. A function may be defined by writing DEF FNy(x) where x and y may be any letter A through Z. y identifies the particular

```
10  REM   THIS PROGRAM FINDS MAGNITUDE AND PHASE ANGLE OF
20  REM   G=K/S(S+A) AS A FUNCTION OF FREQUENCY,W,IN RAD/SEC.
30  REM   FREQUENCY RANGE IS W1 TO W2 IN N EQUAL INTERVALS.
40  REM   K,A,N, AND ALL VALUES OF W MUST BE GREATER THAN ZERO
50  PRINT
60  PRINT "G=K/S(S+A)"
70  PRINT "A=";
80  INPUT A
90  PRINT "K=";
100 INPUT K
110 PRINT "W1=";
120 INPUT W1
130 PRINT "W2=";
140 INPUT W2
150 PRINT "N=";
160 INPUT N
170 LET D=(W2-W1)/N
180 DEF FNG(W)=K/(W*SQR(W**2+A**2))
190 DEF FNA(W)=ATN(W/A)
200 PRINT "FREQ,RAD/SEC","MAGNITUDE","ANGLE,DEG."
210 FOR W=W1 TO W2 STEP D
220 PRINT W,FNG(W),-90-(FNA(W))*57.29578
230 NEXT W
240 PRINT "CHANGE FREQUENCIES? (Y OR N)."
250 INPUT A$
260 IF A$="Y" THEN 330
270 PRINT "CHANGE K? (Y OR N)."
280 INPUT B$
290 IF B$="N" THEN 370
300 PRINT "K=";
310 INPUT K
320 GO TO 180
330 PRINT "CHANGE K? (Y OR N)."
340 INPUT C$
350 IF C$="N" THEN 110
360 GO TO 90
370 END
```

FIGURE 9.2 BASIC program listing for flowchart in Figure 9.1.

9.2 COMPUTER ANALYSIS OF TRANSFER FUNCTIONS

function being defined, while x is a dummy variable; the choice of letter for the dummy variable is purely arbitrary. In line 180 of the listing, the function FNG(W) is defined and can be seen to generate the magnitude of G. Line 190 defines FNA(W) to be arctan ω/a. Actual values for the magnitude and angle of G are calculated and printed in line 220 as the dummy variable W is incremented from W1 to W2 in steps of size (W2 − W1)/N.

Figure 9.3 shows the printed results of a computer run where $a = 5$, $K = 1000$, $\omega_1 = 1$, $\omega_2 = 51$, and $n = 10$. Figure 9.4 shows a Bode plot based on the calculated values of $|G|$ and $\angle G$.

```
G=K/S(S+A)
A=
 5
K=
 1000
W1=
 1
W2=
 51
N=
 10
FREQ,RAD/SEC   MAGNITUDE      ANGLE,DEG.
 1             196.116        -101.310
 6              21.3395       -140.194
11               7.52369      -155.556
16               3.72844      -162.646
21               2.20591      -166.608
26               1.45267      -169.114
31               1.02731      -170.838
36                .764269     -172.093
41                .590509     -173.047
46                .469823     -173.797
51                .382633     -174.401
CHANGE FREQUENCIES? (Y OR N).
N
CHANGE K? (Y OR N).
N
```

FIGURE 9.3 Printout obtained from the program in Figure 9.2.

To illustrate how to use the program iteratively, suppose that $G = 1000/s(s + 5)$ is the open-loop transfer function of a control system and we wish to determine the phase margin. Then we must determine the phase angle at the frequency where $|G| = 1$. By inspecting the tabulated results in the printout (see Figure 9.3), we see that the required frequency must lie between $\omega = 31$ rad/sec (where $|G| = 1.02731$) and $\omega = 36$ rad/sec (where $|G| = 0.764269$). Therefore, we repeat the run with $\omega_1 = 31$, $\omega_2 = 36$, and $n = 10$. This yields the data shown in Figure 9.5. Inspecting Figure 9.5 shows that the required frequency now lies between 31 and 31.5. Performing one more iteration with $\omega_1 = 31$, $\omega_2 = 31.5$, and $n = 10$, we obtain Figure 9.6. From the results of this last run, we settle for $\omega = 31.4$ rad/sec as the unity-gain frequency where the phase angle is $-170.95°$. Our phase margin is, therefore, $180° − 170.95° = 9.05°$.

330 DIGITAL COMPUTER SIMULATION

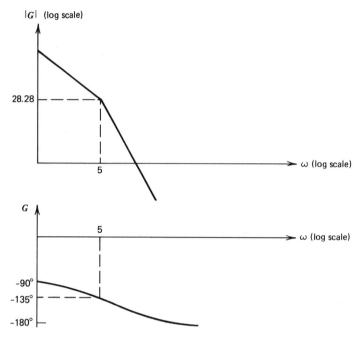

FIGURE 9.4 Bode plot based on Figure 9.3.

```
G=K/S(S+A)
A=
 5
K=
 1000
W1=
 31
W2=
 36
N=
 10
FREQ,RAD/SEC    MAGNITUDE        ANGLE,DEG.
 31              1.02731         -170.838
 31.5000          .995349        -170.981
 32               .964856        -171.119
 32.5000          .935737        -171.254
 33               .907911        -171.384
 33.5000          .881305        -171.511
 34               .855847        -171.634
 34.5000          .831473        -171.754
 35               .808122        -171.870
 35.5000          .785738        -171.983
 36               .764269        -172.093
CHANGE FREQUENCIES? (Y OR N).
N
CHANGE K? (Y OR N).
N
```

FIGURE 9.5 Return of program in Figure 9.2 to find unity-gain frequency.

9.2 COMPUTER ANALYSIS OF TRANSFER FUNCTIONS   331

```
G=K/S(S+A)
A=
 5
K=
 1000
W1=
 31
W2=
 31.5000
N=
 10
FREQ,RAD/SEC    MAGNITUDE        ANGLE,DEG.
 31              1.02731         -170.838
 31.0500         1.02404         -170.852
 31.1000         1.02079         -170.867
 31.1500         1.01756         -170.881
 31.2000         1.01434         -170.895
 31.2500         1.01114         -170.910
 31.3000         1.00795         -170.924
 31.3500         1.00478         -170.938
 31.4000         1.00162         -170.952
 31.4500          .998478        -170.967
 31.5000          .995349        -170.981
CHANGE FREQUENCIES? (Y OR N).
N
CHANGE K? (Y OR N).
N
```

FIGURE 9.6   Final rerun of program; phase margin = 9.05°.

We now modify the program in order to analyze a somewhat more complex transfer function. Let

$$G = \frac{K(s + a)}{s(s + b)(s + c)} \qquad (4)$$

which has a zero at $s = -a$ and poles at $s = 0$, $s = -b$, and $s = -c$. The flowchart for this program will be similar to that in Figure 9.1. The only differences between the programs are the number of input values and the definitions of the functions FNG(W) and FNA(W). In the present case, we must define FNG(W) so that it generates

$$|G| = \frac{K}{\omega}\sqrt{\frac{(\omega^2 + a^2)}{(\omega^2 + b^2)(\omega^2 + c^2)}} \qquad (5)$$

and FNA(W) must generate

$$\angle G = -90 + \arctan \omega/a - \arctan \omega/b - \arctan \omega/c \qquad (6)$$

The program is shown in Figure 9.7.

332 DIGITAL COMPUTER SIMULATION

```
10  REM  THIS PROGRAM FINDS MAGNITUDE AND PHASE ANGLE OF
20  REM  G=K(S+A)/S(S+B)(S+C) AS A FUNCTION OF FREQUENCY,W,IN RAD/SEC,
30  REM  FREQUENCY RANGE IS W1 TO W2 IN N EQUAL INTERVALS.
40  REM  K,A,B,C,N, AND ALL VALUES OF W MUST BE GREATER THAN ZERO.
50  PRINT
60  PRINT "G=K(S+A)/S(S+B)(S+C)"
70  PRINT "ENTER VALUES FOR A,B,C"
80  INPUT A,B,C
90  PRINT "K=";
100 INPUT K
110 PRINT "W1=";
120 INPUT W1
130 PRINT "W2=";
140 INPUT W2
150 PRINT "N=";
160 INPUT N
170 LET D=(W2-W1)/N
180 DEF FNG(W)=(K/W)*SQR((W**2+A**2)/((W**2+B**2)*(W**2+C**2)))
190 DEF FNA(W)=-90+(57.29578)*(ATN(W/A)-ATN(W/B)-ATN(W/C))
200 PRINT "FREQ,RAD/SEC","MAGNITUDE","ANGLE,DEG."
210 FOR W=W1 TO W2 STEP D
220 PRINT W,FNG(W),FNA(W)
230 NEXT W
240 PRINT "CHANGE FREQUENCIES? (Y OR N)."
250 INPUT A$
260 IF A$="Y" THEN 330
270 PRINT "CHANGE K? (Y OR N)."
280 INPUT B$
290 IF B$="N" THEN 370
300 PRINT "K=";
310 INPUT K
320 GO TO 180
330 PRINT "CHANGE K? (Y OR N)."
340 INPUT C$
350 IF C$="N" THEN 110
360 GO TO 90
370 END
```

FIGURE 9.7 Listing of BASIC program that computes |G| and ∠ G for G = K(s + a)/s(s + b)(s + c).

Figure 9.8 shows the results of a computer run for the case

$$G = \frac{500(s + 100)}{s(s + 10)(s + 40)}$$

For this particular run, $\omega_1 = 1$, $\omega_2 = 201$, and $n = 20$.

Figure 9.9 is a sketch of a portion of the polar plot based on the results obtained from Figure 9.8. The tabulated results of Figure 9.8 clearly show that the polar plot crosses the $-180°$ axis with a gain magnitude greater than unity. Note that the angle changes once the $-180°$ axis is crossed: The tabulated values increase and then begin to decrease, confirming that $-180°$ is the asymptotic phase angle that the polar plot approaches as ω approaches infinity. We also observe from the tabulated results that the frequency where the polar plot crosses $-180°$ is somewhere between $\omega = 21$ rad/sec and $\omega = 31$ rad/sec. Suppose that we wish to determine this frequency (ω_{180}) more precisely and to determine the actual gain magnitude at ω_{180}. G might, for example, represent an open-loop transfer function, and we might wish to reduce the gain K sufficiently to make $|G| < 1$ at ω_{180}. From a

9.2 COMPUTER ANALYSIS OF TRANSFER FUNCTIONS

```
G=K(S+A)/S(S+B)(S+C)
ENTER VALUES FOR A,B,C
 100             10            40
K=
 500
W1=
 1
W2=
 201
N=
 20
FREQ,RAD/SEC   MAGNITUDE      ANGLE,DEG.
 1              124.347       -96.5698
 11             7.41485       -146.825
 21             2.31527       -170.376
 31             1.02440       -182.674
 41             .545244       -189.707
 51             .326711       -193.777
 61             .212935       -196.053
 71             .147812       -197.212
 81             .107742       -197.673
 91             8.16357E-02   -197.698
 101            6.38169E-02   -197.455
 111            5.11786E-02   -197.051
 121            4.19222E-02   -196.555
 131            3.49551E-02   -196.012
 141            2.95878E-02   -195.450
 151            2.53691E-02   -194.889
 161            2.19950E-02   -194.339
 171            1.92549E-02   -193.806
 181            1.69997E-02   -193.296
 191            1.51214E-02   -192.810
 201            1.35404E-02   -192.348
CHANGE FREQUENCIES? (Y OR N).
N
CHANGE K? (Y OR N).
N
```

FIGURE 9.8 Results of the program in Figure 9.7.

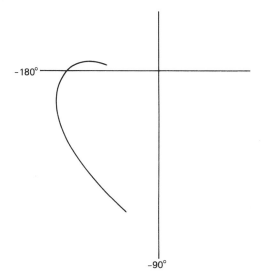

FIGURE 9.9 Polar plot based on Figure 9.8.

knowledge of the value of $|G|$ at ω_{180} when $K = 500$, we could determine what reduction in K would be necessary to achieve stability. We therefore run the program again with $\omega_1 = 21$, $\omega_2 = 31$, and $n = 10$. Figure 9.10 shows that ω_{180} lies between $\omega = 28$ rad/sec and $\omega = 29$ rad/sec. One more iteration with $\omega_1 = 28$, $\omega_2 = 29$, and $n = 10$ shows that $G = -180.017°$ at $\omega = 28.3$ rad/sec. From the results of Figure 9.10, we see that $|G| = 1.24851$ at ω_{180}. We must, therefore, reduce the gain K by a factor of *at least* $1/1.24851 = 0.8$ to achieve a marginally stable system. Of course, we would reduce it even further in practice to realize a specified gain margin.

```
G=K(S+A)/S(S+B)(S+C)
ENTER VALUES FOR A,B,C
 100              10              40
K=
 500
W1=
 21
W2=
 31
N=
 10
FREQ,RAD/SEC   MAGNITUDE      ANGLE,DEG.
    21          2.31527        -170.376
    22          2.10938        -171.959
    23          1.92763        -173.448
    24          1.76651        -174.848
    25          1.62316        -176.168
    26          1.49515        -177.412
    27          1.38046        -178.587
    28          1.27739        -179.696
    29          1.18448        -180.744
    30          1.10050        -181.736
    31          1.02440        -182.674
CHANGE FREQUENCIES? (Y OR N).
Y
CHANGE K? (Y OR N).
N
W1=
 28
W2=
 29
N=
 10
FREQ,RAD/SEC   MAGNITUDE      ANGLE,DEG.
   28           1.27739        -179.696
   28.1000      1.26766        -179.803
   28.2000      1.25804        -179.910
   28.3000      1.24851        -180.017
   28.4000      1.23908        -180.122
   28.5000      1.22975        -180.228
   28.6000      1.22051        -180.332
   28.7000      1.21136        -180.436
   28.8000      1.20231        -180.539
   28.9000      1.19335        -180.642
   29.0000      1.18448        -180.744
CHANGE FREQUENCIES? (Y OR N).
N
CHANGE K? (Y OR N).
N
```

FIGURE 9.10 Results of iterations to find 180° phase-shift frequency.

9.2 COMPUTER ANALYSIS OF TRANSFER FUNCTIONS

We have seen how to compute gain magnitudes and phase angles versus frequency for certain transfer functions containing explicit poles and zeroes. We can generalize the computation, so that it is applicable to any transfer function of the form

$$G = \frac{K(s)^{n_0}(s + a_1)^{n_1}(s + a_2)^{n_2} \ldots (s + a_j)^{n_j}}{(s)^{m_0}(s + b_1)^{m_1}(s + b_2)^{m_2} \ldots (s + b_k)^{m_k}} \tag{7}$$

It is only necessary to define functions in the basic program that calculate

$$|G| = \frac{K(\omega)^{n_0 - m_0}(\omega^2 + a_1^2)^{n_1/2}(\omega^2 + a_2^2)^{n_2/2} \ldots (\omega^2 + a_j^2)^{n_j/2}}{(\omega^2 + b_1^2)^{m_1/2}(\omega^2 + b_2^2)^{m_2/2} \ldots (\omega^2 + b_k^2)^{m_k/2}} \tag{8}$$

or equivalently,

$$|G| = K(\omega)^{n_0 - m_0} \sqrt{\frac{(\omega^2 + a_1^2)^{n_1}(\omega^2 + a_2^2)^{n_2} \ldots (\omega^2 + a_j^2)^{n_j}}{(\omega^2 + b_1^2)^{m_1}(\omega^2 + b_2^2)^{m_2} \ldots (\omega^2 + b_k^2)^{m_k}}}$$

and

$$\angle G = (90)(n_0 - m_0) + n_1 \arctan\left(\frac{\omega}{a_1}\right) + n_2 \arctan\left(\frac{\omega}{a_2}\right)$$
$$+ \ldots + n_j \arctan\left(\frac{\omega}{a_j}\right) - m_1 \arctan\left(\frac{\omega}{b_1}\right) \tag{9}$$
$$- m_2 \arctan\left(\frac{\omega}{b_2}\right) + \ldots - m_k \arctan\left(\frac{\omega}{b_k}\right)$$

We could, in fact, design a universal program for transfer functions in the form of Equation 7 by permitting the user to input the values of $n_0, n_1, \ldots, n_j, m_0, m_1, \ldots, m_k$. This would be a somewhat complex program, and we will not attempt it here. However, it would be an excellent project for the serious student or for a user who must frequently analyze transfer functions of different kinds.

Note that Equation 7 assumes that all poles and zeroes are *known*. In practice, it is more often the case that a known transfer function is in the form of a numerator *polynomial* over a denominator *polynomial*. There are many digital-computer programs that have been developed to find the roots of polynomials, which is precisely what we would have to do in order to express such a transfer function in the form of Equation 7. It is not our intent to cover the various techniques for root solving, since this is properly the realm of the study known as numerical analysis. In any event, it is not necessary to know the poles and zeroes to calculate magnitude and angle as a function of frequency. We will now develop a program that can be used to analyze a very common transfer function with a polynomial denominator, namely, the transfer function with quadradic denominator. Suppose

$$G = \frac{K}{As^2 + Bs + C} \tag{10}$$

Then

$$|G| = \frac{K}{\sqrt{(C - A\omega^2)^2 + (B\omega)^2}} \tag{11}$$

336 DIGITAL COMPUTER SIMULATION

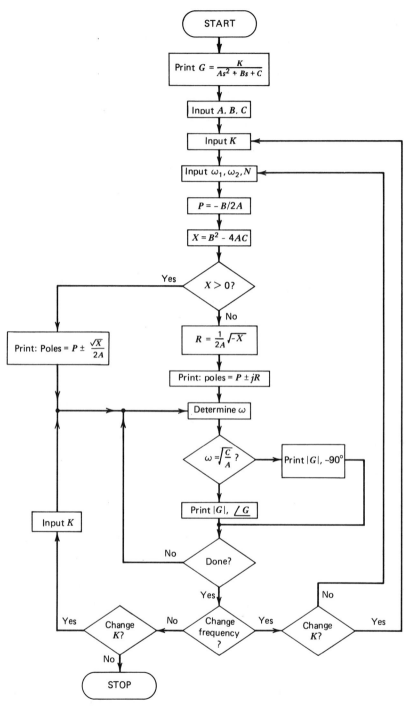

FIGURE 9.11 Flowchart for computing $|G|$ and $\angle G$ with $G = K(As^2 + Bs + C)$.

and

$$\angle G = -\arctan \frac{B\omega}{C - A\omega^2} \tag{12}$$

Our program will define functions that calculate $|G|$ and $\angle G$ according to Equations 11 and 12. We will also include provisions in the program to calculate and print the values of the poles of G. We observe that these poles will be real or complex depending upon the values of $B^2 - 4AC$, so our program must be able to make an appropriate decision on the method of pole calculation to be used. Figure 9.11 shows the flowchart for this program.

This program differs from those we discussed earlier with respect to calculating $\angle G$. Two new complications have arisen: First, we see from Equation 12 that if $\omega = \sqrt{C/A}$ (i.e., if $C = A\omega^2$), we will be dividing by zero, and the computer will generate an error message. Since we know that this value of ω corresponds to the frequency at which $\angle G = -90°$ ($-\arctan \infty = -90°$), we simply include a conditional branch in the program that causes the computer to print $-90°$ instead of the value calculated by Equation 12 whenever such a condition is encountered. This branch is shown in the flowchart. The second complication is somewhat more difficult to circumvent. The arctangent library function (ATN) in most versions of BASIC will only generate angles between $\pm 90°$. We see from Equation 12 that values of ω greater than $\sqrt{C/A}$ result in a negative denominator. Consequently, the ATN function will generate negative angles for ω in this region. But the minus sign in front of Equation 12 (due to the fact that $As^2 + Bs + C$ is in the denominator of G) will then cause the computed angle to be positive. We know that the phase angle of G varies from 0 to $-180°$ as ω varies from 0 to ∞, so we must make some provision to correct the calculated value of G when $\omega > \sqrt{C/A}$. The reader should satisfy him or herself that the correction amounts to subtracting $180°$ from the value calculated by the ATN function whenever $\omega > \sqrt{C/A}$. In order to implement this correction in the program, we will make use of the library functions SGN and INT, which are defined as follows:

$$SGN(x) = \begin{cases} +1 & \text{if } X > 0 \\ 0 & \text{if } X = 0 \\ -1 & \text{if } X < 0 \end{cases}$$

$INT(x)$ = the greatest integer that is less than X

Note that $INT(½) = 0$ and $INT(-½) = -1$. The computation of $\angle G$ is now modified as follows:

$$\angle G = -57.3 \text{ATN}\left(\frac{\omega B}{C - A\omega^2}\right) + 180 \text{INT}\left[\frac{SGN(C - A\omega^2)}{2}\right] \tag{13}$$

(In Equation 13, we have multiplied the ATN function by 57.3 to convert the result to degrees; most ATN library functions produce an angle in radians.) If $\omega < \sqrt{C/A}$, then $(C - A\omega^2) > 0$, $\text{SGN}(C - A\omega^2) = +1$, $\dfrac{\text{SGN}(C - A\omega^2)}{2} = \frac{1}{2}$, and $\text{INT}\left[\dfrac{\text{SGN}(C - A\omega^2)}{2}\right] = 0$, so the angle computed by the ATN function is not modified. On the other hand, if $\omega > \sqrt{C/A}$, then $(C - A\omega^2) < 0$, $\text{SGN}(C - A\omega^2) = -1$, $\dfrac{\text{SGN}(C - A\omega^2)}{2} = -\frac{1}{2}$, and $\text{INT}\left[\dfrac{\text{SGN}(C - A\omega^2)}{2}\right] = -1$. In this case, $-180°$ is added to the angle computed by the ATN function, as we require when $\omega > \sqrt{C/A}$. A listing of the program is shown in Figure 9.12.

```
10  REM  THIS PROGRAM FINDS THE POLES AND THE GAIN AND ANGLE OF
20  REM  G=K/(AS**2+BS+C) AS A FUNCTION OF FREQUENCY,W,IN RAD/SEC.
30  REM  FREQUENCY RANGE IS W1 TO W2 IN N EQUAL INTERVALS.
40  REM  K,A,B,C,AND N MUST BE GREATER THAN ZERO.
50  PRINT
60  PRINT "G=K/(AS**2+BS+C)"
70  PRINT "ENTER VALUES FOR A,B,C"
80  INPUT A,B,C
90  PRINT "K=";
100 INPUT K
110 PRINT "W1=";
120 INPUT W1
130 PRINT "W2=";
140 INPUT W2
150 PRINT "N=";
160 INPUT N
170 LET D=(W2-W1)/N
180 LET P=-B/(2*A)
190 LET X=B**2-4*A*C
200 IF X>0 THEN 240
210 LET R=SQR(-X)/(2*A)
220 PRINT "POLES ARE:";P;"+";"J";R;"AND";P;"-";"J";R
230 GO TO 250
240 PRINT "POLES ARE:";P+SQR(X)/(2*A);"AND";P-SQR(X)/(2*A)
250 DEF FNG(W)=K/SQR((C-A*(W**2))**2+(B*W)**2)
260 DEF FNA(W)=-57.2958*ATN((W*B)/(C-A*(W**2)))+180*INT((SGN(C-A*(W**2)))/2)
270 PRINT "FREQ,RAD/SEC","MAGNITUDE","ANGLE,DEG."
280 FOR W=W1 TO W2 STEP D
290 IF W=SQR(C/A) THEN 450
300 PRINT W,FNG(W),FNA(W)
310 NEXT W
320 PRINT "CHANGE FREQUENCIES? (Y OR N)."
330 INPUT A$
340 IF A$="Y" THEN 410
350 PRINT "CHANGE K? (Y OR N)."
360 INPUT B$
370 IF B$="N" THEN 470
380 PRINT "K=";
390 INPUT K
400 GO TO 250
410 PRINT "CHANGE K? (Y OR N)."
420 INPUT C$
430 IF C$="N" THEN 110
440 GO TO 90
450 PRINT W,FNG(W),"-90.000"
460 GO TO 310
470 END
```

FIGURE 9.12 BASIC program listing for the flowchart in Figure 9.11.

```
G=K/(AS**2+BS+C)
ENTER VALUES FOR A,B,C
 2              40              80
K=
 1000
W1=
 0
W2=
 100
N=
 20
POLES ARE:      -2.25403        AND             -17.7460
FREQ,RAD/SEC    MAGNITUDE       ANGLE,DEG.
 0              12.5000          0
 5               4.94468        -81.4693
 10              2.39457       -106.699
 15              1.41862       -121.661
 20               .929118      -131.987
 25               .649721      -139.479
 30               .476818      -145.097
 35               .363291      -149.429
 40               .285198      -152.850
 45               .229410      -155.610
 50               .188290      -157.878
 55               .157172      -159.771
 60               .133092      -161.372
 65               .114096      -162.743
 70              9.88606E-02   -163.930
 75              8.64614E-02   -164.966
 80              7.62408E-02   -165.879
 85              6.77197E-02   -166.688
 90              6.05433E-02   -167.411
 95              5.44443E-02   -168.060
 100             4.92183E-02   -168.646
CHANGE FREQUENCIES? (Y OR N).
N
CHANGE K? (Y OR N).
N
```

FIGURE 9.13   Results from a run of the program in Figure 9.12; overdamped case.

Figure 9.13 shows the results of a computer run for analyzing

$$G = \frac{1000}{2s^2 + 40s + 80} \quad (14)$$

We see that the poles are real and unequal, so we know that we are dealing with a transfer function that would produce an *overdamped* response to a step input. We observe that the phase angles lie between 0 and $-180°$ and that the gain magnitude decreases steadily with frequency, as we would expect for an overdamped system.

Suppose now that

$$G = \frac{200}{s^2 + 8s + 400} \quad (15)$$

Figure 9.14 shows the results of a run with $\omega_1 = 0$, $\omega_2 = 50$, and $n = 10$. We see that the transfer function has complex poles at $-4 \pm j19.5959$. Since the poles are complex, we know the system is underdamped and that the gain magnitude will rise with increasing ω to some maximum value, then fall off. Inspecting Figure 9.14 shows that $|G|$ increases from $\omega = 0$ to $\omega = 20$, and decreases thereafter. We also observe that $\angle G = -90°$ at $\omega = 20$ rad/sec. This is an expected result, since $\omega_n = 20$ in Equation 15.

Figure 9.14 shows the results of two additional iterations. In this case, we are seeking the maximum value of $|G|$ and the frequency at which it occurs. Since the first run showed that the maximum occurs somewhere between $\omega = 15$ and $\omega = 25$, we use these values for ω_1 and ω_2, respectively, in the second run. The second run shows the maximum to occur between $\omega = 19$ and $\omega = 20$, and using these frequencies in the third run, we find that the maximum is approximately

```
G=K/(A8**2+BS+C)                        CHANGE FREQUENCIES? (Y OR N).
ENTER VALUES FOR A,B,C                  Y
  1       8         400                 CHANGE K? (Y OR N).
K=                                      N
  200                                   W1=
W1=                                      19
  0                                     W2=
W2=                                      20
  50                                    N=
N=                                       10
  10                                    POLES ARE   -4  +8R=  I   19.5959
POLES ARE   -4  +8R=  I   19.5959       FREQ,RAD/SEC  MAGNITUDE    ANGLE,DEG.
FREQ,RAD/SEC  MAGNITUDE    ANGLE,DEG.    19           1.27451      -75.6096
  0           .500000         0          19.1000      1.27551      -77.0309
  5           .530325       -6.08853     19.2000      1.27577      -78.4607
  10          .644157      -14.9314      19.3000      1.27525      -79.8974
  15          .942547      -34.4390      19.4000      1.27397      -81.3393
  20          1.25000      -90.000       19.5000      1.27190      -82.7846
  25          .664364     -138.366       19.6000      1.26905      -84.2316
  30          .360609     -154.359       19.7000      1.26543      -85.6783
  35          .229563     -161.253       19.8000      1.26104      -87.1232
  40          .161039     -165.069       19.9000      1.25589      -88.5643
  45          .120163     -167.509       20.0000      1.25000      -90.0000
  50          9.35561E-02 -169.216       CHANGE FREQUENCIES? (Y OR N).
CHANGE FREQUENCIES? (Y OR N).           N
Y                                       CHANGE K? (Y OR N).
CHANGE K? (Y OR N).                     N
N
W1=
  15
W2=
  25
N=
  10
POLES ARE   -4  +8R=  I   19.5959
FREQ,RAD/SEC  MAGNITUDE    ANGLE,DEG
  15          .942547      -34.4390
  16          1.03807      -41.6336
  17          1.13929      -50.7795
  18          1.22331      -62.1759
  19          1.27451      -75.6096
  20          1.25000      -90.000
  21          1.15653     -103.715
  22          1.02555     -115.514
  23          .890015     -125.034
  24          .767869     -132.510
  25          .664364     -138.366
```

FIGURE 9.14 Iterations of the program in Figure 9.12 to find maximum $|G|$ in the underdamped case.

9.2 COMPUTER ANALYSIS OF TRANSFER FUNCTIONS

1.276 at $\omega = 19.2$ rad/sec. Of course, in this example we could have used Equations 14 and 15 from Chapter 7 to calculate the maximum $|G|$ directly; but our purpose here is to illustrate digital-computer techniques and iterative computations. By way of comparison, Equations 14 and 15 show that

$$|G|_{MAX} = 1.27577 \quad \text{at} \quad \omega_{GMAX} = 19.183 \text{ rad/sec}$$

Programs to analyze more complex systems can be written by using the same general format we have outlined thus far and by defining new magnitude and angle functions that are combinations of the simpler ones. We use the fact that the gain magnitude of transfer functions in series is the product of the individual gain magnitudes, while the phase angle is the sum of the individual phase angles. This is illustrated in Example 9.1.

EXAMPLE 9.1

Write a program that can be used to find a value of K that will cause the control system shown in Figure 9.15 to have a gain margin of 2.

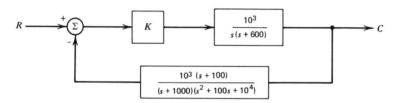

FIGURE 9.15 Find K required for a gain in margin of 2 (Example 9.1).

SOLUTION

The open-loop transfer function is

$$G = \frac{K(s + 100) \times 10^6}{s(s + 600)(s + 1000)(s^2 + 100s + 10^4)}$$

$$= \frac{K(s + 100)}{s(s + 600)(s + 1000)} \cdot \frac{10^6}{(s^2 + 100s + 10^4)}$$

$$= G_1 G_2$$

Having expressed G as the product of two transfer functions that are in general forms that we have already discussed, we can use the previously defined magnitude and angle functions to calculate

$$|G| = |G_1| |G_2|$$

$$\angle G = \angle G_1 + \angle G_2$$

342 DIGITAL COMPUTER SIMULATION

The program shown in Figure 9.16 defines

$$FNI(W) = [FNG(W)] * [FNH(W)]$$

and

$$FNC(W) = FNA(W) + FNB(W)$$

where

$$FNG(W) = |G_1|$$

$$FNH(W) = |G_2|$$

$$FNA(W) = \angle G_1$$

$$FNB(W) = \angle G_2$$

```
10 PRINT "G=K(10**6)(S+100)/S(S+600)(S+1000)(S**2+100S+10000)"
20 PRINT "ENTER VALUES FOR W1,W2,AND N."
30 INPUT W1,W2,N
40 PRINT "K=";
50 INPUT K
60 DEF FNG(W)=(K/W)*SQR((W**2+1E04)/((W**2+36E04)*(W**2+1E06)))
70 DEF FNH(W)=1E06/SQR((1E04-W**2)**2+(100*W)**2)
80 DEF FNI(W)=(FNG(W))*(FNH(W))
90 DEF FNA(W)=-90+(57.29578)*(ATN(W/100)-ATN(W/600)-ATN(W/1000))
100 DEF FNB(W)=-57.29578*(ATN((W*100)/(1E04-W**2)))+180*INT((SGN(1E04-W**2))/2)
110 DEF FNC(W)=FNA(W)+FNB(W)
120 PRINT "W,RAD/SEC","MAGNITUDE","ANGLE,DEG."
130 FOR W=W1 TO W2 STEP (W2-W1)/N
140 IF W=10 THEN 180
150 PRINT W,FNI(W),FNC(W)
160 NEXT W
170 GO TO 200
180 PRINT W,FNI(W),FNA(W)-90
190 GO TO 160
200 PRINT "CHANGE FREQUENCIES? (Y OR N)>"
210 INPUT A$
220 IF A$="Y" THEN 20
230 PRINT "CHANGE K? (Y OR N)."
240 INPUT B$
250 IF B$="Y" THEN 40
260 END
```

FIGURE 9.16 Program for solution of Example 9.1.

Figure 9.17 shows the results of several runs. We arbitrarily start with $K = 100,000$, $\omega_1 = 50$, and $\omega_2 = 200$. Inspecting the results of this first run reveals that ω_{180} lies between 125 and 140 rad/sec. In the next two runs, we iterate until ω_{180} is found at $\omega = 138.05$. The magnitude at this frequency is seen to be $|G| = 12.0327$. Since we wish to have a gain margin of two, we now change K to $\dfrac{1}{2(12.0327)} \times 100,000 = 4155.34$. The last run confirms that $|G|$ is now approximately 0.5 at ω_{180}, and we have, therefore, achieved the desired gain margin.

```
G=K(10**6)(S+100)/S(S+600)(S+1000)(S**2+100S+10000)
ENTER VALUES FOR W1,W2, AND N.
  50            200              10
K=
 100000
W,RAD/SEC     MAGNITUDE      ANGLE,DEG.
  50           41.1507        -104.751
  65           34.8939        -115.258
  80           30.0494        -129.281
  95           24.9150        -145.033
 110           19.6651        -159.748
 125           15.1260        -171.781
 140           11.6365        -181.080
 155            9.07854       -188.264
 170            7.21495       -193.962
 185            5.84052       -198.642
 200            4.80769       -202.620
CHANGE FREQUENCIES? (Y OR N).
Y
ENTER VALUES FOR W1,W2, AND N.
 125            140              10
K=
 100000
W,RAD/SEC     MAGNITUDE      ANGLE,DEG.
 125           15.1260        -171.781
 126.500       14.7296        -172.826
 128           14.3439        -173.843
 129.500       13.9691        -174.834
 131           13.6050        -175.799
 132.500       13.2515        -176.739
 134           12.9084        -177.653
 135.500       12.5755        -178.544
 137           12.2528        -179.412
 138.500       11.9398        -180.257
 140           11.6365        -181.080
CHANGE FREQUENCIES? (Y OR N).
Y
ENTER VALUES FOR W1,W2, AND N.
 137            138.500          10
K=
 100000
W,RAD/SEC     MAGNITUDE      ANGLE,DEG.
 137           12.2528        -179.412
 137.150       12.2210        -179.497
 137.300       12.1894        -179.583
 137.450       12.1579        -179.668
 137.600       12.1264        -179.753
 137.750       12.0951        -179.837
 137.900       12.0638        -179.922
 138.050       12.0327        -180.006
 138.200       12.0016        -180.090
 138.350       11.9707        -180.174
 138.500       11.9398        -180.257
CHANGE FREQUENCIES? (Y OR N).
Y
ENTER VALUES FOR W1,W2, AND N.
 138.050        138.100           1
K=
 4155.34
W,RAD/SEC     MAGNITUDE      ANGLE,DEG.
 138.050        .499999       -180.006
 138.100        .499569       -180.034
CHANGE FREQUENCIES? (Y OR N).
N
CHANGE K? (Y OR N).
N
```

FIGURE 9.17 Iterations of the program in Figure 9.16; $K = 4155.34$ (Example 9.1).

9.3 COMPUTING TIME-DOMAIN RESPONSES

Given the LaPlace transform $G(s)$ of a transfer function and the transform of the input to it $R(s)$, the time-domain response is, of course, the inverse transform of the product $G(s)R(s)$. See Figure 9.18. If we know the poles of $G(s)$ and $R(s)$,

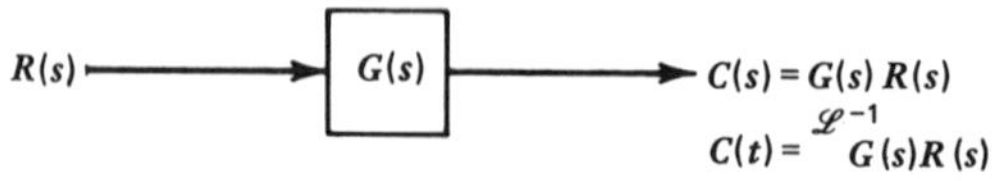

```
R(s) ──────▶ │ G(s) │ ──────▶ C(s) = G(s) R(s)
                                C(t) = 𝓛⁻¹ G(s)R(s)
```

FIGURE 9.18 The time-domain response is an inverse transform.

then it is a relatively easy matter to find $C(t)$. We may use the method of partial fractions and express the response as a sum of terms, each of whose coefficients is determined by the partial-fraction methods discussed in Chapter 4. Suppose, for example, we wish to find the inverse transform of the general form (Equation 4) that we considered in Section 9.2

$$\frac{K(s + a)}{s(s + b)(s + c)}$$

Expanding in partial fractions, we have

$$\frac{K(s + a)}{s(s + b)(s + c)} = \frac{A_1}{s} + \frac{A_2}{s + b} + \frac{A_3}{s + c}$$

We find A_1, A_2, and A_3 in the usual way

$$A_1 = \left.\frac{K(s + a)}{(s + b)(s + c)}\right|_{s=0} = \frac{aK}{bc} \tag{16}$$

$$A_2 = \left.\frac{K(s + a)}{s(s + c)}\right|_{s=-b} = \frac{K(-b + a)}{-b(-b + c)} \tag{17}$$

$$A_3 = \left.\frac{K(s + a)}{s(s + b)}\right|_{s=-c} = \frac{K(-c + a)}{-c(-c + b)} \tag{18}$$

We can write a program that calculates A_1, A_2, and A_3 according to Equations 16, 17, and 18 and then compute the response

$$A_1 + A_2 e^{-at} + A_3 e^{-bt} \tag{19}$$

A BASIC program that will perform these calculations is shown in Figure 9.19.

9.3 COMPUTING TIME-DOMAIN RESPONSES

Since the program is quite straightforward and follows the general format of those we have discussed earlier, a flowchart is not shown. Note that the coefficients A_1, A_2, and A_3 are computed in lines 100, 110, and 120. The time-domain function expressed by Equation 19 is evaluated in line 160 by defining the function FNI(T). We include the usual provisions for iteration by permitting changes in the values of t at which Equation 19 is evaluated. Note also that we do not permit the values of B and C to be equal; Equations 17 and 18 would cause division by zero if this were the case. And, of course, Equation 19 would not be the correct inverse transform if B were equal to C, since then we would have a pole of order two. It is left as an exercise to modify the program so as to permit $B = C$. (Include a conditional branch that would cause the evaluation of a different time-domain function when the values that are input for B and C are equal.)

```
10 REM   THIS PROGRAM FINDS THE INVERSE OF G=K(S+A)/S(S+B)(S+C).
20 REM   K,A,B,AND C MUST BE POSITIVE. B CANNOT = C. THE INVERSE IS
30 REM   COMPUTED AT N EQUAL TIME INTERVALS BETWEEN T1 AND T2.
40 PRINT
50 PRINT "G=K(S+A)/S(S+B)(S+C)"
60 PRINT "ENTER K,A,B,C"
70 INPUT K,A,B,C
80 PRINT "ENTER T1,T2,N"
90 INPUT T1,T2,N
100 LET A1=(K*A)/(B*C)
110 LET A2=(K*(-B+A))/((-B)*(-B+C))
120 LET A3=(K*(-C+A))/((-C)*(-C+B))
130 PRINT "THE INVERSE TRANSFORM IS:"
140 PRINT A1;"+";A2;"EXP(";-B;"T)";"+";A3;"EXP(";-C;"T)"
150 PRINT
160 DEF FNI(T)=A1+A2*EXP(-B*T)+A3*EXP(-C*T)
170 PRINT "TIME,SEC","INVERSE"
180 FOR T=T1 TO T2 STEP (T2-T1)/N
190 PRINT T,FNI(T)
200 NEXT T
210 PRINT "CHANGE TIME POINTS? (Y OR N)."
220 INPUT A$
230 IF A$="Y" THEN 80
240 END
```

FIGURE 9.19 Listing of a BASIC program that finds the inverse of $F(s) = K(s + a)$ $/s(s + b)(s + c)$.

Figure 9.20 shows the results of a computer run of the program in Figure 9.19 for the function we examined earlier

$$F(s) = \frac{500(s + 100)}{s(s + 10)(s + 50)}$$

In this case, we used $t_1 = 0$ sec, $t_2 = 0.5$ sec, and $n = 20$. The time-domain function is seen to begin at 0 and rise gradually towards its final value of 100.

346 DIGITAL COMPUTER SIMULATION

```
G=K(S+A)/S(S+B)(S+C)
ENTER K,A,B,C
500             100             10              50
ENTER T1,T2,N
0               .500000         20
THE INVERSE TRANSFORM IS:
100  +   -112.500 EXP( -10  T)  +   12.5000 EXP( -50  T)
TIME,SEC        INVERSE
0               0
2.50000E-02     15.9662
5.00000E-02     32.7914
7.50000E-02     47.1527
.100000         58.6978
.125000         67.7323
.150000         74.9048
.175000         80.4524
.200000         84.7753
.225000         88.1427
.250000         90.7655
.275000         92.8081
.300000         94.3990
.325000         95.6379
.350000         96.6028
.375000         97.3543
.400000         97.9395
.425000         98.3953
.450000         98.7502
.475000         99.0267
.500000         99.2420
CHANGE TIME POINTS? (Y OR N).
N
```

FIGURE 9.20   Results of a run of the program in Figure 9.19.

EXAMPLE 9.2

Use the program in Figure 9.19 to determine the time-domain response e_o of the operational-amplifier circuit shown in Figure 9.21 when the input is a 10-V step.

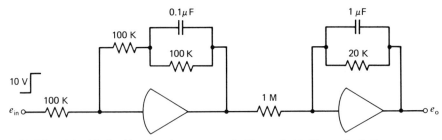

FIGURE 9.21   Find $e_o(t)$ using the program in Figure 9.19 (Example 9.2).

SOLUTION

If we let Z_1 and Z_2 represent the input impedances to each amplifier ($Z_1 = 100$ K and $Z_2 = 1$ M in this case), while Z_{f1} and Z_{f2} represent the impedances in the feedback paths of each amplifier, then we know

$$E_o(s) = \left(\frac{-Z_{f1}}{Z_1}\right)\left(\frac{-Z_{f2}}{Z_2}\right)E_1(s)$$

$$= \left(\frac{Z_{f1}Z_{f2}}{Z_1 Z_2}\right)E_1(s) \tag{20}$$

9.3 COMPUTING TIME-DOMAIN RESPONSES

Now

$$Z_{f1} = 10^5 + \frac{10^5/10^{-7}s}{10^5 + (1/10^{-7}s)} = 10^5 + \frac{10^7}{s + 100} = \frac{10^5(s + 200)}{s + 100}$$

and

$$Z_{f2} = \frac{2 \times 10^4/10^{-6}s}{2 \times 10^4 + (1/10^{-6}s)} = \frac{10^6}{s + 50}$$

Thus, Equation 20 becomes

$$E_o(s) = \left[\frac{10^5(s + 200)}{10^5(s + 100)} \frac{10^6}{10^6(s + 50)}\right] E_1(s)$$

and since $E_1(s) = 10/s$, this may be written

$$E_o(s) = \frac{10(s + 200)}{s(s + 100)(s + 50)} \tag{21}$$

Figure 9.22 shows the results of a computer run of the program in Figure 9.19, with $K = 10$, $a = 200$, $b = 100$, $c = 50$, $t_1 = 0$ sec, $t_2 = 0.1$ sec, and $n =$

```
G=K(S+A)/S(S+B)(S+C)
ENTER K,A,B,C
 10          200              100            50
ENTER T1,T2,N
 0           .100000          20
THE INVERSE TRANSFORM IS!
 .400000   +   .200000  EXP( -100  T)  +  -.600000  EXP( -50  T)

TIME,SEC          INVERSE
 0                 0
 5.00000E-03       5.40257E-02
 1.00000E-02       .109657
 1.50000E-02       .161206
 2.00000E-02       .206339
 2.50000E-02       .244514
 3.00000E-02       .276079
 3.50000E-02       .301775
 4.00000E-02       .322462
 4.50000E-02       .338982
 5.00000E-02       .352097
 5.50000E-02       .362461
 6.00000E-02       .370624
 6.50000E-02       .377036
 7.00000E-02       .382064
 7.50000E-02       .386000
 8.00000E-02       .389078
 8.50000E-02       .391482
 9.00000E-02       .393359
 9.50000E-02       .394824
 .100000           .395966
CHANGE TIME POINTS? (Y OR N).
N
```

FIGURE 9.22 Results of a computer run of the program in Figure 9.19 for Equation 21 (Example 9.2).

20. In comparison with the results in Figure 9.20, we note that the response rises towards its final value (0.4) more quickly, which is due to the poles (-100 and -50) being further from the origin than in the previous case. Recall that poles close to the origin *dominate* the response in the sense that they are responsible for long time constants and correspondingly long times to dampen out.

Now let us consider a typical control-systems problem, one that readily lends itself to computer solution by inverse-transform methods. Suppose we have the control system whose block diagram is shown in Figure 9.23.

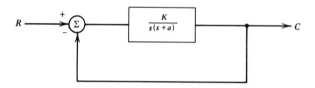

FIGURE 9.23 Control system whose response is to be computed.

This system was studied extensively in Chapter 7 in the context of its root locus, that is, in terms of the locations of its closed-loop poles as a function of the open-loop gain K. We now wish to write a program that will generate the response of the system $C(t)$ to a step input $Ru(t)$. The closed-loop transfer function is

$$G' = \frac{G}{1 + G} = \frac{K/s(s + a)}{1 + K/s(s + a)} = \frac{K}{s^2 + as + K} \qquad (22)$$

Therefore, since $R(s) = R/s$, the response is

$$C(s) = \frac{R}{s} G'(s) = \frac{KR}{s(s^2 + as + K)} \qquad (23)$$

Recall that the response will be over-, under-, or critically damped, depending on the magnitude of K, specifically on whether or not $(a^2 - 4K)$ is greater than, less than, or equal to zero, respectively. Therefore, our program must first test $(a^2 - 4K)$ to determine which of three possible time-domain functions should be evaluated. If $(a^2 - 4K)$ is less than zero, we know the response will be a constant term plus a damped sinusoid. Recall that

$$C(t) = A_1 + \frac{M}{\omega} e^{-\alpha t} \sin(\omega t + \theta) \qquad (24)$$

where

$$A_1 = \frac{RK}{(s^2 + as + K)}\bigg|_{s=0} = R \qquad (25)$$

$$M\angle\theta = K(s)\bigg|_{s=-\alpha+j\omega} = \frac{RK}{s}\bigg|_{s=-\alpha+j\omega} \quad (26)$$

and $-\alpha$ and $\omega =$ real imaginary parts of the complex roots of $s^2 + as + K$. In the present case, the roots of the quadratic are

$$-\frac{a}{2} \pm \frac{\sqrt{a^2 - 4K}}{2}$$

Since we are considering the underdamped case, $a^2 < 4K$, we have

$$\alpha = \frac{a}{2} \qquad \omega = \frac{\sqrt{4K - a^2}}{2} \quad (27)$$

Thus from Equation 26,

$$M\angle\theta = \frac{RK}{(a/2) + j(\sqrt{4K - a^2}/2)} \quad (28)$$

$$M = \frac{RK}{\sqrt{(a^2/4) + (4K - a^2)/4}} = R\sqrt{K} \quad (29)$$

$$\theta = -\arctan\left[\frac{(4K - a^2)/2}{-a/2}\right]$$

$$= \arctan\left(\frac{4K - a^2}{a}\right) - \pi \quad (30)$$

Using Equations 25, 27, and 29 in Equation 24, we obtain

$$C(t) = R + \frac{R\sqrt{K}}{\sqrt{4K - a^2}/2}\, e^{-(a/2)t} \sin\left(\frac{\sqrt{4K - a^2}}{2}t + \theta\right)$$

$$= R + 2R\sqrt{\frac{K}{4K - a^2}}\, e^{-(a/2)t} \sin\left(\frac{\sqrt{4K - a^2}}{2}t + \theta\right) \quad (31)$$

350 DIGITAL COMPUTER SIMULATION

where θ is given by Equation 30. We will implement Equation 31 in our program to calculate the response in the underdamped case.

In the case where $a^2 - 4K > 0$, we know that the quadratic expression $s^2 + as + K$ has real, unequal factors

$$C(s) = \frac{RK}{s(s^2 + as + K)} = \frac{RK}{s(s + P_1)(s + P_2)} \qquad (32)$$

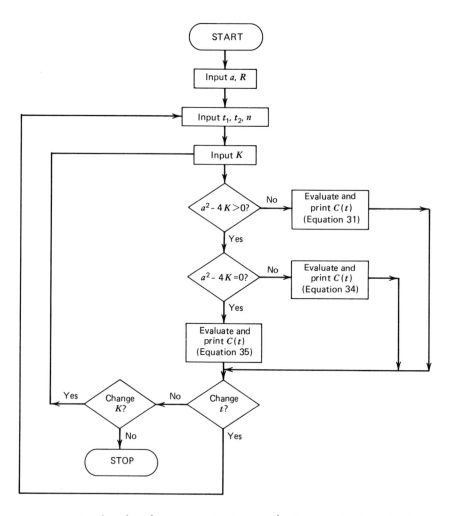

FIGURE 9.24 Flowchart for program to compute the response in Figure 9.23.

where

$$P_1 = \frac{a}{2} + \frac{\sqrt{a^2 - 4K}}{2}$$

$$P_2 = \frac{a}{2} - \frac{\sqrt{a^2 - 4K}}{2}$$

In this case, we can determine the inverse by applying a partial fraction expansion

$$C(s) = \frac{R}{s} + \frac{A_1}{s + P_1} + \frac{A_2}{s + P_2} \quad (33)$$

Evaluating the coefficients in the usual way shows that

$$A_1 = \frac{RK}{-P_1(-P_1 + P_2)}$$

$$A_2 = \frac{RK}{-P_2(-P_2 + P_1)}$$

The inverse transform is then

$$C(t) = R + A_1 e^{-P_1 t} + A_2 e^{-P_2 t} \quad (34)$$

The program will evaluate Equation 34 whenever $a^2 - 4K > 0$.
Finally, if $a^2 - 4K = 0$, the critically damped case, we have

$$C(s) = \frac{RK}{s[s + (a/2)]^2} = R + \frac{A_1}{[s + (a/2)]} + \frac{A_2}{[s + (a/2)]^2}$$

where

$$A_1 = \left.\frac{d(RK/s)}{ds}\right|_{s=-a/2} = \left.\frac{-RK}{s^2}\right|_{s=-a/2} = \frac{-RK}{(a^2/4)} = \frac{-RK}{K} = -R$$

$$A_2 = \left.\frac{RK}{s}\right|_{s=-a/2} = \frac{RK}{-(a/2)} = \frac{-2RK}{a} = -R\sqrt{K}$$

Thus,

$$C(t) = R - Re^{-(a/2)t} - R\sqrt{K}\, te^{-(a/2)t} \quad (35)$$

The program will evaluate Equation 35 whenever the system is critically damped. Figure 9.24 shows a flowchart for the program.

Figure 9.25 is a listing of the program. Note that the functions that are defined to evaluate C(t) in the three possible cases are on line 210 (FNI), line 370 (FNJ), and line 460 (FNK); they correspond to the underdamped, overdamped, and critically damped cases, respectively. Even though only one of these functions will be

```
10 REM   THIS PROGRAM FINDS THE CLOSED-LOOP TIME-DOMAIN
20 REM   RESPONSE OF A SYSTEM WITH OPEN-LOOP TRANSFER
30 REM   FUNCTION K/S(S+A) WHEN THE INPUT IS AN R-UNIT STEP.
40 REM   A MUST BE >0.  COMPUTATIONS ARE BETWEEN T1 AND T2 IN N INTERVALS.
50 PRINT "G=K/S(S+A)"
60 PRINT "ENTER A,R"
70 INPUT A,R
80 PRINT "ENTER T1,T2,N"
90 INPUT T1,T2,N
100 PRINT "K=";
110 INPUT K
120 LET D=A**2-4*K
130 IF D>=0 THEN 270
140 LET S=-D
150 LET W=(SQR(S))/2
160 LET Z=ATN((SQR(S))/A)-3.1416
170 LET M=2*R*SQR(K/S)
180 PRINT "THE RESPONSE IS:"
190 PRINT R;"+";M;"EXP(";-A/2;"T)";"SIN(";W;"T+";Z*57.29578;")"
200 PRINT
210 DEF FNI(T)=R+M*(EXP((-A/2)*T))*SIN((W*T)+Z)
220 PRINT "TIME,SEC","VALUE OF RESPONSE"
230 FOR T=T1 TO T2 STEP (T2-T1)/N
240 PRINT T,FNI(T)
250 NEXT T
260 GO TO 500
270 IF D=0 THEN 420
280 LET P=(SQR(D))/2
290 LET P1=(A/2)+P
300 LET P2=(A/2)-P
310 LET A1=(R*K)/((-P1)*(-P1+P2))
320 LET A2=(R*K)/((-P2)*(-P2+P1))
330 PRINT "THE RESPONSE IS:"
340 PRINT R;"+";A1;"EXP(";-P1;"T)";"+";A2;"EXP(";-P2;"T)"
350 PRINT
360 PRINT "TIME,SEC","VALUE OF RESPONSE"
370 DEF FNJ(T)=R+A1*EXP(-P1*T)+A2*EXP(-P2*T)
380 FOR T=T1 TO T2 STEP (T2-T1)/N
390 PRINT T,FNJ(T)
400 NEXT T
410 GO TO 500
420 PRINT "THE RESPONSE IS:"
430 PRINT R;"-";R;"EXP(";-A/2;"T)";"-R*SQR(K)";"T";"EXP(";-A/2;"T)"
440 PRINT
450 PRINT "TIME,SEC","VALUE OF RESPONSE"
460 DEF FNK(T)=R-R*EXP((-A/2)*T)-R*(SQR(K))*T*EXP((-A/2)*T)
470 FOR T=T1 TO T2 STEP (T2-T1)/N
480 PRINT T,FNK(T)
490 NEXT T
500 PRINT "CHANGE TIMES? (Y OR N)."
510 INPUT A$
520 IF A$="Y" THEN 80
530 PRINT "CHANGE K? (Y OR N)."
540 INPUT B$
550 IF B$="Y" THEN 100
560 END
```

FIGURE 9.25   BASIC program listing for the flowchart in Figure 9.24.

evaluated during any single computer run, they must have different designations (I, J, and K) as shown, or else the computer will generate an error message.

EXAMPLE 9.3

Figure 9.26 shows the results of a computer solution to the problem in Example

```
G=K/S(S+A)
ENTER A,R
 10             1
ENTER T1,T2,N
 0              1              15
K=
 16
THE RESPONSE IS
 1   +     .333333   EXP(  -8  T)  +    -1.33333  EXP(  -2  T)

TIME,SEC          VALUE OF RESPONSE
 0                 6.93889E-17
 6.66667E-02       2.86510E-02
 .133333           9.34801E-02
 .200000           .173539
 .266667           .257286
 .333333           .338605
 .400000           .414482
 .466667           .483650
 .533333           .545804
 .600000           .601151
 .666667           .650146
 .733333           .693353
 .800000           .731358
 .866667           .764732
 .933333           .794006
 1.00000           .819665
CHANGE TIMES? (Y OR N).
N
CHANGE K? (Y OR N).
Y
K=
 25
THE RESPONSE IS:
 1   -     1  EXP( -5  T)  -5  T   EXP( -5  T)

THE RESPONSE IS:
 0                 0
 6.66667E-02       4.46249E-02
 .133333           .144305
 .200000           .264241
 .266667           .384940
 .333333           .496332
 .400000           .593994
 .466667           .676760
 .533333           .745227
 .600000           .800852
 .666667           .845413
 .733333           .880713
 .800000           .908422
 .866667           .930007
 .933333           .946713
 1.00000           .959572
```

FIGURE 9.26 Results of computer runs of the program in Figure 9.25 (Example 9.3). (Continued on next page.)

6.16, where we were required to find the response to a unit-step input of a system for which $G = K/s(s + 10)$, using $K = 16$, $K = 25$, and $K = 50$. The responses are seen to be over, critically, and underdamped, respectively. For each of the three values of gain K, we requested values of the response between $t = 0$ and $t = 1$ sec, in 15 equal intervals. From the tabulated results, we note that the response

354 DIGITAL COMPUTER SIMULATION

```
CHANGE TIMES? (Y OR N).
N
CHANGE K? (Y OR N).
Y
K=
50
THE RESPONSE IS:
1  +    1.41421   EXP( -5  T)  SIN(  5  T+  -135.000  )

TIME, SEC           VALUE OF RESPONSE
0                   7.34644E-06
6.66667E-02         8.84668E-02
.133333             .279031
.200000             .491673
.266667             .681791
.333333             .830070
.400000             .933258
.466667             .996864
.533333             1.03002
.600000             1.04226
.666667             1.04182
.733333             1.03493
.800000             1.02583
.866667             1.01705
.933333             1.00982
1.00000             1.00455
CHANGE TIMES? (Y OR N).
N
CHANGE K? (Y OR N).
N
```

FIGURE 9.26 *(continued)*

rises toward the value 1.00 more quickly as the gain is increased. In the case where $K = 50$, we see the response oscillating around 1.00, an oscillation that will eventually damp out. (Note that the very small calculated response values at $t = 0$, where we know the response should be exactly zero, are due to round-off error.)

EXAMPLE 9.4

Use the program listed in Figure 9.25 to determine the time required for the voltage on the capacitor to reach 12 V after the switch is closed in the circuit shown in Figure 9.27. (If the response is underdamped, find the *first* time instant that the capacitor voltage reaches 12 V.)

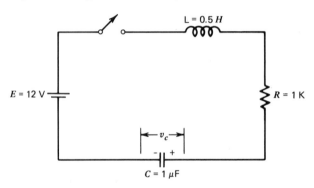

FIGURE 9.27 Find the time required for v_c to reach 12 V (Example 9.4).

SOLUTION

We know from the voltage-divider rule that

$$V_c(s) = \frac{1/Cs}{Ls + R + (1/Cs)} \frac{E}{s} = \frac{E/s}{LCs^2 + RCs + 1} = \frac{E/LC}{s[s^2 + (R/L)s + (1/LC)]}$$

Substituting the component values from the circuit, we find

$$V_c(s) = \frac{2.4 \times 10^7}{s(s^2 + 2 \times 10^3 s + 2 \times 10^6)}$$

By comparison with Equation 22, we see that in order to use the program in Figure 9.25, we must let

$$a = 2 \times 10^3$$
$$K = 2 \times 10^6$$
$$KR = 2.4 \times 10^7 \Rightarrow R = 12$$

Figure 9.28 shows the results of a run using these values, with $t_1 = 0$, $t_2 = .01$, and $n = 10$. Examining the results reveals that the response is underdamped, $v_c(t) = 12 + 16.9706e^{-1000t} \sin(1000t - 135°)$. We also see that $v_c(t)$ reaches 12 V at some point between $t = .002$ and $t = .003$. Consequently, the run is repeated with $t_1 = .002$, $t_2 = .003$, and $n = 20$. We see that $v_c(t)$ reaches 12 V in approximately 2.35 msec.

```
G=K/S(S+A)
ENTER A,R
 2000           12
ENTER T1,T2,N
 0              1.00000E-02    10
K=
 2000000
THE RESPONSE IS:
 12    +     16.9706    EXP(  -1000   T)   SIN(   1000   T+  -135.000  )

TIME,SEC         VALUE OF RESPONSE
 0                8.81572E-05
 1.00000E-03      5.90008
 2.00000E-03      11.1991
 3.00000E-03      12.5071
 4.00000E-03      12.3100
 5.00000E-03      12.0546
 6.00000E-03      11.9798
 7.00000E-03      11.9846
 8.00000E-03      11.9966
 9.00000E-03      12.0007
 1.00000E-02      12.0008
CHANGE TIMES? (Y OR N).
Y
ENTER T1,T2,N
 2.00000E-03     3.00000E-03    20
K=
 2000000
THE RESPONSE IS:
 12    +     16.9706    EXP(  -1000   T)   SIN(   1000   T+  -135.000  )
```

FIGURE 9.28 Solution to Example 9.4 ($T = 2.35$ msec).

```
TIME,SEC         VALUE OF RESPONSE
2.00000E-03      11.1991
2.05000E-03      11.3414
2.10000E-03      11.4734
2.15000E-03      11.5953
2.20000E-03      11.7075
2.25000E-03      11.8104
2.30000E-03      11.9044
2.35000E-03      11.9900
2.40000E-03      12.0674
2.45000E-03      12.1372
2.50000E-03      12.1996
2.55000E-03      12.2552
2.60000E-03      12.3043
2.65000E-03      12.3472
2.70000E-03      12.3844
2.75000E-03      12.4163
2.80000E-03      12.4431
2.85000E-03      12.4653
2.90000E-03      12.4831
2.95000E-03      12.4970
3.00000E-03      12.5071
CHANGE TIMES? (Y OR N).
N
CHANGE K? (Y OR N).
N
```

FIGURE 9.28 (continued)

9.4 OPTIMIZATION BY AUTOMATIC ITERATIVE COMPUTATION

Iterative computation is the computational method used to find the solution to a problem by repeatedly performing the same calculation or series of calculations. Each time the calculations are performed (that is, for each iteration), the constants are changed in such a way that the calculations yield a result that is closer to the true solution than the result of the previous iteration. Iterative computation typically includes some provision for the result of each iteration to be tested to determine if it is close enough to the true solution, that is, to determine if it meets some accuracy criterion.

A simple example of an iterative computation is the procedure followed to find the square root of a number using a calculator that has no square-root function. We would make an initial guess at the root, square the guessed value, then see how close the result was to the number whose root we wished to find. If the result were larger than our number, we would reduce our guess by a certain amount and try again. Thus, for example, if we wished to find $\sqrt{5}$, we might make an initial guess of 3. Since $3 \times 3 = 9$ is too big, we try 2; since $2 \times 2 = 4$ is too small, we try 2.5. But $2.5 \times 2.5 = 6.25$ is again too big, so we try 2.3 and find that $2.3 \times 2.3 = 5.29$, and so forth. Notice that each iteration gets us closer and closer to the true value of $\sqrt{5}$. If we decide that we want a root value that produces the number $5 \pm .02$ when squared, then we would continue iterating until we found $2.24 \times 2.24 = 5.0176$.

We have already seen several examples in this chapter of iterative computations using a computer. In each of these examples, we were able to examine the results

9.4 OPTIMIZATION BY AUTOMATIC ITERATIVE COMPUTATION

of a given iteration, then manually enter new data to produce a result that was closer to the solution we were seeking. In Example 9.1, we iterated 3 times in order to find $\omega_{180} = 138.05$. These computations, though iterative, were not automatic, since it was necessary for the programmer to examine the results of each iteration and make a judgment about what new values of the constants should be used for the next iteration. Naturally, the programmer is expected to choose new values that produce results that are closer to the desired solution than those of the previous iteration. When the task of deciding what new values should be used to produce a better solution is left to the computer rather than to the user, the iterations are said to be *automatic*. The computer can be programmed to perform automatic iterations by using instructions that examine the results of each computation, compare these results to some criterion, adjust the constants in an appropriate way, and branch to repeat the computation if the criterion has not been met. When writing a program to perform automatic iterations, the programmer must be careful to avoid putting the computer into an infinite loop. This can be the result of specifying a criterion that cannot be met or of causing the constants to be adjusted in such a way that each iteration produces results that are further from the desired solution instead of closer.

The method of automatic iterations is an effective means for optimizing the design of a control system. As we know, there are numerous trade-offs that must be considered when specifying the constants of a system. For example, the desirable effects of increased open-loop gain (increased bandwidth, decreased response time) are generally offset by reduced stability margins. Other criteria that might be considered in the optimization process include error coefficients, percent overshoot, and settling time (the time required for oscillatory ringing to dampen out). The computer can be programmed to perform automatic iterations that change such factors as gain and locations of zeroes in order to meet specified values of any of these criteria.

We will illustrate automatic iterative computation by developing a BASIC program that finds the value of gain K required to realize a gain margin selected by the user. Recall that gain margin is the increase in open-loop gain that will cause the open-loop gain to be unity at the frequency where the open-loop phase shift is 180°. We will consider the system whose open-loop transfer function is

$$G = \frac{K(s + a)}{s(s + b)(s + c)} \tag{36}$$

This transfer function might represent the third and most accurate approximation of a servomotor in a system that has been compensated by the addition of a zero at $s = -a$. Instead of working with the gain margin per se, we will allow the user to input the actual value of open-loop gain desired at the frequency where the phase shift is 180°. Thus, for example, if a gain margin of 2 is desired, the user would specify an open-loop gain of 0.5 at the frequency where the phase shift is 180°. The program will then calculate the value of frequency-independent gain K required to meet this criterion. Iterations will be required to find the frequency where the phase shift is 180°, after which the program will calculate the value of gain at that frequency. Once this gain is calculated, it is a simple matter to determine what additional gain is required to obtain the user-specified gain.

358   DIGITAL COMPUTER SIMULATION

The strategy we will use to determine the frequency at which the phase angle is 180° (ω_{180}) is to continually increment the values of ω at which the phase angle is calculated. We know that the total phase must eventually reach $-180°$. The first problem we must consider is the choice of an initial value (the first guess) for ω. Clearly, we would not wish to choose a first value for ω that would produce a phase shift in excess of 180°, since our strategy is to increase the value of ω in each iteration. For a given ω, the phase angle in Equation 36 is found from

$$\Theta = -90 + \arctan\frac{\omega}{a} - \arctan\frac{\omega}{b} - \arctan\frac{\omega}{c} \qquad (37)$$

Let $x = \min(b, c)$ and $y = \max(b, c)$. From Equation 37, we see that the phase at $\omega = x$ would be

$$\Theta_x = -90 + \arctan\frac{x}{a} - \arctan\frac{x}{b} - \arctan\frac{x}{c} \qquad (38)$$

Now, if $a < x$, then $\frac{x}{a} > 1$, $\frac{x}{b} \leq 1$, and $\frac{x}{c} > 1$. Consequently, $\arctan\frac{x}{a} > 45°$, and $\arctan\frac{x}{b} + \arctan\frac{x}{c} \leq 90°$. Hence, ω_x cannot be $-180°$, and we can choose $\omega = x$ for our initial value of ω, provided $a < x$.

Suppose now that $a > y$. Then $a > y > x$. Without loss of generality, we may assume $x = b$ and $y = c$. Then Equation 37 becomes

$$\Theta_x = -90 + \arctan\frac{b}{a} - \arctan 1 - \arctan\frac{b}{c}$$

Since $c > b$, $\arctan\frac{b}{c} < 45°$, and Θ_x again cannot be $-180°$. Thus, we may also choose $\omega = x$ for our first value of ω when $a > y$.

Finally, suppose $x < a < y$. Again without loss of generality, suppose $x = b$ and $y = c$. Then Equation 37 evaluated at $\omega = a$ would give us

$$\Theta_a = -90 + \arctan 1 - \arctan\frac{a}{b} - \arctan\frac{a}{c}$$

$$= -45 - \left(\arctan\frac{a}{b} + \arctan\frac{a}{c}\right)$$

In order for Θ_a to equal $-180°$, the expression in parentheses would have to equal 135°. But since $a < c$, $\arctan\frac{a}{c} < 45°$ and since $\arctan\frac{a}{b} < 90°$, we see that this is not possible. Hence, for the case $x < a < y$, we are safe in starting our iterative procedure at $\omega = a$.

Having established the procedure for determining the initial choice of ω, we may now turn attention to the details of the iterative procedure itself. As noted earlier, the total phase Θ of our transfer function must ultimately approach 180°. However, depending on the relative magnitudes of a, b, and c, Θ may cross through $-180°$

9.4 OPTIMIZATION BY AUTOMATIC ITERATIVE COMPUTATION

(be more negative than $-180°$) before it approaches $-180°$. The latter case will occur when the value of a in the transfer function is less than b or c (that is, when the zero is closer to the origin than one of the poles). If we designed our iterative procedure to seek a phase of $-180°$ and we had this situation, then the procedure could never terminate. Thus, we will choose as our criterion for terminating the iterations a phase angle of $-180° \pm 1°$. A $1°$ accuracy is certainly acceptable in the practical situation.

The next question we must address is how large should the increment in ω be each time a new iteration is performed. This clearly should be a function of the values of a, b, and c. If, for example, we had $a = 1000$, $b = 1$, and $c = 500$ (so that our initial value of ω was 1), we would not wish to increment ω by, say, 0.01, as we might wish to do if $b = 1$ and $c = 1.1$. Another potential problem emerges as we consider increment size. Suppose that Θ is changing so rapidly that two successive iterations yield values that are first too small and then too large. For example, supose that one iteration produced $-175°$, and our increment size was such that the next iteration produced $-185°$. We would clearly have passed through the $-180°$ point without satisfying our $\pm 1°$ accuracy criterion. This possibility tells us that we must either keep the increment size small or include some provision for *decrementing* (reducing the size of) ω so that we can go back and home in on ω_{180} whenever an iteration produces a phase angle greater than $180°$. We will choose the latter course of action, since there is no reliable way of making the increment size small enough to meet every possible situation and since we would waste a lot of computer time performing iterations with small increments if it turned out that we had the case where the phase never actually goes through $-180°$ but approaches it asymptotically.

To resolve the conflicting requirements that the two possible situations impose on increment size (large increments in ω if ω_{180} is never actually reached versus small increments in the vicinity of ω_{180} if that frequency is crossed), we propose a twofold solution. First, to avoid an excessive number of iterations should we have the case where ω_{180} is never reached, we will increment the size of the increments in ω. We will choose an initial increment of

$$D = \frac{y - \omega_I}{10}$$

where ω_I is the initial value of ω, selected according to the method we have discussed previously, and $y = \max(b, c)$. We will retain this value of the increment in the iterations until the value of ω reaches y, then, the increment value will be increased to equal y. This increment value will be retained until ω reaches $10y$, at which time the increment value will be increased by a factor of 10 to $10y$. We will allow this process to continue, increasing the value of the increment by 10 each time the value of ω reaches $10y$, $100y$, $1000y$, and so forth. To ensure that this process does not continue indefinitely, we will impose a limitation on the size of N in the increment value of Ny. (In the program that we will presently examine, N is chosen to be 10,000.) After each iteration, we will examine the calculated phase angle to determine if it is within $1°$ of $-180°$ and also to determine if it has exceeded

$-180°$. If it is within the 1° accuracy criterion, we will branch to compute the gain required. If, on the other hand, the angle exceeds $-180°$ (by more than 1°), we will implement the second part of our solution to this problem, namely, an iterative loop that "homes in" on ω_{180}, as we have already discussed. In this loop, we will reduce the increment size as required to find the frequency at which the phase is $-180° \pm 1°$. If the phase actually crosses $-180°$, we know that we can eventually find ω_{180} by alternately increasing values of ω that produce phase angles less than 180° and decreasing values of ω that produce phase angles greater than 180°. When the ω_{180} frequency is found, we will again branch to calculate the required gain.

The value of gain K required to obtain a user-specified gain of M at ω_{180} is easily found from

$$K = \frac{M}{FNG(\omega_{180})}$$

where $FNG(\omega_{180})$ is the value of gain calculated from

$$|G| = \frac{1}{\omega_{180}} \sqrt{\frac{(\omega_{180})^2 + a^2}{[(\omega_{180})^2 + b^2][(\omega_{180})^2 + c^2]}}$$

Figure 9.29 shows a flowchart that summarizes the logic implemented by the program. Note that we include a provision to print only the phase angle finally

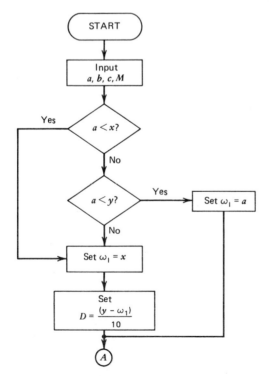

FIGURE 9.29 Flowchart of a program that performs automatic iterations.

9.4 OPTIMIZATION BY AUTOMATIC ITERATIVE COMPUTATION   361

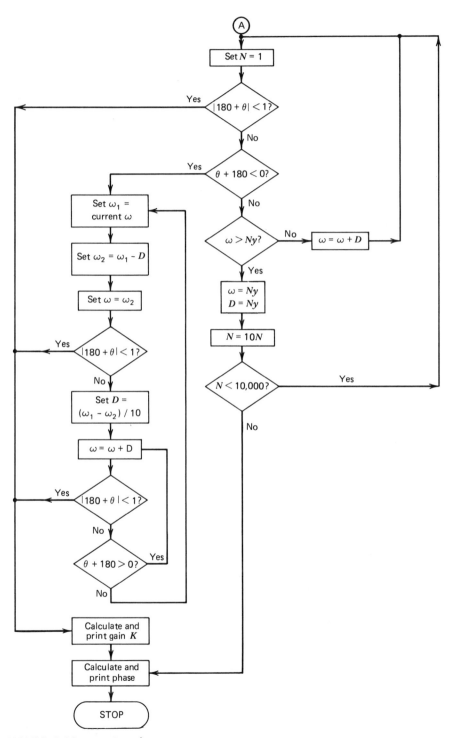

FIGURE 9.29 (continued)

```
10 REM     THIS PROGRAM FINDS THE OPEN LOOP GAIN K
20 REM     NECESSARY TO OBTAIN A DESIRED GAIN MARGIN.
30 REM     USER INPUTS THE CLOSED LOOP GAIN DESIRED AT
40 REM     THE FREQUENCY WHERE PHASE =-180 DEGREES.
50 PRINT
60 PRINT "G=K(S+A)/S(S+B)(S+C),   A,B,C>0."
70 PRINT "A,B,C=";
80 INPUT A,B,C
90 PRINT "ENTER CLOSED-LOOP GAIN DESIRED AT -180 DEG. PHASE"
100 INPUT M
110 LET X=MIN(B,C)
120 LET Y=MAX(B,C)
130 IF A<X THEN 150
140 IF A<Y THEN 170
150 LET W=X
160 GO TO 180
170 LET W=A
180 LET D=(Y-W)/10
190 DEF FNA(W)=-90+57.2958*(ATN(W/A)-ATN(W/B)-ATN(W/C))
200 DEF FNG(W)=(1/W)*(SQR((W**2+A**2)/((W**2+B**2)*(W**2+C**2))))
210 LET N=1
220 IF ABS(FNA(W)+180)<1 THEN 410
230 IF FNA(W)+180<0 THEN 320
240 IF W>N*Y THEN 270
250 LET W=W+D
260 GO TO 220
270 LET W=N*Y
280 LET D=N*Y
290 LET N=10*N
300 IF N<10000 THEN 220
310 GO TO 430
320 LET W1=W
330 LET W2=W1-D
340 LET W=W2
350 IF ABS(FNA(W)+180)<1 THEN 410
360 LET D=(W1-W2)/10
370 LET W=W+D
380 IF ABS(FNA(W)+180)<1 THEN 410
390 IF FNA(W)+180>0 THEN 370
400 GO TO 320
410 LET K=M/FNG(W)
420 PRINT "THE REQUIRED GAIN IS K=";K
430 PRINT "AT W=";W;"RAD/SEC, PHASE IS=";FNA(W);"DEGREES"
440 END
```

FIGURE 9.30 BASIC program listing for the flowchart in Figure 9.29.

reached if it should turn out that we must iterate until $N > 10,000$ without finding a phase angle equal to $-180° \pm 1°$. Figure 9.30 is a listing of the program. Note the use of the ABS (absolute value) instruction to determine whether Θ is within $\pm 1°$ of 180°. This check is made in lines 220, 350, and 380.

Figure 9.31 shows the results of three computer runs of the program. In each run, the values of b and c are the same, $b = 50$ and $c = 100$, and the desired gain at ω_{180}, 0.5, is the same. The zero of the transfer function is changed for each run: $a = 200$ in the first run, $a = 75$ in the second, and $a = 10$ in the third. Thus, the zero is moved closer to the origin in each succeeding run. Note the increased bandwidth that results from moving the zero closer to the origin. The frequency ω_{180} increases from 140 rad/sec to 9000 rad/sec. This program illustrates the process of automatic iteration. It is easy to see how we might extend the program to include provisions for further optimization with respect to some other criterion.

9.5 CONTINUOUS SYSTEM MODELING PROGRAM (CSMP)   363

```
G=K(S+A)/S(S+(S+C)),   A,B,C>0.
A,B,C=
  200           50            100
ENTER CLOSED-LOOP GAIN DESIRED AT -180 DEG. PHASE
 .500000
THE REQUIRED GAIN IS K=       7333.59
AT W=      140  RAD/SEC, PHASE IS      -179.817    DEGREES


G=K(S+A)/S(S+(S+C)),   A,B,C>0.
A,B,C=
  75            50            100
ENTER CLOSED-LOOP GAIN DESIRED AT -180 DEG. PHASE
 .500000
THE REQUIRED GAIN IS K=       1.25017E+07
AT W=     5000   RAD/SEC, PHASE IS     -179.141    DEGREES


G=K(S+A)/S(S+(S+C)),   A,B,C>0.
A,B,C=
  10            50            100
ENTER CLOSED-LOOP GAIN DESIRED AT -180 DEG. PHASE
 .500000
THE REQUIRED GAIN IS K=       4.05031E+07
AT W=     9000   RAD/SEC, PHASE IS     -179.109    DEGREES
```

FIGURE 9.31   Results of three runs of the program in Figure 9.30.

For example, we might wish to have a gain margin that falls in some range (say, not less than 2.0) and yet have a response time that does not exceed a predetermined value. We could define response time as the time required for the system to reach 63.2% of its final value (one time constant) in response to a step input and then have the program calculate this time by computing the inverse transform (time-domain response) of the system when subjected to a step input. If the calculated response time is not less than or equal to the prescribed value, the program could alter the location of the zero (at $s = -a$) in some systematic way and recheck the gain margin and resulting response time. Proceeding in this fashion, we might ultimately find a value of gain K and zero location a that satisfy both requirements on gain margin and response time. Of course, it is possible that the prescribed values of gain margin and response time could not be met simultaneously for a given system, in which case the program would be written to print a message to that effect. It would certainly be possible in this situation to write the program in such a way that it would print the *optimum* combination of gain and zero location, that is, the combination that produces maximum gain margin and minimum response time.

9.5 CONTINUOUS SYSTEM MODELING PROGRAM (CSMP)*

As mentioned previously, there are several digital-computer program packages that have been specifically designed to solve problems in control systems, network analysis, and differential equations. These programs employ languages that are oriented

---

*Portions of this section are by permission of *Systems/360 Continuous System Modeling Program User's Manual*, copyright © 1969 by International Business Machines Corporation.

to the conventions and terminology of those fields and are, therefore, easily learned by users familiar with the fields. Furthermore, these program packages are very versatile in the sense that they can be used to find solutions to a wide variety of problems, many of which would be far too complex to be solved by using conventional analysis techniques. The programs we have discussed thus far in this chapter were developed to solve specific problems in specific system configurations, whereas the large program packages can be used in virtually any problem configuration with a minimum of programming effort. On the other hand, using these programs requires access to a large-scale computer, considerable computer memory capacity, and, of course, purchasing the software package itself.

Among such program packages is the Continuous System Modeling Program (CSMP) developed by the IBM Corporation for use on its System/360. We will discuss the language and programming format of CSMP to the extent that we can write programs capable of solving basic control-systems problems of the level of complexity we have encountered heretofore.

The statements written by a programmer using CSMP can be classified as belonging in one of three different program segments: *initial, dynamic,* and *terminal.* Statements are essentially FORTRAN in nature in that constants, variables, and mathematical operations are expressed using FORTRAN-like conventions. The initial segment of the program is used to define constants that are used in the program (e.g., CONSTANT PI = 3.1426), to define parameter values (values that are automatically changed to specified values when the program is automatically rerun), and to specify certain other program capabilities with which we need not be concerned. IBM defines a variety of statement types that may be used in CSMP, but we will not attempt to classify or even to list all the options and capabilities of the program package, since it will not be necessary for solving the straightforward problems we will consider. One statement that must be included tells the computer the period of time (*problem* time, not computer or computational time) that the simulation should be allowed to run. For example, if the solution to a particular problem turned out to be $y = 5(1 - e^{-10t})$, we would probably want a total time equal to approximately five time constants or 0.5 sec. Of course, if we have no prior knowledge whatsoever of the nature of the solution, then a certain amount of trial and error will be required to specify a time necessary to obtain a representative portion of the solution. This time specification may be accomplished in the initial segment by the statement

$$\text{TIMER FINTIM} = t$$

where t is the total time in seconds.

Since the ultimate solution to any problem solved by CSMP will be a list of time points with the corresponding values of the variable whose solution we are seeking at those time points, we will also specify the desired interval between time points. This allows us to control the *resolution* of the solution and may be specified by another TIMER statement in the terminal section (or by the same timer statement that specifies FINTIM, separated from FINTIM by a comma). When we want a printed plot of the output values, as we will, the required time interval is specified

9.5 CONTINUOUS SYSTEM MODELING PROGRAM (CSMP)

by OUTDEL = Δt. Thus, we would write

$$\text{TIMER OUTDEL} = \Delta t$$

or

$$\text{TIMER FINTIM} = t, \quad \text{OUTDEL} = \Delta t$$

where Δt is the required time interval between solution values. Again, a certain amount of trial and error may be necessary to achieve a value of Δt that yields satisfactory resolution.

The dynamic segment of the program contains the FORTRAN statements that define the actual problem to be solved. This segment is very easy to construct, since it simply consists of a series of statements that list the transfer functions and the inputs and output of each component of the system being simulated. These statements may be written in any order. It is only necessary to be certain to include every input, output, and transfer function present, irrespective of which components are connected to which. Transfer functions and driving functions are specified in certain formats. In Tables 9-1 and 9-2, we list only those that we will require for the types of control-systems problems we are considering.

TABLE 9-1

Transfer Function	Format
1. $1/s$	$y = \text{INTGRL}(ic, x)$ $y = $ output $x = $ input $ic = $ initial condition $= y(0)$
2. s	$y = \text{DERIV}(ic, x)$ $y = $ output $x = $ input $ic = $ initial condition $= x(0)$
3. $1/(ps + 1)$	$y = \text{REALPL}(ic, p, x)$ $y = $ output $x = $ input $ic = $ initial condition $= y(0)$
4. $(p_1 s + 1)/(p_2 s + 1)$	$y = \text{LEDLAG}(p_1, p_2, x)$ $y = $ output $x = $ input
5. $1/(s^2 + 2p_1 p_2 s + p_2^2)$	$y = \text{CMPXPL}(ic_1, ic_2, p_1, p_2, x)$ $y = $ output $x = $ input $ic_1 = y(0)$ $ic_2 = \dot{y}(0)$

Note that in some cases it may be necessary to rearrange the form of a given transfer function in order to make it fit one of the formats of Table 9-1. For example, a transfer function

$$G(s) = \frac{1}{(s + 20)}$$

would have to be written in the equivalent form

$$G(s) = \frac{1}{20} \times \frac{1}{s/20 + 1}$$

in order to fit entry number 3 in the table. In this case, it would also be necessary to multiply by 1/20 to maintain equivalence. Thus, a block in the system with form shown in Figure 9.32a would be redrawn as the two equivalent blocks shown in Figure 9.32b. The FORTRAN statements defining this block in the dynamic segment would be written

X1 = .05*X

Y = REALPL(0,.05,X1)

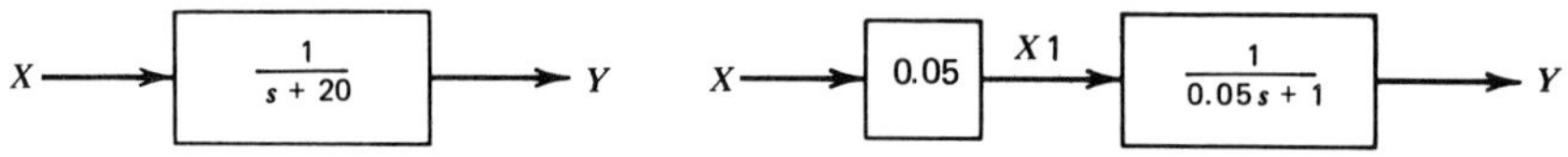

FIGURE 9.32 CSMP representation of a block.

In a similar way, transfer functions containing more than one pole or zero would be broken down into component blocks that fit the formats of Table 9-1 and that when taken together are equivalent to the original transfer function.

EXAMPLE 9.5

Write the FORTRAN statements that would be required to specify the transfer function

$$G(s) = \frac{100(s + 10)}{s(s^2 + 8s + 16)}$$

SOLUTION

We first break G(s) into component blocks as shown in Figure 9.33. The second of these component blocks must be rewritten to fit the form of entry number 4 in Table 9-1. Thus,

$$s + 10 = 10 \times \left(\frac{s}{10} + 1\right)$$

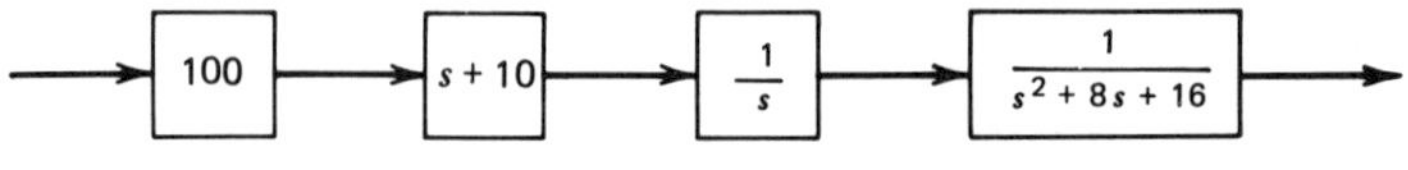

FIGURE 9.33 Decomposition of a transfer function into component blocks (Example 9.5).

Note that we will use $p_2 = 0$ in this case. In order to fit the last block to format

9.5 CONTINUOUS SYSTEM MODELING PROGRAM (CSMP)

of entry number 5, we observe that $p_2^2 = 16$ and $2p_1 p_2 = 8$. Thus, $p_2 = 4$ and $p_1 = 1$.

Noting that the factor 10 that we removed from the term $(s + 10)$ is multiplied by the factor (100) in the original breakdown to produce an overall multiplying factor of $(10)(100) = 1000$, we may now redraw the block diagram of our transfer function as shown in Figure 9.34. The variable names (X, X1, etc.) that we

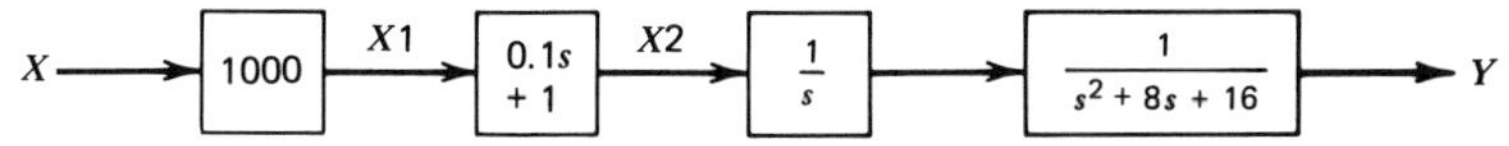

X → [1000] → X1 → [0.1s + 1] → X2 → [1/s] → [1 / (s² + 8s + 16)] → Y

FIGURE 9.34 CSMP representation of Figure 9.33 (Example 9.5).

choose for inputs and outputs are completely arbitrary and need only conform to the usual FORTRAN requirements (except that no distinction need be made between fixed and floating-point variables). The complete set of FORTRAN statements defining our transfer function may now be written as follows:

$$X1 = 1000*X$$

$$X2 = \text{LEDLAG}(.1, 0., X1)$$

$$X3 = \text{INTGRL}(0., X2)$$

$$Y = \text{CMPXPL}(0., 0., 1.0, 4.0, X3)$$

Again, these statements may be written in any order.

Table 9-2 lists the formats required to specify various types of driving functions. We can cause the computer to generate other driving functions that are combinations of those listed in Table 9-2 by writing a set of statements expressing the desired combinations.

TABLE 9-2

Function	Format
1. Step occurring at $t = p$ $$y = \begin{cases} 0 & t < p \\ 1 & t \geq p \end{cases}$$	$y = \text{STEP}(p)$
2. Ramp with unity slope beginning at $t = p$ $$y = \begin{cases} 0 & t < p \\ t - p & t \geq p \end{cases}$$	$y = \text{RAMP}(p)$
3. Inpulse function $y = \delta(t - p_1) + \delta(t - p_1 - p_2) +$ $\delta(t - p_1 - 2p_2) + \delta(t - p_1 - 3p_2)$ $+ \cdots$ $=$ a sequence of unity-weight impulse functions beginning at $t = p_1$ sec and separated by p_2 sec.	$y = \text{IMPULS}(p_1, p_2)$

TABLE 9-2 Con't.

Function	Format
4. Pulse functions (triggered one shot) y = unity-height pulse with minimum width p sec triggered when $x > 0$ and remaining high while $x > 0$.	y = PULSE(p, x)
5. Sine wave $y = \sin[p_2(t - p_1) + p_3]$ = sine wave beginning at $t = p_1$, frequency p_2 rad/sec, and phase angle p_3 rad.	y = SINE(p$_1$, p$_2$, p$_3$)

EXAMPLE 9.6

Write a set of statements in CSMP format to generate a 2 Hz square wave that alternates between ±5 V. The square wave is to be designated SQ2.

SOLUTION

We will first generate a 2-Hz sine wave $V = \sin 4\pi t$ and use this to trigger the pulse generator: V1 = PULSE (0., V). Multiplying V1 by 10 will thus produce a 2-Hz square wave that alternates between 0 and +10 V. We then generate the required wave by adding a −5-V dc level to this square wave:

$$V = \text{SINE}(0., 12.566, 0.)$$
$$V1 = \text{PULSE}(0., V)$$
$$V2 = 10.0 * V1$$
$$V3 = \text{STEP}(0.)$$
$$V4 = -5.0 * V3$$
$$SQ2 = V2 + V4$$

When constructing the dynamic segment of a CSMP program, a good procedure to follow is to first make a block diagram where each block represents one of the transfer functions in Table 9-1 and where all inputs and outputs, including driving functions, have been labeled. It is then an easy task to write the list of statements that define the problem.

EXAMPLE 9.7

Write the dynamic segment of the CSMP program which defines the system shown in Figure 9.35 when it is subjected to a 5-V step input at $t = 0$.

9.5 CONTINUOUS SYSTEM MODELING PROGRAM (CSMP)

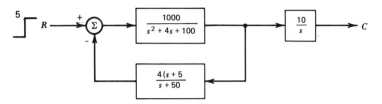

FIGURE 9.35 Write the dynamic segment of a CSMP program for the control system (Example 9.6).

SOLUTION

We must first modify the transfer function of the block in the feedback loop so that it fits the form of entry number 4 in Table 9-1

$$\frac{4(s + 5)}{(s + 50)} = \frac{4(5)[(s/5) + 1]}{50[(s/50) + 1]} = \frac{0.4(0.2s + 1)}{(0.02s + 1)}$$

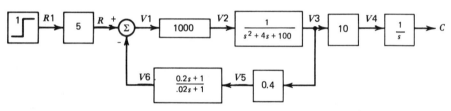

FIGURE 9.36 CSMP representation of Figure 9.35 (Example 9.6).

We then redraw our block diagram and assign variable names as shown in Figure 9.36. The statements defining this system are now written as follows:

$$R1 = STEP(0.)$$
$$R = 5.0*R1$$
$$V1 = R - V6$$
$$V2 = 1000.*V1$$
$$V3 = CMPXPL(0.,0.,.2,10.,V2)$$
$$V5 = .4*V3$$
$$V6 = LEDLAG(.2,.02,V5)$$
$$V4 = 10*V3$$
$$C = INTGRL(0.,V4)$$

We can readily see how powerful CSMP may be for solving very complex systems. A system involving multiple feedback paths as well as "feed-forward" paths and numerous summing and differencing junctions can easily be modeled by a set of statements similar to those in Example 9.6. The program has the additional ca-

370 DIGITAL COMPUTER SIMULATION

pability of simulating virtually any mathematical function of the variables, including many nonlinear functions, and can be driven by a variety of signals, including random noise.

For our purposes, the terminal segment of the program need consist only of a statement that causes the variable of interest to be plotted. The statement

PRTPLOT x

will cause the variable named x to be plotted on the same printout with the computed values of x at the specified time intervals.

We can now illustrate a complete CSMP program. The program format must conform to certain requirements. First, the initial, dynamic, and terminal segments must be identified as such by a statement (or card) containing INITIAL, DYNAMIC, or TERMINAL at the beginning of each segment. Definitions of constants in the initial segment must be preceded by the word CONSTANT, though more than one constant can be defined in the same CONSTANT statement. Each constant definition is separated by a comma, and, as mentioned previously, statements concerning simulation time (FINTIM) and computational interval (OUTDEL) must be preceded by the word TIMER. The last two statements in the program must be END and STOP.

We have described relatively few of the options and capabilities of CSMP and yet have enough to simulate some very complex systems. In Example 9.8, we will illustrate the application of CSMP to one of the control-systems problems we studied earlier.

EXAMPLE 9.8

Write CSMP programs to determine the response of the control system shown in Figure 9.37 with $V = 10$, $\omega_1 = 10$, and $K_2 = 8$ for each of the cases $K1 = 4$, $K1 = 8$, and $K1 = 40$. The input is a unit step in each case.

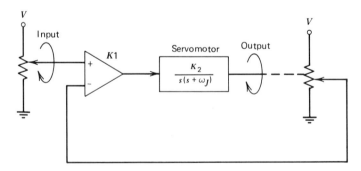

FIGURE 9.37 Write a CSMP program to determine the response (Example 9.8).

SOLUTION

We first draw a block diagram that allows us to identify the transfer function of each component in the system, as shown in Figure 9.38.

9.5 CONTINUOUS SYSTEM MODELING PROGRAM (CSMP)

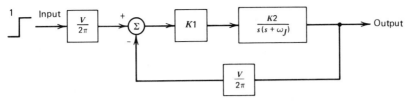

FIGURE 9.38 Block diagram of Figure 9.37 (Example 9.8).

We now redraw the block in a form suitable for CSMP modeling and label inputs and outputs of each component. See Figure 9.39.

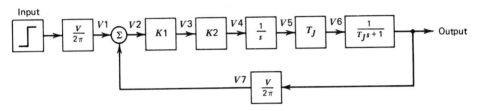

FIGURE 9.39 CSMP representation of Figure 9.38 (Example 9.8).

In Figure 9.40, we have expressed $(1/s + \omega_1)$ in its equivalent form

$$T_1\left(\frac{1}{T_1 s + 1}\right) \quad \text{where} \quad T_1 = 1/\omega_1.$$

The program for the case $K1 = 4$ is shown in Figure 9.40.

```
INITIAL
TIMER FINTIM=2.0
CONSTANT V=10.0, TWOPI=6.2831852
CONSTANT K1=4.0, K2=8.0, TJ=.1
DYNAMIC
INPUT=STEP(0.)
V1=(V/TWOPI)*INPUT
V2=V1-V7
V3=K1*V2
V4=K2*V3
V5=INTGRL(0.,V4)
V6=TJ*V5
OUTPUT=REALPL(0.,TJ,V6)
V7=(V/TWOPI)*OUTPUT
TERMINAL
TIMER OUTDEL=.02
PRTPLOT OUTPUT
END
STOP
```

FIGURE 9.40 CSMP Program for $K1 = 4$ (Example 9.8).

Figure 9.40 does not show the job-control language (JCL) statements normally required to run a FORTRAN program nor any other special statement required to access the CSMP package. These will, of course, generally be required and their

372 DIGITAL COMPUTER SIMULATION

form will depend upon the user's computer facility. In any event, when the program is loaded and executed properly, the printed output will appear as shown in Figure 9.41.

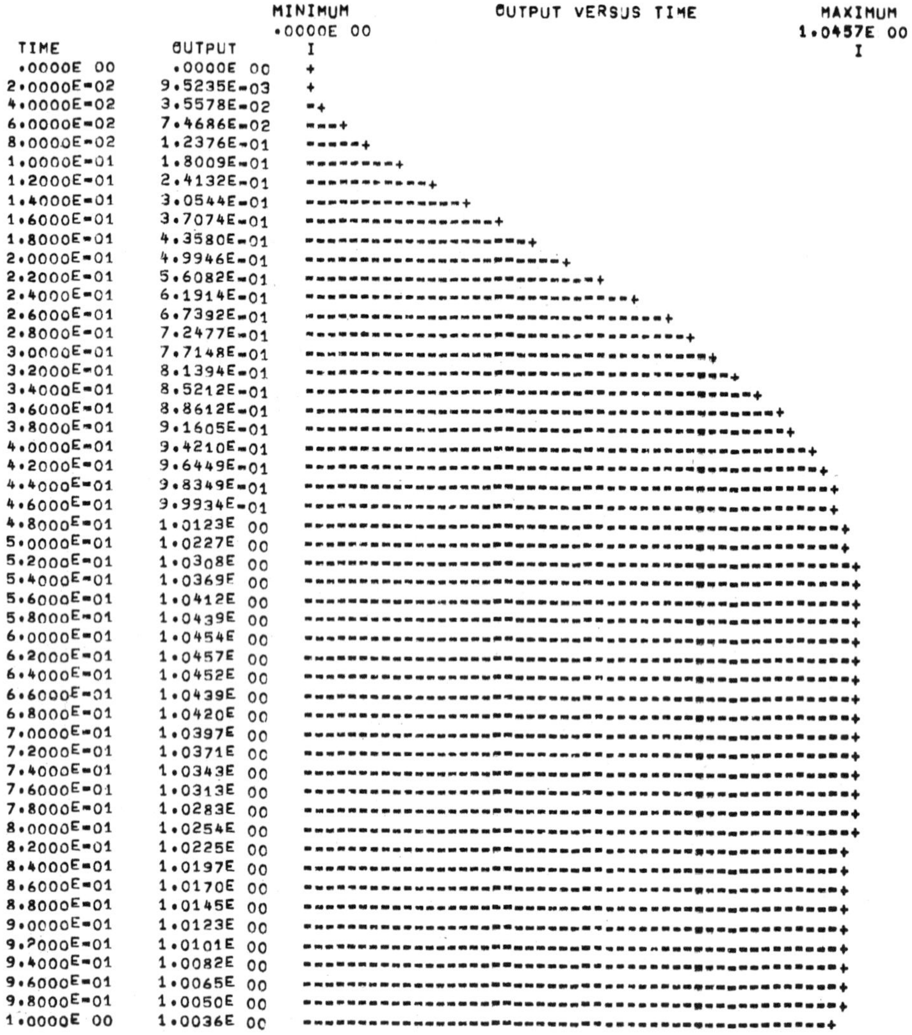

```
                        MINIMUM                OUTPUT VERSUS TIME              MAXIMUM
                        .0000E 00                                              1.0457E 00
    TIME        OUTPUT        I                                                    I
   .0000E 00   .0000E 00      +
  2.0000E-02   9.5235E-03     +
  4.0000E-02   3.5578E-02     -+
  6.0000E-02   7.4686E-02     ---+
  8.0000E-02   1.2376E-01     -----+
  1.0000E-01   1.8009E-01     --------+
  1.2000E-01   2.4132E-01     -----------+
  1.4000E-01   3.0544E-01     ---------------+
  1.6000E-01   3.7074E-01     -------------------+
  1.8000E-01   4.3580E-01     ----------------------+
  2.0000E-01   4.9946E-01     --------------------------+
  2.2000E-01   5.6082E-01     -----------------------------+
  2.4000E-01   6.1914E-01     --------------------------------+
  2.6000E-01   6.7392E-01     -----------------------------------+
  2.8000E-01   7.2477E-01     --------------------------------------+
  3.0000E-01   7.7148E-01     ----------------------------------------+
  3.2000E-01   8.1394E-01     ------------------------------------------+
  3.4000E-01   8.5212E-01     --------------------------------------------+
  3.6000E-01   8.8612E-01     ----------------------------------------------+
  3.8000E-01   9.1605E-01     ------------------------------------------------+
  4.0000E-01   9.4210E-01     -------------------------------------------------+
  4.2000E-01   9.6449E-01     --------------------------------------------------+
  4.4000E-01   9.8349E-01     ---------------------------------------------------+
  4.6000E-01   9.9934E-01     ----------------------------------------------------+
  4.8000E-01   1.0123E 00     -----------------------------------------------------+
  5.0000E-01   1.0227E 00     -----------------------------------------------------+
  5.2000E-01   1.0308E 00     ------------------------------------------------------+
  5.4000E-01   1.0369E 00     ------------------------------------------------------+
  5.6000E-01   1.0412E 00     ------------------------------------------------------+
  5.8000E-01   1.0439E 00     ------------------------------------------------------+
  6.0000E-01   1.0454E 00     ------------------------------------------------------+
  6.2000E-01   1.0457E 00     ------------------------------------------------------+
  6.4000E-01   1.0452E 00     ------------------------------------------------------+
  6.6000E-01   1.0439E 00     ------------------------------------------------------+
  6.8000E-01   1.0420E 00     ------------------------------------------------------+
  7.0000E-01   1.0397E 00     ------------------------------------------------------+
  7.2000E-01   1.0371E 00     ------------------------------------------------------+
  7.4000E-01   1.0343E 00     ------------------------------------------------------+
  7.6000E-01   1.0313E 00     ------------------------------------------------------+
  7.8000E-01   1.0283E 00     ------------------------------------------------------+
  8.0000E-01   1.0254E 00     -----------------------------------------------------+
  8.2000E-01   1.0225E 00     -----------------------------------------------------+
  8.4000E-01   1.0197E 00     -----------------------------------------------------+
  8.6000E-01   1.0170E 00     -----------------------------------------------------+
  8.8000E-01   1.0145E 00     -----------------------------------------------------+
  9.0000E-01   1.0123E 00     -----------------------------------------------------+
  9.2000E-01   1.0101E 00     -----------------------------------------------------+
  9.4000E-01   1.0082E 00     ----------------------------------------------------+
  9.6000E-01   1.0065E 00     ----------------------------------------------------+
  9.8000E-01   1.0050E 00     ----------------------------------------------------+
  1.0000E 00   1.0036E 00     ----------------------------------------------------+
```

FIGURE 9.41 CSMP printout for $K1 = 4$ (Example 9.8).

Although FINTIM = 2.0, only the first second of OUTPUT is displayed in Figure 9.41, since the response is very nearly constant between $t = 1$ and $t = 2$ sec. We see that the response is very slightly underdamped, the peak overshoot being 1.0457 at $t = 0.62$ sec.

Figure 9.42 shows the CSMP program for the case $K1 = 8$, and Figure 9.43 shows the results of the run. Again, only the first second of OUTPUT is displayed.

```
INITIAL
TIMER FINTIM=2.0
CONSTANT V=10.0, TWOPI=6.2831852
CONSTANT K1=8.0, K2=8.0, TJ=.1
DYNAMIC
INPUT=STEP(0.)
V1=(V/TWOPI)*INPUT
V2=V1-V7
V3=K1*V2
V4=K2*V3
V5=INTGRL(0.,V4)
V6=TJ*V5
OUTPUT=REALPL(0.,TJ,V6)
V7=(V/TWOPI)*OUTPUT
TERMINAL
TIMER OUTDEL=.02
PRTPLOT OUTPUT
END
STOP
```

FIGURE 9.42 CSMP program for $K1 = 8$ (Example 9.8).

Inspecting Figure 9.43 (page 374) clearly reveals that the response is less damped than that of the previous case as we might expect, since the gain $K1$ has been doubled. The peak overshoot is now 1.166 and occurs at $t = 0.36$ sec. Note the decrease in time required for the output to reach the unity-input value.

Finally, Figure 9.44 (page 375) is the CSMP program for the case $K1 = 40$, and Figure 9.45 shows the first one second of OUTPUT for this case. The response in this case is seen to be quite underdamped. The oscillatory (ringing) output is readily apparent. The first peak overshoots to 1.488 at $t = 0.14$ sec, and then undershoots to 0.772. If greater resolution were desired in the vicinity of this ringing (if, for example, we wished to measure the ringing frequency or the exponential decay time of the envelope), we could alter the CSMP program to make FINTIM equal to 1 sec and OUTDEL equal to, say, .005 sec.

374 DIGITAL COMPUTER SIMULATION

```
                        MINIMUM           OUTPUT VERSUS TIME        MAXIMUM
                        .0000E 00                                   1.1666E 00
TIME         OUTPUT        I                                           I
.0000E 00    .0000E 00     +
2.0000E-02   1.9015E-02    +
4.0000E-02   7.0686E-02    ---+
6.0000E-02   1.4719E-01    -------+
8.0000E-02   2.4122E-01    -----------+
1.0000E-01   3.4612E-01    ----------------+
1.2000E-01   4.5607E-01    ---------------------+
1.4000E-01   5.6611E-01    --------------------------+
1.6000E-01   6.7218E-01    -------------------------------+
1.8000E-01   7.7108E-01    ------------------------------------+
2.0000E-01   8.6047E-01    ----------------------------------------+
2.2000E-01   9.3880E-01    --------------------------------------------+
2.4000E-01   1.0052E 00    -----------------------------------------------+
2.6000E-01   1.0593E 00    --------------------------------------------------+
2.8000E-01   1.1015E 00    ----------------------------------------------------+
3.0000E-01   1.1323E 00    -----------------------------------------------------+
3.2000E-01   1.1526E 00    ------------------------------------------------------+
3.4000E-01   1.1636E 00    -------------------------------------------------------+
3.6000E-01   1.1666E 00    -------------------------------------------------------+
3.8000E-01   1.1630E 00    -------------------------------------------------------+
4.0000E-01   1.1540E 00    ------------------------------------------------------+
4.2000E-01   1.1409E 00    ------------------------------------------------------+
4.4000E-01   1.1251E 00    -----------------------------------------------------+
4.6000E-01   1.1075E 00    ----------------------------------------------------+
4.8000E-01   1.0891E 00    ---------------------------------------------------+
5.0000E-01   1.0708E 00    --------------------------------------------------+
5.2000E-01   1.0532E 00    -------------------------------------------------+
5.4000E-01   1.0369E 00    ------------------------------------------------+
5.6000E-01   1.0221E 00    ------------------------------------------------+
5.8000E-01   1.0092E 00    -----------------------------------------------+
6.0000E-01   9.9831E-01    -----------------------------------------------+
6.2000E-01   9.8945E-01    ----------------------------------------------+
6.4000E-01   9.8259E-01    ----------------------------------------------+
6.6000E-01   9.7761E-01    ----------------------------------------------+
6.8000E-01   9.7436E-01    ----------------------------------------------+
7.0000E-01   9.7264E-01    ----------------------------------------------+
7.2000E-01   9.7224E-01    ----------------------------------------------+
7.4000E-01   9.7293E-01    ----------------------------------------------+
7.6000E-01   9.7450E-01    ----------------------------------------------+
7.8000E-01   9.7672E-01    ----------------------------------------------+
8.0000E-01   9.7939E-01    ----------------------------------------------+
8.2000E-01   9.8234E-01    ----------------------------------------------+
8.4000E-01   9.8540E-01    -----------------------------------------------+
8.6000E-01   9.8845E-01    -----------------------------------------------+
8.8000E-01   9.9136E-01    -----------------------------------------------+
9.0000E-01   9.9407E-01    -----------------------------------------------+
9.2000E-01   9.9651E-01    -----------------------------------------------+
9.4000E-01   9.9863E-01    -----------------------------------------------+
9.6000E-01   1.0004E 00    ------------------------------------------------+
9.8000E-01   1.0019E 00    ------------------------------------------------+
1.0000E 00   1.0030E 00    ------------------------------------------------+
```

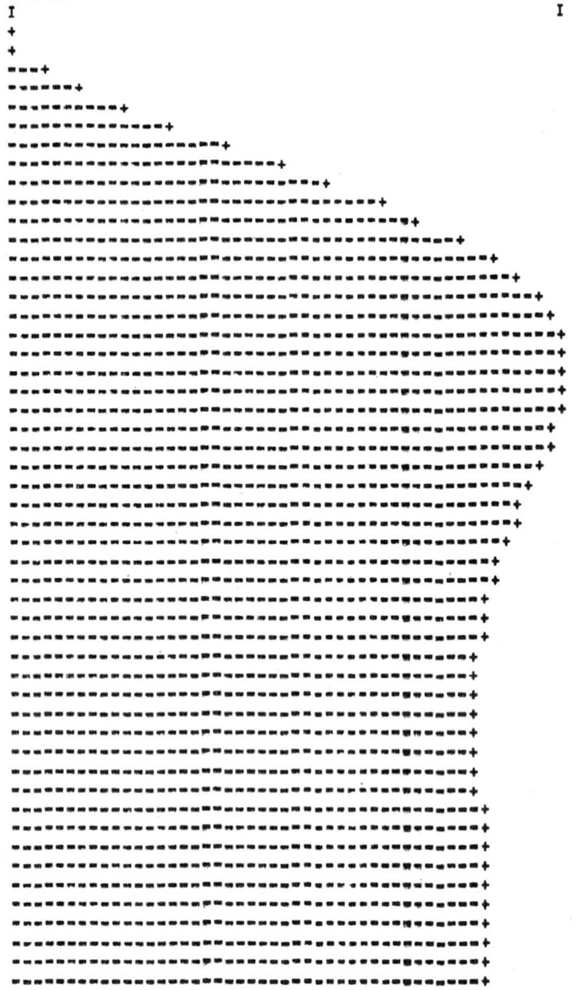

FIGURE 9.43 CSMP printout for $K1 = 8$ (Example 9.8).

9.5 CONTINUOUS SYSTEM MODELING PROGRAM (CSMP)

```
INITIAL
TIMER FINTIM=2.0
CONSTANT V=10.0, TWOPI=6.2831852
CONSTANT K1=40, K2=8, TJ=.1
DYNAMIC
INPUT=STEP(0.)
V1=(V/TWOPI)*INPUT
V2=V1-V7
V3=K1*V2
V4=K2*V3
V5=INTGRL(0.,V4)
V6=TJ*V5
OUTPUT=REALPL(0.,TJ,V6)
V7=(V/TWOPI)*OUTPUT
TERMINAL
TIMER OUTDEl=.02
PRTPLOT OUTPUT
END
STOP
```

FIGURE 9.44   CSMP program for $K1 = 40$ (Example 9.8).

```
                        MINIMUM         OUTPUT VERSUS TIME        MAXIMUM
                        .0000E 00                                 1.4888E 00
TIME        OUTPUT      I                                         I
.0000E 00   .0000E 00   +
2.0000E-02  9.3807E-02  ---+
4.0000E-02  3.3510E-01  ------------+
6.0000E-02  6.5335E-01  -------------------------+
8.0000E-02  9.7684E-01  -------------------------------------+
1.0000E-01  1.2459E 00  -----------------------------------------------+
1.2000E-01  1.4216E 00  ------------------------------------------------------+
1.4000E-01  1.4888E 00  ---------------------------------------------------------+
1.6000E-01  1.4552E 00  --------------------------------------------------------+
1.8000E-01  1.3450E 00  ---------------------------------------------------+
2.0000E-01  1.1922E 00  --------------------------------------------+
2.2000E-01  1.0322E 00  --------------------------------------+
2.4000E-01  8.9532E-01  ---------------------------------+
2.6000E-01  8.0227E-01  ------------------------------+
2.8000E-01  7.6198E-01  ----------------------------+
3.0000E-01  7.7220E-01  -----------------------------+
3.2000E-01  8.2192E-01  -------------------------------+
3.4000E-01  8.9495E-01  ---------------------------------+
3.6000E-01  9.7381E-01  ------------------------------------+
3.8000E-01  1.0431E 00  ---------------------------------------+
4.0000E-01  1.0921E 00  -----------------------------------------+
4.2000E-01  1.1154E 00  ------------------------------------------+
4.4000E-01  1.1136E 00  ------------------------------------------+
4.6000E-01  1.0915E 00  -----------------------------------------+
4.8000E-01  1.0567E 00  ----------------------------------------+
5.0000E-01  1.0180E 00  --------------------------------------+
5.2000E-01  9.8305E-01  -------------------------------------+
5.4000E-01  9.5750E-01  ------------------------------------+
5.6000E-01  9.4429E-01  -----------------------------------+
5.8000E-01  9.4358E-01  -----------------------------------+
6.0000E-01  9.5325E-01  ------------------------------------+
6.2000E-01  9.6965E-01  ------------------------------------+
6.4000E-01  9.8859E-01  -------------------------------------+
6.6000E-01  1.0062E 00  --------------------------------------+
6.8000E-01  1.0194E 00  ---------------------------------------+
7.0000E-01  1.0268E 00  ---------------------------------------+
7.2000E-01  1.0279E 00  ---------------------------------------+
7.4000E-01  1.0238E 00  ---------------------------------------+
7.6000E-01  1.0161E 00  --------------------------------------+
7.8000E-01  1.0069E 00  --------------------------------------+
8.0000E-01  9.9807E-01  -------------------------------------+
8.2000E-01  9.9122E-01  -------------------------------------+
8.4000E-01  9.8720E-01  -------------------------------------+
8.6000E-01  9.8623E-01  -------------------------------------+
8.8000E-01  9.8794E-01  -------------------------------------+
9.0000E-01  9.9153E-01  -------------------------------------+
9.2000E-01  9.9600E-01  -------------------------------------+
9.4000E-01  1.0004E 00  --------------------------------------+
9.6000E-01  1.0039E 00  --------------------------------------+
9.8000E-01  1.0061E 00  --------------------------------------+
1.0000E 00  1.0068E 00  --------------------------------------+
```

FIGURE 9.45   CSMP printout for $K1 = 40$ (Example 9.8).

REFERENCES

1. **Carnahan, Brice**, and **James O. Wilkes**. *Digital Computing and Numerical Analysis.* Wiley, 1973.
2. **Gottfried, Byron S.** *Programming with BASIC.* McGraw-Hill, 1975.
3. **IBM.** *System/360 Continuous System Modeling Program.* Program Number 360A-CX-16X. IBM Publication H20-0367-3.
4. **Speckhart, Frank H.**, and **Walter L. Green**. *A Guide to Using CSMP.* Prentice-Hall. 1976.
5. **Stephenson, Robert E.** *Computer Simulation for Engineers.* Harcourt Brace Jovanovich, 1971.

EXERCISES

9.1 Modify the program in Figure 9.2 so that it will automatically print messages A MUST BE GREATER THAN ZERO or N MUST BE GREATER THAN ZERO if the user inputs values for a or n that are either zero or negative in value. In each case, cause the program to branch back to the appropriate input line after printing the message, so that the user can immediately revise the input.

9.2 Use the program in Figure 9.2 to determine the phase margin of the system in Figure 9.46.

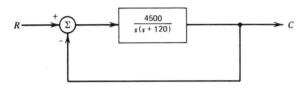

FIGURE 9.46

9.3 Use the program in Figure 9.2 to determine the gain of the system in Exercise 9.2 when the phase angle is $-135°$.

9.4 Modify the program in Figure 9.2 so that it may be used to produce data for a plot of log $|G|$ versus log ω.

9.5 Write a BASIC program that computes the gain magnitude and phase shift of the transfer function

$$G = \frac{K}{(s + a)(s + b)(s + c)}$$

Use this program to find the frequency at which $G = -180°$ when $K = 800$, $a = 15$, $b = 30$, and $c = 100$. If G is an open-loop transfer function, do these values result in a stable closed-loop system?

9.6 Use the program in Figure 9.7 to construct a Bode plot for the open-loop transfer function of the system in Figure 9.47. Plot exact values of log $|G|$ versus log ω using log-log paper and sketch the asymptotic plot on the same paper. Why is there a significant discrepancy between exact and asymptotic values of $|G|$ in the vicinity of $\omega = 8.5$ to 10 rad/sec? In what situations do you conclude that computer simulation has a distinct advantage over asymptotic analysis?

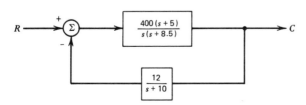

FIGURE 9.47

378 DIGITAL COMPUTER SIMULATION

9.7 Use the program in Figure 9.7 to determine the gain and phase margins of the system in Figure 9.48.

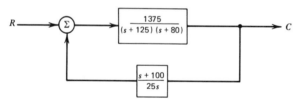

FIGURE 9.48

9.8 Use the program in Figure 9.12 to determine the maximum value of $|G|$ when

$$G = \frac{5000}{s^2 + 12s + 832}$$

9.9 Modify the program in Figure 9.12 so that it prints out the damping factor ζ and the damped natural frequency $\omega_n\sqrt{1 - \zeta^2}$ whenever ζ is less than unity.

9.10 Write a BASIC program that computes $|G|$ and $\angle G$ when

$$G = \frac{K(s + a)^2}{s^2(s + b)(s^2 + cs + d)}$$

9.11 Use the program in Figure 9.12 to determine the value of K that will cause the system in Figure 9.49 to be critically damped.

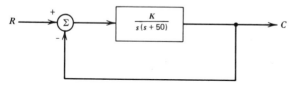

FIGURE 9.49

9.12 Modify the program in Figure 9.19 so that it has the additional capability of computing and printing the inverse transform for the case $B = C$.

9.13 Use the program in Figure 9.19 to determine the time-domain response of the operational amplifier circuit in Figure 9.50 when the input is a 5-V step.

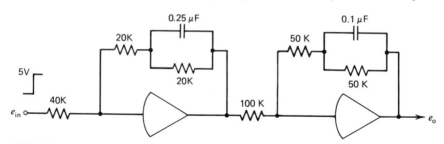

FIGURE 9.50

9.14 Use the program in Figure 9.25 to solve Exercise 9.11.

9.15 Use the program in Figure 9.25 to generate data for a plot of K versus peak-transient response of the system in Figure 9.51.

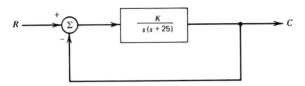

FIGURE 9.51

9.16 Write a BASIC program that can be used to find the value of K required in the system in Figure 9.23 in order to achieve a user-specified damping factor ξ.

9.17 Write a BASIC program that will compute the time-domain response of the system whose closed-loop transfer function is shown in Figure 9.52 when it is subjected to an R-unit-step input.

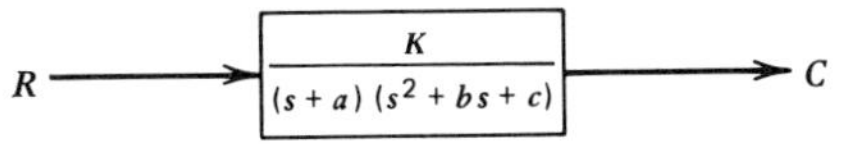

$$R \longrightarrow \boxed{\frac{K}{(s+a)(s^2+bs+c)}} \longrightarrow C$$

FIGURE 9.52

9.18 Convert each of the following transfer functions to a set of equivalent transfer functions that can be expressed in the CSMP format of Table 9-1.

(a) $G(s) = \dfrac{800s}{(s+10)(s+100)}$

(b) $G(s) = \dfrac{(s+1)}{5s(12s^2+24s+96)}$

(c) $G(s) = \dfrac{160(s+80)}{s^2(s+125)^2}$

9.19 Write the CSMP format for specifying each of the following driving functions:

(a) VIN = a unit-step input occurring 50 msec after $t = 0$.
(b) THETA = a unity-amplitude cosine function with frequency 36 Hz.
(c) INPUT = a 100-Hz square wave that alternates between $+1$ V and 0 V.

9.20 Write a set of statements in CSMP format that will produce each of the following driving signals:

(a) ANGLE = a 10-V pulse beginning at $t = 2.0$ sec and 40 msec wide.
(b) IN = a sine wave with peak-to-peak value 10 V, frequency 100 Hz, phase angle 30°, and a dc level (offset) of $+15$ V.
(c) R = a ramp voltage that reaches 50 mV in 1 sec.

380 DIGITAL COMPUTER SIMULATION

(d) I = a sequence of alternately positive and negative impulse functions, each with weight 10 and separated by 1-sec intervals.

9.21 Write the dynamic segment of the CSMP program required to model each of the following systems in Figure 9.53:

(a) $R = 10u(t)$.

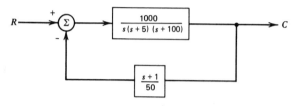

(b) $R = 6$-V step occurring at $t = 1$ sec.

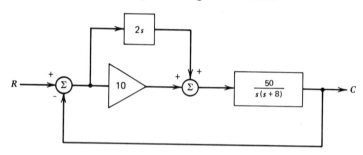

(c) R = series of unit-weight impulses occurring at 40-msec intervals.

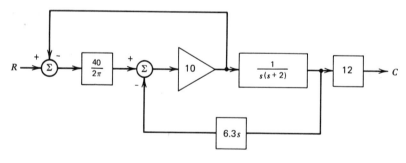

FIGURE 9.53

CHAPTER

10
MICROPROCESSOR-BASED CONTROL SYSTEMS

10.1 ROLE OF THE MICROPROCESSOR

The digital revolution in electronics, spawned by the development of large and very large scale integrated circuits (LSI and VLSI) and by the development of microprocessors (which many believe will have an impact on electronics comparable to the invention of the transistor), has already resulted in the replacement of many traditionally analog control systems by cheaper, faster, and more accurate digital systems. There is abundant evidence of this incursion of digital technology into what was once the exclusive realm of continuous systems: industrial process control, navigation systems, weapons guidance, tracking systems, and a host of others. Furthermore, the availability, versatility, and low cost of digital-logic circuitry has resulted in the introduction of feedback control loops in many devices that were previously open loop. Among these are appliances, automobile ignitions, TV games, measuring instruments, and traffic controllers. The microprocessor has emerged as the central control element in these systems, the device that accepts feedback data, processes and compares it with prescribed values, and issues the appropriate control output(s). There is a great proliferation of these devices on the market now, and we will treat only one representative type in any detail. We will also present a survey of the types of support components that are generally required to design a complete microprocessor-based control system. It should be noted that many of the peripheral devices we will discuss in this chapter are now being fabricated on the microprocessor chip itself. Even though we may discuss them subsequently as if they were separate entities, the reader should realize that in many cases these peripherals are physically together on one integrated circuit chip with a microprocessor. This trend towards integrating devices on a single chip has resulted in systems with very low "chip counts" (that is, a small total number of chips) and reduced fabrication costs as well as ever smaller physical sizes. We hear, for example, of the one-chip microcomputer—a microprocessor that shares a chip with memory and usually with some limited input/output interface circuitry.

A microprocessor may be thought of as the *central processing unit* (CPU) of a computer. It is the device that contains the circuitry capable of performing logical and arithmetic operations on data, of issuing and responding to control signals, and of performing these functions in an orderly sequence as directed by a program stored in memory. We do not, however, wish to view the microprocessor as part of a computer. We are interested in the microprocessor's role as a *controller*. There is a subtle difference in viewpoint here: The term "computer" connotes a device whose main role is problem solving or data processing, while the term "controller" emphasizes the device's capability to issue control signals in response to externally generated feedback signals or data. The difference is not in the device itself, but in the way it is used. When the microprocessor is used as a controller, it is quite true that it may perform many of the same functions as a computer: It must make comparisons and perhaps perform computations and must certainly be programmed to perform in an orderly fashion. Indeed, in some complex control applications, the microprocessor may be called upon to perform some rather lengthy computations, and the distinction in roles may not then be so clear. Even in these situations, though, it is appropriate to refer to the device as a controller, since a computer is generally regarded as a device that is routinely programmed to solve a variety of problems and produce answers, while the controller's ultimate output is a signal that somehow alters the state of the system under control.

In its role as a controller, the microprocessor is said to be *dedicated*; that is, it is programmed to perform a certain task or series of tasks in a wholly predictable manner. The system designer prepares a program that allows for every contingency. Certain inputs to the microprocessor will invariably result in certain outputs. The program is thus similarly dedicated to the particular application. In contrast, when the microprocessor serves as the CPU of a microcomputer, as for example in the now widely popular home computer, the programmer may change the tasks it performs at will simply by changing the program that governs the sequence of operations the CPU must follow.

10.2 THE ANALOG/DIGITAL INTERFACE

Before studying the characteristics and capabilities of the microprocessor itself, we should note that a great many microprocessor-based control systems are used to control the magnitude of an analog-type variable (position, pressure, temperature, etc.) and, therefore, must operate with analog-type input, output, and feedback signals. Consider for example a position control system where the input is the angular rotation of a potentiometer shaft and the output is the angular rotation of a motor shaft. If a microprocessor controller were incorporated in this system, it would be programmed to compare the magnitude of a binary number proportional to the input with the magnitude of a binary number proportional to the output. It would then generate a binary number equal to the difference between these two, and this number would in turn cause the output shaft to rotate in the correct direction until such time as the difference (error) became zero. These are, of course, the same functions that are performed by the continuous-position control systems studied in earlier chapters except, in this case, the comparison of input to output is accomplished using binary arithmetic. Clearly then, we require a means for converting the analog inputs and outputs to digital form as well as a means for converting the digital output of the microprocessor to analog form. In other words, we need *analog-to-digital* (A/D) converters and *digital-to-analog* (D/A) converters to form an *interface* between the analog and digital components of the system. See Figure 10.1.

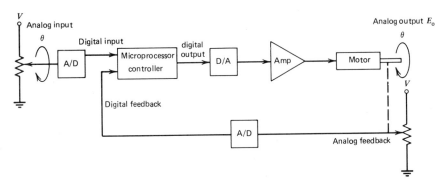

FIGURE 10.1 Analog/Digital interfaces required in a microprocessor-based position-control system.

One widely used method for converting a binary word in parallel form to an analog voltage is a precision-resistor network connected as shown in Figure 10.2. Here we assume an eight-bit binary input with bits D0 through D7.

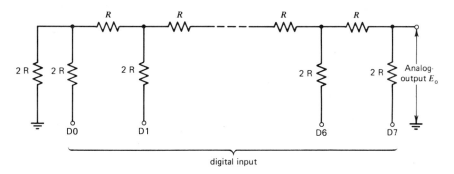

FIGURE 10.2 An R-2R ladder D/A converter.

For obvious reasons, the network shown in Figure 10.2 is called an *R-2R ladder network*. To understand its operation, suppose that binary 0 equals 0 V (ground) and that binary 1 equals +5 V. If all bits D0 through D7 are 0, then clearly E_o = 0. Suppose D7 = 1 and all other bits equal 0; the equivalent circuit then appears as shown in Figure 10.3.

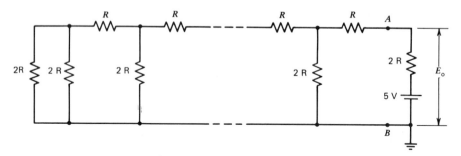

FIGURE 10.3 Equivalent circuit of the R-2R ladder with input 10000000.

Note that in Figure 10.3 the total equivalent resistance to the left of the terminals A–B is 2 R. (Starting at the left-hand side of the diagram, 2 R||2 R = R, which is in series with R. The resultant 2 R is in parallel with another 2 R and so forth.) The circuit is therefore equivalent to that shown in Figure 10.4. We see in this

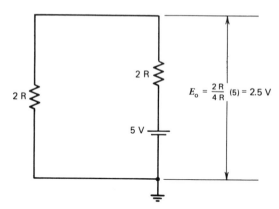

FIGURE 10.4 Simplified circuit equivalent to Figure 10.3.

case that E_o = 2.5 V. The analog output corresponding to the binary number 10000000 is therefore ½(5) = 2.5 V. By a similar circuit analysis, it is easy to show that the analog output corresponding to a 01000000 is ¼(5) = 1.25 V. In general, if the binary number contains a single binary 1 and it is in position D_n (n between 0 and 7), then the output is $(1/2^{8-n})5$ V. By applying the principle of superposition, we can find the output corresponding to any binary input. This is illustrated in example 10.1.

EXAMPLE 10.1

Find the analog output of the D/A converter illustrated in Figure 10.3 when the binary input is 11010000. Assume that binary 1 and 0 are +5 and 0 V, respectively.

SOLUTION

We have 1's in bit positions D4, D6, and D7. The output voltage resulting from each of these acting alone is

$$n = 4 \quad \left(\frac{1}{2^4}\right)5 = \frac{5}{16} \text{ V}$$

$$n = 6 \quad \left(\frac{1}{2^2}\right)5 = \frac{5}{4} \text{ V}$$

$$n = 7 \quad \left(\frac{1}{2}\right)5 = \frac{5}{2} \text{ V}$$

By superposition, the output resulting from all inputs acting simultaneously is therefore

$$\frac{5}{16} + \frac{5}{4} + \frac{5}{2} = \left(\frac{13}{16}\right)5 = 4.0625 \text{ V}$$

Each time the binary number applied to the converter is incremented by 1, the analog output increases by $(1/2^8)5 = 5/256 = .0195$ V. Thus, if the binary input were supplied from a binary counter that counted up from 00000000 through 11111111, the analog output would be a "staircase" voltage, as shown in Figure 10.5.

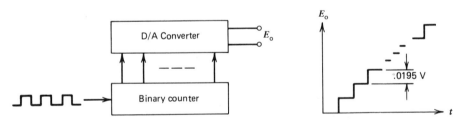

FIGURE 10.5 Output of D/A converter driven by a binary counter.

Digital-to-analog converters are available in integrated-circuit packages that include active circuitry in addition to the R-2R ladder network. One version of these devices is designed to operate with a reference voltage that may be time varying. The analog output is thus proportional to the product of an analog input and a binary input, and the device is called a *multiplying D/A converter*. Appendix B contains specifications and application information for one such device, the Analog Devices AD7520/7521. Note the use of a 10 K–20 K ladder network. Note also how the device is designed to operate in conjunction with an operational amplifier to act as a four-quadrant multiplier (both analog and digital inputs may be either positive or negative). When used in this manner, a binary input consisting of all 1's is the maximum negative input, and all 0's correspond to the maximum positive input. This is called *bipolar* operation.

Many electronic analog-to-digital (A/D) converters incorporate a D/A converter as a necessary component in their conversion method; they include the *counter* type, the *dual-slope integrating* type, and the *successive approximation* type of A/D converter. The first two of these methods use a binary counter that is driven by a fixed-frequency clock signal and produce a binary count that is proportional to the magnitude of the analog input. The time required to perform the conversion thus depends on the magnitude of the analog input. The successive approximation method, which is the most widely used method of A/D conversion and which is the only one we will discuss in detail, has a conversion time that is independent of the magnitude of the analog input.

The method of successive approximation uses a D/A converter to convert its own binary output to an equivalent analog voltage. This analog voltage is compared by using a voltage comparator to the analog input of the converter. The bits of the binary output are then adjusted one at a time, beginning with the most significant bit, until the voltage comparator determines that the binary output is equal to the analog input. Figure 10.6 illustrates the method.

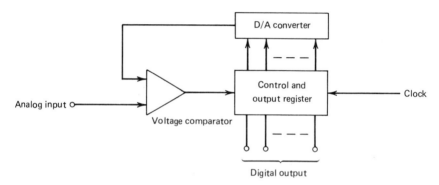

FIGURE 10.6 Block diagram of the successive-approximation method of A/D conversion.

To simplify the explanation of this technique, let us suppose we have only a four-bit device (with output $D_3D_2D_1D_0$). Further, suppose that an analog input of 8 V produces the binary output 1000, while 4 V produces 0100, and so forth. If the input is, say, 5 V, then the method of successive approximation would proceed as follows:

1. The initial binary output is 0000. The most significant bit D_3 is set to 1, so the D/A converter produces a voltage proportional to 8 V. This 8-V level is compared to the 5-V analog input.
2. Since $5 < 8$, bit D_3 is reset to 0, and bit D_2 is set to 1. The binary output is now 0100, and so the D/A converter produces a 4-V level, which is compared to the 5-V input.
3. Since $5 > 4$, bit D_2 is left in its 1 state, and bit D_1 is set to 1. The binary number is now 0110 or 6 V.
4. Since $6 > 5$, bit D_1 is reset to 0, and bit D_0 is set to 1, resulting in 0101 or 5 V.
5. Since $0101 = 5 =$ the input voltage, the conversion is complete.

The comparisons and adjustments of bits take place under the synchronization of the clock, which is shown connected to the device in Figure 10.6. Since one comparison and one bit adjustment takes place during each clock period, the total conversion time depends only on the number of bits of the output rather than on the magnitude of the input. On the average, the method of successive approximation is faster than either the counter or dual-slope integrating types of converters.

The speed with which an A/D converter can perform conversions is a very important characteristic. In practice, the analog input is changing with time. Clearly, this would be the case in a control system such as the one illustrated in Figure 10.1, where the feedback signal, for example, would continue to change until such time as the correct angular position of the motor shaft was reached. The A/D converter must be able to produce a binary output in a period of time that is negligible compared to the time during which significant changes in the analog signal may occur. Putting it another way, the conversion frequency (number of conversions per second) must be greater than the highest frequency component of the analog input.

The Shannon sampling theorem tells us just how great the conversion frequency must be provided we know the maximum frequency component of the analog input. According to this theorem, a signal must be sampled at a rate equal to *at least twice* the frequency of the maximum-frequency component of the signal. Failure to sample (convert) the signal at a rate at least this high causes the introduction of spurious frequency components in the output. This is called *alias* error, and the unwanted frequency component is called an alias frequency. Suppose, for example, that an A/D converter is capable of performing one conversion in 0.1 msec. Its conversion rate is thus $1/10^{-4} = 10,000$ conversions per second and, therefore, should not be used for analog signals with frequency components greater than 5 kHz.

Another important specification for the A/D converter is its *resolution*, the smallest change in input voltage that it can detect (and that can thus cause a change in the binary output). This is equal to the maximum rated analog input voltage (full-scale input) divided by 2^n, where n is the number of bits of the binary output. In practice, resolution is often specified as a percentage, which is equivalent to the ratio of the weight of the least significant bit ($2^0 = 1$) to the weight of the most significant bit (2^n). For example, the resolution of a ten-bit A/D converter might be expressed as $(1/2^{10}) \times 100\% = 0.1\%$. If the maximum-analog input voltage, or reference voltage, of the converter were 15 V, then this resolution would be equivalent to $10^{-3} \times 15 = 15$ mV. Since a knowledge of the number of bits is sufficient information to determine the percent resolution, the resolution is frequently expressed simply as "n-bits." The ten-bit converter in the foregoing example could be said to have ten-bit resolution.

EXAMPLE 10.2

Use the specifications for the DATEL Model ADC-EH8B1 analog-to-digital converter provided in Appendix B to answer the following questions:

(a) What is the resolution in bits, percent, and volts when used in the unipolar mode?

(b) What is the conversion time and conversion rate?

(c) What is the theoretical maximum-frequency component in the analog input?

SOLUTIONS

(a) The ADC-EH8B1 is an eight-bit converter and, therefore, has eight-bit resolution or $1/256 \times 100\% = 0.4\%$. When operated in the unipolar mode, its full-scale input is 10 V, and so the resolution is 0.04 V or 40 mV.

(b) From the data sheet, the conversion time is 4 μsec. The conversion rate is, therefore, $1/(4 \times 10^{-6}) = 250{,}000$ conversions/sec.

(c) By the Shannon sampling theorem, the maximum-frequency component of the analog input is 250 kHz/2 = 125 kHz.

Figure 10.1 shows the analog feedback signal of a position-control system being developed by a potentiometer. The figure shows that this voltage is the input to an A/D converter, which produces the digital (binary) number necessary for processing by the microprocessor. A more efficient and reliable means of producing a digital code proportional to the rotary position of a shaft is by using an *optical-shaft encoder*. Mounted on the shaft of this device is a disk that contains a pattern of transparent and opaque segments. A fixed light source directs light rays onto a narrow portion of the disk and photosensitive light sensors on the opposite side are illuminated or darkened according to the pattern. As the shaft rotates, the pattern opposite the light source changes and so does the combination of illuminated light sensors. The pattern is designed so that equal increments of rotation create unique combinations of illuminated sensors. The electrical outputs of these sensors are buffered and amplified, so a unique binary output code is generated for each increment of rotation.

The encoder we have just described is called an *absolute*, or *whole-word* encoder, since it produces a unique n-bit binary-code word for each of 2^n shaft positions. Its resolution is thus $360°/2^n$. The *incremental* encoder, in contrast, simply produces a serial train of pulses proportional to total shaft rotation. The pulses are used to drive a binary counter whose total count is then proportional to total shaft rotation. Clearly, this type of encoder could also be used to digitize the angular velocity of a shaft simply by counting the number of pulses per unit time.

Appendix B contains technical data on the Litton Encoder line of absolute shaft encoders. Note that models with resolutions from 10 to 13 bits are available. This corresponds to resolutions from $360°/2^{10} = 0.352°$ to $360°/2^{13} = 0.044°$. In the latter case, this means that a unique binary code will be generated for every 0.044°, or 2.64 min, of rotation.

Note also that in the Litton Encoder line models are available with outputs in Gray, binary, or 8-4-2-1 BCD code. The binary and BCD coded models have the advantage that the numbers are in a form that can be processed directly by a CPU that has been programmed to perform binary or BCD arithmetic. The disadvantage of these codes is that a small misalignment of the light sensors relative to the optical pattern can result in codes that correspond to radically different positions when, in fact, only a small angular rotation occurs. Using Gray Code eliminates this problem, since Gray Code has the characteristic that only *one* bit changes in the transition from any number to any adjacent number. For example, in binary code, the transition from 0111 to 1000 results in a change in all four bits, whereas in Gray Code this same transition is represented by going from 0100 to 1100. The disadvantage of Gray Code is that it is not a weighted code (bit positions carry no arithmetic weight as in binary), and it is, therefore, not practical to perform arithmetic computations using this code. On the other hand, the CPU that receives feedback in this code can be programmed to convert it to binary before processing.

10.3 MICROPROCESSOR SYSTEM DESIGN CONSIDERATIONS

Whether the microprocessor functions as the CPU of a microcomputer or as a system controller, there are certain other components that must be present in the overall system and with which the microprocessor must communicate in order for it to function at all. Chief among these is a memory; we will defer discussing the organization and types of memories used in microprocessor control systems for the moment and concentrate on what we mean by communication between the microprocessor and its peripheral components. The microprocessor is, of course, a digital device and, therefore, receives and generates digital logic signals, which are typically +5 V and 0 V, corresponding to binary 1 and binary 0, respectively. A variety of digital signals are transmitted to and from the microprocessor by way of *buses*. A bus is simply a group of signal lines that have some functional characteristic in common. The *data bus*, for example, consists of the parallel lines on which digital *data words* (a data word is typically eight bits or one *byte*) are transmitted. We refer to an eight-bit microprocessor as one that has an eight-bit data bus, and such a microprocessor chip would have eight pins assigned as data pins. The data pins and data-bus lines are usually identified by the designations D0, D1, . . ., D7. The *address bus* (typically 16 bits) consists of the lines on which memory addresses are transmitted. Thus, 16 pins on the microprocessor chip (A0 through A15) are dedicated to the address bus. In a typical sequence of operations, the microprocessor *reads* a data word stored in a certain memory address by transmitting that address to memory via the address bus and then receives the data word from memory via the data bus. Similarly, the microprocessor *writes* in memory by first transmitting a memory address (the address in which data is to be stored) on the address bus and then transmitting the data word to that memory address via the data bus. In either case, we say that memory has been *accessed* by the CPU and that communication has taken place via the buses. Figure 10.7 illustrates this process.

Note that in figure 10.7 we have used "fat" arrows to represent buses, in other words, to represent a parallel collection of signal lines. Note also that the data bus has arrowheads on both ends, implying that data words can be transmitted from the CPU to memory or vice versa. This is, of course, necessary if we are to be able to both read and write in memory, and such a bus is said to be *bidirectional*. The address bus on the other hand is *unidirectional*, since there is no need for memory to transmit addresses to the CPU.

A third type of signal present in a microprocessor system is the *control* signal. There may be several such control signals in a system, some of which are inputs

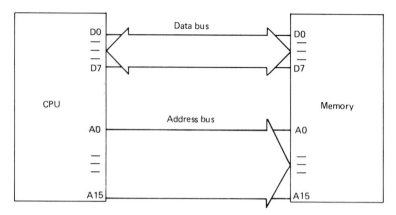

FIGURE 10.7 Block diagram of the bus structure connecting memory and microprocessor CPU.

to the CPU to effect its control and others that are generated by the CPU to control external devices. The types and functions of the control signals constitute what is probably the greatest variation in the design of various microprocessors. Some authors and manufacturers lump the control signals together as a group and refer to them as the *control bus*, although there is frequently such diversity in the functions of the various control signals that this terminology may be misleading. Some of the functions that may be performed by control signals generated by the CPU include *enabling* (activating) memory, enabling other input/output (I/O) devices, enabling read or write operations, and *acknowledging* the recognition and acceptance of other control signals that have been transmitted *to* the CPU. Control signals generated by external devices and transmitted to the CPU to control it include *resetting* (starting the CPU in its program), *hold* and *wait*, which are methods of suspending the normal sequence of CPU operations, and *interrupt*, which causes the CPU to temporarily abandon its present tasks in favor of a more urgent one. We will elaborate on these functions when we discuss microprocessor characteristics in greater detail in a later section. Of course, the CPU must also receive dc power and have a ground connection, so certain pins are allocated for this purpose. Most modern CPUs require only a single dc voltage, usually +5 V. Finally, the CPU is synchronized by a *clock* signal, which is essentially a sequence of recurring pulses with a frequency somewhere in the interval between the manufacturer's specified minimum and maximum operating frequency.

Figure 10.8 shows a generalized block diagram of the CPU in a system that includes I/O devices as well as memory. Input devices, by definition, produce inputs *to* the CPU, while output devices receive data *from* the CPU. In other words, the CPU is always the point of reference when distinguishing between input and output. In Figure 10.8, we notice that memory and the I/O devices share the same data bus. Of course, only one device can receive or transmit data on the data bus at any given time, and this time sharing by the bus is called *time multiplexing*. The various control signals ensure that only one device has *control* of the bus at any given time. It should be noted that many systems also time multiplex the

10.3 MICROPROCESSOR SYSTEM DESIGN CONSIDERATIONS

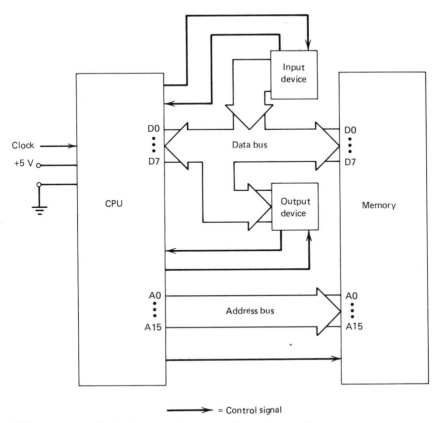

FIGURE 10.8 Block diagram of a CPU with memory and I/O devices.

address bus. At this point, we might legitimately question the wisdom of having more than one device electrically connected to the same lines. For example, if both memory and an input device share the same data bus (i.e., if both have their data lines physically connected to the same points) and if the input device drives these lines by placing an eight-bit data byte on the bus, what is to prevent those logic levels from adversely affecting, or being affected by, the memory lines? Suppose bit D0 of the input device is +5 V while bit D0 of memory is 0 V? Since these points are electrically connected, what is to prevent that memory line from shorting out the +5 V produced on D0 by the input device?

The answers to the questions just posed lie in the fact that devices that share a common bus are *tri-state* or *three-level logic* devices. By this we mean that the outputs of these devices can be put into a third, or *high impedance* state, in addition to the normal logic-zero and logic-one states. When the input or output of such a device is in the high impedance state, it is effectively disconnected, or open circuited, so its presence is not felt by any other lines connected to it. It is just as if there were a set of switches inside the device that is opened to disconnect the device from its pins. When the device is *enabled*, or *selected*, these switches close, and the generation of normal logic levels can resume. These ideas are illustrated in Figure 10.9.

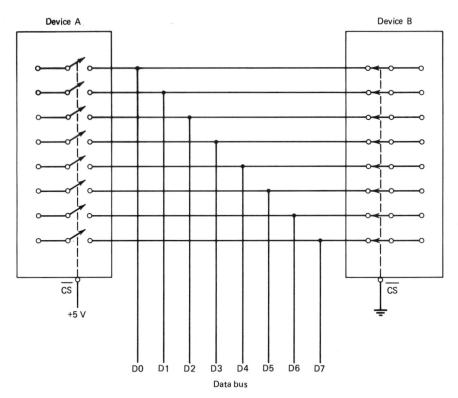

FIGURE 10.9 Illustration of the three-level logic concept; device A is in the high impedance state.

In Figure 10.9, device A is in a high impedance state while device B is enabled, or selected. As far as the data bus is concerned, only device B is connected to it, and as far as device B is concerned, only the data bus is connected to it. There is no way for current to flow in or out of device A, and so its presence is unknown to any other devices sharing the data bus. Device A is said to be *disabled*.* Of course, the switches shown in Figure 10.9 are fictitious and serve only to illustrate a point, but the effect is the same. Note that each device in Figure 10.9 has an input labeled $\overline{CS}$. This is the *chip-select* input, and the bar above the letters CS indicates that it is an *active-low* input. When this input is made low, as shown on device B, the device is enabled, or selected, and it operates in the normal fashion, while connecting $\overline{CS}$ to a logic-one level as shown on device A, causes it to enter the high impedance state. Chip-select inputs are usually but not always active low. Chip-select inputs are the inputs used by the control signals in the microprocessor system to ensure that only one device has control of a bus at any given time.

EXAMPLE 10.3

A microprocessor generates active low control signals designated $\overline{I/OREQ}$, $\overline{READ}$, and $\overline{WRITE}$, which go low (logic zero) when the device is requesting an I/O device,

---

Inhibited and *floated* are other terms used to describe this situation.

10.3 MICROPROCESSOR SYSTEM DESIGN CONSIDERATIONS

is ready to read, or is ready to write, respectively. Draw a logic diagram that combines these signals in such a way that an active low enabling signal is produced at the proper time for an input device. Repeat for an output device.

SOLUTION

We first construct a truth table, remembering that active low means that the signal is true, or that the action is to be taken, when the logic level is zero. For the input device, we need only consider the signals $\overline{\text{I/OREQ}}$ and $\overline{\text{READ}}$, since the CPU can only read an input device.

$\overline{\text{I/OREQ}}$	$\overline{\text{RD}}$	$\overline{\text{CS}}$
0	0	0
0	1	1
1	0	1
1	1	1

We see that the input device will only be selected ($\overline{\text{CS}}$ low) when both $\overline{\text{I/OREQ}}$ and $\overline{\text{READ}}$ are low. The truth table corresponds to the logical OR function, and hence, we use an OR gate to combine $\overline{\text{I/OREQ}}$ and $\overline{\text{READ}}$. Similarly, $\overline{\text{I/OREQ}}$ and $\overline{\text{WRITE}}$ are combined in an OR gate to enable the output device. This is illustrated in Figure 10.10. (The data bus is not shown.)

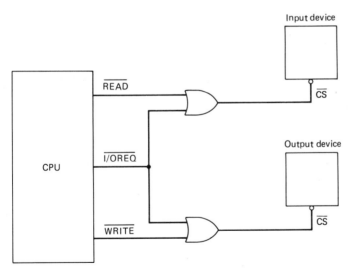

FIGURE 10.10 Combining control signals to enable I/O devices (Example 10.3).

We have devoted considerable space to the trilevel logic concept because it is an exceptionally important characteristic of a microprocessor system. The control-systems designer must have a thorough understanding of this concept in order to develop the logic circuitry required to ensure that the correct devices are enabled or inhibited at the correct times. Furthermore, the microprocessor itself has the capability of placing its own buses in a high impedance state (when commanded to do so by an externally generated control signal). Thus, the CPU may effectively disconnect its data pins D0 through D7, its address pins A0 through A15, and its control-signal pins from the system buses. When this occurs, the CPU is said to have *floated* its data, address, or control lines. A typical example of a situation where these lines must be floated occurs when an external device (a device other than the CPU itself) must have access to memory. This external device may, for example, have to write data into memory, which means that the CPU must yield control of the data and address buses to the device. This is called *direct-memory access* (DMA), and is an important capability of every microprocessor.

Let us turn now to the functions and characteristics of memory. Physically, the memory in microprocessor control systems is almost always composed of one or more integrated circuit chips. We will discuss the various physical configurations and characteristics presently, but let us first classify memory types according to function. In this context, memory is used for two purposes: to store a program and to store data; accordingly, we speak of *program memory* and *data memory*. The student should realize that in a normal sequence of operations, the CPU will continually read the contents of successive addresses in program memory in order to continually *fetch* instructions that tell it what tasks to perform. The sequential nature of this procedure is significant. Unless told otherwise by an instruction it fetches from memory, the CPU will invariably return to the next memory address in sequence to fetch its next instruction. Program memory thus contains a set of instructions that direct the CPU to perform the desired tasks in the desired order. Suppose, for example, that we wished to program the CPU to add two numbers stored in two memory addresses and store the sum in a third memory address. The program would consist first of an instruction causing the CPU to read the contents of the address where the first number is stored; then to read the contents of the address where the second number is stored; then to perform the addition; and then to store the sum (write the sum) in the third memory address. The CPU would execute this program by reading and performing the instructions in the order that they are stored in program memory. Note that the CPU does not normally write into program memory. Data memory, on the other hand, is the section of memory used to store the quantities that the CPU will use in its computations or the quantities that result from those computations. In our example, the two numbers to be added as well as their sum might be stored in data memory. The contents of data memory are subject to change, while the contents of program memory (in control-systems applications) are not changed. The same program could be used to add two numbers and store their sum no matter what numbers may have been, or may later be, stored in data memory. Thus, the CPU must generally be able to both read and write in data memory.

Physically, program memory and data memory may or may not be of the same types, though in control-systems applications program memory is usually *Read Only Memory* (ROM), while data memory is *Random Access Memory* (RAM), sometimes called read-write memory. As mentioned, the CPU need only read program memory in order to fetch its instructions; hence the use of ROM for storing programs. A further important characteristic of ROM is that it is *nonvolatile*, which means that its contents are not lost when power is removed. This is certainly a desirable feature in control systems applications: It would be most inconvenient if it were necessary to reprogram memory in the field every time the system was shut down. RAM can be both read and written into, but it is *volatile*: Its contents are lost if power is removed. (Understand that we speak only of integrated circuit memory—other memory types, not generally used in microprocessor control systems, may have nonvolatile RAM.)

We have mentioned that the typical microprocessor address bus has 16 address lines. This means that the microprocessor is capable of generating, and therefore accessing, $2^{16} = 65{,}536$ different memory addresses. In practice, it is rare that a memory of such large size is required. In some practical control-systems applications, less than 1000 total-program and data-memory addresses are required. Whatever the requirement, integrated-circuit memory capacity is almost always some power of 2, since n address lines are capable of specifying 2^n addresses. If a particular application required 1000 total memory addresses, we would use only ten of the 16 available address lines and provide a memory with $2^{10} = 1024$ addresses. If we could find a single memory chip capable of storing 1024 eight-bit words, it would have ten address pins and eight data pins, and we would connect the ten address pins to the address bus (say, A0 through A9) and the eight data pins to the data bus. Figure 10.11 shows these connections.

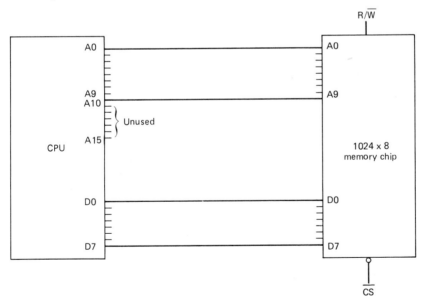

FIGURE 10.11 Address and data-pin connections for a 1024 × 8 memory chip.

Note that in Figure 10.11 the memory chip is shown with an input labeled $\overline{R/W}$. This is a typical read/write control input. Provided that $\overline{CS}$ is low, then $\overline{R/W}$ high allows the memory chip to be read, that is, causes data to flow from memory to the CPU, while $\overline{R/W}$ low allows the memory chip to be written into.

It is more often the case that a memory of the required capacity must be assembled using more than one integrated-circuit chip. A memory chip that can store M N-bit words is said to be organized M by N, or $M \times N$, where M is some power of 2. Thus, the chip in Figure 10.11 would be referred to as 1024×8. It is conventional to round off the number M and express it in terms of thousands, so this chip might also be called $1 \text{ K} \times 8$. Table 10-1 lists powers of 2 and the abbreviated way they are usually specified. Note that 32 K and 64 K are not actually the values of 2^{15} and 2^{16} rounded off to the nearest one thousand, but this is the way they are abbreviated in practice (though one occasionally hears "65 K").

Suppose we wished to construct a $1 \text{ K} \times 8$ memory using $1 \text{ K} \times 4$ memory chips. Then clearly we would need two such chips, and they would be connected as shown in Figure 10.12.

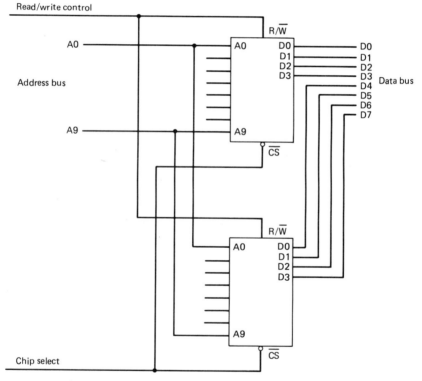

FIGURE 10.12 Address and data-pin connections for two $1 \text{ K} \times 4$ memory chips.

Referring to Figure 10.12, we see that when a certain ten-bit address appears on the address bus, the same address will be accessed in both chips and that one chip contains the least significant four bits (D0–D3) of the data word stored at that address, while the other chip contains the most significant four bits (D4–D7).

10.3 MICROPROCESSOR SYSTEM DESIGN CONSIDERATIONS

TABLE 10-1

n	2^n	Abbreviated Form
8	256	¼ K
9	512	½ K
10	1,024	1 K
11	2,048	2 K
12	4,096	4 K
13	8,192	8 K
14	16,384	16 K
15	32,768	32 K
16	65,536	64 K

EXAMPLE 10.4

How many 2 K × 1 chips would be required to construct a 16-K memory?

SOLUTION

Each 2 K × 1 chip stores 2048, or 2-K, bits. But one data byte requires eight bits of storage. Consequently, each 2 K of eight-bit storage requires 8 2 K × 1 chips. Since we wish 16 K of storage, we will need

$$\left(\frac{16 \text{ K}}{2 \text{ K}}\right) \times 8 = 8 \times 8 = 64 \text{ chips}$$

We will conclude our discussion of memory by considering one other way that memory types can be classified. Integrated-circuit memory is either *static* or *dynamic*. From the programmer's standpoint, it makes no difference which type is used, but the system designer must be aware of this difference and may need to make special provisions to accommodate one type or the other. The basic storage mechanism in dynamic memory is the very small gate-to-source capacitance of a field-effect transistor. Since this capacitance eventually discharges, dynamic memory must be periodically recharged, or *refreshed*. We will not treat memory-refresh techniques in any detail but will be aware that a system using dynamic memory must have the capability of providing refresh signals periodically. Static memory, on the other hand, will retain its contents indefinitely (so long as power is connected). With this obvious advantage of static memory, we might correctly suppose that dynamic memory is used because it has certain offsetting advantages. The fact is that due to the simplicity of construction of the basic storage cell in dynamic memory, the dynamic-memory chip has a much larger storage capacity than its static counterpart. Consequently, a memory of given size can be assembled using fewer dynamic chips than one of the same size using static chips. Also, dynamic memory generally consumes much less power than static memory.

We turn our attention now to the means by which the microprocessor accesses and communicates with I/O devices other than memory. In a microprocessor-based control system, input devices provide the CPU with data that represent the present state of the system under control. If, for example, the system is designed to control temperature, an input device would furnish a data byte representing the temperature at the point where control is to be maintained. Similarly, an input device might furnish bytes representing pressure, flow rate, speed, position, voltage, or any other controlled quantity. Input devices thus furnish the *feedback* signals of the control system. The feedback signal might be as simple as a single pulse, as, for example, in a traffic-controller application when a vehicle passes over a pressure-actuated switch under a street. Output devices (which will also require interfacing) consist of the devices that are used to change the state of the system in accordance with output generated by the CPU. The output device may be as simple as a switch that is opened or closed by pulses generated by the CPU, or it may be an actuator that produces some displacement, voltage, or other quantity in direct proportion to the magnitude of the data byte received from the CPU.

Frequently, the CPU must output data to electro-optical devices, which display the current state of the system or other quantities of interest (such as elapsed time, cumulative flow, etc.). In some control systems, the CPU must communicate with peripheral devices similar to those used in computer systems: teletypewriters, video terminals, magnetic disks or tapes, and so forth. The extent to which these types of devices are present in a given system depends upon the complexity of the system and the degree to which an operator is able to override or alter the characteristics of the system. We would not expect to find such elaborate I/O equipment associated with a microprocessor-controlled traffic light but might well expect it to be associated with the reactor-control apparatus in a nuclear power plant.

There are at least two different philosophies among microprocessor designers and manufacturers about the mode by which the CPU may gain access to an I/O device. The first of these is called *memory-mapped I/O* and can be implemented using virtually any microprocessor. The essence of this technique is that each I/O device is assigned a memory address and is, therefore, accessed in exactly the same way as any other memory address. If the program stored in memory directs the CPU to input a data byte from a memory address (which happens to correspond to the address assigned to an input device), then the CPU issues that address on its address bus just as it would for any other memory address and then waits to receive data on its data bus from that address. Unknown to the CPU, we have arranged for that data to come from an input device rather than from a memory location. Of course, this arrangement we make amounts to installing a *decoder* on the address bus; and when the decoder detects the presence of the address assigned to the input device, it enables the operation of that device. The input device, in turn, places its data byte on the data bus, and the CPU never knows the difference! Needless to say, we would not assign an I/O device any memory address that had already been assigned to an actual location in memory, for in that case, the system would attempt to place two data bytes simultaneously on the data bus whenever that address appeared on the address bus. Example 10.5 illustrates decoding and address assignment for memory-mapped I/O in a typical application.

10.3 MICROPROCESSOR SYSTEM DESIGN CONSIDERATIONS 399

EXAMPLE 10.5

A microprocessor control system has an 8-K memory that has been assigned the hexadecimal addresses 0000 through 0FFF. The CPU must be able to access an input device and an output device using the memory-mapped I/O technique. Assign addresses to the devices and show how they would be wired into the system buses.

SOLUTION

The memory addresses that are assigned to the 8 K of memory are expressed in binary as follows:

$$A_{15}\, A_{14}\, A_{13}\, A_{12}\, A_{11}\, A_{10}\, A_9\, A_8\, A_7\, A_6\, A_5\, A_4\, A_3\, A_2\, A_1\, A_0$$
$$0\ 0\ 0\ 0\ 0\ 0\ 0\ 0\ 0\ 0\ 0\ 0\ 0\ 0\ 0\ 0$$

through

$$0\ 0\ 0\ 0\ 1\ 1\ 1\ 1\ 1\ 1\ 1\ 1\ 1\ 1\ 1\ 1$$

It is clear that address bits A_{12} through A_{15} are always 0 whenever an actual memory location is addressed. Consequently, we can use these bits to specify device addresses. We have numerous choices; for example, we might choose $A_{14}=1$, $A_{15}=0$ to select the input device, and $A_{14}=1$, $A_{15}=1$ to select the output device. The devices would then be assigned binary addresses

$$A_{15}\ \ldots\ \ldots\ A_0$$
$$0\ 1\ X\ X\ X\ X\ X\ X\ X\ X\ X\ X\ X\ X\ X\ X$$
$$1\ 1\ X\ X\ X\ X\ X\ X\ X\ X\ X\ X\ X\ X\ X\ X$$

where X can be either 1 or 0. If we choose all the X's to be 0, the hex address of the input device would be 4000 and that of the output device C000. We would still have to ensure that true memory address 0000 was not accessed when one of these device addresses appeared on the address bus, since bits A_0 through A_{11} are all 0 in both cases. To avoid this conflict, bit A_{14} could be used to disable memory, as shown in Figure 10.13.

It is recommended that the reader construct truth tables corresponding to the logic shown in Figure 10.13 to confirm that the correct components are enabled for the correct address-bit configurations.

Another solution to the problem in Figure 10.13 would be to assign the input and output devices the following addresses:

$$A_{15}\ \ldots\ \ldots\ A_0$$
$$1\ 0\ 0\ 0\ 0\ 0\ 0\ 0\ 0\ 0\ 0\ 0\ 0\ 0\ 0\ 0 = 8000_{16}$$
$$0\ 1\ 0\ 0\ 0\ 0\ 0\ 0\ 0\ 0\ 0\ 0\ 0\ 0\ 0\ 0 = 4000_{16}$$

respectively. The logic required for decoding would then be the somewhat simpler arrangement shown in Figure 10.14.

400 MICROPROCESSOR-BASED CONTROL SYSTEMS

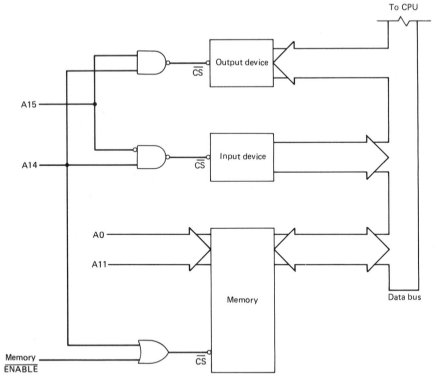

FIGURE 10.13 Bus connections for memory-mapped I/O (Example 10.5).

The second method used by some microprocessors to access an I/O device involves using special instructions for input and output and generating special control signals that can be used to enable I/O devices whenever one of these instructions is executed. Thus, when the CPU fetches an INPUT or OUTPUT instruction from program memory, it responds by causing a particular control-signal output to become active (i.e., to go low or high) and then issues a *device code*, usually on the address bus, that identifies the particular I/O device to be selected. Of course, the INPUT or OUTPUT instruction must in some way include the device code desired, so that the CPU can, in turn, reproduce that code on its address bus.

EXAMPLE 10.6

A microprocessor-based control system is being designed to control the linear velocity of a conveyor belt and the rate at which a certain automated operation takes place on the production line of a manufacturing plant. Two transducers, T1 and T2, produce eight-bit data bytes representing the belt velocity and the rate of the automated operation. Each rate is controlled by the speed of an electric motor that rotates at a speed determined by the magnitude of an eight-bit data byte furnished to its respective motor controller. Motor controller MC1 controls the speed of the belt-velocity motor, and MC2 controls the speed of the motor regulating the automated operation. Assuming the CPU produces the same control signals as in

10.3 MICROPROCESSOR SYSTEM DESIGN CONSIDERATIONS

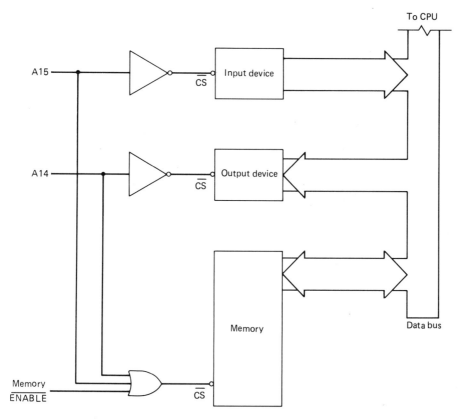

FIGURE 10.14 An alternative method for the memory-mapped I/O in Example 10.5.

Example 10.3, assign device codes to the transducers and motor controllers, and show how these would be wired into the system. Assume the CPU issues device codes on the least significant eight bits of its address bus and that all devices are enabled by active-low signals.

SOLUTION

We wish to assign device codes in such a way that a minimum amount of decoding is necessary. Accordingly, we choose to assign bit A0 to T1, A1 to T2, A2 to MC1, and A3 to MC2 and to use device codes that result in one and only one address bit being low when a particular device code is specified. The following assignment of codes meets these requirements:

$$
\begin{array}{rccccccccl}
& A7 & & & & & & & A0 & \\
T1: & 1 & 1 & 1 & 1 & 1 & 1 & 1 & 0 & = FE_{16} \\
T2: & 1 & 1 & 1 & 1 & 1 & 1 & 0 & 1 & = FD_{16} \\
MC1: & 1 & 1 & 1 & 1 & 1 & 0 & 1 & 1 & = FB_{16} \\
MC2: & 1 & 1 & 1 & 1 & 0 & 1 & 1 & 1 & = F7_{16} \\
\end{array}
$$

(Note that bits A4 through A7 are arbitrary.) Following Example 10.3, we may then wire the system as shown in Figure 10.15.

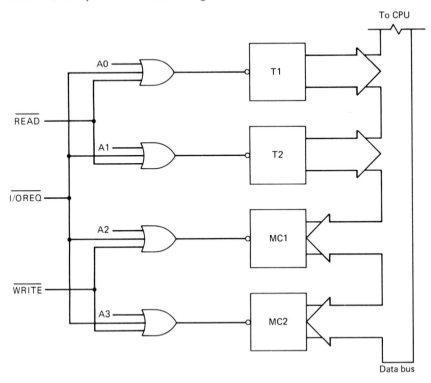

FIGURE 10.15   Bus connections for a system controlling conveyor-belt velocity and an automated-manufacturing operation (Example 10.6).

Though not shown in Figure 10.15, we need a means of ensuring that memory is not enabled at the same time that an I/O device is enabled. For this purpose, we would use an active-low $\overline{\text{MREQ}}$ (memory request) control signal generated by the CPU (or by external logic). This control signal goes low whenever a memory access is required and is high at all other times. Thus, it would normally be connected to a $\overline{\text{CS}}$ input on memory and would disable memory (go high) whenever $\overline{\text{I/OREQ}}$ went low. Note also that we could have assigned the same device codes to MC1 and MC2 that we used for T1 and T2, since only one of $\overline{\text{READ}}$ and $\overline{\text{WRITE}}$ would be active when we wished to access either an input or output device, respectively. Since eight address bits are available to specify a device code and since separate control signals are available for enabling input devices and output devices, we could use this method with up to $2^8 = 256$ different input devices and 256 different output devices. Of course, decoding the device codes becomes a more complex problem as the number of I/O devices increases; rarely would we need 512 I/O devices in a practical control system.

10.4 MICROPROCESSOR ARCHITECTURE AND PROGRAMMING FOR CONTROL SYSTEMS APPLICATIONS

In this section, we intend to look inside a typical microprocessor. We will study the architecture (the arrangement and operation of its component parts) in enough detail to understand the interface and programming requirements of a microprocessor-based control system. In contrast with one popular approach to introducing microprocessor theory, we do not intend to invent a hypothetical microprocessor that attempts to be representative of the many different types that are now available. Instead, we will discuss a specific microprocessor, with the result that the diligent student will have the knowledge of at least one real microprocessor currently in use. We choose the Mostek Z-80 microprocessor for this purpose, since it does share many of the characteristics of a rather broad class of microprocessors, and it is widely used in industry. In any event, the student will find that the transition to learning and understanding other microprocessors is a relatively easy task, once the basic concepts of any one device have been mastered. The specifications for the Z-80A microprocessor, an upgraded version of the original Z-80, may be found in Appendix B.

An important part of every microprocessor is its *register* complement. A register is simply a set of storage devices, such as flip-flops, used for storing binary numbers. In the microprocessor, registers are used to store data bytes and addresses; thus, it contains both eight-bit and 16-bit registers. Unlike memory storage, the contents of the various registers will generally change many times as a program is executed, since the program will cause register contents to be loaded, cleared, incremented, or shifted as necessary to accomplish the tasks it is designed to perform. One very important register is designated the *accumulator*. The accumulator will usually contain one of the data bytes (*operands*) of every arithmetic and logic operation as well as the results of each such operation. In a typical example, if two binary numbers are to be added, then one of the numbers is first loaded into the accumulator; after the addition has been performed, the resulting sum is found in the accumulator. In an eight-bit microprocessor, the accumulator is an eight-bit register. Other eight-bit registers, *general purpose registers*, may also be present and may be identified by letters (B, C, etc.) or similar designations. Sixteen-bit registers are used to store memory addresses for a variety of purposes, which we will discuss presently. The control-systems designer must be intimately acquainted with the microprocessor registers and thoroughly understand their functions in order to be able to write a program or interface the microprocessor with the other components of the system. The register complement of the Z-80 microprocessor is shown in Figure 10.16.

We note in Figure 10.16 that the Z-80 has an eight-bit accumulator (frequently referred to as the ACC or A register), general-purpose eight-bit registers designated B, C, D, E, H, and L and 16-bit registers labeled *program counter* (PC), *stack pointer* (SP), and *index registers* (IX and IY). There is also an alternate set of eight-bit registers with prime designations and two special-purpose eight-bit registers I and R.

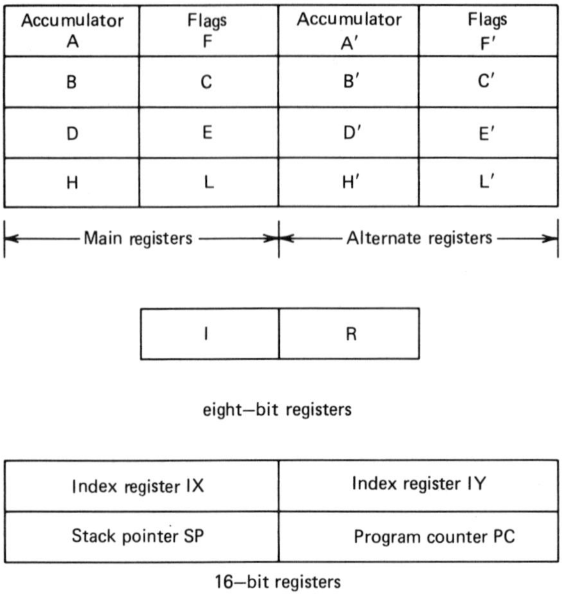

Accumulator A	Flags F	Accumulator A′	Flags F′
B	C	B′	C′
D	E	D′	E′
H	L	H′	L′

|←——— Main registers ———→|←——— Alternate registers ———→|

I	R

eight–bit registers

Index register IX	Index register IY
Stack pointer SP	Program counter PC

16–bit registers

FIGURE 10.16 Register complement of the Z-80 microprocessor.

We will introduce Z-80 programming techniques by discussing the instructions that are available for loading the eight-bit registers, for transferring data between registers, and for moving data to and from memory. It is not our intention to discuss or even list every Z-80 instruction. We will concentrate on the instructions that are most frequently used in control-systems applications and leave it to the student to consult manufacturer's literature for a detailed study of the instruction complement. (In this chapter, we will discuss enough instructions in enough detail to illustrate programming concepts for a number of common control-system applications; but the student should be aware that there may well be more *efficient* ways of programming to accomplish a given task if he or she uses the full instruction complement. An efficient program is one that requires a minimum of storage locations in program memory.) Every instruction has an *operation code* (op-code) consisting of at least one eight-bit byte. The op-code identifies the task that a given instruction accomplishes and is abbreviated when programming by use of a mnemonic (memory aid); for example, the mnemonic for load is LD. The operand(s) of a given instruction are identified by (but not necessarily equal to) the bytes that immediately follow the op-code. Recall that we previously referred to the operands of an add instruction as the binary numbers that were to be added. A complete instruction may require one, two, three, or four bytes to specify, depending on how many bytes are required to identify the op-code and the operand(s). Accordingly, one instruction may occupy one, two, three, or four successive addresses in program memory. The Z-80 instructions that are used to transfer bytes from one eight-bit register to another are one-byte instructions and are written in the following format

LD d,s

10.4 MICROPROCESSOR ARCHITECTURE AND PROGRAMMING

TABLE 10-2

Destination Register	Source Register						
	A	B	C	D	E	H	L
A	7F	78	79	7A	7B	7C	7D
B	47	40	41	42	43	44	45
C	4F	48	49	4A	4B	4C	4D
D	57	50	51	52	53	54	55
E	5F	58	59	5A	5B	5C	5D
H	67	60	61	62	63	64	65
L	6F	68	69	6A	6B	6C	6D

Where s stands for the *source* register and d stands for the *destination* register. Thus,

$$\text{LD B,C}$$

is an instruction, which when executed, would cause the contents of the C register to be transferred, or loaded into, the B register. The source-register contents (contents of C in this case) are unchanged by all such instructions. Symbolically, we write

$$(C) \longrightarrow (B)$$

where the parentheses denote "contents of" and the arrow means "are transferred to." Every instruction has a unique sequence of bits that identifies its particular opcode; but op-codes are usually expressed in octal or hexadecimal form to save space. For example, the op-codes for LD A,B and LD H,L are $78)_{16}$ and $65)_{16}$, respectively. Table 10-2 shows the hexadecimal op-codes for every possible load (LD) instruction that transfers a byte from one register to another.

Eight-bit registers can also be loaded by *immediate*-type instructions. These are two byte instructions in which the byte immediately following the op-code is treated as the data to be loaded into the specified register. Thus, for example, in the instruction

$$\text{LD A}$$
$$\text{FF}$$

$FF)_{16}$ is the data loaded into the accumulator. Symbolically,

$$FF)_{16} \longrightarrow (A)$$

The hexadecimal op-codes for the load-immediate instruction are shown in Table 10-3.

TABLE 10-3

			Load Immediate			
A	B	C	D	E	H	L
3E	06	0E	16	1E	26	2E

When writing a program, we will always list one byte per line in the same way that the bytes would ultimately be stored in program memory. We will also list the memory address at which each byte is to be stored. Since each memory address requires 16 bits, or two bytes, to specify, we will need four hexadecimal digits to list each address.

EXAMPLE 10.7

(a) Write the program listed below in hexadecimal form.
(b) What are the contents of the accumulator after the program is executed?

Address	Contents
0000	LD B
0001	52
0002	LD C,B
0003	LD D,C
0004	LD H,D
0005	LD A,H
0006	HALT

(Note that the last instruction in every program must be HALT, which has hex op-code 76.)

SOLUTION

(a)

Address	Contents
0000	06
0001	52
0002	48
0003	51
0004	62
0005	7C
0006	76

(b) The instruction stored at addresses 0000 and 0001 loads the B register with 52_{16}. The sequence that takes place thereafter is: (B) → (C); (C) → (D); (D) → (H); (H) → (A). Thus, the accumulator ultimately contains 52_{16}. Of course, this program accomplishes nothing useful. It is simply presented as an exercise in program interpretation.

Although the H and L registers are classified as general purpose, they can be used in one very special way. The Z-80 has certain instructions that interpret the contents of the H and L *register pair* as a memory address. (Note that two eight-bit registers are needed to specify one 16-bit address.) These so-called *indirect-addressing* instructions treat the H register contents as the high order (most significant) eight-bits of an address and the L register contents as the low order eight-bits of the address. In these cases, the H and L register pair is said to *point* to a memory

10.4 MICROPROCESSOR ARCHITECTURE AND PROGRAMMING

TABLE 10-4

Source = ((HL))	Destination						
	A	B	C	D	E	H	L
	7E	46	4E	56	5E	66	6E

Destination = ((HL))	Source						
	A	B	C	D	E	H	L
	77	70	71	72	73	74	75

location. The operand of an indirect addressing instruction is contained in the memory address pointed to by the H and L register pair. Therefore, using these instructions, we can load any register from any memory location or transfer the contents of any memory location to any register. Of course, our program must load the H and L register pair with the memory location to be affected before the instruction causing the transfer to or from memory is executed. Table 10-4 lists the hexadecimal op-codes for the indirect-load instructions.

In Table 10-4, the symbol ((HL)) means the contents of the memory location pointed to by the H and L register pair (the contents of the contents of HL). The mnemonics for these one-byte instructions are written in the following formats:

$$\text{LD (HL), s}$$
$$\text{or}$$
$$\text{LD d, (HL)}$$

which mean, symbolically

$$(s) \rightarrow ((HL))$$
$$\text{and}$$
$$((HL)) \rightarrow (d)$$

respectively.

EXAMPLE 10.8

What are the contents of the H, L, and A registers after the following program is executed?

Address	Contents
0000	LD L
0001	25
0002	LD H
0003	01
0004	LD A
0005	FF
0006	LD (HL), A
0007	LD L, (HL)
0008	HALT

SOLUTION

The sequence of operations is as follows:

$$25 \rightarrow (L)$$
$$01 \rightarrow (H)$$
$$FF \rightarrow (A)$$
$$(A) \rightarrow ((HL)), \text{ that is, } FF \rightarrow (0125)$$
$$((HL)) \rightarrow (L), \text{ that is, } (0125) \rightarrow (L), \text{ that is, } FF \rightarrow (L)$$

Thus, H contains $01)_{16}$, L contains $FF)_{16}$, and A contains $FF)_{16}$. Again, this is not a particularly useful program.

Although there are a number of other eight-bit load instructions in the Z-80 complement, we have covered a sufficient number for our purposes and turn now to the jump instructions. We hinted earlier that unless directed to do otherwise by an instruction fetched from memory, the CPU always reads its next byte from the next sequential memory address. The instructions that cause the CPU to go elsewhere in memory to fetch its next instruction are called *jump*, or *branch*, instructions. The simplest of these is the *unconditional jump*, which is a three-byte instruction: an op-code followed by two bytes that specify the next memory address to be accessed. The mnemonic (hex op-code C3) is JP, and the format is as follows:

JP
ll
hh

where ll is the hex code for the low order eight-bits of the next address to be read, and hh is the high order eight-bits of that address. Note carefully the sequence in which the low and high order address bytes are written: Low order is always first.

EXAMPLE 10.9

What does the following program accomplish?

Address	Contents
0000	LDA
0001	00
0002	JP
0003	00
0004	00
0005	HALT

SOLUTION

Very little. After the accumulator is loaded with 00, the CPU fetches its next instruction from 0000. But this again causes the accumulator to be loaded with 00, so there is no change in the accumulator contents. The process continues indefi-

10.4 MICROPROCESSOR ARCHITECTURE AND PROGRAMMING

nitely, and the HALT instruction is never executed; in other words, the program is in an infinite loop.

The CPU contains a set of flip-flops known as *flags*. These flags are automatically set or reset when the results of arithmetic and logic operations meet certain specified conditions. For example, the ZERO flag is a flip-flop that is set whenever the result of an operation is equal to zero. It must be remembered, though, that only certain instructions *affect* each flag, the instructions that cause arithmetic and logical operations to be performed. The zero flag is not affected, for example, by any of the LD instructions. The programmer should consult the manufacturer's list of instruction specifications if there is ever any doubt about whether or not a given instruction affects a particular flag. It must further be remembered that if an instruction does affect one of the flags, then that flag will always be either set or reset according to the outcome when the instruction is executed. For example, if the CPU adds two binary numbers and the sum is *not* zero, the ZERO flag will be reset. Any flag that is set or reset will remain set or reset until the next instruction affecting that flag is executed. The Z-80 CPU contains six flags: the ZERO (Z), CARRY (C), SIGN (S), HALF-CARRY (H), ADD-SUBTRACT (N), and PARITY-OVERFLOW (P/V) flags. We will use only the first three of these in our subsequent work. The ZERO flag has already been described. The CARRY flag is set whenever a carry occurs out of the most significant bit of the result of an arithmetic or logical operation, and reset otherwise. For example, C would be set equal to 1 when the CPU performed the addition $FF_{16} + FF_{16}$. The SIGN flag is set whenever the result of an arithmetic or logic operation produces a 1 in the most significant position, and reset otherwise. It is called the SIGN flag because in two's-complement arithmetic, the most significant bit is the sign bit (equal to 1 when the number is negative).

A *conditional-jump* instruction causes the CPU to branch to a specified address if a certain condition determined by the state of the corresponding flag is satisfied. Thus, for example, the instruction JUMP IF NON-ZERO (JP NZ) would cause the CPU to jump to the address specified by the next two bytes of the instruction (low address byte first), provided that the ZERO flag were reset. Otherwise, the CPU takes the next instruction in sequence. The conditional-jump instructions and their corresponding hex op-codes are listed in Table 10-5.

TABLE 10-5

Instruction	Mnemonic	Hex op-code
Jump if Zero	JP Z	CA
Jump if Non-zero	JP NZ	C2
Jump if Carry	JP C	DA
Jump if No Carry	JP NC	D2
Jump if Positive	JP P	F2
Jump if Negative	JP N	FA

The contents of each eight-bit register and the contents of each memory location may be *incremented* (contents increased by 1) or *decremented* (contents decreased by 1) by using appropriate Z-80 instructions. The mnemonic for increment is INC and that for decrement is DEC; these increment and decrement instructions require only one byte to specify and are written in formats illustrated by the following:

INC L

DEC (HL)

INC L increments the contents of the L register, and DEC (HL) decrements the contents of the memory location pointed to by the H and L register pair. The increment and decrement instructions affect the SIGN and ZERO flags. The hex op-codes for these instructions are shown in Table 10-6.

One very common task that the microprocessor is used to perform in control-systems applications is that of generating a fixed *time delay*. In many situations, a fixed time must be allowed to elapse between the input or output of data or control signals. For these purposes, we write a program segment that though accomplishing nothing else useful is known to require a certain amount of time to execute. For example, if the system receives a certain input, it may be necessary to delay the system's response to this input by, say, 0.5 sec. Whenever that input was received, we would have the CPU jump to that set of instructions that required exactly 0.5 sec to execute. The last instruction in this time-delay program would cause the CPU to jump back into the main program and proceed to generate the required output (response) signal. In this manner, the CPU can serve as a time-delay relay in addition to performing its other duties. This is a good example of how we can put into practice one important guideline in designing microprocessor-based control systems: replace hardware with software. Software (programming) is generally more economical and results in smaller chip counts, smaller size, and less power consumption than any hardware used to accomplish the same task.

In order to write a time-delay program, we must, of course, know how much time the CPU requires to execute each instruction. By consulting manufacturers' specifications, we can learn exactly how many *clock periods* (or *clock cycles* or *states*) each instruction requires. By knowing the clock frequency of our system, we can convert a certain number of clock periods to an equivalent time duration. The number of clock periods required to execute the instructions we have studied so far are listed in Table 10-7. The most efficient way to write a time-delay program is by using the conditional-jump instructions to cause the CPU to loop through the same instructions over and over again, as many times as required to consume the prescribed time. Example 10.10 illustrates this principle.

TABLE 10-6

	A	B	C	D	E	H	L	(HL)
INC	3C	04	0C	14	1C	24	2C	34
DEC	3D	05	0D	15	1D	25	2D	35

10.4 MICROPROCESSOR ARCHITECTURE AND PROGRAMMING

TABLE 10-7

Instruction	Clock Periods
LD d, s	4
LD d, (HL)	7
LD (HL), s	7
LD r (immediate load data) (register r)	7
all JP instructions	10
INC r, DEC r	4
INC (HL), DEC (HL)	11
HALT	4

EXAMPLE 10.10

By counting the number of bolts on a conveyor belt that travel past a certain point, a microprocessor is used to control the rate at which a certain machine bolt is manufactured. A photoelectric cell generates a pulse that, in turn, increments an electronic counter every time one bolt passes. The microprocessor must input the count from the counter every second in order to determine the number of bolts per second being manufactured. Write a program segment that will create the required time delay, ±10 msec. The CPU clock frequency is 100 kHz.

SOLUTION

The period of each clock cycle is $T = 1/10^5 = 0.01$ msec. We must, therefore, write a program that takes 1 sec/0.01 msec = 100,000 clock periods to execute, ±1,000 clock periods. The program that is shown starting at address 0050 requires 100,555 clock periods, or 1.00555 sec, to execute and, therefore, satisfies the requirements.

Address	Contents	Clock Periods
0050	LD B	7
0051	1C	
0052	LD C	7
0053	FF	
0054	DEC C	4
0055	JP NZ	10
0056	54	
0057	00	
0058	DEC B	4
0059	JP NZ	10
005A	52	
005B	00	

A flowchart for this program is shown in Figure 10.17.

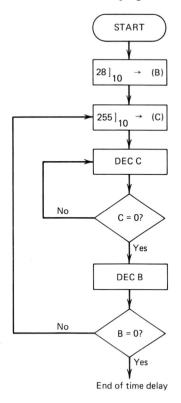

FIGURE 10.17   Flowchart for a Z-80 1-sec-time delay program (Example 10.10).

After B is loaded with $1C)_{16} = 28)_{10}$ and C is loaded with $FF)_{16} = 255)_{10}$, C is decremented 255 times, which uses up $255 \times 14 = 3570$ clock periods; that is, C is counted down from 255 to zero. Then B is decremented; it now contains $27)_{10}$. Then C is reloaded with $255)_{10}$ and counted down again, using up another 3570 clock periods. This process continues until B has been counted down to zero; that is, C is loaded and counted down $28)_{10}$ times, for a total of $28 \times 3570 = 99{,}960$ clock periods. Since C is loaded 28 times, this takes up $28 \times 7 = 196$ clock periods; B is loaded once, which adds another 7, and B is decremented 28 times for a total of $4 \times 28 = 112$ clock periods. The conditional jump at 0059 adds a total of $10 \times 28 = 280$ periods. Thus, the total number of clock periods is $99{,}960 + 196 + 7 + 112 + 280 = 100{,}555$ clock periods, or 1.00555 sec.

We have referred on several occasions to the *logic operations* that the CPU can perform. These operations are extremely useful in control-systems applications, and we will study the nature and use of the instructions that implement logic operations in some detail. (There is at least one one-bit microprocessor used in control-systems applications that accomplishes all its tasks using essentially only logic operations.) The Z-80 can perform three Boolean logic operations: AND, OR, and EXCLU-

SIVE OR (XOR). These operations are performed *bit by bit* on two eight-bit data bytes. For example, we know that 1 AND 1 = 1, 1 AND 0 = 0, and 0 AND 0 = 0. This logic is performed on each pair of corresponding bits of each of two data bytes whenever the AND instruction is executed. One of the bytes must always be in the accumulator, while the other may be in any eight-bit register or in any memory location. The one-byte AND instructions are written in the format

$$\text{AND r}$$

or

$$\text{AND (HL)}$$

and in a similar format for OR and XOR. The result of each logic operation appears in the accumulator after the instruction is executed. Symbolically, the results of each logic instruction are expressed as shown (where $\wedge$ means AND, $\vee$ means OR, and $\veebar$ means EXCLUSIVE OR):

AND r	$(A) \wedge (r) \rightarrow (A)$
AND (HL)	$(A) \wedge ((HL)) \rightarrow (A)$
OR r	$(A) \vee (r) \rightarrow (A)$
OR (HL)	$(A) \vee ((HL)) \rightarrow (A)$
XOR r	$(A) \veebar (r) \rightarrow (A)$
XOR (HL)	$(A) \veebar ((HL)) \rightarrow (A)$

EXAMPLE 10.11

If the accumulator contains 9D and the E register contains E3, what are the contents of the accumulator after each of the following instructions is performed?

(a) AND E
(b) OR E
(c) XOR E

SOLUTION

(a) (A) = 9D = 10011101
   (E) = E3 = 11100011
   (A) $\wedge$ (E) = 10000001 = 81 $\rightarrow$ (A)

(b) (A) = 9D = 10011101
   (E) = E3 = 11100011
   (A) $\vee$ (E) = 11111111 = FF $\rightarrow$ (A)

(c) (A) = 9D = 10011101
   (E) = E3 = 11100011
   (A) $\veebar$ (E) = 01111110 = 7E $\rightarrow$ (A)

Logic operations can also be performed on the accumulator contents and the immediately following byte. These immediate type instructions are of course two-byte instructions (op-code plus immediate data). Table 10-8 shows the hex op-codes for the various logic instructions.

414 MICROPROCESSOR-BASED CONTROL SYSTEMS

TABLE 10-8

	IMMED	A	B	C	D	E	H	L	(HL)
AND	E6	A7	A0	A1	A2	A3	A4	A5	A6
OR	F6	B7	B0	B1	B2	B3	B4	B5	B6
XOR	EE	AF	A8	A9	AA	AB	AC	AD	AE

EXAMPLE 10.12

A microprocessor-based control system is used to control the starting and stopping of a pump that pumps fluid into three reservoirs numbered 1, 2, and 3. Whenever one reservoir is filled, the pump must be shut off. The system contains circuitry that will cause the number of a reservoir to be stored in memory location 0050 when that reservoir is filled. If no reservoirs are filled, 0050 contains 00. Write a program segment, beginning at address 0000, that checks to see if any reservoir is filled and if so, branches to address 0100 (where we can assume there is a program segment that causes a shut-down signal to be generated).

SOLUTION

Address	Contents
0000	LD A
0001	03
0002	LD H
0003	00
0004	LD L
0005	50
0006	AND (HL)
0007	JZ
0008	06
0009	00
000A	JP
000B	00
000C	01

We load the accumulator with 03 and load the HL register pair so that it points to memory location 0050. The result of the AND (HL) instruction will be zero as long as the contents of 0050 remain 00 (no reservoir filled). Since logic instructions affect the ZERO flag, the CPU in this case will jump back and repeat the AND instruction. This process continues indefinitely, until a 1 shows up in either D0, D1, or both, corresponding to reservoir numbers 01, 02, or 03 being filled. In that case, the result of the AND instruction is not zero (it is 01, 02, or 03), and the CPU branches to 0100 as required.

Example 10.12 illustrates a commonly used technique called *masking*. We say that in Example 10.12 bits D3 through D7 have been "masked out" of a data byte when that byte is ANDed with 03. Since masking allows the CPU to respond

TABLE 10-9

Instruction	Mnemonic	Operation Performed
Rotate Left Circular	RLC	CY ← D7 ← ← ← D0 (circular)
Rotate Right Circular	RRC	CY ← D7 → → → D0 (circular)
Rotate Left	RL	CY ← D7 ← ← ← D0 (through CY)
Rotate Right	RR	CY → D7 → → → D0 (through CY)

according to the states of certain bits, the technique is widely used for checking the status of external devices. We might, for example, assign bit positions D0 through D7 to eight different external devices, stipulating that the device cause its assigned bit position to be 1 if the device is on, and 0 if it is off. Using masking, we could then check the status of any one or a combination of devices. Masking can also be used to alter a data byte by changing specific bits in the byte. The Z-80 is unusual because it has an additional group of instructions that can be used to set, reset, or test any bit of any byte in any register or in any memory location. The mnemonics are SET, RES, and BIT, respectively. There are 250 different op-codes for performing all these possible bit operations, so we will not list them here.

The Z-80 instruction complement includes 72 instructions that may be used for *shifting* data bytes in a variety of ways. Many of these include shifting into or out of the CARRY flag; that is, the C flip-flop is often treated as if it were an additional bit position in a register or memory location. The shift instructions that cause bits to be recirculated are referred to as rotate instructions. Since these rotate instructions do affect the C flag, conditional jumps (JP C or JP NC) can be used following the execution of the rotate instructions if desired. We will discuss only four of the nine different ways in which the bits in a register or memory location can be shifted or rotated. Table 10-9 illustrates these instructions by showing schematically how each bit is shifted. A single execution of one of these instructions causes each bit to shift *one* position in the direction and manner illustrated. The carry flag is designated CY, instead of C, to avoid confusion with the C register. To write a rotate instruction, simply write the appropriate mnemonic, followed by a register designation or by (HL). With the exception of accumulator rotations, these instructions have two-byte op-codes, so we will not attempt to list them completely; the 72 different hex op-codes may be found in the manufacturer's specifications. Example 10.13 illustrates one use of rotate instructions.

EXAMPLE 10.13

Figure 10.18 is a diagram of the intersection of two streets, where pressure activated switches are installed at each of the four corners. A microprocessor is used to control the flow of traffic through the intersection by sequencing the traffic signal in accordance with pulses received from the switches. As shown in Figure 10.18, switches S1, S2, S3, and S4 control bit positions D7, D6, D5, and D4 of the accumulator. Assume a bit position is set to 1 whenever the corresponding switch is activated. Whenever a pulse is received from S1 or S3, indicating traffic on

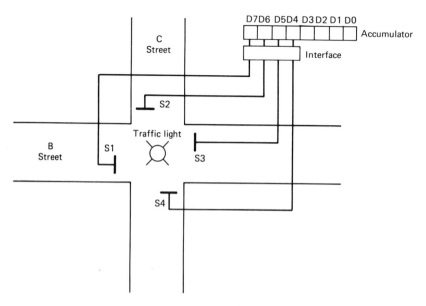

FIGURE 10.18 A traffic intersection with pressure switches (Example 10.13).

B street, the B register is to be incremented. Whenever a pulse is received from S2 or S4, indicating traffic on C street, the C register is to be incremented. Write a program that continually monitors and counts the traffic on B and C streets. Assume that the accumulator can only be loaded by switch activations during the time it is being monitored (i.e., assume that the microprocessor disables the interface circuitry between the switches and the accumulator whenever a new entry is detected).

SOLUTION

A flowchart for the program is shown in Figure 10.19.

10.4 MICROPROCESSOR ARCHITECTURE AND PROGRAMMING

The program is:

Address	Contents	Comments
0000	LD B	Initialize B
0001	00	
0002	LD C	Initialize C
0003	00	
0004	LD A	Clear A
0005	00	
0006	AND A	Check A for any 1-bits
0007	F0	
0008	JP Z	No 1s; go back and check again
0009	06	
000A	00	
000B	RL A	Check D7
000C	JP C	Jump if D7 = 1
000D	1B	
000E	00	
000F	RL A	D7 = 0; check D6
0010	JP C	Jump if D6 = 1
0011	17	
0012	00	
0013	RL A	D6 = 0; check D5
0014	JP C	Jump is D5 = 1
0015	1B	
0016	00	
0017	INC C	Increment C
0018	JP	and go back to checking A
0019	04	
001A	00	
001B	INC B	Increment B
001C	JP	and go back to checking A
001D	04	
001E	00	

Note the masking operation performed by the AND A instruction using the immediate byte F0. If the presence of a 1 is detected in any bit position as a result of this operation, the program proceeds to rotate the accumulator left until it determines which bit contains the 1 and then increments the appropriate traffic counter.

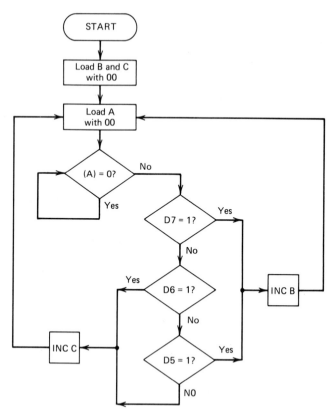

FIGURE 10.19 Flowchart for a program that counts traffic (Example 10.13).

The program shown in Example 10.13 is somewhat unrealistic in that it contains no instructions to accomplish the actual updating of the accumulator whenever a switch is activated; that is, there are no programming provisions for reading new data into the accumulator. We will rectify this situation shortly when we discuss the INPUT and OUTPUT instructions. But let us first finish our discussion of arithmetic and logic operations by mentioning the ADD and SUBTRACT (SUB) instructions. We have mentioned previously that the accumulator is the principal register involved in arithmetic instructions. The ADD and SUB instructions cause the contents of a specified register or of a memory location (or an immediate byte) to be added or subtracted from the contents of the accumulator and the result to be placed in the accumulator. Arithmetic operations are performed using two's-complement arithmetic and affect the overflow (P/V) flag: This flag is set if the result of an add or subtract instruction exceeds the eight-bit capacity of the accumulator to represent it. There are also variations of these instructions involving the carry flag and a special instruction for handling BCD arithmetic, but we will not have occasion to use them and will not list their op-codes here. It should be noted that there are no "multiply" or "divide" instructions in the Z-80 complement. This absence of a hardware capability to perform multiplication and division is typical of most microprocessors. If there is a requirement for such operations, as in com-

10.4 MICROPROCESSOR ARCHITECTURE AND PROGRAMMING

TABLE 10-10

	A	B	C	D	E	H	L	(HL)	IMMED
ADD	87	80	81	82	83	84	85	86	C6
SUB	97	90	91	92	93	94	95	96	D6

puter applications, the programmer must write software using the available instructions to implement multiplication and division. The op-codes for ADD and SUB are shown in Table 10-10.

In Section 10.3, we described two methods that are used by the CPU to access the input/output devices of a system. We covered the memory-mapped I/O technique in detail at that time. We now know that it would be possible to transfer data between any register and an I/O device simply by executing

$$\text{LD r, (HL)}$$
or
$$\text{LD (HL), r}$$

provided the HL register pair is pointing to an address that we have assigned to the I/O device.

We mentioned that the second technique for I/O access involved using special INPUT and OUTPUT instructions. We will conclude our discussion of Z-80 software by studying two such instructions. These are two-byte instructions where the second byte is the eight-bit device code that has been assigned to the device to be accessed. The Z-80 has a number of variations of the basic input and output operations, but the only two instructions we will consider are

$$\text{IN (DB)} \quad \text{and} \quad \text{OUT (D3)}$$
$$\text{dd} \quad \quad \quad \text{dd}$$

where dd is the device code. When the IN instruction is executed, a data byte is transmitted from the device specified by dd to the accumulator register in the CPU. Similarly, when OUT is executed, a data byte is transferred from the accumulator to the specified output device. Thus, the accumulator is the point of destination and the point of origin within the CPU for data transfers between the CPU and I/O devices. Executing these instructions causes the appropriate control signals to go active ($\overline{\text{I/OREQ}}$ and $\overline{\text{READ}}$ or $\overline{\text{WRITE}}$, as described in Section 10.2) and they, in turn, are used to enable the specified device, as previously described. Recall that the CPU places the eight-bit device code on the low order eight bits of the address bus so that external logic can select the specified device at the correct time. The time required for an input device to place a byte on the data bus once the device has been selected or the time required for an output device to accept a byte supplied by the CPU is called the *access time* of the device, and it is generally longer than the access time of solid-state memory. For this reason, the Z-80 automatically inserts a *wait state* in its normal sequence of operations whenever an IN or OUT instruction is executed. This wait state amounts to a pause equal in duration to one clock period, during which the CPU does nothing. It is simply a short delay to give an I/O device time to set up. For a clearer understanding of this and other refinements in the timing and synchronization of the CPU with external devices, the control-systems designer must consult the manufacturer's timing diagrams. This is essential both for understanding CPU operation and for designing system-interface logic.

We should now modify the program in Example 10.13 so that it correctly reflects the means by which the accumulator contents are updated. An IN instruction should be written between the instruction that clears the accumulator (stored at 0004 and 0005) and the AND instruction at 0006 and 0007. The conditional-jump instruction should then branch the program back to the IN instruction. The student should redraw the flowchart and rewrite the program to convince him- or herself that this modification is correct.

EXAMPLE 10.14

A microprocessor is to be used to control the flow of data in a telecommunications system. Blocks of 50 sequential data bytes are received from a remote transmitter at random time intervals. These bytes are to be relayed (retransmitted) by an on-site transmitter whenever the transmitter is not busy with other transmissions. The receiver signifies that it is receiving data by storing FF in an eight-bit status port that has been assigned device code 01. It will then transfer the received data bytes to the CPU via a data port (input device) assigned number 02. The on-site transmitter signals that it is ready to transmit by storing FF in status port number 03. It will then receive the data bytes to be transmitted via data port number 04. The system is illustrated in Figure 10.20.

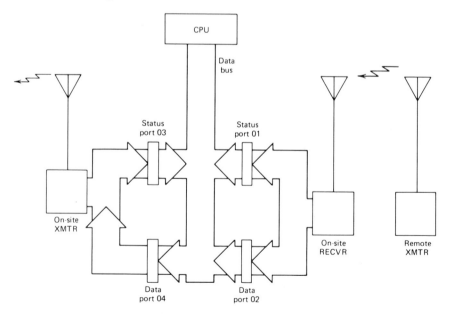

FIGURE 10.20 A microprocessor-controlled telecommunications system (Example 10.14).

10.4 MICROPROCESSOR ARCHITECTURE AND PROGRAMMING

Assuming that the remote transmitter will not transmit a new block of data bytes until the on-site transmitter has finished transmitting the current block, write a program that regulates this data flow.

SOLUTION

A flowchart for the program is shown in Figure 10.21.

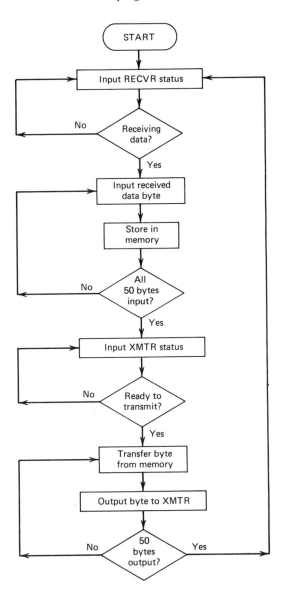

FIGURE 10.21 Flowchart for the microprocessor-controlled telecommunications system (Example 10.14).

422 MICROPROCESSOR-BASED CONTROL SYSTEMS

The program is:

Address	Contents	Comments
0000	LD B	Initialize B register with # of bytes:
0001	32	$50)_{10} = 32)_{16}$
0002	LD H	Point HL register pair to memory address 00AA
0003	00	
0004	LD L	
0005	AA	
0006	IN	Input RECVR status from status port 01
0007	01	
0008	SUB	Check to see if equal to FF (data received)
0009	FF	
000A	JP NZ	If not, go back and check status again
000B	06	
000C	00	
000D	IN	Input data byte from data port 02
000E	02	
000F	LD (HL), A	Transfer byte to memory
0010	INC L	Point HL to next address
0011	DEC B	B = number of bytes yet to go
0012	JP NZ	If all 50 bytes not yet input, go back and input
0013	0D	another
0014	00	
0015	LD B	Number of bytes to be transmitted
0016	32	
0017	LD H	Point HL to memory where bytes are stored
0018	00	
0019	LD L	
001A	AA	
001B	IN	Input status of XMTR
001C	03	
001D	SUB	Check to see if XMTR is ready to transmit
001E	FF	
001F	JP NZ	If not, jump back and check again
0020	1B	
0021	00	
0022	LD A,(HL)	Transfer data byte from memory
0023	OUT	
0024	04	
0025	INC L	Point HL to next address
0026	DEC B	Total number of bytes yet to go
0027	JP NZ	If all 50 bytes not output, go back and output
0028	22	another
0029	00	

002A	JP		Finished this block; go back and wait for next
002B	00		block to be received
002C	00		

One very commonly used technique is illustrated by the program in Example 10.14. Note that the CPU remains in a loop while awaiting notification that an external device is ready. This loop is the continuous repetition of the IN instruction (while checking a status port) until such time as the input byte signifies a ready status. Note also that this program provides a good example of an application requiring program memory, which could be ROM, and data memory, which must be RAM (read/write memory).

We conclude by reiterating an earlier assertion, namely, that the program in Example 10.14 could be written in a more efficient manner had we used the full range of capabilities of the Z-80 instruction complement. The Z-80 has a number of instructions that are capable of moving *blocks* of data between the CPU and I/O devices. It is also possible to load the HL register pair using a single instruction and to perform jumps using an addressing mode that requires only a single byte to specify the address jumped to. Using all of these capabilities would considerably reduce the length of the program. The Z-80 has other powerful instructions that we have not had to use but deserve mention. A single instruction can be used to move data stored in a block of memory locations to any other block of memory. There is also a *block search* instruction that causes the CPU to search a memory block of any size for a data byte that matches (is equal to) a byte previously loaded into the accumulator. Finally, there is a set of instructions that can be used to perform 16-bit arithmetic operations between sets of register pairs.

10.5 INTERRUPTS

In many microprocessor-based control systems there are devices or system components that must gain the attention of the CPU for one reason or another. We have seen several examples in previous sections; some other examples include the following:

1. A human operator wishes to override the system and depresses a key to interrupt normal system operation.
2. A hazardous condition is detected, and the CPU must immediately jump to another program segment to correct it.
3. Conditions indicating an imminent power failure are detected, and the CPU must jump to a program segment that generates the control signals necessary to switch in a standby power source.
4. An input device has new data that must replace current data stored in data memory.

In Example 10.14, we saw how to write a program that causes the CPU to continually check the contents of a status port to see if a device needs attention. If a system has several such devices, any one of which might require attention from the CPU, we could easily write a program that continually checks the devices in an endless sequence, that is, interrogates each device in turn and then jumps to an appropriate program segment if any particular device is "ready," or needs the services of the CPU. This technique is called *polling* and has one serious disadvantage. Clearly, when the CPU is engaged in polling, it can do nothing else. In many practical systems, it is necessary for the CPU to process data, make comparisons, generate time delays, and to perform many such tasks necessary to maintain control of system operation.

There is another way for external devices to gain the attention of the CPU without requiring it to spend all of its time polling. A CPU has at least one input that when activated will cause the CPU to temporarily abandon its current task and yield to the demands of an external device. This process is called an *interrupt*. By activating the interrupt input of the CPU (making it high or low, as the case may be), we cause the CPU to jump to a special *interrupt-service subroutine*, a program segment that has been written to serve the needs of the interrupting device. When the CPU completes executing this subroutine, it will automatically return to the point in the program (we will call it the main program) where it was interrupted and resume execution where it left off. An example may help clarify this concept.

Suppose that a microprocessor control system is used to automatically brake a railway train. The system continously monitors train velocity and by using a radar feedback, computes the rate at which the train is approaching any obstacle in its track. If this rate of approach ever exceeds a certain value, the microprocessor automatically causes the brakes to be applied. Thus, the main task of the CPU is to compute rate of approach and to compare this rate with a maximum allowable

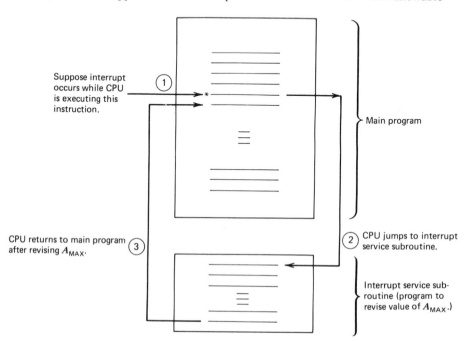

FIGURE 10.22 Sequence of CPU operations resulting from an interrupt.

value, say, A_{MAX}. If the train enters a rainstorm and the track becomes wet, then it would be necessary to reduce the maximum allowable rate of approach A_{MAX} due to the increased braking distance required. The system would have a moisture sensor that triggers an interrupt in the presence of rain. No matter where or when this interrupt might occur, the CPU would immediately jump to a program segment that revises the value of A_{MAX}. Having set in the new value of A_{MAX}, the processor would then return to the main program and resume calculating the approach rate. This sequence is illustrated in Figure 10.22.

Since the CPU may be interrupted *anywhere* in its main program, we might wonder how it is always able to return to precisely the same location where the interrupt did occur upon completing the interrupt-service subroutine. The answer is that activating the interrupt input automatically causes the CPU *to store in memory the address to which it must return*. This address is, in fact, the address of the next instruction in the program following the instruction being executed when the interrupt occurs. (The CPU will always finish executing an instruction it has already started before it allows itself to be interrupted, that is, before it *honors* that interrupt.) The next question is, Where in memory does the CPU store the address to which it must return? And the answer to this question is that the programmer must set aside, that is, reserve, an area in read/write memory where that address can be stored, and he or she must tell the CPU where that area of memory is. The area of memory so reserved is called the *stack*, and the programmer informs the CPU of the location of the stack by means of an instruction that laods the *stack-pointer*. The stack-pointer is a 16-bit register in the CPU that contains the first address of the stack, that is, that points to the stack. Since the stack is an area of memory reserved for storing an address and since memory locations can only contain eight-bit bytes, it is clear that the stack must consist of at least two memory locations to be able to contain a 16-bit address.

The Z-80 mnemonic for loading the stack-pointer is LD SP (hex op-code 31). The format for writing this three-byte instruction is

LD SP
ll
hh

where ll is the low order eight bits of the first address of the stack, and hh is the high order eight bits of that address. For example, if we wanted to locate the stack beginning at address 00A5, we would write

LD SP
A5
00

This instruction shold always be located at the beginning of the main program, since an interrupt might occur anywhere and the CPU must know where the stack is to be found. The Z-80 stack is loaded from the bottom up; that is, given that the first address of the stack is pointed to by the stack-pointer, the next address is the address that is one less than the contents of the stack pointer. In our example, the first address of the stack is 00A5 and the next address is 00A4. When the CPU honors an interrupt, it stores the high byte of the return address in 00A4 and the low byte of that address in 00A3. Note that what we called the first address of the stack is not used! When this occurs, the stack pointer is automatically decremented by two, so that it now points to 00A3. The address pointed to by the stack pointer is called the *top of the stack*. In this example, the top of the stack before the interrupt occurred was 00A5; and after the interrupt was honored, the top of the stack became 00A3.

To continue this illustration, let us suppose that the CPU is executing a one-byte instruction located at address 0057 of the main program when an interrupt occurs. Since the stack pointer contains 00A5 at this point, the CPU stores 00 in 00A4 and 58 in 00A3. When the CPU finishes executing the interrupt-service subroutine, it retrieves the return address from the stack, low byte first then high byte, and the stack pointer is incremented by 2. Thus, when the CPU returns to the main program, the top of the stack is once again 00A5. Of course, the CPU must have some way of knowing when it has reached the end of the interrupt-service subroutine; that is, it must be informed that now is the time to retrieve the return address and to return to the main program. The CPU knows it has reached the end of an interrupt service subroutine when it encounters a RETURN (RET) instruction. Thus, the programmer must make certain that the last instruction in his subroutine is this one-byte instruction RET (op-code C9). Executing RET automatically causes the CPU to retrieve the return address from the stack and to jump to that address in the main program. The only question remaining in this mechanism is, How does the CPU know where to find the interrupt-service subroutine when it is interrupted? We will discuss the answer to this question in more detail later, but for now, suffice it to say that it is generally the responsibility of the interrupting device to inform the CPU exactly where it must jump to in order to find the interrupt service subroutine. Thus, a system must have external-logic circuitry that transmits a jump address to the CPU whenever an interrupt occurs. (This is one disadvantage of the interrupt system compared to polling.)

The first encounter with the complexities of the interrupt sequence and the ensuing stack operations can be somewhat perplexing. Part of the difficulty lies in the abundance of new terminology associated with these operations. A good procedure for mastering these concepts is to memorize the terminology, read over the foregoing material as many times as necessary to feel comfortable with it, and study Example 10.15 carefully.

EXAMPLE 10.15

The program that follows is used in a microprocessor control system to input data bytes from port number 01, add one to each byte, and output each byte to port number 02. Every time an interrupt occurs during this process, the CPU must increment the B register by 1, then resume its original task. Thus, the interrupt service subroutine counts the number of interrupts that occur. Assume that the CPU automatically jumps to address 0066, where the interrupt-service subroutine begins, whenever an interrupt does occur.

10.5 INTERRUPTS

Address	Contents	Comments
0000	LD SP	Set top of the stack to 00BB
0001	BB	
0002	00	
0003	LD B	Set B register
0004	00	count to zero
0005	IN	Input a byte
0006	01	(port 01) → (A)
0007	INC A	Add one to it
0008	OUT	Output result
0009	02	to port 02
000A	JP	Jump back to input next byte
000B	05	
000C	00	
⋮	⋮	
0066	INC B	Increment count
0067	RET	Return to main program

If an interrupt occurs while the CPU is executing the instruction stored at 0005 and 0006, answer the following questions:

(a) What is the top of the stack before the interrupt occurs?
(b) What are the contents of 00BB, 00BA, and 00B9 when the CPU jumps to 0066?
(c) What is the top of the stack while the instruction stored at 0066 is being executed?
(d) What are the contents of the stack pointer after the CPU returns to the main program?
(e) What are the contents of 00B9 after the CPU returns to the main program?

SOLUTION

Figure 10.23 illustrates the sequence of events that takes place when the interrupt occurs.

(a) Before the interrupt occurs, the top of the stack is 00BB. The stack pointer always points to the top of the stack.

(b) The contents of 00BB are unknown; this location is not actually used. 00BA contains the high byte of the return address (00), and 00B9 contains the low byte (07).

(c) The stack pointer is decremented by 2, so it now points to the new top of the stack, 00B9.

(d) When the CPU returns, the stack pointer is incremented by 2, so it again points to 00BB.

(e) None of these operations affect what was previously stored in any memory location, so 00B9 in the stack still contains 07.

428 MICROPROCESSOR-BASED CONTROL SYSTEMS

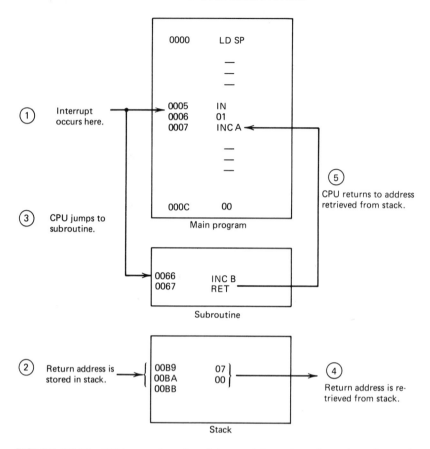

FIGURE 10.23 CPU operations involving an interrupt-service subroutine and a stack (Example 10.15).

The interrupt procedure we have outlined in Figure 10.23 will be repeated any time another interrupt occurs except that, of course, the return address stored in the stack will probably be different from 0007, depending on where in the main program the interrupt takes place. (What happens if the interrupt occurs while the CPU is in the subroutine?) This after all is the whole idea behind the interrupt technique: The CPU knows where to return in the main program no matter when the interrupt occurs.

Many microprocessors (including the Z-80) recognize two types of interrupts: a *maskable interrupt*, and a *nonmaskable interrupt*. A maskable interrupt is one that the CPU can ignore under certain conditions. For example, in a Z-80 system, whenever an interrupt of the maskable type is honored, all future interrupt requests will be ignored unless there is an instruction in the program that directs the CPU to "enable interrupts." The instruction mnemonic is EI and has op-code F8. There is also an instruction, "disable interrupts" (DI, op-code F3) that directs the CPU to ignore all interrupt requests of the maskable variety. Thus, through software, the programmer has control over whether or not an interrupt of this type will be honored by the CPU. This capability, in fact, defines the maskable interrupt: one that can be masked out or ignored. An example of a situation where the EI and DI instructions would be valuable occurs when there is a program segment whose execution is so critical that it has priority over any possible interrupts. The programmer would begin this segment with DI and end it with EI. Also, as we have already noted, once a maskable interrupt is honored, subsequent interrupts will be ignored unless an EI instruction is executed. Therefore, the EI instruction should appear somewhere in the interrupt-service subroutine that serves the maskable interrupt. Furthermore, when the CPU is *reset*, that is, when it begins executing its main program, interrupts are automatically disabled, so an EI instruction must appear somewhere early in the main program if maskable-interrupt requests are anticipated.

The nonmaskable interrupt, as the name implies, cannot be ignored by the CPU. This type of interrupt is reserved for the highest priority interrupts, such as, for example, when an imminent power failure or a safety hazard is detected. The Z-80 CPU has two separate active-low inputs for maskable and nonmaskable interrupts: $\overline{\text{INT}}$ and $\overline{\text{NMI}}$, respectively. It should be noted that it is possible for a system to have many different devices that are capable of interrupting the CPU and for each interrupting device to have its own special interrupt-service subroutine stored in program memory. This is possible even though the CPU has only two interrrupt inputs, but external logic is required to *arbitrate* multiple interrupt requests. Special chips are available for this purpose. These permit the designer to assign priorities among the various interrupting devices, and in many cases, the chips are designed to identify for the CPU exactly which device among several is requesting an interrupt.

The reader may have detected an apparent flaw in the interrupt scheme as we have described it so far. Suppose an interrupt-service subroutine requires using one or more of the same registers that are used in the main program. If the CPU is to return to the main program and resume computation exactly as it left off when interrupted, then obviously all of the register contents must be exactly the same as they were when the interrupt occurred. We cannot allow the service subroutine to alter these contents. Most microprocessors resolve this dilemma by means of a special set of instructions that temporarily store register contents on the stack. In a Z-80 system, these are called push instructions, and they are used to store contents of register *pairs* on the stack. Thus, each push instruction affects two additional memory addresses in the stack and causes the stack pointer to be decremented by 2 each time one of them is executed. The register pairs are defined to be B-C, D-E, H-L, and A-F, where F is an 8-bit byte containing the six flag states (and two unused bit positions). The high order register of each pair (B, D, H, or A) is stored in the stack first, followed by the low order register. If the programmer realizes that it is possible for an interrupt to occur somewhere in the main program where the register contents are vital to the continued correct execution of that program, then he or she must place however many push instructions are required at the beginning of the interrupt service subroutine in order to preserve those contents. Another set

430 MICROPROCESSOR-BASED CONTROL SYSTEMS

TABLE 10-11

	AF	BC	DE	HL
PUSH	F5	C5	D5	E5
POP	F1	C1	D1	E1

of special instructions is used to restore register contents, that is, to transfer register contents from the stack to the CPU at the end of the service subroutine. The Z-80 refers to these as pop instructions, and they, too, operate on register pairs. Each push instruction has its counterpart in a pop instruction. The Z-80 stack is an example of a procedure known as "last-in, first-out" (LIFO). That is, data is popped off the stack in the reverse order that it was pushed onto the stack. Thus, pop instructions must be written in the reverse order of their push counterparts. The push and pop instructions are summarized in Table 10-11.

EXAMPLE 10.16

The program that follows is interrupted while executing the OUT instruction at 0009. The interrupt-service subroutine begins at 0066. Assume the contents of port 01 were $20)_{16}$ and the contents of port 02 were $1A)_{16}$ just before the interrupt occurred.

Address	Contents	Comments
0000	LD SP	Top of stack = 0050
0001	50	
0002	00	
0003	IN	Input byte from port 01
0004	01	
0005	LD E,A	Store it in E
0006	IN	Input byte from port 02
0007	02	
0008	ADD E	Add it to (E)
0009	OUT	Output sum to port 03
000A	03	
000B	JP	Repeat
000C	03	
000D	00	
⋮	⋮	
0066	PUSH AF	Store A, F on stack
0067	PUSH DE	Store D, E on stack
0068	IN	Input byte from port 04
0069	04	
006A	LD E,A	Store it in E
006B	IN	Input byte from

006C	05	port 05
006D	ADD E	Add it to (E)
006E	OUT	Output sum to
006F	06	port 06
0070	POP DE	Restore D, E
0071	POP AF	Restore A, F
0072	RET	Return

Two points to notice in this program: (1) Even though F and D are not used, we must PUSH the register pairs AF and DE in order to preserve A and E; and (2) the POP instructions occur in the reverse order of the PUSH instructions. The sequence of stack operations is as follows:

1. The stack pointer is loaded with 0050, so the top of the stack is initially 0050.
2. When the interrupt occurs at 0009, the high order address of the return instruction (00) is stored at 004F, and the low order address (0B) is stored at 004E. The top of the stack is now 004E.
3. Since A contains $20)_{16} + 1A)_{16} = 3A)_{16}$, the PUSH instruction at 0066 stores $3A)_{16}$ at 004D and unknown flag contents at 004C. The top of the stack is now 004C.
4. The contents of D are unknown, but E contains $1A)_{16}$, so the PUSH DE instruction causes D to be stored at 004B, and $1A)_{16}$ to be stored at 004A. The top of the stack is now 004A.
5. The POP DE instruction at 0070 restores the contents of the D and E registers and increments the stack pointer by 2, so the top of the stack is now 004C.
6. The POP AF instruction restores the contents of A and F, and the top of the stack becomes 004E.
7. The RET instruction is executed, the CPU returns to 000B, and the top of the stack is once again 0050.

Figure 10.24 shows all the stack addresses that were affected and their contents when the CPU returns to the main program. As usual, POP and RET instructions do not alter stack contents. (In case the reader may be wondering why there are no EI instructions in the programs of the preceding examples, we are assuming that the interrupts are of the Z-80 nonmaskable type, which always cause the CPU to jump to 0066 to find the interrupt-service subroutine.)

	Stack Address	Contents	
	004A	20	DE contents
Top of the stack at	004B	(D)	
various times during	004C	(F)	AF contents
program execution	004D	3A	
	004E	0B	Return address
	004F	00	
	0050	XX	

FIGURE 10.24 Stack addresses and contents (Example 10.16).

The Z-80 has one programming feature that eliminates the need for PUSH and POP instructions in many small systems. In Figure 10.16, we noted the existence of an *alternate* set of registers, designated A′, B′, and so forth. It is possible to *exchange* registers (A with A′, B with B′, etc.) using certain Z-80 instructions. When an exchange operation is performed, the CPU has a whole new set of registers to work with, and the contents of the original set are unchanged. Thus, when the CPU enters an interrupt-service subroutine, an exchange operation will preserve current register contents, while still permitting register operations using the alternate set. At the conclusion of the subroutine, another exchange is performed, and the CPU returns to the main program with its original register contents.

Many systems are designed to honor one or more interrupt requests even when the CPU is executing the interrupt-service subroutine of a previously honored interrupt. Such systems are said to have a *nested interrupt* capability. In these situations, the CPU abandons its current subroutine, loads the stack in unused locations above the current stack contents, and jumps to the new subroutine designed to service the new interrupt. On completing the new interrupt-service subroutine, the CPU returns to the subroutine it was executing when the new interrupt occurred. We can visualize how it would be possible to have interrupts interrupting interrupts interrupting interrupts, and so on, to a virtually unlimited level of nesting. Of course, the greater the nesting, the larger the stack must be, and the programmer must be certain to reserve enough memory space to ensure that the stack does not eventually grow into program or data memory. The designer may exercise control over interrupt nesting at any given time through use of the masking instructions (EI and DI) and external logic that sets priorities on the various interrupting devices and denies low priority interrupt requests when higher priority interrupts are being serviced.

We will now take up the means by which an interruting device is able to communicate to the CPU the address to which it must jump in order to find the correct interrupt-service subroutir.e. We have already noted parenthetically that a nonmaskable Z-80 interrupt request will always cause the CPU to jump to 0066, so the programmer simply writes the nonmaskable interrupt-service subroutine beginning at that address and no further accommodations are necessary. The Z-80 will respond to maskable-interrupt requests (when its $\overline{\text{INT}}$ input is made low) in one of three different ways, depending on the *interrupt mode* it happens to be in. The interrupt modes are set by the programmer using the instructions IM0, IM1, and IM2: "Set interrupt mode zero, . . . mode one, . . . and mode two," respectively. If the programmer does not specify the mode, the Z-80 will assume mode 0. (We will discuss only modes 0 and 1 in detail.) When the CPU recognizes a mode-0 interrupt, it finishes executing the instruction in progress as previously described. The CPU then issues a unique combination of control-signal outputs that signify that it has *acknowledged* (honored) the interrupt. At this time, it becomes the responsibility of the external, interrupting device (or logic circuitry controlled by it) to respond to the acknowledgement by placing a special eight-bit code on the data bus, which the CPU translates into a jump address. There are eight such codes, each referred to as a RESTART instruction, and we say that the interrupting

TABLE 10-12

Restart Instruction Mnemonic	Op-Code	Jump Address (Hex)
RST 0	C7	0000
RST 8	CF	0008
RST 16	D7	0010
RST 24	DF	0018
RST 32	E7	0020
RST 40	EF	0028
RST 48	F7	0030
RST 56	FF	0038

device *jams* a RESTART onto the data bus. (This is the only time that the CPU receives an instruction from a source other than memory.) The eight RESTART instructions, their hex op-codes, and the addresses they cause the CPU to jump to, are summarized in Table 10-12.

Since there are eight different RESTART instructions, the Z-80 can serve eight different interrupting devices when it is in interrupt mode 0. Each device would have its own subroutine, each beginning at one of the addresses shown in Table 10-12. Note, however, that there are only eight addresses between the first instruction of each subroutine. This is rarely sufficient space to write a typical interrupt-service subroutine, so the programmer frequently writes an unconditional jump instruction (to an uncluttered region of memory) at each RESTART address.

Interrupt mode 1 is useful when there is only one device in the system capable of requesting a maskable interrupt. When a mode-1 interrupt is honored, the CPU automatically jumps to 0038 to find the interrupt-service subroutine. Thus, no external-logic circuitry is required to jam a byte on the data bus. Interrupt mode 1 is similar to the nonmaskable interrupt in this respect but is, of course, maskable.

In interrupt mode 2, external-logic circuitry jams a byte on the data bus, and this byte is combined with the contents of the I register (see Figure 10.16) to form a 16-bit address at which the CPU can find still another address. We will not describe mode-2 interrupts in any further detail other than to say that this is the most versatile of the modes and makes it possible to locate interrupt service subroutines *anywhere* in memory.

Before concluding our discussion of interrupts, we should mention that the stack operations that have been described are also very useful in many applications that do not involve interrupts. It is frequently the case that the CPU must execute the same series of instructions many times during the course of a program. The most efficient way to prepare such a program is to write the series of instructions as a self-contained *subroutine* somewhere in program memory. The CPU can then exit the main program and jump to the same subroutine each time this series of instructions must be performed. On completing the subroutine, the CPU returns to the main program and resumes executing where it left off. Since the return address will, in general, be different each time the CPU completes the subroutine, the

434 MICROPROCESSOR-BASED CONTROL SYSTEMS

stack may be used in the same way that it is used with interrupts to store the return address. The instruction that causes the CPU to store the return address in the stack and jump to a subroutine is the three-byte instruction

$$\begin{array}{c} \text{CALL} \\ \text{ll} \\ \text{hh} \end{array}$$

where ll is the low address byte of the first instruction in the subroutine and hh is the high address byte. The hex op-code for CALL is CD. The CALL and RETURN instructions we have described are unconditional, but it should be noted that the Z-80 also has conditional versions, which CALL or RETURN depending on one of the condition flags. As an example, we might want to call a subroutine that computes the square root of a number generated in the main program if and only if that number is positive. We could use the conditional call: CALL P (call if sign positive) for this purpose. Example 10.17 illustrates using the CALL and RETURN instructions in conjunction with a subroutine.

EXAMPLE 10.17

A microprocessor is used to monitor the salinity (salt content) of a solution used in a certain chemical process. The salinity is measured by passing an electrical current through two electrodes that are immersed in the solution: the greater the salinity, the greater the current flow. The current flows through a resistor of fixed

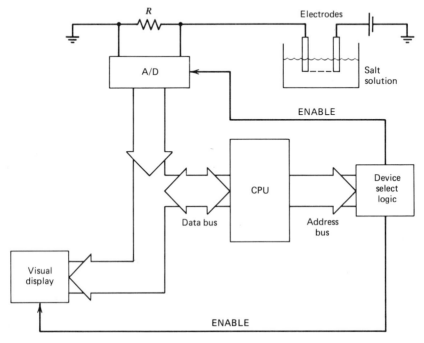

FIGURE 10.25 A salinity monitor (Example 10.17).

value, so the voltage drop across the resistor is proportional to the salinity of the solution. An A/D converter converts the voltage drop to an eight-bit number that is sampled by the microprocessor every 3 sec. The procedure is as follows: Sample the A/D converter, wait 1 sec, output the number to a visual display, wait 2 sec, then sample the A/D converter again. The A/D converter has been assigned device code 07 and the visual-display device, code A3. The system is illustrated in Figure 10.25. Assuming a 100-kHz clock frequency, write a program which implements the procedure described.

SOLUTION

We will use the one-second time-delay program from Example 10.10 as a subroutine beginning at address 0050. Rather than repeat the entire program, we simply identify it as "1-sec time-delay subroutine."

Address	Contents	Comments
0000	LD SP	Set top of stack
0001	AA	to 00AA
0002	00	
0003	IN	Input A/D
0004	07	converter
0005	CALL	Call one
0006	50	second time delay
0007	00	subroutine
0008	OUT	Output to
0009	A3	visual display
000A	CALL	Call one
000B	50	second time delay
000C	00	subroutine
000D	CALL	Call one
000E	50	second time delay
000F	00	again (total delay approximately 2 sec)
0010	JP	Jump back to
0011	03	input A/D converter
0012	00	again
⋮		
0050		Beginning of 1-sec time-delay subroutine
⋮		
005B		End of 1-sec time-delay subroutine
005C	RET	

10.6 DESIGN EXAMPLE—A SORTING ROBOT

This section is devoted to illustrating the step-by-step procedure to follow when designing a complete microprocessor control system. For illustration purposes, we will choose a realistic situation that requires the application of many of the hardware and software techniques discussed in the preceding sections. Our first step is to prepare a concise statement of the problem. This statement will include all the tasks that must be performed to solve the problem and will list all of the constraints that may bear on it.

A parcel-distribution facility must sort packages into three different size categories: small, medium, and large. The size category of a package is determined by the length of its longest dimension (height, width, or depth). Packages must be sorted so that each is placed into one of the three different bins according to its size. When any one bin has been filled to a total weight (including bin) of 100 lb, an empty bin must replace it. The packages enter the facility on a conveyor belt where they have been placed so that their longest dimension is vertical. There are three additional conveyor belts that transport bins: one belt for each type of bin. The physical arrangement of this belt system is illustrated in Figure 10.26.

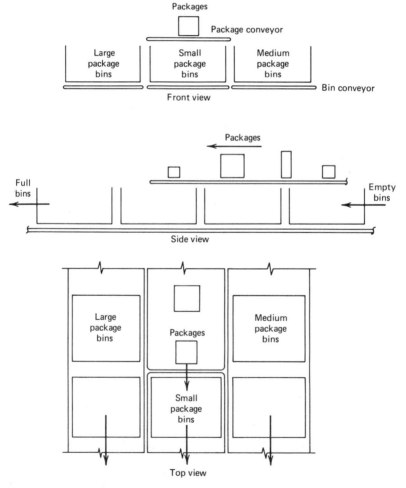

FIGURE 10.26 Conveyor arrangement in an automatic package-sorting system.

10.6 DESIGN EXAMPLE—A SORTING ROBOT

A robot with two arms will be installed beside the conveyor system. The purpose of the robot is to push medium packages into medium package bins, pull large packages into large package bins, and allow small packages to travel to the end of the belt where they fall into small package bins. This arrangement is illustrated in Figure 10.27.

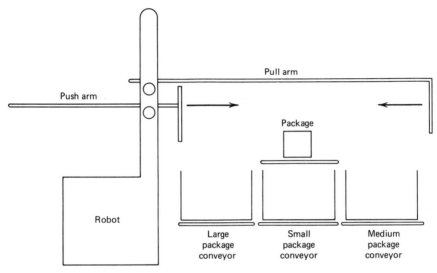

FIGURE 10.27 Push and pull arms of a robotic package sorter.

A vertical bank of three photoelectric cells is installed beside the package conveyor for the purpose of determining package size. Light beams crossing the path of oncoming packages are used to illuminate the cells. (See Figure 10.28.)

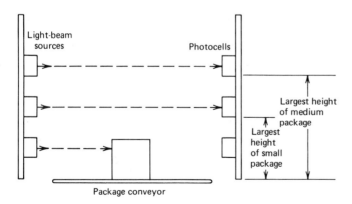

FIGURE 10.28 Photoelectric sensor used to detect package size.

The photocell system is designed to load binary 1's into the least significant three bits of an eight-bit port according to the scheme shown in Figure 10.29.

Bit position	D7	D6	D5	D4	D3	D2	D1	D0	Significance
	X	X	X	X	X	0	0	0	No beams interrupted
Contents	X	X	X	X	X	0	0	1	Beam 1 interrupted (small package)
	X	X	X	X	X	0	1	1	Beams 1 and 2 interrupted (medium package)
	X	X	X	X	X	1	1	1	All beams interrupted (large package)

FIGURE 10.29 Codes produced by photocells to indicate package sizes (X = Don't Care).

There is a weight sensor beneath each bin while it is in the loading position that produces a high output when the total weight of the bin is less than 100 lb, and a low output when the weight reaches 100 lb.

A microprocessor control system is to be designed to perform the following tasks:

1. Actuate the push arm of the robot for 1 sec whenever a medium-size package passes in front of it. This is to be accomplished by delivering a start signal to the push-arm motor controller, followed 1 sec later by a stop signal. (The start signal causes the push arm to travel to its full extension in ½ sec, then retract in the next ½ sec. It will resume cycling so long as a start signal is present, stopping only when a stop signal is received.)
2. Actuate the pull arm for 1 sec whenever a large package passes. The start-stop scheme is the same as for the push arm.
3. Actuate neither arm if
   (a) No package is present
   (b) A small package passes by
4. If the weight of any one of the three bins being loaded reaches 100 lb, the microprocesssor must
   (a) Stop the package conveyor belt
   (b) Advance the belt carrying the full bin one position, so that an empty bin moves into its place. This is accomplished by generating a start signal, followed in 1 sec by a stop signal, which is delivered to the appropriate belt-motor controller.
   (c) Restart the package conveyor belt and resume sorting.

Assume that all motor controllers have a latching eight-big port and that FF stored in the port will start the associated motor, while 00 will stop it. Our next step is to construct a block diagram showing all the components involved in the system and all data-bus, address-bus, and control-signal interconnections. This is shown in Figure 10.30.

Constructing the block diagram stimulates the thought processes that produce ideas for overall system design and assist in preliminary planning. At this stage, we do not want to concern ourselves with details of interface circuitry or programming, but rather wish to formulate an overall strategy for solving the problem. Figure 10.30 reflects some of the broad decisions we have made about the problem at hand. Note that we have decided to regard the six motor controllers in the system

10.6 DESIGN EXAMPLE—A SORTING ROBOT

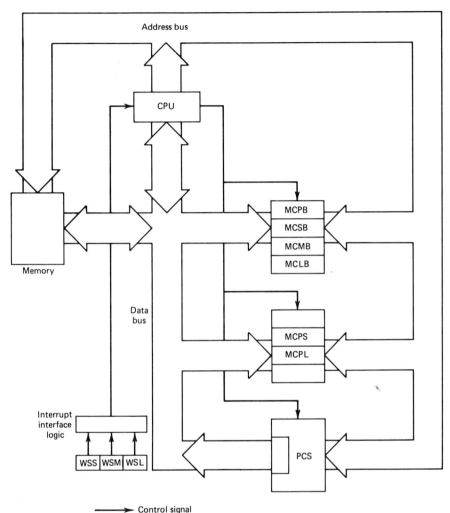

MCPB = package belt motor controller
MCSB = small-bin motor controller
MCMB = medium-bin motor controller
MCLB = large-bin motor controller
PCS = photoelectric sensors

MCPS = push-arm motor controller
MCPL = pull-arm motor controller
WSS = small-bin weight sensor
WSM = medium-bin weight sensor
WSL = large-bin weight sensor

FIGURE 10.30 Block diagram showing control, data, and address buses for the package-sorting system.

as output devices and transmit start-stop signals to them via the data bus. We will, therefore, need to assign device codes and provide for decoding logic on the address bus. Also, we have decided to use the weight-sensor outputs as interrupt signals to the CPU. Thus, whenever a bin is filled, the CPU will interrupt its sorting tasks, stop the package conveyor, and advance the correct bin conveyor. Interrupt interface logic will be required in order to identify for the CPU which bin conveyor needs

440 MICROPROCESSOR-BASED CONTROL SYSTEMS

advancing, that is, which weight sensor caused the interrupt. Finally, we treat the photocell port as an input device, since the CPU must interpret the photocell status code (see Figure 10.29) in order to determine package size.

Our next step is to construct a flowchart that shows the logic and sequence of overall system operation. Again, this is a thought stimulating process that may lead to revisions in the hardware decisions made earlier. At this point, we are comparing trade-offs between hardware and software but keeping in mind the guideline that software should generally be used whenever possible to replace hardware. A flowchart is shown in Figure 10.31.

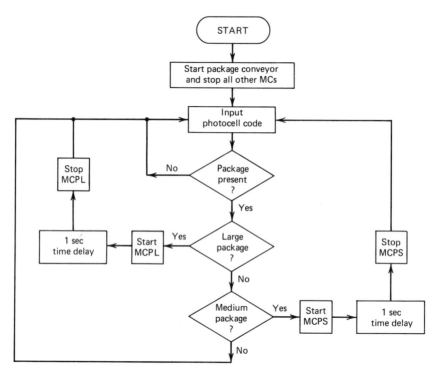

FIGURE 10.31 Flowchart of microprocessor operations in the package-sorting system.

At this point, we should note that we have made some implicit assumptions concerning the mechanics of the system. For example, there are no provisions in the flowchart for stopping the package conveyor when a push arm or pull arm is activated. In order for a package to be pushed or pulled properly into a bin, we are, therefore, making one or more of the following assumptions: (1) The package-conveyor speed is very slow compared to the push- or pull-arm speed; (2) The medium- and large-package bins are situated sufficiently in front of the direction of package travel, so that packages will fall into them properly; (3) The contact surfaces of the robot arms are sufficiently large compared to package sizes, so that contact will be maintained while packages are moving forward; (4) Package bins are

10.6 DESIGN EXAMPLE—A SORTING ROBOT

sufficiently large compared to package sizes, so that any deflection caused by simultaneous forward package movement and push or pull forces will not affect the fall of the packages into the bins; (5) The packages on the package conveyor are spaced so that no new package will completely pass through the photocell detectors during the time that another package is being pushed or pulled off the conveyor.

If we cannot make any of these assumptions, it may be necessary to stop the package conveyor each time the presence of a medium- or large-size package is detected in front of the robot. We will proceed with our design example under the assumption that this is not necessary, but the foregoing serves to illustrate an important aspect of the design procedure. Designers must always be aware (and beware) that they inevitably make some assumption about system characteristics whenever they propose a solution to a design problem. In the practical world, system designers cannot afford the luxury of making arbitrary assumptions that just happen to make a particular design work. To help ensure that all contingencies are accounted for, an excellent approach for designers at all steps along the way is to ask themselves what if? In other words, consider every possible event or system characteristic that might create a problem or cause the system to fail, just as expert bridge players consider every possible card distribution that could cause them to lose their contract, before beginning play of the cards.

Before getting into the details of the program, we must make certain hardware decisions. For example, we must assign device codes to the six output devices and to the one input device. As discussed in Section 10.3, we wish to make these assignments so that a minimum amount of decoding logic is necessary. We can eliminate the need for distinguishing between input and output devices in our decoding logic (i.e., eliminate the need for the $\overline{\text{READ}}$ and $\overline{\text{WRITE}}$ control signals) by assigning each of the seven devices a unique code. In this way, the only Z-80 control signal we will require is $\overline{\text{I/OREQ}}$, since it goes active low when either an input or an output device is requested. To eliminate extensive address-bus decoding, we choose the seven device codes so that one and only one address bit is low in each device code. A device-code assignment scheme that meets this criterion is illustrated in Figure 10.32.

Device	Binary Device Code								Hex Code
	A7	A6	A5	A4	A3	A2	A1	A0	
MCPB	1	1	1	1	1	1	1	0	FE
MCSB	1	1	1	1	1	1	0	1	FD
MCMB	1	1	1	1	1	0	1	1	FB
MCLB	1	1	1	1	0	1	1	1	F7
MCPS	1	1	1	0	1	1	1	1	EF
MCPL	1	1	0	1	1	1	1	1	DF
PCS	1	0	1	1	1	1	1	1	BF

FIGURE 10.32 Assignment of device codes to I/O devices (see also Figure 10.30).

Assuming active-low enabling signals, the decoding logic for MCPB, whose device code causes only A0 to be low, is shown in the truth table and logic diagram of Figure 10.33.

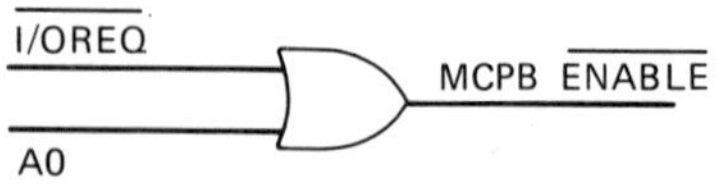

$\overline{I/OREQ}$	A0	$\overline{MCPB\ ENABLE}$
0	0	0
0	1	1
1	0	1
1	1	1

FIGURE 10.33 Logic required to enable one I/O device.

Decoding for the other six devices is similar, the only difference being the particular address bit used. Thus, all device-code decoding can be accomplished with two-quad two-input OR-gate chips.

Before designing the memory-interface circuitry, we must know how much memory will be required; we must, therefore, begin working on the program. Since we will need three interrupt-service subroutines, we will design the system to recognize mode-0 interrupts and will, therefore, have to locate the first instruction of each subroutine in one of the RESTART addresses (see Table 10.11). The hardware implication of this decision is that we must have at least some memory located in address space that has high address byte $(00)_{16}$, since all RESTART addresses have this high byte. The next question is, How much memory space will be required for each interrupt-service subroutine? We cannot know for certain at this early stage, but it is possible to make some preliminary estimates based on the design decisions thus far. We know that an interrupt-service subroutine must perform the following tasks in the sequence listed: (1) stop the package conveyor; (2) start the appropriate bin conveyor; (3) stop the bin conveyor after 1 sec; and (4) start the package conveyor again.

We might also expect that we will need some instructions in the subroutine to preserve the register contents of the main program, so space must be allowed in the subroutine for some PUSH- and POP-type instructions. This would be a good time to pose a what-if question. For example, what if an interrupt occurs while a push-arm or pull-arm maneuver is being executed? In fact, it is likely that an interrupt *will* occur shortly after a medium or large package has been pushed or pulled off the conveyor, for it is then that the weight sensor of a just-filled bin will be triggered. It is, therefore, likely that the pull or push arm of the robot will be retracting when an interrupt occurs. But since the interrupt will cause the CPU to abandon the time-delay routine used for timing the interval between starting and stopping the arm movement, the arm will continue to retract and is liable to reextend during the 1 sec required to advance the bin conveyor. (Recall that the arm motion continues in endless cycles until a STOP command is received.) There

10.6 DESIGN EXAMPLE—A SORTING ROBOT

are two ways of resolving this potential problem: (1) Since the bin-advance time is 1 sec, which equals the *period* of a push-arm or pull-arm cycle, we can rely on the arm being in the exact same position when the CPU returns from the interrupt as it was when the interrupt occurred; or (2) we can stop the push-arm or pull-arm motion at the beginning of the interrupt-service subroutine and start it again at the end of the subroutine. The latter choice seems preferable, since any minor differences between arm-cycle and bin-advance time are liable to accumulate over many cycles of operation and eventually introduce serious errors in arm position. Of course, it will not be necessary to incorporate these start-stop provisions in the interrupt-service subroutine used to advance the small package bins, since no robot-arm motion is involved.

A trial writing of the interrupt-service subroutines for the medium- and large-package interrupts reveals that they require more memory space than is available between the RESTART addresses. We, therefore, write an unconditional jump instruction at the first address of RESTART 8 for the medium-package interrrupt and at the first address of RESTART 16 for the large-package interrupt. The interrupt-service subroutine for the small-package interrupt can begin at the first address of RESTART 24. (At the first address of RESTART 0, 0000, we will write an unconditional jump to the main program, since resetting the CPU causes program execution to begin at 0000.)

The small-package interrupt-service subroutine follows.

Address	Contents	Comments
0018	PUSH—	Preserve registers
0019	PUSH—	(to be determined later)
001A	XOR A	Load MC stop code (00) into A
001B	OUT	Stop package conveyor
001C	FE	
001D	LD A	Load MC start code
001E	FF	into A
001F	OUT	Start small bin
0020	FD	conveyor
0021	CALL	Jump to 1-sec time
0022	—	delay (address to be
0023	—	determined later)
0024	XOR A	Load MC stop code into A
0025	OUT	Stop small bin
0026	FD	conveyor
0027	LD A	Load MC start code
0028	FF	into A
0029	OUT	Start package conveyor
002A	FE	
002B	POP—	Restore registers
002C	POP—	
002D	EI	Enable future interrupts
002E	RET	Return to main program

We will leave a few memory addresses for possible future modification of this subroutine and begin the medium-bin interrupt subroutine at 0035. Thus, at the first address of RESTART 8, 0008, we write the unconditional jump

0008	JP	
0009	35	
000A	00	

The medium bin subroutine follows.

Address	Contents	Comments
0035	PUSH—	Preserve registers
0036	PUSH—	(to be determined later)
0037	LD A	Load MC stop code
0038	00	into accumulator
0039	OUT	Stop package conveyor
003A	FE	
003B	OUT	Stop push arm
003C	EF	
003D	LD A	Load MC start code
003E	FF	into accumulator
003F	OUT	Start medium bin
0040	FB	conveyor
0041	CALL	Jump to 1-sec time
0042	—	delay (address to be
0043	—	determined later)
0044	LD A	Load MC stop code
0045	00	into accumulator
0046	OUT	Stop medium bin
0047	FB	conveyor
0048	LD A	Load MC start code
0049	FF	into accumulator
004A	OUT	Start package conveyor
004B	FE	
004C	OUT	Start push arm
004D	EF	
004E	POP—	Restore registers
004F	POP—	
0050	EI	Enable future interrupts
0051	RET	Return to main program

Again leaving a few addresses for possible future modification, we begin the large-bin interrupt subroutine at 0055. This subroutine will be the same as the preceding one except that it will stop and start the pull arm instead of the push arm and start and stop the large-bin instead of medium-bin conveyor. Thus, it is only necessary

10.6 DESIGN EXAMPLE—A SORTING ROBOT

to change the appropriate device codes in the previous subroutine, and we will not rewrite it here. At the first address of RESTART 16, 0010, we write the unconditional jump

0016	JP
0017	55
0018	00

The large-bin interrupt subroutine will require the same amount of memory as the medium-bin subroutine and will, therefore, occupy memory addresses 0055 through 0071. Therefore, we begin the main program at 0075 (and write an unconditional jump to 0075 at address 0000).

Address	Contents	Comments
0075	EI	Enable interrupts
0076	LD SP	Set top of stack to address
0077	—	to be determined later
0078	—	
0079	XOR A	Load MC stop code into A
007A	OUT	Stop small bin conveyor
007B	FD	
007C	OUT	Stop medium bin conveyor
007D	FB	
007E	OUT	Stop large bin conveyor
007F	F7	
0080	OUT	Stop push arm
0081	EF	
0082	OUT	Stop pull arm
0083	DF	
0084	LD A	Load MC start code into A
0085	FF	
0086	OUT	Start package conveyor
0087	FE	
0088	IN	Input photocell code
0089	BF	
008A	LD B,A	Temporarily store code in B
008B	AND	Check for presence of 1
008C	07	in D0, D1, or D2
008D	JP Z	No package present; keep
008E	88	checking photocell code
008F	00	
0090	LD A,B	Retrieve photocell code
0091	AND	Check for presence of 1 in D2
0092	04	
0093	JP NZ	Large package present

Address	Contents	Comments
0094	9F	Jump to actuate pull arm
0095	00	
0096	LD A,B	Retrieve photocell code
0097	AND	Check for presence of 1 in D1
0098	02	
0099	JP NZ	Medium package present
009A	AD	Jump to actuate push arm
009B	00	
009C	JP	Small package present;
009D	88	no arms need actuating;
009E	00	jump to check photocells again
009F	LD A	Load A with MC start code
00A0	FF	
00A1	OUT	Start pull arm
00A2	DF	
00A3	CALL	Jump to 1-sec time delay
00A4	—	
00A5	00	
00A6	LD A	Load A with MC stop code
00A7	00	
00A8	OUT	Stop pull arm
00A9	DF	
00AA	JP	Check photocell code again
00AB	88	
00AC	00	
00AD	LD A	Load A with MC start code
00AE	FF	
00AF	OUT	Start push arm
00B0	EF	
00B1	CALL	Jump to 1-sec time delay
00B2	—	
00B3	00	
00B4	LD A	Load A with MC stop code
00B5	00	
00B6	OUT	Stop push arm
00B7	EF	
00B8	JP	Check photocell code again
00B9	88	
00BA	00	

10.6 DESIGN EXAMPLE—A SORTING ROBOT

Assuming a 100-kHz clock, we can use the same 1-sec time-delay routine that was used in Example 10.10. This routine can begin at address 00C0.

Address	Contents
00C0	LD B
00C1	1C
00C2	LD C
00C3	FF
00C4	DEC C
00C5	JP NZ
00C6	C4
00C7	00
00C8	DEC B
00C9	JP NZ
00CA	C2
00CB	00
00CC	RET

We may now fill in the bytes we temporarily omitted when writing the subroutines and main program. Since the main program and the time-delay program use the A, B, and C registers, the AF and BC register pairs should be pushed and popped in the interrupt-service subroutines. The address of the time-delay subroutine 00C0 can now be written at those locations in the interrupt subroutines and main program where the time delay is called. Finally, the top of the stack can safely be placed at 0150, so the stack-pointer load instruction at 0076 can be finalized. The program, including subroutines, can be stored in the memory space from 0000 through 00FF, a total of 256 contiguous locations, some of which are unused. Figure 10.34 summarizes the program.

Note that the stack is the only area of memory that cannot be ROM. The CPU must be able to write into the stack, but no other instructions in the program require writing into memory. We have placed the top of the stack in an address space that has high byte 01 and will use one small RAM memory chip solely for the stack. To store the rest of the program, we may use a Texas Instruments 745470/71 256 × 8 static PROM. The access time of this chip is 50 nanoseconds (nsec), and since we are using a 100-kHz clock (one period = .01 msec), we will have no timing problems with this choice. In practice, a PROM is generally used for system development and prototype testing, since the contents can be erased and reprogrammed with relative ease. If the system is to be produced in large quantities, the final program is usually in mask programmable ROM, that is, custom programmed by the chip manufacturer to the user's specifications. The chip we choose for the stack is the 6810 static RAM, which is organized 128 × 8 and has access time 450 nsec.

Figure 10.35 shows how the two memory chips would be interconnected and tied to the system buses. The 6810 RAM is connected to the address bus in such a way that it is accessed by the CPU for memory addresses 0100 through 017F.

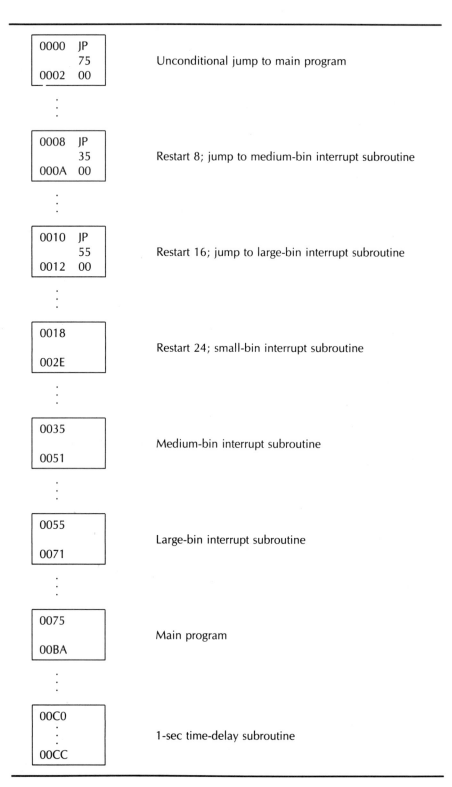

FIGURE 10.34 Summary of program blocks.

10.6 DESIGN EXAMPLE—A SORTING ROBOT

The 6810 has both active high and active low chip-select inputs, and we use one of the active high inputs (CSO) to enable the chip whenever address bit A8 is high. Address bit A7 is not connected, since it is always zero in the address space 0100 through 017F. The Z-80 CPU produces an active low $\overline{\text{MREQ}}$ control signal that goes low whenever a memory read or write operation is to be performed. This signal is OR'ed with $\overline{\text{READ}}$ to enable the ROM and tied directly to an active-low chip-select ($\overline{\text{CS1}}$) on the RAM to enable it. R/W is connected directly to the $\overline{\text{WRITE}}$ control output from the Z-80, since this signal is low when the CPU writes and high when it reads.

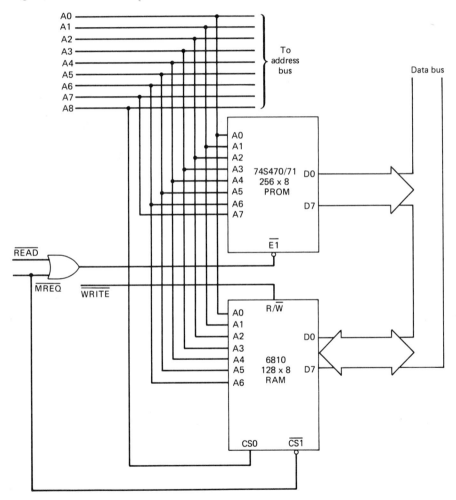

FIGURE 10.35 RAM and PROM memory interconnections.

The only interface logic left to consider is that between the interrupt generating weight sensors and the CPU. We know that it is necessary to jam a RST 8, RST 16, or RST 24 instruction on the data bus when one of the weight sensors detects a bin weighing 100 lb. As mentioned earlier, there are a number of versatile, commercially available chips that can be used to perform this function as well as many other functions, and it is not our intention to treat any one such device in detail. A relatively simple and inexpensive scheme that will serve our needs is based on the 8214 Priority Interrupt Control Unit and the Intel 8212 eight-bit I/O port. The 8214 can be configured to establish priorities among eight different interrupt-request inputs ($\overline{R0}$ through $\overline{R7}$) and issue a three-bit code (A0, A1, A2) identifying which one of the eight possible interrupt requests is to be honored. We have only three interrupting devices and no need to assign priorities, since we know that only one weight sensor will be triggered and only one full bin advanced at any given time. There is no possibility of a bin becoming filled while another full bin is being replaced, so we will not concern ourselves with the details of priority arbitration by the 8214.

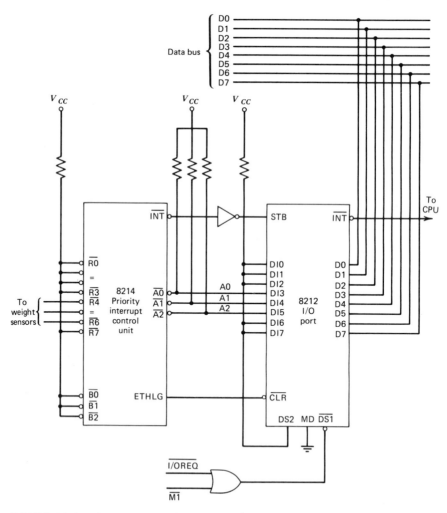

FIGURE 10.36 Interrupt control and interfacing.

The 8212 eight-bit I/O port is used to latch a RESTART instruction op-code onto the data bus at the appropriate time (when the CPU acknowledges an interrupt). The three-bit code (A0, A1, A2) produced by the 8214 is an input to the 8212, which, in turn, outputs the corresponding RESTART code to the data bus. The scheme is based on the fact that bits 0, 1, 2, 6, and 7 are binary 1 in *every* RESTART op-code. (To verify this, see Table 10-12.) An abbreviated wiring diagram, showing only the pin connections of significance to our application, is shown in Figure 10.36.

The following points should be noted from Figure 10.36: (1) Interrupt request inputs $\overline{R0}$, $\overline{R1}$, $\overline{R2}$, $\overline{R3}$, and $\overline{R7}$ to the 8214 are all tied high; we use only $\overline{R4}$, $\overline{R5}$, and $\overline{R6}$, since these cause RST 24, RST 16, and RST 8, respectively, to be jammed on the data bus; (2) $\overline{B0}$, $\overline{B1}$, and $\overline{B2}$ are all tied high; this causes the 8214 to recognize *all* interrupt requests; (3) Inputs DI0, DI1, DI2, DI6, and DI7 to the 8212 are all tied high; as already mentioned, these bits are all 1 in every RST op-code; (4) The Z-80 acknowledges an interrupt by causing $\overline{I/OREQ}$ and $\overline{M1}$ (which we have not discussed) to go low simultaneously; when this occurs, the 8212 is enabled by virtue of the connection to $\overline{DS1}$ (device select); (5) The 8212 is latched by the $\overline{INT}$ connection to its STB (strobe) input; (6) The 8212 produces an $\overline{INT}$ interrupt signal to the CPU; thus when the CPU acknowledges the interrupt, the 8212 is enabled and the correct RST op-code is placed on the data bus.

This concludes our design example. We must again remind the reader that we have barely scratched the surface of microprocessor-based control systems. The great proliferation of microprocessor types and support devices precludes even a survey, much less an exhaustive treatment, of this subject in the space we have allocated. We have not, for example, even mentioned the important area of series I/O interface, which is widely used in systems that transmit and receive data from many types of remote I/O devices: printers, teletypewriters, video terminals, and auxiliary memory, to mention but a few. Be aware that a custom-made chip designed to serve nearly every support function for every type of microprocessor is probably available. These are referred to as microprocessor families, though support chips in some families are designed to operate satisfactorily with CPUs outside the family. The serious designer of a microprocessor-based control system is best advised to consult the technical literature published by manufacturers. Microprocessor manufacturers today provide excellent documentation of their products, and no single textbook is capable of duplicating these detailed specifications, application notes, and operating characteristics for every family now in existence.

REFERENCES

1. **Bibbero, Robert J.** *Microprocessors in Instruments and Control.* Wiley, 1977.
2. **Boyce, Jefferson C.** *Microprocessor and Microcomputer Basics.* Prentice-Hall, 1979.
3. **Goody, Roy W.** *Microcomputer Fundamentals, A Laboratory Approach.* Science Research Associates, 1980.

4. **Hilburn, J. L.**, and **Paul M. Julich.** *Microcomputers/Microprocessors: Hardware, Software, and Applications.* Prentice-Hall, 1976.
5. **Hnatek, Eugene R.** *A User's Handbook of Semiconductor Memories.* Wiley 1977.
6. **Hunter, Ronald P.** *Automated-Process Control System—Concepts and Hardware.* Prentice-Hall, 1978.
7. **Lenk, John D.** *Handbook of Microprocessors, Microcomputers, and Minicomputers.* Prentice-Hall, 1979.
8. **Leventhal, Lance A.** *Introduction to Microprocessors: Software, Hardware, Programming.* Prentice-Hall, 1978.
9. **Lin, Wen C.**, ed. *Microprocessors: Fundamentals and Applications.* IEEE Press, 1976.
10. **McGlynn, Daniel R.** *Microprocessors, Technology, Architecture, and Applications.* Wiley, 1976.
11. **MOSTEK.** *Microcomputer Data Book.* MOSTEK Corporation, 1979.
12. **Nashelsky, Louis.** *Introduction to Digital-Computer Technology.* 2d ed. Wiley, 1977.
13. **Osborne, Adam.** *An Introduction to Microcomputers, Volume 0—The Beginners' Book.* Osborn/McGraw-Hill, 1977.
14. **Porat, Dan I.**, and **Arpand Barna.** *Introduction to Digital Techniques.* Wiley, 1979.
15. **Tocci, Ronald J.**, and **Lester P. Laskowski.** *Microcomputers and Microprocessors: Hardware and Software.* Prentice-Hall, 1979.
16. **William, Gerald E.** *Digital Technology.* Science Research Associates, 1977.

EXERCISES

10.1 Assuming that logic $1 = +10$ V and logic $0 = 0$ V (ground), calculate the analog output voltage of the R-2R ladder network in a ten-bit D/A converter when the input is:

(a) 1000000000
(b) 0101010101
(c) 1110000000

10.2 A 12-bit A/D converter that uses the successive-approximation technique is driven by a 6-MHz clock. What is its conversion time and conversion rate?

10.3 What is the maximum permissible frequency component of the analog input to an A/D converter with a conversion time of 50 nsec?

10.4 Using the specifications for the Datel ADC-EH10B2 A/D converter provided in Appendix B, answer the following questions (assuming the device is operated with input range 0 to $+10$ V):

(a) What is the resolution in bits, percent, and volts?
(b) Could this converter be used to convert an analog-input signal with maximum frequency 200 kHz? Explain.

10.5 What minimum resolution in bits would be required for an absolute optical shaft encoder designed to produce a unique binary code for each 5 min of angular-shaft rotation?

10.6 A Z-80 microprocessor is being used to replace an 8080 microprocessor in an existing control system. The 8080 system produces the following control signals: $\overline{\text{MEMR}}$ (memory read), $\overline{\text{MEMW}}$ (memory write), $\overline{\text{I/ORD}}$ (I/O read), and $\overline{\text{I/OWR}}$ (I/O write). Derive the logic that should be used to combine the Z-80 control signals ($\overline{\text{READ}}$, $\overline{\text{WRITE}}$, $\overline{\text{MREQ}}$, and $\overline{\text{I/OREQ}}$) in such a way that they produce the equivalent 8080 control signals.

10.7 A static ROM chip is organized 4 K × 4. How many pins on the chip are assigned to (a) address bits; (b) data bits?

10.8 Find the total number of 1 K × 1 memory chips that would be required to construct a memory for eight-bit data bytes if the memory size is to be

(a) 8 K
(b) 32 K
(c) the maximum possible size for use with a microprocessor with 16 address bits.

10.9 A microprocessor control system has a 16K memory that has been assigned hexadecimal addresses 0000 through 3FFF. Using the memory-mapped I/O technique, assign addresses to two different output devices and show how their $\overline{\text{CS}}$ inputs would be connected into the system.

10.10 What hexadecimal device code has been assigned to the input device shown in Figure 10.37?

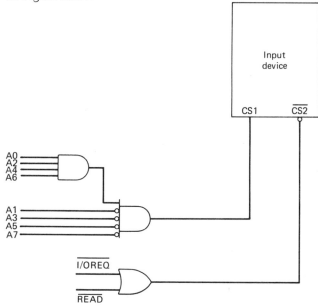

FIGURE 10.37

10.11 Given the following Z-80 program:

Address	Contents
0000	LD A
0001	55
0002	LD B
0003	20
0004	LD A,B
0005	LD B,A
0006	HALT

What are the contents of A and B after the HALT is executed?

10.12 Write a Z-80 program segment, using only the LOAD instructions, which exchanges the contents of the H and L registers.

10.13 Given the following Z-80 program:

Address	Contents
0000	LD H
0001	03
0002	LD L
0003	AA
0004	LD B
0005	AF
0006	LD (HL),B
0007	LD A,(HL)
0008	LD C,A
0009	LD B,C
000A	HALT

After the HALT is executed, what are the contents of A, B, C, H, L, and memory address 03AA?

10.14 Given the following Z-80 program that begins at address 0050:

Address	Contents
0050	26
0051	00
0052	2E
0053	0A
0054	75
0055	2D
0056	C2
0057	54
0058	00
0059	76

(a) Write the program using op-code mnemonics.
(b) Describe briefly what this parogram accomplishes.

10.6 EXERCISES

10.15 Beginning at address 0000, write a Z-80 program that loads 30 into 0030, 31 into 0031, and so forth, up to FF in 00FF. Use *only* the LD, INC, and conditional JP instructions. (Hint: A register containing FF will contain 00 after it is incremented once.)

10.16 How many clock periods would be required in a time-delay program designed to consume 1 min if the clock frequency is 2.5 MHz?

10.17 Assuming a 0.5-MHz clock frequency, determine how long the following program takes to execute:

Address	Contents
0000	LD E
0001	0F
0002	LD D
0003	A0
0004	DEC D
0005	JP NZ
0006	04
0007	00
0008	DEC E
0009	JP NZ
000A	02
000B	00
000C	HALT

10.18 What are the contents of the accumulator after the instruction XOR A is executed?

10.19 Given the following Z-80 program:

Address	Contents
0000	LD A
0001	57
0002	LD B
0003	3A
0004	AND B
0005	LD C,A
0006	OR C
0007	LD D,A
0008	XOR D
0009	HALT

What are the hexadecimal contents of A, B, C, and D after the HALT is executed?

10.20 In Example 10.13, suppose the traffic-control system is modified so that the pressure-actuated switches S1, S2, S3, and S4 load binary 1's into bit positions D0, D2, D4, and D6, respectively. Rewrite the program in the example so that it counts traffic on B and C streets with this modification incorporated.

10.21 Write a program that inputs a byte from input device 01, outputs it to output device 01, then inputs from input device 02 and outputs to output device 02, and so forth, in sequence, until it inputs from input device AF and outputs to output device AF. The program should then automatically start over and repeat this sequence indefinitely.

10.22 Given the following Z-80 program:

Address	Contents
0000	LD SP
0001	F0
0002	00
0003	LD B
0004	00
0005	IN
0006	01
0007	OR B
0008	JP NZ
0009	05
000A	00
⋮	⋮
0066	PUSH AF
0067	IN
0068	01
0069	OUT
006A	75
006B	POP AF
006C	RET

Input device 01 contains $(4D)_{16}$. Suppose a single nonmaskable interrupt occurs while the CPU is executing the instruction at address 0005. (A nonmaskable interrupt automatically causes the CPU to jump to address 0066 to find the interrupt-service subroutine.) Answer the following questions:

(a) What is the top of the stack after the instruction at 0003 is executed?

(b) What are the contents of the stack pointer after the interrupt is honored and the CPU jumps to 0066 but before the instruction at 0066 is executed?

(c) What is the top of the stack just after the PUSH AF instruction is performed?

(d) What are the contents of the stack pointer just after the POP AF instruction is performed?

(e) What are the contents of the A and B registers just after the RET instruction is performed?

(f) What are the contents of the stack pointer just after the RET instruction is performed?

(g) What are the contents of memory addresses 00EC through 00EF just after the RET instruction is performed?

10.23 A microprocessor control system has six different I/O devices that are capable of generating interrupt requests. The system is designed to permit nested interrupts (any device may interrupt the CPU while it is performing the interrupt-service subroutine of another device). If the main program and every interrupt-service subroutine require the use of the E and L registers, how much memory space should be allocated for the stack?

10.24 Many logic devices have the characteristic that they react to *open* inputs in the same way they would if the inputs were high or logic one. Such devices are said to "float their inputs high." Assuming the Z-80 floats open data-bus inputs high, what would happen to the system described in the design example in Section 10.6 if the interrupt interface circuitry in Figure 10.36 were omitted entirely? That is, assuming the rest of the system circuitry is as described, what would happen when a weight sensor requested an interrupt if there were no provisions to jam a RESTART instruction on the data bus? Explain the reasons for your answer.

ANSWERS TO EXERCISES

10.1 (a) 5 V
 (b) 3.33 V
 (c) 8.75 V

10.2 2 μsec; 500,000 conversions/sec

10.3 10 MHz

10.4 (a) 12 bits; .025%; 2.5 mV
 (b) Yes. Conversion rate is 500,000 samples per second, so by the Shannon sampling theorem, input frequencies up to 250 kHz are permissible.

10.5 13 bits

10.6 See Figure 10.38.

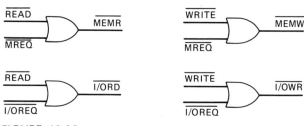

FIGURE 10.38

10.7 (a) 12
 (b) 4

10.8 (a) 64
 (b) 256
 (c) 512

458 MICROPROCESSOR-BASED CONTROL SYSTEMS

10.10 $55)_{16}$

10.11 $20)_{16}$

10.12
0000   LD A,H
0001   LD H,L
0002   LD L,A

10.13 AF, AF, AF, 03, AA, AF

10.14 (a)
0050   LD H   0055   DEC L
0051   00   0056   JP NZ
0052   LD L   0057   54
0053   0A   0058   00
0054   LD (HL),L   0059   HALT

(b) This program stores 0A in 000A, 09 in 0009, 08 in 0008, and so forth, down to 01 in 0001.

10.15
0000   LD H
0001   00
0002   LD L
0003   30
0004   LD (HL),L
0005   INC L
0006   JP NZ
0007   04
0008   00
0009   HALT

10.16 150×10^6

10.17 67.852 msec

10.18 $00)_{16}$

10.19 (A) = 00; (B) = 3A; (C) = 12; (D) = 12

10.22 (a) 00F0   (e) (A) = 4D; (B) = 00
       (b) 00EE   (f) 00F0
       (c) 00EC   (g) (00EC) = flags; (unknown)
       (d) 00EE   (00ED) = 4D
                             (00EE) = 07
                             (00EF) = 00

10.23 $36)_{10}$ addresses, not counting the initial top of the stack.

10.24 The CPU would jump to address 0038, since the open data bus would be treated as if all binary 1's were present when a RESTART instruction was expected. But $FF)_{16}$ is the op-code for RST 56.

APPENDIX

A

ADVANCED TOPICS IN LAPLACE TRANSFORMS

A.1 REPRESENTATION OF DELAYED TIME FUNCTIONS

In many electrical, electronic, and electromechanical systems, inputs may occur at points in time *after* $t = 0$. A system may, for example, be subjected to several driving signals that occur one after the other in time. We might want to analyze the response of such a system beginning at $t = 0$ when the first input is applied and also take into account the effects of inputs that occur at later times. Such inputs are said to be *delayed* by an amount of time equal to the time (after $t = 0$) when they occur.

Many other systems contain components that intentionally or unintentionally delay the propagation of signals through them. That is, they contain devices whose outputs appear only after the lapse of a certain period of time from the time the inputs occur. An example of such a device is a *delay line*. Figure A.1 illustrates a 10-msec delay line whose input is $5 \sin(2\pi \times 100t)$. The output, like the input,

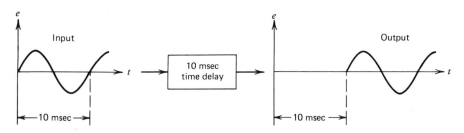

FIGURE A.1   Sinusoidal input and output waveforms in a 10-msec delay line.

460 ADVANCED TOPICS IN LAPLACE TRANSFORMS

FIGURE A.2   The unit-step function u(t).

is a 5-V peak sine wave but begins 10 msec (or one full period) after the initial application of the input.

The concept of delayed signals is also applicable to the mathematical analysis of systems that are driven by, or that develop, nonlinear waveforms. For example, the clock signal in a digital system can be viewed as a series of pulses that occur

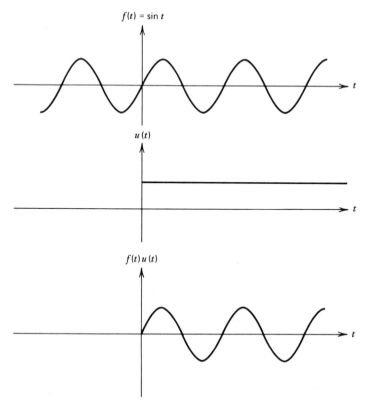

FIGURE A.3   The product of a unit step and a sine wave equals a sine wave switched on at $t = 0$.

A.1 REPRESENTATION OF DELAYED TIME FUNCTIONS

one after the other, that is, that are delayed in time with respect to one another. The inputs and outputs of the various logic gates, counters, shift registers, and so forth, of a digital system can all be considered as signals that are switched in at various times after $t = 0$.

In order to use the techniques of LaPlace transformation to analyze the behavior of systems with delayed inputs or of systems containing delay lines, we must learn how to represent delayed time functions and, eventually, how to transform them into the frequency or s-domain. Toward this end, let us first review the concept of the unit-step function $u(t)$. The unit step is defined mathematically by

$$u(t) = \begin{cases} 1 \text{ for } t \geq 0 \\ 0 \text{ for } t < 0 \end{cases}$$

This function is sketched in Figure A.2.

The unit step is seen to have a "jump" at $t = 0$, that is, an instantaneous change in value from 0 to 1 occurring at $t = 0$. (Such a jump is called a discontinuity in the function.) The unit step is useful when we wish to express mathematically that a certain time-domain function is suddenly applied to a system; that is, switched into the circuit at $t = 0$. Strictly speaking, the mathematical expression for a time-domain function $f(t)$ such as $f(t) = 10t$, $f(t) = 5 \sin 100 t$, and so forth, is defined (has values) for *negative* t as well as positive t. But since we are typically only interested in system behavior for $t \geq 0$, we want to write $f(t)$ in such a way that it is clear that $f(t) = 0$ for all negative values of t. We can accomplish this by writing $f(t)u(t)$. Since $u(t)$ is zero for all negative t, the *product* $f(t)u(t)$ is likewise 0 for all negative t. Furthermore, since $u(t) = 1$ for all $t \geq 0$, the product $f(t)u(t)$ exactly equals $f(t)$ for all positive t. This is illustrated in Figure A.3.

EXAMPLE A.1

A *ramp* voltage $v(t)$ that increases linearly with time at the rate of 6 V/sec is switched into a circuit at $t = 0$. Express the function $v(t)$ mathematically.

SOLUTION

If we wrote $v(t) = 6t$, we would have the correct representation for $t > 0$; but this expression would produce negative voltages for $t < 0$. Hence, we write

$$v(t) = 6tu(t)$$

If a time-domain is switched into a circuit at $t = a$ sec instead of at $t = 0$, we must modify our representation. The rule for expressing the fact that a time-domain function $f(t)$ has been delayed by a certain time a may be stated as follows: Replace t by $(t - a)$ everywhere that t occurs in the undelayed function. Thus, a function $f(t)u(t)$ delayed by a sec is written $f(t - a)u(t - a)$.

EXAMPLE A.2

Write an expression for the output of the 10-msec delay line shown in Figure A.1.

SOLUTION

The input to the delay line is $5 \sin (2\pi \times 100t)u(t)$. The output is, therefore, $[5 \sin (2\pi \times 100)(t - 0.01)]u(t - 0.01)$. Note that in Example A.2 we replaced t by $(t - 0.01)$ everywhere that t occurred in the original expression, including $u(t)$.

EXAMPLE A.3

A 12-V dc voltage is applied to a circuit at $t = 2.5$ sec. Write an expression for the voltage.

SOLUTION

A 12-V dc source switched in at $t = 0$ would be represented by $12u(t)$, which has value 12 for all $t > 0$ and value zero for all $t < 0$. Hence, if the switching is delayed by 2.5 sec, we have

$$e(t) = 12u(t - 2.5)$$

This is sketched in Figure A.4.

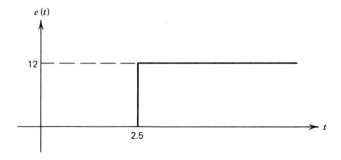

FIGURE A.4 Waveform representing a 12-V source switched in at $t = 2.5$ sec: $e(t) = 12u(t - 2.5)$ (Example A.3).

Occasionally, we will wish to represent functions by expressions such as $f(t - a)u(t)$ or $f(t)u(t - a)$. We must be very careful to distinguish between the meanings of these expressions and the meaning of $f(t - a)u(t - a)$. The expression $f(t - a)u(t)$ represents a function that has been delayed by a sec when it is switched in at $t = 0$.

EXAMPLE A.4

Write an expression for the voltage whose graph is shown in Figure A.5.

A.1 REPRESENTATION OF DELAYED TIME FUNCTIONS

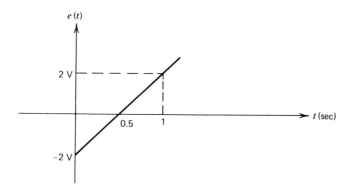

FIGURE A.5 Write an expression for the voltage ramp (Example A.4).

SOLUTION

The function whose graph is shown in Figure A.5 is a ramp with slope $2/(1 - 0.5)$ = 4 V/sec. Consider first the functions $4t$ and $4(t - 0.5)$. They are shown in Figure A.6. Clearly, if $4(t - 0.5)$ is multiplied by $u(t)$, the result would be zero for all $t < 0$ and would reproduce $4(t - 0.5)$ for all $t > 0$. Hence, the function shown in Figure A.5 is $e(t) = 4(t - 0.5)u(t)$.

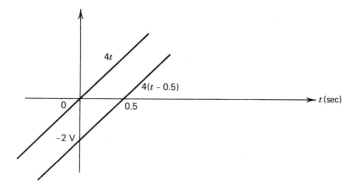

FIGURE A.6 The ramp $4t$ delayed by 0.5 sec and multiplied by $u(t)$ (Example A.4).

EXAMPLE A.5

Sketch the following voltages versus time:

(a) $e_1(t) = 4(t - 0.5)u(t - 0.5)$
(b) $e_2(t) = 4tu(t - 0.5)$

SOLUTION

(a) $e_1(t)$ is a 4-V/sec ramp, $4tu(t)$, that has been delayed by 0.5 sec. Hence, it appears as shown in Figure A.7(b).

464 ADVANCED TOPICS IN LAPLACE TRANSFORMS

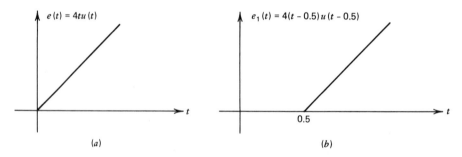

FIGURE A.7 The ramp 4t delayed by 0.5 sec (Example A.5a).

(b) $e_2(t)$ is a 4-V/sec ramp that has been multiplied by a unit step equal to zero for all $t < 0.5$ sec. Hence, it appears as shown in Figure A.8(b). Note the jump, or discontinuity, in $e_2(t)$.

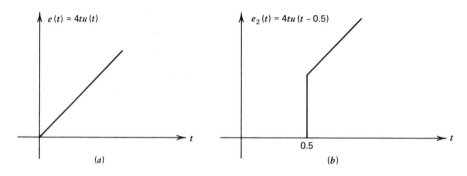

FIGURE A.8 The ramp 4t multiplied by the delayed step $u(t - 0.5)$ (Example A.5b).

EXAMPLE A.6

Write an expression for the voltage shown in Figure A.9.

FIGURE A.9 Write an expression for e(t) (Example A.6).

A.1 REPRESENTATION OF DELAYED TIME FUNCTIONS

SOLUTION

We first write an expression for $e(t)$ as if it were defined for all values of t. Referring to Figure A.10, we see that this function has slope $(4 - 3)/10^{-3} = 1000$ V/sec.

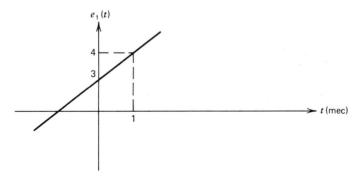

FIGURE A.10 The function in Figure A.9 when defined for all time t.

Its intercept on the vertical axis is 3; and, therefore, from the slope-intercept form of analytic geometry, its equation is

$$e_1(t) = 10^3 t + 3 \text{ V}$$

Since the desired function is zero for all $t < 0$, we have

$$e(t) = e_1(t)u(t)$$

or

$$e(t) = (10^3 t + 3)u(t)$$

We may use the technique for representing delayed-time functions to represent *pulses*. A rectangular pulse, for example, can be thought of as a positive step function added to a *negative* step function that occurs at some later time. Figure A.11 shows a 10-V pulse that begins at $t = 0$ and ends at $t = 0.1$ sec. This

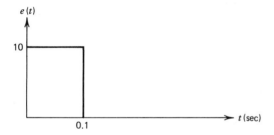

FIGURE A.11 A 10-V rectangular pulse with width 0.1 sec.

pulse may be represented by

$$e(t) = 10u(t) - 10u(t - 0.1)$$

Note that $e(t) = 0$ for $t < 0$, $e(t) = 10$ for $0 \leq t < 0.1$ and $e(t) = 10 - 10 = 0$ for $t \geq 0.1$. In effect, we have added two step functions, one of which is negative, to obtain the desired result. This is illustrated in Figure A.12.

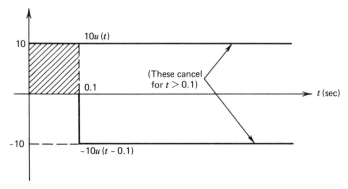

FIGURE A.12 The pulse in Figure A.11 represented as the sum of a step and a delayed negative step.

EXAMPLE A.7

Write an expression for a 15-V pulse that begins at $t = 2$ msec and is 48 msec wide.

SOLUTION

In this case, we wish to subtract a 15-V step beginning at $(48 + 2) = 50$ msec from a 15-V step that begins at $t = 2$ msec. Hence, we have

$$e(t) = 15u(t - 0.002) - 15u(t - 0.05)$$

See Figure A.13.

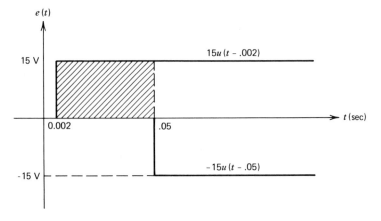

FIGURE A.13 A 15-V delayed pulse (Example A.7).

A.1 REPRESENTATION OF DELAYED TIME FUNCTIONS

When ramp functions of different slopes are added, the result is another ramp whose slope is the *sum* of the slopes of those added. This holds true even if some of the ramps have negative slopes, in which case they subtract from the net slope of the result. Figure A.14 illustrates the result of adding three ramp functions, two of which are delayed. Figure A.14 shows how we might represent a triangular pulse.

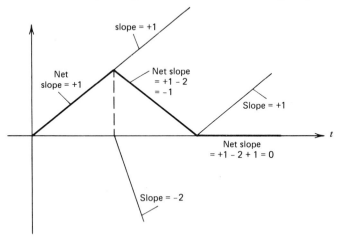

FIGURE A.14 A triangular pulse may be represented as the sum of three ramp functions.

EXAMPLE A.8

Write an expression for the triangular voltage pulse shown in Figure A.15.

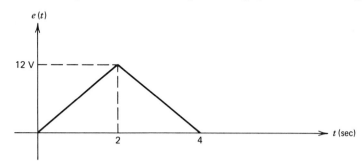

FIGURE A.15 Write an expression for the triangular voltage pulse (Example A.8).

SOLUTION

From $t = 0$ to $t = 2$ sec, $e(t)$ is a ramp with slope $+6$: $6tu(t)$. From $t = 2$ to $t = 4$ sec, the function must have *net* slope -6. Hence, we must add in a negative-going ramp with slope -12 beginning at $t = 2$: $6tu(t) - 12(t - 2)u(t - 2)$. Finally, for $t \geq 4$, the net slope must be zero, so we must add a positive ramp with slope $+6$ beginning at $t = 4$. The required function is then

$$e(t) = 6tu(t) - 12(t - 2)u(t - 2) + 6(t - 4)u(t - 4)$$

EXAMPLE A.9

Write an expression to represent the waveform shown in Figure A.16.

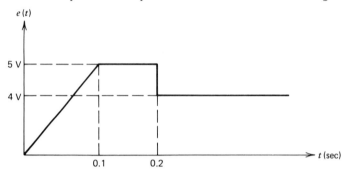

FIGURE A.16 Write an expression for the voltage waveform (Example A.9).

SOLUTION

From $t = 0$ to $t = 0.1$, $e(t) = 50tu(t)$. From $t = 0.1$ to $t = 0.2$, the net slope is zero, so a ramp with slope -50 must be added beginning at $t = 0.1$. For $t \geq 0.2$, the slope is still zero, but the net voltage is 4, indicating that a -1-V step is added beginning at $t = 0.2$. The required function is then

$$e(t) = 50tu(t) - 50(t - 0.1)u(t - 0.1) - u(t - 0.2)$$

A.2 REPRESENTATION OF NONLINEAR PERIODIC FUNCTIONS

We can use the techniques developed in the previous section to write expressions for nonlinear periodic waveforms, such as square waves, sawtooth waveforms, and so on. In essence, we will write an expression that represents the sum of an infinite number of pulses, each of the required shape (square, triangular, etc.), and each delayed in time from the preceding one by an amount equal to the period. For this purpose, we will use the conventional mathematical representation of an infinite sum.

When an infinite number of terms such as $A_1, A_2, A_3, \ldots$ are added together, we can express this summation as follows:

$$\sum_{n=1}^{\infty} A_n = A_1 + A_2 + \ldots$$

The subscript n, called the index, can be any letter or symbol and can appear with the terms in a number of different ways. Suppose, for example, we wished to write an expression for the infinite summation

$$(1)^1 + \left(\frac{1}{2}\right)^2 + \left(\frac{1}{3}\right)^3 + \ldots$$

A.2 REPRESENTATION OF NONLINEAR PERIODIC FUNCTIONS

We could write this sum as

$$\sum_{n=1}^{\infty} \left(\frac{1}{n}\right)^n$$

Substituting $n = 1$ into $(1/n)^n$ gives us the first term of our sum, $(1/1)^1$, while $n = 2$ gives $(1/2)^2$, and so on. The symbol Σ simply means that we are to add *all* the terms obtained by substituting $n = 1, n = 2, \ldots$.

EXAMPLE A.10

Write an expression for the infinite sum

$$f(t) = 2tu(t) + 4(t - 1)u(t - 1) + 6(t - 2)u(t - 2) + \ldots$$

SOLUTION

When writing an expression for an infinite sum, it is usually necessary to study the terms in order to discover the *pattern* that relates successive terms. In this example, we see that the expression represents an infinite sum of ramps, each with slope two greater than that of the preceding one, and each delayed by 1 sec more than the preceding one. We concentrate first on a method for representing the slopes. We require an expression that equals 2 when n is 1, 4 when n is 2, and so on. Clearly, the expression $2n$ will do. As for the time delays, we need an expression that equals 0 when $n = 1$, 1 when $n = 2$, and so on; $(n - 1)$ is obviously such an expression. Hence, we write

$$f(t) = \sum_{n=1}^{\infty} 2n[t - (n - 1)]u[t - (n - 1)]$$

or equivalently,

$$f(t) = \sum_{n=1}^{\infty} 2n(t - n + 1)u(t - n + 1)$$

It is sometimes convenient to allow the index to range from zero (or any other integer) to infinity rather than from one to infinity. The function is this example could equally well have been expressed by

$$f(t) = \sum_{n=0}^{\infty} (2n + 2)(t - n)u(t - n)$$

Frequently, we must alternately add and subtract terms in an infinite sum. That is, the algebraic sign joining terms must alternate between plus and minus as the index increases. This sign alternation can be expressed by multiplying each term by $(-1)^n$ or $(-1)^{n+1}$.

EXAMPLE A.11

Write an expression for the infinite sum

$$f(t) = u(t - 3) - 3u(t - 5) + 5u(t - 7) - 7u(t - 9) + \ldots$$

SOLUTION

In this example the integers we must represent are all *odd* numbers. For any integer value of n, $(2n + 1)$ or $(2n - 1)$ will represent an odd number. We note that the time delay of each step function is the odd number that is two greater than the magnitude of the step and that the signs between terms are alternating. We may, therefore, represent $f(t)$ by

$$f(t) = \sum_{n=1}^{\infty} (-1)^{n+1}(2n - 1)u[t - (2n + 1)]$$

or

$$f(t) = \sum_{n=1}^{\infty} (-1)^{n+1}(2n - 1)u(t - 2n - 1)$$

As noted earlier, we are now in a position to represent a periodic waveform as an infinite sum of pulses. As our first example, we will consider the square wave.

EXAMPLE A.12

Write an expression for a square wave that alternates between $+5$ V and 0 V with a period of 0.1 sec, as shown in Figure A.17.

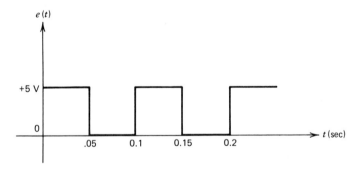

FIGURE A.17 Write an expression for the square wave (Example A.12).

SOLUTION

The first term in our expression will clearly be a 5-V step at $t = 0$. The first positive pulse will then be generated by subtracting a 5-V step beginning at $t = 0.05$ sec, that is, delayed by 0.05 sec. The next pulse is generated by adding a 5-V step beginning at $t = 0.1$ sec and then subtracting a 5-V step beginning at $t = 0.15$

A.3 DELAYED AND NONLINEAR PERIODIC WAVES

sec and so forth. Thus,

$$e(t) = 5u(t) - 5u(t - .05) + 5u(t - 0.1) - 5u(t - 0.15) + \ldots$$

or

$$e(t) = \sum_{n=0}^{\infty} (-1)^n 5u(t - 0.05n)$$

Note that $(-1)^0 = +1$, since *any* number raised to the zero power is equal to one.

EXAMPLE A.13

Sketch the function represented by

$$e(t) = 0.1tu(t) + \sum_{n=1}^{\infty} (-1)^n 0.2(t - n)u(t - n)$$

SOLUTION

The function begins as a ramp with slope 0.1. (The infinite sum contributes nothing until $t = 1$, since its first term is delayed by 1 sec.) At $t = 1$ sec, the initial ramp has reached a value of $(0.1)(1) = 0.1$ V, and at that instant a ramp with slope -0.2 is switched in. Hence, the net slope after $t = 1$ is $(0.1) - (0.2) = -0.1$; this continues until $t = 2$. At $t = 2$, $e(t) = 0.1 - (0.1)(1) = 0$. Also at $t = 2$, a ramp with slope $+0.2$ is switched in, yielding a net slope now of $+0.1$. Thus, $e(t)$ begins to rise again from zero with a slope of $+0.1$. This process is seen to continue, with the slope alternating between ± 0.1 every second. The function $e(t)$ thus has the appearance of a triangular wave with period 2 sec, as shown in Figure A.18.

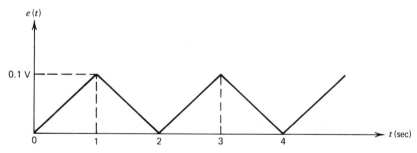

FIGURE A.18 A triangular wave with period 2 sec.

A.3 LAPLACE TRANSFORMS OF DELAYED AND NONLINEAR PERIODIC WAVES

In the two previous sections, we have developed mathematical techniques for representing time-domain functions that are delayed in time or that can be rep-

472 ADVANCED TOPICS IN LAPLACE TRANSFORMS

resented as sums of time-delayed functions. In this section, we will learn how to transform such functions into the frequency or s- domain. The fundamental theorem relating the transform of a time-domain function to its transform when it is delayed in time can be summarized as follows:

$$\text{If } \mathcal{L}[f(t)u(t)] = F(s)$$
$$\text{then } \mathcal{L}[f(t - a)u(t - a)] = e^{-as}F(s)$$

That is, we multiply the transform of a function by e^{-as} in order to obtain the transform of the function when it has been delayed by a sec. Delayed functions are often referred to as shifted functions, since they can be visualized as having been shifted to the right on the time axis, and the theorem just given is sometimes called the shifting theorem. From the shifting theorem, it is, of course, also true that

$$\mathcal{L}^{-1}e^{-as}F(s) = f(t - a)u(t - a)$$

EXAMPLE A.14

Write the LaPlace transform for the output of the delay line shown in Figure A.1.

SOLUTION

The input waveform is $5 \sin(2\pi \times 100t)$. Thus, the transform of this (undelayed) waveform from Table 4-1 is

$$E(s) = \frac{5(200\pi)}{s^2 + (200\pi)^2} = \frac{10^3 \pi}{s^2 + 4\pi^2 \times 10^4}$$

Since the output is delayed from the input by $a = 0.01$ sec, then from the shifting theorem the transform of the output is

$$\frac{10^3 \pi e^{-0.01s}}{s^2 + 4\pi^2 \times 10^4}$$

EXAMPLE A.15

Find the inverse transform of $F(s) = (e^{-s}/s) + 2e^{-2s}/(s + 4)$.

SOLUTION

Since

$$\mathcal{L}^{-1} 1/s = u(t)$$

and

$$\mathcal{L}^{-1} \frac{2}{s + 4} = 2e^{-4t}$$

we have

$$\mathcal{L}^{-1} F(s) = u(t - 1) + 2e^{-4(t-2)}u(t - 2)$$

A.3 DELAYED AND NONLINEAR PERIODIC WAVES

Since we now know how to find the transforms of delayed time functions, we can find the transforms of pulses and periodic waves, which, as we have seen, can be represented by sums of delayed functions. We use the fact that the transform of a sum is the sum of the transforms.

EXAMPLE A.16

Find the LaPlace transform of the pulse shown in Figure A.13.

SOLUTION

We saw that this pulse can be represented in the time domain by

$$e(t) = 15u(t - 0.002) - 15u(t - 0.05)$$

Hence, from the shifting theorem, its transform is

$$E(s) = \frac{15e^{-0.002s}}{s} - \frac{15e^{-0.05s}}{s}$$

EXAMPLE A.17

Find the LaPlace transform of the infinite series of impulses whose weights are alternately ± 5, as shown in Figure A.19.

SOLUTION

The series of impulses can be written in the time domain as

$$f(t) = 5\delta(t) - 5\delta(t - 10^{-3}) + 5\delta(t - 2 \times 10^{-3}) - \ldots$$

$$= \sum_{n=0}^{\infty} (-1)^n 5\delta(t - n \times 10^{-3})$$

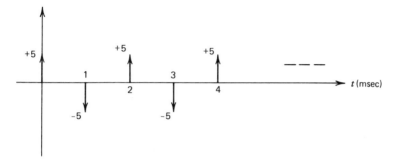

FIGURE A.19 Find the LaPlace transform of the series of impulses (Example A.17)

474 ADVANCED TOPICS IN LAPLACE TRANSFORMS

The LaPlace transform is then expressed by

$$F(s) = 5 - 5e^{-10^{-3}s} + 5e^{-2\times 10^{-3}s} - \ldots$$

$$= \sum_{n=0}^{\infty} (-1)^n 5e^{-n\times 10^{-3}s}$$

Figure A.19 is the waveform that we would obtain at the output of a differentiator when the input is a square wave that alternates between 0 and +5 V with period 2 msec. We now have the means of transforming such a wave and, hence, of analyzing circuits and systems that contain such signals by using LaPlace transform methods.

EXAMPLE A.18

Determine the transform of the output of an integrator whose input is

$$e_{in} = 6u(t) + \sum_{n=1}^{\infty} (-1)^n 12u(t - 2n)$$

SOLUTION

The input waveform is a square wave that alternates between ± 6 V with period 4 sec. The transform of e_{in} may be found by applying the shifting theorem

$$E_{in}(s) = \frac{6}{s} + \sum_{n=1}^{\infty} (-1)^n \frac{12e^{-2ns}}{s}$$

We know that integration corresponds to division by s, so the transform of the integrator's output may be written

$$E_o(s) = \frac{E_{in}(s)}{s} = \frac{6}{s^2} + \sum_{n=1}^{\infty} (-1)^n \frac{12}{s^2} e^{-2ns}$$

As a point of interest, let us find the inverse transform of the output. Since $\mathcal{L}^{-1} 1/s^2 = t$, we find

$$\mathcal{L}^{-1}E_o(s) = \mathcal{L}^{-1}\left[\frac{6}{s^2} + \sum_{n=1}^{\infty} (-1)^n \frac{12}{s^2} e^{-2ns}\right]$$

$$= 6tu(t) + \sum_{n=1}^{\infty} (-1)^n 12(t - 2n)u(t - 2n)$$

Thus, $e_o(t)$ is a summation of ramps that has net slope $+6$ from $t = 0$ to $t = 2$, -6 from $t = 2$ to $t = 4$, $+6$ from $t = 4$ to $t = 6$, and so on. In other words, $e_o(t)$ is a triangular wave with period 4 sec, as we might expect from integrating a square wave. The input and output are sketched in Figure A.20.

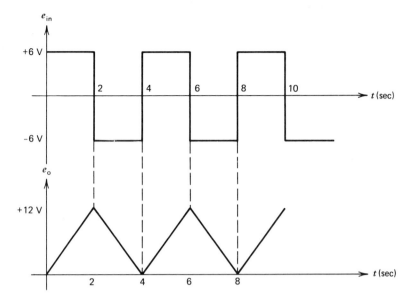

FIGURE A.20 Square-wave input and triangular output of an integrator (Example A.18).

A.4 LAPLACE TRANSFORM ANALYSIS OF CIRCUITS WITH DELAYED AND NONLINEAR PERIODIC WAVES

Having developed the mathematical techniques necessary to represent delayed and periodic waves in both the time and frequency domains, we are now in a position to apply these techniques to solving circuits. The approach we use is precisely the one we used in Chapter 5, namely, transform the circuit parameters and applied signals, solve for the unknown(s), and find the inverse transform(s). As a first example, we will study an RC-network driven by a voltage pulse.

EXAMPLE A.19

The RC-network shown in Figure A.21 is driven by a 5-V pulse that begins at $t = 0$ and is 100 msec wide. Find an expression for the voltage across the resistor as a function of time.

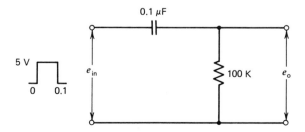

FIGURE A.21 Find e_o versus time using LaPlace transforms (Example A.19).

SOLUTION

From the voltage-divider rule, we know

$$E_o(s) = \left[\frac{10^5}{10^5 + 1/(10^{-7}s)}\right]E_{in}(s) = \left(\frac{s}{s+100}\right)E_{in}(s)$$

Now

$$e_{in}(t) = 5u(t) - 5u(t-0.1)$$

so

$$E_{in}(s) = \frac{5}{s} - \frac{5e^{-0.1s}}{s}$$

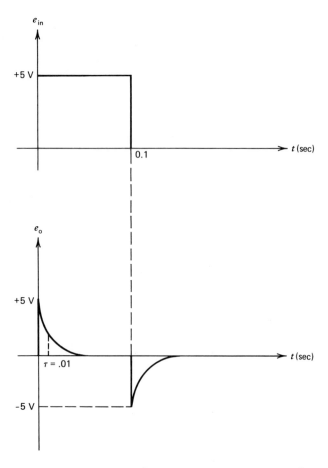

FIGURE A.22 Input and output in Figure A.21 (Example A.19).

A.4 CIRCUITS WITH DELAYED AND NONLINEAR PERIODIC WAVES

Therefore,

$$E_o(s) = \left[\frac{s}{s+100}\right]\left[\frac{5}{s} - \frac{5e^{-0.1s}}{s}\right]$$

$$= \frac{5}{s+100} - \left[\frac{5}{s+100}\right]e^{-0.1s}$$

Notice that we have separated that part of the transform that contains the unshifted term $(5/s + 100)$ from the part that contains the shifted term $(5/s + 100)e^{-0.1s}$. We can thus find the inverse transforms of these two parts separately and combine the results

$$\mathcal{L}^{-1}[E_o(s)] = 5e^{-100t}u(t) - 5e^{-100(t-0.1)}u(t-0.1)$$

The input and output waves are sketched in Figure A.22. These results are not unexpected, since we recognize the RC-network as a circuit that approximates a differentiator when driven by a pulse whose width is long compared to the RC time constant τ. The point is that we now have the ability to obtain an exact mathematical expression for the output voltage regardless of the time constant or pulse width.

EXAMPLE A.20

Find the voltage across the capacitor in the network in Figure A.23 when the input is a square wave that alternates between ± 10 V with a period of 50 μ sec.

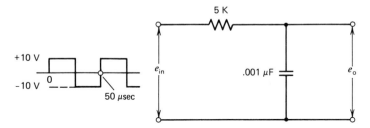

FIGURE A.23 Find e_o versus time using LaPlace transforms (Example A.20).

SOLUTION

By the voltage-divider rule,

$$E_o(s) = \left[\frac{1/(10^{-9}s)}{5 \times 10^3 + 1/(10^{-9}s)}\right]E_{in}(s) = \left(\frac{2 \times 10^5}{s + 2 \times 10^5}\right)E_{in}(s)$$

Now

$$e_{in}(t) = 10u(t) - 20u(t - 25 \times 10^{-6}) + 20u(t - 50 \times 10^{-6}) - \cdots$$

$$= 10u(t) + \sum_{n=1}^{\infty}(-1)^n 20u(t - 25n \times 10^{-6})$$

478 ADVANCED TOPICS IN LAPLACE TRANSFORMS

Therefore,

$$E_{in}(s) = \left(\frac{10}{s}\right) - \left(\frac{20}{s}\right)e^{-25 \times 10^{-6}s} + \left(\frac{20}{s}\right)e^{-50 \times 10^{-6}s} - \cdots$$

$$= \left(\frac{10}{s}\right) + \sum_{n=1}^{\infty} -1^n(20/s)e^{-25n \times 10^{-6}s}$$

$$E_o(s) = \left(\frac{2 \times 10^5}{s + 2 \times 10^5}\right)\left[\frac{10}{s} + \sum_{n=1}^{\infty} (-1)^n \frac{20}{s} e^{-25n \times 10^{-6}s}\right]$$

$$= \left(\frac{2 \times 10^5}{s + 2 \times 10^5}\right)\frac{10}{s} - \left(\frac{2 \times 10^5}{s + 2 \times 10^5}\right)\frac{20}{s} e^{-25 \times 10^{-6}s} + \cdots$$

Thus, using pair 7 from Table 4-1,

$$e_o(t) = 10(1 - e^{-2 \times 10^5 t})u(t) - 20[1 - e^{-2 \times 10^5(t - 25 \times 10^{-6})}] \cdot u(t - 25 \times 10^{-6}) + \cdots$$

$$= 10(1 - e^{-2 \times 10^5 t})u(t) + \sum_{n=1}^{\infty} (-1)^n 20 \cdot [1 - e^{-2 \times 10^5(t - 25n \times 10^{-6})}]u(t - 25n \times 10^{-6})$$

The input and output functions are sketched in Figure A.24.

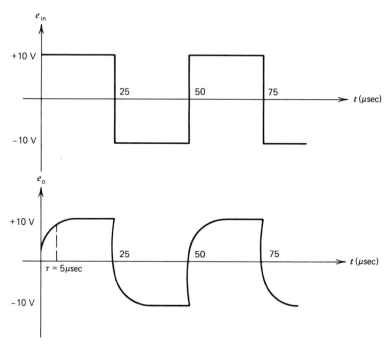

FIGURE A.24 Input and output waveform of Figure A.23 (Example A.20).

If there is an ideal differentiator (with transfer function of the form Ks) in a system that is subjected to waveforms containing discontinuities, or "jumps," then

A.4 CIRCUITS WITH DELAYED AND NONLINEAR PERIODIC WAVES

we must be careful to include the differentiation of these jumps in our analysis. Recall that differentiation of a step function produces an impulse function with weight, or "area," equal to the size of the step

$$\frac{dAu(t)}{dt} = A\delta(t)$$

In a similar way, differentiating *any* waveform that contains a jump produces an impulse with weight equal to the size of the jump. This is illustrated in Example A.21.

EXAMPLE A.21

Find the output of a differentiator whose input is the voltage shown in Figure A.25.

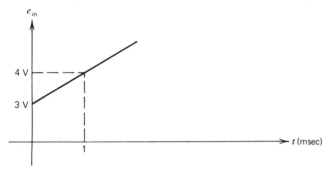

FIGURE A.25 Find the output of a differentiator with input e_{in} (Example A.21).

SOLUTION

The voltage e_{in} in Figure A.25 is a ramp with slope $(4 - 3)/10^{-3} = 1000$ V/sec. It has a 3-V jump at $t = 0$. Since the derivative of a ramp with slope 1000 is the constant 1000 and since the derivative of a 3-V jump is $3\delta(t)$, we have

$$e_o(t) = \frac{de_{in}(t)}{dt} = [3\delta(t) + 10003]u(t)$$

This is sketched in Figure A.26.

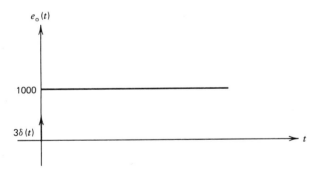

FIGURE A.26 Output of a differentiator with input shown in Figure A.25 (Example A.21).

480 ADVANCED TOPICS IN LAPLACE TRANSFORMS

EXAMPLE A.22

Use LaPlace transforms to find the output of a differentiator with transfer function $100s$ when the input is the pulse shown in Figure A.27.

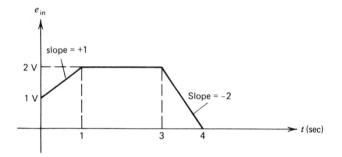

FIGURE A.27 Find the output of a differentiator with input e_{in} (Example A.22):

SOLUTION

$$e_{in}(t) = (t + 1)u(t) - (t - 1)u(t - 1) - 2(t - 3)u(t - 3)$$
$$+ 2(t - 4)u(t - 4)$$
$$= tu(t) + u(t) - (t - 1)u(t - 1) - 2(t - 3)u(t - 3)$$
$$+ 2(t - 4)u(t - 4)$$

$$E_{in}(s) = \frac{1}{s^2} + \frac{1}{s} - \frac{1}{s^2}e^{-s} - \frac{2}{s^2}e^{-3s} + \frac{2}{s^2}e^{-4s}$$

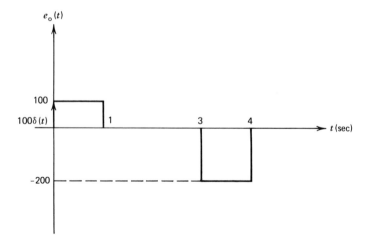

FIGURE A.28 Output of a differentiator with input shown in Figure A.27 (Example A.22).

A.4 CIRCUITS WITH DELAYED AND NONLINEAR PERIODIC WAVES

Since $E_o(s) = (100s)E_{in}(s)$, we have

$$E_o(s) = \frac{100}{s} + 100 - \frac{100}{s}e^{-s} - \frac{200}{s}e^{-3s} + \frac{200}{s}e^{-4s}$$

Then

$$e_o(t) = 100u(t) + 100\delta(t) - 100u(t-1) - 200u(t-3) + 200u(t-4)$$

This output is sketched in Figure A.28.

Note that we could have obtained Figure A.28 directly by observing that the derivative of the input must consist of an impulse due to the jump at $t = 0$ and constant portions due to the constant slopes (all multiplied by the gain 100 of the differentiator). Example A.22 shows how we may obtain the mathematical expression corresponding to this result. If a waveform contains jumps at points of time other than $t = 0$ and if the waveform is differentiated, then of course the resulting impulse functions occur at the same points in time as the jumps.

All of the circuit-analysis theorems discussed in Chapter 5 can be used to solve circuits involving nonlinear or delayed waveforms. In Example A.23, we will see how Thevenin's theorem can be applied to a circuit in which the voltage source is a sawtooth waveform.

EXAMPLE A.23

Find the transformed Thevenin equivalent circuit with respect to the terminals a–b in Figure A.29 if the voltage source is the sawtooth waveform shown.

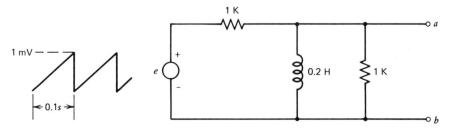

FIGURE A.29 Find a transformed Thevenin equivalent circuit at terminals a–b (Example A.23).

SOLUTION

The Thevenin equivalent impedance is found in the usual way after replacing the voltage source with a short circuit

$$Z_{TH}(s) = 1\text{ K} \| (1\text{ K}) \| (0.2s)$$

$$= \frac{(0.2s)(500)}{0.2s + 500}$$

$$= \frac{500s}{s + 2.5 \times 10^3}$$

The Thevenin equivalent voltage is the voltage that appears across the terminals a–b, which is the same as the voltage across Z_{EQ} in Figure A.30.

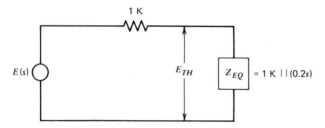

FIGURE A.30   The Thevenin equivalent voltage appears across z_{EQ} (Example A.23).

$$Z_{EQ} = \frac{200s}{0.2s + 1\text{ K}} = \frac{10^3 s}{s + 5 \times 10^3}$$

Therefore,

$$E_{TH} = \frac{Z_{EQ}}{1\text{ K} + Z_{EQ}} E(s) = \frac{10^3 s/(s + 5 \times 10^3)}{10^3 + 10^3 s/(s + 5 \times 10^3)} E(s)$$

$$= \frac{10^3 s}{2 \times 10^3 s + 5 \times 10^6} E(s) = \frac{0.5s}{s + 2.5 \times 10^3} E(s)$$

Now the applied voltage e may be regarded as a ramp with slope $10^{-3}/10^{-1} = .01$ V/sec from which an infinite sum of delayed steps is subtracted, as shown in Figure A.31. Thus, we may write

$$e(t) = .01 t u(t) - \sum_{n=1}^{\infty} 10^{-3} u(t - 0.1n)$$

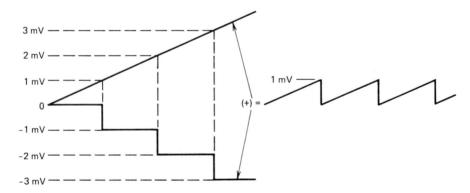

FIGURE A.31   Representing the sawtooth in Figure A.29 as the difference of a ramp and a sum of delayed steps (Example A.23).

A.4 CIRCUITS WITH DELAYED AND NONLINEAR PERIODIC WAVES

Therefore,

$$E(s) = \frac{.01}{s^2} - \sum_{n=1}^{\infty} 10^{-3} \frac{e^{-0.1ns}}{s}$$

Consequently,

$$E_{TH}(s) = \left(\frac{0.5s}{s + 2.5 \times 10^3}\right)\left(\frac{.01}{s^2} - \sum_{n=1}^{\infty} 10^{-3} \frac{e^{-0.1ns}}{s}\right)$$

$$= \frac{5 \times 10^{-3}}{s(s + 2.5 \times 10^3)} - \frac{0.5 \times 10^{-3}}{s + 2.5 \times 10^3} \sum_{n=1}^{\infty} e^{-0.1ns}$$

As a final example, we will apply the method of superposition to find the voltage in a circuit to which two pulses delayed in time with respect to each other are applied.

EXAMPLE A.24

In the circuit shown in Figure A.32, e_1 is a 5-V pulse beginning at $t = 0$ and 10 msec wide, while e_2 is a -10-V pulse beginning at $t = 5$ msec and 10 msec wide. Find the voltage across the capacitor as a function of time.

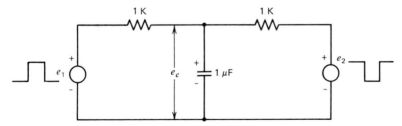

FIGURE A.32 Find the voltage across the capacitor using superposition and LaPlace transforms (Example A.24).

SOLUTION

We first find the voltage across the capacitor Ec_1 due to e_1 with e_2 shorted out.

$$Ec_1(s) = \left(\frac{Z_{EQ}}{1\text{ K} + Z_{EQ}}\right)E_1(s)$$

where

$$Z_{EQ} = \frac{(1\text{ K}) 1/10^{-6}s}{1\text{ K} + (1/10^{-6}s)} = \frac{10^6}{s + 10^3}$$

Thus,

$$Ec_1(s) = \left[\frac{10^6/(s + 10^3)}{10^3 + 10^6/(s + 10^3)}\right]E_1(s) = \frac{10^3}{s + 2 \times 10^3}E_1(s)$$

Now
$$e_1(t) = 5u(t) - 5u(t - .01)$$

So
$$E_1(s) = (5/s) - (5/s)e^{-.01s}$$

and therefore,
$$Ec_1(s) = \frac{10^3}{s + 2 \times 10^3}\left[\frac{5}{s} - \frac{5}{s}e^{-.01s}\right]$$

$$= \frac{5 \times 10^3}{s(s + 2 \times 10^3)} - \frac{5 \times 10^3 e^{-.01s}}{s(s + 2 \times 10^3)}$$

Inverting, we find
$$e_{c_1}(t) = \frac{(5 \times 10^3)}{(2 \times 10^3)}[1 - e^{-2 \times 10^3 t}]u(t) - \frac{5 \times 10^3}{2 \times 10^3}[1 - e^{-2 \times 10^3(t-.01)}]u(t - .01)$$

In a similar way, by shorting out e_1 and using the fact that
$$e_2(t) = -10u(t - .005) + 10u(t - .015)$$

and
$$E_2(s) = \frac{-10}{s}e^{-.005s} + \frac{10}{s}e^{-.015s}$$

we find
$$E_{c_2}(s) = \frac{-10 \times 10^3 e^{-.005s}}{s(s + 2 \times 10^3)} + \frac{10 \times 10^3 e^{-.015s}}{s(s + 2 \times 10^3)}$$

Then
$$e_{c_2}(t) = \frac{-10 \times 10^3}{2 \times 10^3}e^{-2 \times 10^3(t-.005)}u(t - .005)$$

$$+ \frac{10 \times 10^3}{2 \times 10^3}e^{-2 \times 10^3(t-.015)}u(t - .015)$$

Finally,
$$e_c(t) = e_{c_1}(t) + e_{c_2}(t)$$
$$= 2.5(1 - e^{-2 \times 10^3 t})u(t) - 2.5[1 - e^{-2 \times 10^3(t-.01)}]u(t - .01)$$
$$- 5[1 - e^{-2 \times 10^3(t-.005)}]u(t - .005) + 5[1 - e^{-2 \times 10^3(t-.015)}]u(t - .015)$$

The voltage across the capacitor as a function of time thus resembles that shown in Figure A.33.

A.4 CIRCUITS WITH DELAYED AND NONLINEAR PERIODIC WAVES 485

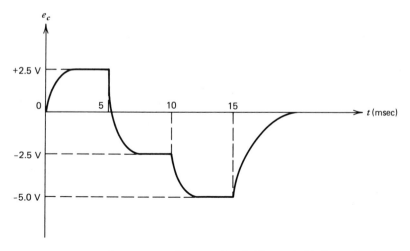

FIGURE A.33 Voltage across the capacitor in Figure A.32 (Example A.24).

APPENDIX B
SPECIFICATIONS AND DATA SHEETS FOR PRODUCTS CITED IN THE EXAMPLES

Vernitech: Model 112, Linear Motion/Infinite Resolution Potentiometer

Courtesy of Vernitech Corporation.

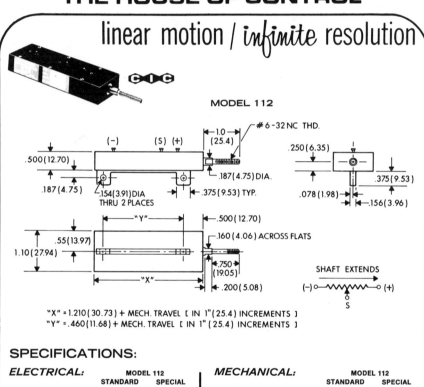

488 SPECIFICATIONS AND DATA SHEETS

CIC/V (Vernitech): Potentiometer Type/Infinite Resolution Pressure Transducers, Models 1000

Courtesy of Vernitech Corporation.

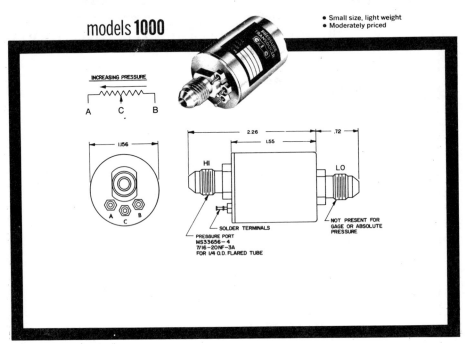

- Small size, light weight
- Moderately priced

ELECTRICAL SPECIFICATIONS:

Pressure Range	Any range to suit from 0-10 to 0-500 PSI Gage, Absolute, or Differential
Terminal Resistance	50Ω to 10KΩ ±10%
Independent Linearity	To ±0.5% F.S.
Repeatability	±0.1% F.S.
Power Rating	0.20 Watts
Life	To 10 million cycles, depending upon application
Temperature Sensitivity	±0.02%/°C
Response Time	20 milliseconds to respond to 63% of step pressure input

MECHANICAL SPECIFICATIONS:

Overpressure	20%	Higher Available
Case Pressure	200 PSIA	
Weight	2 ounces	
ENVIRONMENTAL: (Typical)	MODEL 1000	MODEL 1100
Temperature Range	−55°C to +85°C	
Acceleration Error	±0.05%/G to 50G	±0.05%/G to 20G
Pressure Media	Absolute and Differential: air or oil, non-corrosive.	

Gage: any media compatible with Ni-Span-C & Aluminum

TYPICAL STATIC AND DYNAMIC ERROR BANDS (Based on ±0.5% Linearity)

Pressure Range (PSI)	0-10 to 0-25	0-30 to 0-100	0-125 to 0-500
Static Error Band	±1.8%	±1.3%	±1.0%
Dynamic Error Band	±0.8%	±0.7%	±0.7%

SPECIFICATIONS AND DATA SHEETS 489

Vernitech: Model 15010, Multi-turn/Infinite Resolution Potentiometer

Courtesy of Vernitech Corporation.

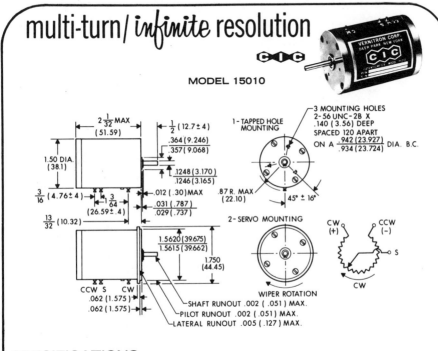

SPECIFICATIONS:

ELECTRICAL:	STANDARD	SPECIAL
Resistance Range ±20%	10KΩ	5K-100KΩ
Independent Linearity or Conformity	±1.0%	To ±0.1%
Electrical Function Angle ±10°	3600°	—
Electrical Contact Angle	Same as Mechanical Rotation	
Power Dissipation (at 25°C.)	2 Watts	—
Operating Temperature Range	−55°C to +125°C	—
Dielectric Strength	750V RMS	—

MECHANICAL:	STANDARD	SPECIAL
Mechanical Rotation +110° −0°	3600°	—
Starting Torque in-Oz. (Max.)	1.0	—
Max. Weight in Oz.	3.25	—
Stop Strength (in-Oz.)	75	—
Moment of Inertia	0.6 gcm^2	—

Note: Supplied with thermoset plastic cover only.

*Taps are available in 2 types. See page 42.
**Of Peak-to-Peak excitation.
For Optional Electrical Characteristics, see page 42.

For Optional Mechanical Configurations, see page 43.

490 SPECIFICATIONS AND DATA SHEETS

Vernitech: Precision Gearhead, Size 8
Speed Reducer, Size 8

Courtesy of Vernitech Corporation.

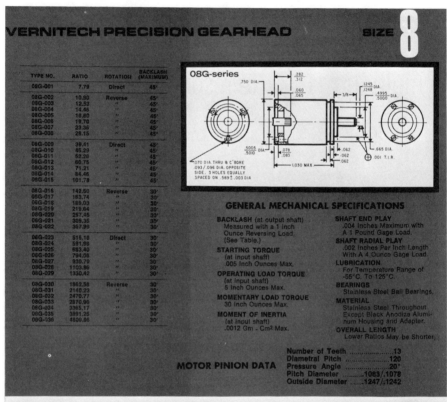

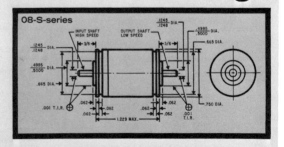

SPECIFICATIONS AND DATA SHEETS 491

Electro-Craft: E-586 BVC Control
               E-586 MG

Courtesy of Electro-Craft Corporation.

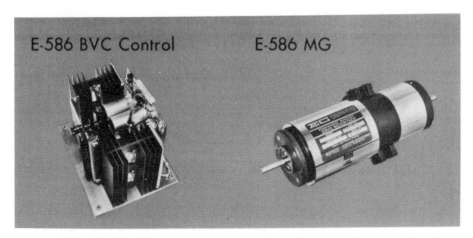

E586-BI DIRECTIONAL VELOCITY OR
POSITION CONTROL

STANDARD FEATURES:

100:1 speed range 50 to 5000 rpm
Closed-loop completely transistorized velocity feedback amplifier
Adjustable torque limit
Single turn-center tap-speed control potentiometer
Position sensor potentiomer (bi-directional position system only)
115/230 VAC, 50/60 Hz fuse and circuit breaker protection

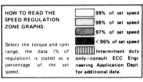

SPEED REGULATION ZONES
E-586 BVC CONTROL

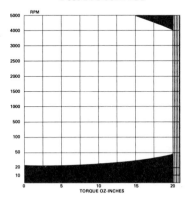

492 SPECIFICATIONS AND DATA SHEETS

Electro-Craft: E586-BPC Bi-Directional Position Control System

Courtesy of Electro-Craft Corporation.

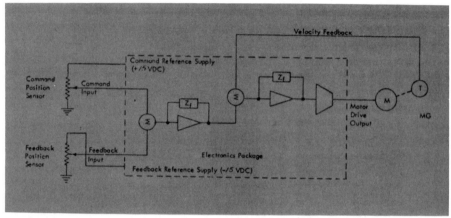

E586-BPC BI-DIRECTIONAL POSITION CONTROL SYSTEM

Brief Description:

The E586-BPC system is a position control system employing a high-gain, closed-loop electronic control. The input command can be any variable resistance or ungrounded reference DC voltage. The motor angular position of a potentiometer or any other error voltage source. Velocity feedback is employed to control the travel rate while the motor is "homing in" on the desired position.

System Specifications:

The *position accuracy* of the system is limited by two variables: (1) the drift characteristics of the reference power supplies, and (2) the accuracy of the command and feedback signals. Using the potentiometer supplied with the standard E586-BPC, system accuracy is approximately 1% per revolution or $3.6°$. Using improved power supply regulation and sensors, the accuracy can easily be improved to .1% per revolution or $.36°$. For high accuracy systems, please consult an Electro-Craft representative.

The *system travel speed* is controlled by velocity feedback from an integral tachometer and closed-loop servo. The standard E-586-BPC controls the motor to run up to 5,000 rpm during travel to the "target" position.

Command input can be accomplished in either of two ways: (1) a position setting on a potentiometer, using the E586-BPC reference power supply voltage, or (2) an ungrounded command voltage of 0 to +15 volts DC. Care must be exercised to insure that the input command range is not greater than the feedback sensor output range.

The *feedback input* requirements are similar to the command requirements and can be either a position on a potentiometer, using the feedback reference voltage source, or an ungrounded 0 to −15 volts DC signals.

Electro-Craft: M-100/M-110 Tachometer

Courtesy of Electro-Craft Corporation.

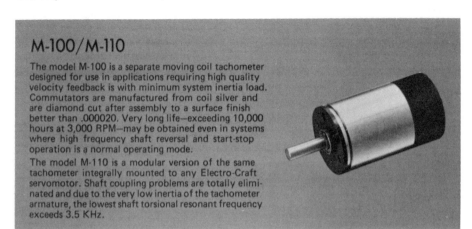

M-100/M-110

The model M-100 is a separate moving coil tachometer designed for use in applications requiring high quality velocity feedback is with minimum system inertia load. Commutators are manufactured from coil silver and are diamond cut after assembly to a surface finish better than .000020. Very long life—exceeding 10,000 hours at 3,000 RPM—may be obtained even in systems where high frequency shaft reversal and start-stop operation is a normal operating mode.

The model M-110 is a modular version of the same tachometer integrally mounted to any Electro-Craft servomotor. Shaft coupling problems are totally eliminated and due to the very low inertia of the tachometer armature, the lowest shaft torsional resonant frequency exceeds 3.5 KHz.

HOW TO ORDER

To order, you must supply sales model number. For complete tachometer order

M-100 — _____ — S
        tach winding   for special requirements
        (C winding   add S and specify
        supplied unless
        otherwise
        specified)

For modular tachometer order

_____ — _____ /110— _____ — S
motor model winding model   tach winding   for special

The M-110 tachometer can be supplied with any Electro-Craft motor.

Typical oscilloscope picture of M-100/M-110 tachometer output showing AC ripple component

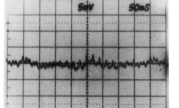

STANDARD RIPPLE FILTER SCHEMATIC

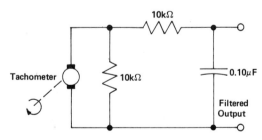

494 SPECIFICATIONS AND DATA SHEETS

Electro-Craft: Performance Specifications
Winding Variations

Courtesy of Electro-Craft Corporation.

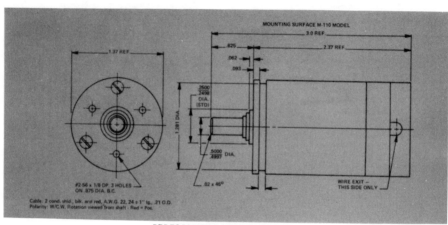

PERFORMANCE SPECIFICATIONS

Parameter	Value	Units
Linearity	.2	% Max. Deviation from Perfect Linearity
Ripple		
M-100	1.5	Max Percentage Peak to Peak AC
M-110	2	Component in DC output at 167 to 6,000 RPM
Ripple Frequency	19	Cycles/Revolution
Speed Range	1-6,000	RPM
Temperature Coefficent	−.01	%/°C
Armature Inertia		
M-100	9×10^{-5}	oz. in. sec^2
M-110	7×10^{-5}	oz. in. sec^2
Insulation Resistance	10	Megohm
Friction Torque	.25	oz. in., max.
Torsional Resonant Frequency When Coupled to an Electro-Craft MCM Servomotor	3.5	KHz, min.
Rated Operating Life	10,000	Hours at 3,000 RPM
Operating Temperature Range	0-155	°C

WINDING VARIATIONS

Winding Constants	Winding		
	A	B	C
Output Voltage Gradient (Volts/KRPM)	3.0	2.5	1
Armature Resistance (ohms)	150	150	150
Armature Inducance (mhy)	4	4	4
Recommended Minimum Load Impedance (ohms)	10,000	10,000	10,000

SPECIFICATIONS AND DATA SHEETS

Electro-Craft: E-508/E-512 Series, FHP D.C. permanent magnet motor

Courtesy of Electro-Craft Corporation.

E-508/E-512

The E-508/E-512 series is a moderate cost of FHP D.C. permanent magnet motors designed for a broad range of industrial applications, such as tape transport reel drives or position control systems. Reliability and long life have been designed into these motors by the use of corrosion-resistant shaft and housing materials and permanently lubricated ball bearings. Pictured at the right is the E-508 and the longer E-512.

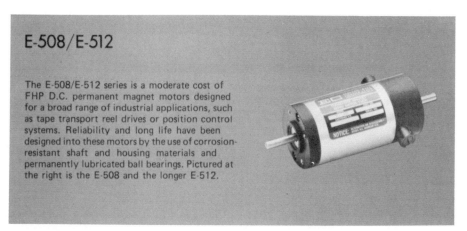

HOW TO ORDER

To order, you must supply sales model number. Order codes are shown below in color

E – _____ – _____ – S
   motor model   motor winding   for special

	Winding	Torque Constant (K_T) (oz.in./amp ± 10%)	Voltage Constant (K_E) (Volts/KRPM) ± 10%	Winding Resistance (Ohms) at 25°C ± 15% (+0.40 ohms for terminal resistance)	Armature Inductance (Millihenries)	Max. Pulse Current (Amps) (to avoid demagnetization)
508	A	2.60	1.92	.85	1.13	24
	B	3.28	2.42	1.34	1.57	19
	C	4.13	3.05	2.12	2.28	15
	D	5.20	3.84	3.36	3.40	12
510	A	5.88	4.35	1.24	3.39	24
	B	7.41	5.48	1.99	4.92	19
	C	9.34	6.91	3.15	7.31	15
	D	11.76	8.70	4.99	11.10	12
512	A	8.68	6.42	1.64	5.65	24
	B	10.94	8.09	2.64	8.41	19
	C	13.78	10.19	4.18	12.68	15
	D	17.36	12.84	6.63	19.47	12

MODEL 508

Peak torque (oz. in.) 62
Max. rated continuous torque at stall (oz. in.) 10
Armature inertia (oz. in. sec.^{2}) .0015
Damping factor (oz. in./KRPM) .1
Thermal resistance (°C./watt) 6.84
  (Armature-to-ambient)
Electrical time constant (msec) .90
Mechanical time constant (msec) 29
Motor weight 1 lb. 2 oz.
Static friction torque 3 oz. in.

MODEL 510

Peak torque 9(oz. in.) 140
Max. rated continuous torque at stall (oz. in.) 19
Armature inertia (oz. in. sec.^{2}) .0038
Damping factor (oz. in./KRPM) .1
Thermal resistance (°C./watt) 5.0
  (Armature-to-ambient)
Electrical time constant (msec) 2.06
Mechanical time constant (msec) 24.7
Motor weight 1 lb. 8 oz.
Static friction torque 3 oz. in.

MODEL 512

Peak torque (oz. in.) 208
Max. rated continuous torque at stall (oz. in.) 25
Armature inertia (oz. in. sec^2) .0062
Damping factor (oz. in./KRPM) .2
Thermal resistance (°C./watt) 4.18
  (Armature-to-ambient)
Electrical time constant (msec) 2.77
Mechanical time constant (msec) 24
Motor weight 2 lb. 8 oz.
Static friction torque 3 oz. in.

496 SPECIFICATIONS AND DATA SHEETS

Electro-Craft: Series Motor Ratings, Construction Features, Speed Torque Curves

Courtesy of Electro-Craft Corporation.

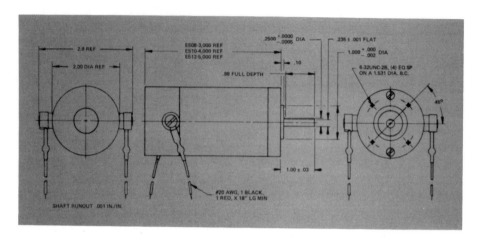

SERIES MOTOR RATINGS

Maximum rated armature temperature 155°C
Maximum no load speed 6 KRPM
Rated brush life 5,000 hours @ 1 KRPM
Maximum friction torque 3 oz. in.
Maximum shaft radial load 10 lbs.,
    0.5 in. from front bearing continuous
    for a minimum of 1,000 hours

CONSTRUCTION FEATURES

Winding insulation Class F
ABEC Class 1 ball bearings
PVC lead insulation
Totally enclosed housing
Available options:
Integrally mounted tachometer (see pp. 54-59)
Integrally mounted optical encoder (see pp. 38-41)

SPEED TORQUE CURVES

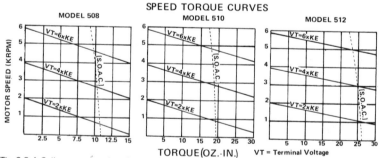

The S.O.A.C. line represents the safe operating area, continuous with unit mounted on 10"x10"x¼" heat sink.

SPECIFICATIONS AND DATA SHEETS

Electro-Craft: E-530/E-532 Series, Motor-Tach package

Courtesy of Electro-Craft Corporation.

E-530/E-532

The E-530/E-532 series incorporates the inherent low ripple, low inertia, and long life characteristics of a moving coil tachometer in an economical motor-tach package.

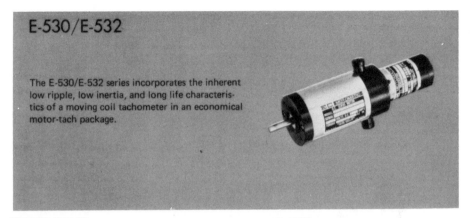

HOW TO ORDER

To order you must supply sales model number. Order codes are shown below in color.

```
E –_____–_____–_____–S
    motor-tach type  motor winding   tach winding   for Special
```

	Motor Winding	Torque Constant (K_T) (oz. in./amp ± 10%)	Voltage Constant (K_E) (Volts/KRPM) ± 10%	Winding Resistance (Ohms) at 25°C ± 15% (+0.40 ohms for terminal resistance)	Armature Inductance (Millihenries)	Max. Pulse Current (Amps) (to avoid demagnetization)
530	A	10.02	7.41	1.24	3.39	24
	B	12.63	9.34	1.99	4.92	19
	C	15.91	11.76	3.15	7.31	15
	D	20.04	14.82	4.99	11.10	12
531	A	11.77	8.70	1.44	4.53	24
	B	14.83	10.96	2.31	6.67	19
	C	18.69	13.81	3.66	9.99	15
	D	23.54	17.40	5.81	15.28	12
532	A	14.81	10.95	1.64	5.65	24
	B	18.66	13.80	2.64	8.41	19
	C	23.51	17.38	4.18	12.68	15
	D	29.63	21.90	6.63	19.47	12

MODEL E-530
Peak torque (oz. in.) 240
Max. rated continuous torque at stall (oz. in.) 29
Armature intertia (including tach) (oz. in. sec^2) .0039
Damping factor (oz. in./KRPM) .1
Thermal resistance (°C/watt) 5.0
  (Armature-to-ambient)
Electrical time constant (msec) 2.06
Mechanical time constant (msec) 8.8
Motor weight 2 lb. 1 oz.
Static Friction Torque 3 oz. in.

MODEL E-531
Peak torque (oz. in.) 282
Max. rated continuous torque at stall (oz. in.) 38
Armature inertia (including tach) (oz. in. sec^2) .0051
Damping factor (oz. in./KRPM) .15
Thermal resistance (°C/watt) 4.60
  (Armature-to-ambient)
Electrical time constant (msec) 2.46
Mechanical time constant (msec) 9.4
Motor weight 2 lb. 8 oz.
Static Friction Torque 3 oz. in.

MODEL E-532
Peak torque (oz. in.) 355
Max. rated continuous torque at stall (oz. in.) 50
Armature inertia (including tach) (oz. in. sec^2) .0063
Damping factor (oz. in./KRPM) .2
Thermal resistance (°C/watt) 4.18
  (Armature-to-ambient)
Electrical time constant (msec) 2.77
Motor weight 3 lb. 0 oz.
Static Friction Torque 3 oz. in.

498 SPECIFICATIONS AND DATA SHEETS

Electro-Craft: Series Motor Ratings, Construction Features, Tachometer Winding Variations, Speed Torque Curves

Courtesy of Electro-Craft Corporation.

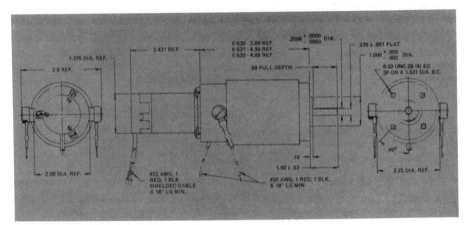

SERIES MOTOR RATINGS
Maximum rated armature temperature 155°C
Maximum no load speed 6 KRPM
Rated brush life 5,000 hours @ 1 KRPM
Maximum friction torque 3 oz. in.
Maximum shaft radial load 10 lbs.
  0.5 in. from front bearing continuous for a minimum of 1,000 hours

CONSTRUCTION FEATURES
Winding insulation Class F
ABEC Class 1 ball bearings
PVC lead insulation
Totally enclosed housing
Available options:
Integrally mounted tachometer (see pp. 54-59)
Integrally mounted optical encoder (see pp. 38-41)

TACHOMETER WINDING VARIATIONS

Winding Constants	A	B	C	D
Output Voltage Gradient (Volts/KRPM)	3.0	7.0	14.2	21
Armature Resistance (ohms) (includes temperature compensation network)	570	570	720	950
Armature Inductance (mhy)	6.2	33	138	255
Optimum Load Impedance (ohms)	5,000	5,000	5,000	10,000

(See pp. 54-61 for more detailed tachometer information)

SPEED TORQUE CURVES

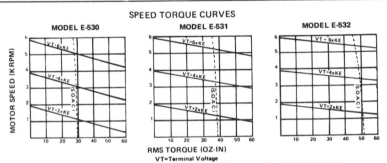

RMS TORQUE (OZ-IN)
VT=Terminal Voltage

The S.O.A.C. line represents the safe operating area, continuous with unit mounted on 10"x10"x¼" heat sink.

Analog Devices, Inc.:   Low Cost, Laser Trimmed, Precision IC Op Amp (AD517)

Courtesy of Analog Devices, Inc.

Low Cost, Laser Trimmed, Precision IC Op Amp
AD517

FEATURES
Low Input Bias Current: 1nA max (AD517L)
Low Input Offset Current: 0.25nA max (AD517L)
Low V_{os}: 25µV max (AD517L), 150µV max (AD517J)
Low V_{os} Drift: 0.5µV/°C (AD517L)
Internal Compensation
Low Cost

PRODUCT DESCRIPTION

The AD517 is a high accuracy monolithic op amp featuring extremely low offset voltages and input currents. Analog Devices' thermally-balanced layout and superior IC processing combine to produce a truly precision device at low cost.

The AD517 is laser trimmed at the wafer level (LWT) to produce offset voltages less than 25µV and offset voltage drifts less than 0.5µV/°C unnulled. Superbeta input transistors provide extremely low input bias currents of 1nA max and offset currents as low as 0.25nA max. While these figures are comparable to presently available BIFET amplifiers at room temperature, the AD517 input currents decrease, rather than increase, at elevated temperatures. Open-loop gain in many IC amplifiers is degraded under loaded conditions due to thermal gradients on the chip. However, the AD517 layout is balanced along a thermal axis, maintaining open-loop gain in excess of 1,000,000 for a wide range of load resistances.

The input stage of the AD517 is fully protected, allowing differential input voltages of up to ±V_S without degradation of gain or bias current due to reverse breakdown. The output stage is short-circuit protected and is capable of driving a load capacitance up to 1000pF.

The AD517 is well suited to applications requiring high precision and excellent long-term stability at low cost, such as stable references, followers, bridge instruments and analog computation circuits.

The circuit is packaged in a hermetically sealed TO-99 metal can, and is available in three performance versions (J, K, and L) specified over the commercial 0 to +70°C range; and one version (AD517S) specified over the full military temperature range, −55°C to +125°C. The AD517S is also available with full processing to the requirements of MIL-STD-883A, Level B.

PRODUCT HIGHLIGHTS
1. Offset voltage is 100% tested and guaranteed on all models. Testing is performed using a controlled-temperature drift bath following a 5 minute warm-up period.
2. The AD517 exhibits extremely low input bias currents without sacrificing CMRR (over 100dB) or offset voltage stability.
3. The AD517 inputs are protected (to ±V_S), preventing offset voltage and bias current degradation due to reverse breakdown of the input transistors.
4. Internal compensation is provided, eliminating the need for additional components (often required by high accuracy IC op amps).
5. The AD517 can directly replace 725, 108, and AD510 amplifiers. In addition, it can replace 741-type amplifiers if the offset-nulling potentiometer is removed.
6. Thermally-balanced layout insures high open-loop gain independent of thermal gradients induced by output loading, offset nulling, and power supply variations.
7. Every AD517 is baked for 48 hours at +150°C, temperature cycled from −65°C to +200°C, and subjected to a high G shock test to assure reliability and long-term stability.

Analog Devices, Inc.: Specifications

Courtesy of Analog Devices, Inc.

SPECIFICATIONS (typical @ +25°C and ±15V dc unless otherwise noted)

MODEL	AD517J	AD517K	AD517L	AD517S[1]
OPEN LOOP GAIN				
$V_O = \pm 10V$, $R_L \geqslant 2k\Omega$	10^6 min	*	*	*
T_{min} to T_{max}	500,000 min	*	*	250,000
OUTPUT CHARACTERISTICS				
Voltage @ $R_L \geqslant 2k\Omega$, T_{min} to T_{max}	±10V min	*	*	*
Load Capacitance	1000pF	*	*	*
Output Current	10mA min	*	*	*
Short Circuit Current	25mA	*	*	*
FREQUENCY RESPONSE				
Unity Gain, Small Signal	250kHz	*	*	*
Full Power Response	1.5kHz	*	*	*
Slew Rate, Unity Gain	0.10V/µs	*	*	*
INPUT OFFSET VOLTAGE				
Initial Offset, $R_S \leqslant 10k\Omega$	150µV max	50µV max	25µV max	**
vs. Temp., T_{min} to T_{max}	3.0µV/°C max	1.0µV/°C max	0.5µV/°C max	**
vs. Supply	25µV/V max	10µV/V max	**	**
(T_{min} to T_{max})	40µV/V max	15µV/V max	**	20µV/V max
INPUT OFFSET CURRENT				
Initial	1nA max	0.75nA max	0.25nA max	**
T_{min} to T_{max}	1.5nA max	1.25nA max	0.4nA max	2nA max
INPUT BIAS CURRENT				
Initial	5nA max	2nA max	1nA max	**
T_{min} to T_{max}	8nA max	3.5nA max	1.5nA max	10nA max
vs. Temp, T_{min} to T_{max}	±20pA/°C	±10pA/°C	±4pA/°C	**
INPUT IMPEDANCE				
Differential	15MΩ‖1.5pF	20MΩ‖1.5pF	**	**
Common Mode	$2.0 \times 10^{11} \Omega$	*	*	*
INPUT NOISE				
Voltage, 0.1Hz to 10Hz	2µV p-p	*	*	*
f = 10Hz	35nV/√Hz	*	*	*
f = 100Hz	25nV/√Hz	*	*	*
f = 1kHz	20nV/√Hz	*	*	*
Current, f = 10Hz	0.05pA/√Hz	*	*	*
f = 100Hz	0.03pA/√Hz	*	*	*
f = 1kHz	0.03pA/√Hz	*	*	*
INPUT VOLTAGE RANGE				
Differential or Common Mode max Safe	±V_S	*	*	*
Common Mode Rejection, $V_{in} = \pm 10V$	94dB min	110dB min	**	**
Common Mode Rejection, T_{min} to T_{max}	94dB min	100dB min	**	**
POWER SUPPLY				
Rated Performance	±15V	*	*	*
Operating	±(5 to 18)V	*	*	±(5 to 22)V
Current, Quiescent	4mA max	3mA max	**	**
TEMPERATURE RANGE				
Operating Rated Performance	0 to +70°C	*	*	−55°C to +125°C
Storage	−65°C to +150°C	*	*	*

NOTES
*Specifications same as AD517J
**Specifications same as AD517K
Specifications subject to change without notice.
[1] The AD517S is available fully processed and screened to the requirements of MIL-STD-883A, Level B. Consult factory for pricing.

SPECIFICATIONS AND DATA SHEETS 501

Analog Devices, Inc.: Typical Performance Curves

Courtesy of Analog Devices, Inc.

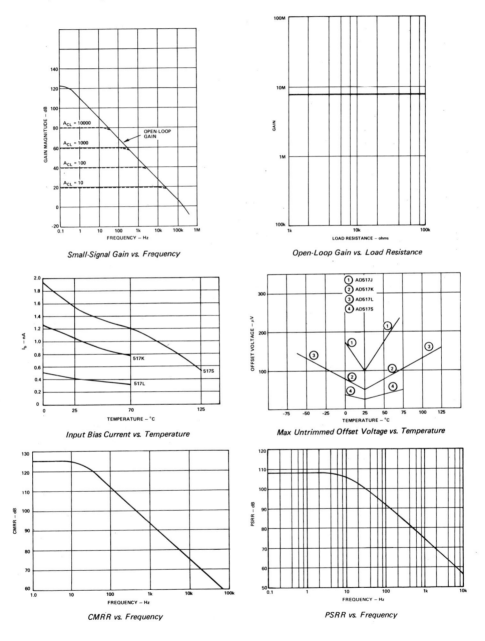

Analog Devices, Inc.: Typical Performance Curves (continued)

Courtesy of Analog Devices, Inc.

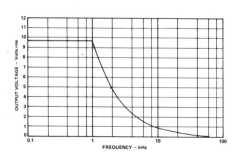

Maximum Undistorted Output vs. Frequency (Distortion ≤ 1%)

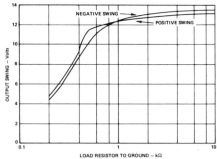

Output Voltage vs. Load Resistance

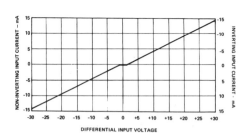

Input Current vs. Differential Input Voltage

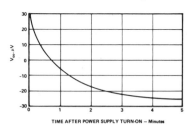

Warm-Up Offset Voltage Drift

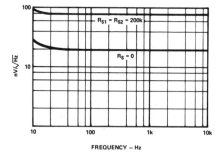

Total Input Noise Voltage vs. Frequency

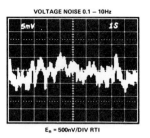

Low Frequency Voltage Noise (0.1 to 10Hz)

Analog Devices, Inc.: Accurate, Wideband, Multiplier, Divider, Square Rooter (Model 429)

Courtesy of Analog Devices, Inc.

Accurate, Wideband, Multiplier, Divider, Square Rooter

MODEL 429

FEATURES:
1.0%/0.5% Accuracy without Trimming (429A/B)
Low Drift to 1.0mV/°C max
Wideband — 10MHz
0.2% Nonlinearity max (429B)
External Amplifiers not Required

APPLICATIONS:
Fast Divider
Modulation and Demodulation
Phase Detection
Instrumentation Calculations
Analog Computer Functions
Adaptive Process Control
Trigonometric Computations

GENERAL DESCRIPTION

The model 429, an extremely fast multiplier/divider, should be considered if bandwidth, temperature coefficient, or accuracy are critical parameters. Based on the transconductance principle to achieve high speed, the model 429 offers a unique combination of features, those being ½% max error (429B) and 10MHz small signal bandwidth.

Both models 429A and 429B are internally trimmed achieving max errors of 1.0% and 0.5% respectively. By fine trimming the offset and feedthrough with external trim potentiometers typical performance may be improved to 0.5% for the 429A and 0.2% for the 429B.

In addition to high accuracy and high bandwidth, the model 429 offers exceptionally good stability for changes in ambient temperature. Model 429B is 100% temperature tested in order to guarantee an overall accuracy temperature coefficient of only 0.04%/°C max. Additionally, offset drift is held to only 1mV/°C max. To satisfy OEM requirements of low cost, the 429 uses transconductance principles with the latest design techniques and components to achieve guaranteed performance at competitive prices.

MULTIPLICATION ACCURACY

Multiplication accuracy is generally specified as a percentage of full scale output. This implies that error is independent of signal level. However, for signal levels less than 2/3 of full scale, error tends to decrease roughly in proportion to the input signal. A good approximation of error behavior is:

$f(X, Y) \cong |X| \epsilon_x + |Y| \epsilon_y$, where ϵ_x and ϵ_y are the fractional nonlinearities specified for the X and Y inputs

EXAMPLE: For Model 429A, $\epsilon_x = 0.5\%$, $\epsilon_y = 0.3\%$. What maximum error can one expect for x = 5V, y = 1V, providing the offset is zeroed out? Can one get less by interchanging inputs?
1. Nominal output is XY/10 = (5)(1)/10 = 500mV
2. Expected error is (5) (0.5%) + (1) (0.3%) = 28mV, 5.6% of output (0.28% of F.S.)
3. Interchanging inputs (1) (0.5%) + (5) (0.3%) = 20mV, 4.0% of output (0.20% of F.S.)

Compare this with the overly conservative error predicted by the overall 1% of full scale specification: 100mV, or 20% of output.

FREQUENCY RELATED SPECIFICATIONS

Accuracy, and its components, feedthrough, linearity, gain, (and phase shift) are frequency dependent. Feedthrough is constant up to 100kHz for the Y input, and up to 400kHz for the X input. Beyond these frequencies it rises at approximately a 6dB/octave rate due to distributed capacitive coupling. A plot of typical feedthrough vs. frequency is shown in Figure 1. For this measurement one input is driven with a 20V p-p sine wave while the other input is grounded and the feedthrough is measured at the output. This error will decrease roughly in proportion to the input signal, and will also vary with temperature (about 0.01%/°C of the nonzero input). Low frequency feedthrough error can be further reduced from the internally trimmed limits by the use of optional external potentiometers.

Non-linearity likewise increases with frequency at a 6dB/octave rate above the break frequency. With the Y input driven

Analog Devices, Inc.: Specifications

Courtesy of Analog Devices, Inc.

SPECIFICATIONS (typical @ +25°C and ±15V dc unless otherwise noted)

MODEL	429A	429B
MULTIPLICATION CHARACTERISTICS		
Output Function	XY/10	*
Error, with Internal Trim, at +25°C	±1% max	±0.5% max
Error, with External Trim, at +25°C	±0.7%	±0.3%
Avg vs. Temp (-25°C to +85°C)	±0.05%/°C	±0.04%/°C max
Avg vs. Supply	±0.05%/%	*
SCALE FACTOR		
Initial Error at +25°C	0.5%	0.25%
Avg vs. Temp (-25°C to +85°C)	0.03%/°C	0.02%/°C
Avg vs. Supply	0.03%/%	
OUTPUT OFFSET		
Initial at +25°C (Adjustable to Zero)	±20mV max	±10mV max
Avg vs. Temp (-25°C to +85°C)	±2mV/°C	±1mV/°C max
Avg vs. Supply	±1mV/%	*
NONLINEARITY		
X Input (X = 20V p–p 50Hz, Y = ±10V)	0.5% max	0.2% max
Y Input (Y = 20V p–p 50Hz, X = ±10V)	0.3% max	0.2% max
FEEDTHROUGH		
X = 0, Y = 20V p–p, 50Hz	50mV p–p, max	20mV p–p, max
With External Trim	16mV p–p	10mV p–p
Y = 0, X = 20V p–p, 50Hz	100mV p–p, max	30mV p–p, max
With External Trim	50mV p–p	20mV p–p
BANDWIDTH		
–3dB	10MHz	*
Full Power Response	2MHz min	*
Slew Rate	120V/μs min	*
1% Amplitude Error	300kHz min	*
1% Vector Error (0.57°)	50kHz min	*
Differential Phase Shift ($\theta_x - \theta_y$)	1° @ 1MHz	*
Small Signal Rise Time 10–90%	40ns	*
Settling to ±1% (±10V step)	500ns	*
Overload Recovery	0.2μs	*
OUTPUT NOISE		
5Hz to 10kHz	0.6mV rms	*
5Hz to 10MHz	3.0mV rms	*
OUTPUT CHARACTERISTICS		
Voltage, 1kΩ load	±11V min	*
Current	±11mA min	*
Load Capacitance	0.01μF max	*
INPUT RESISTANCE		
X Input	10kΩ±5%	*
Y Input	11kΩ±2%	*
Z Input	27kΩ±10%	*
INPUT BIAS CURRENT		
Input X, Y, Z	±100nA	*
Z	±20μA	*
MAXIMUM INPUT VOLTAGE		
For Rated Accuracy	±10.5V	*
Maximum Safe	±16V	*
WARM UP		
To Rated Specifications	1 second	*
POWER SUPPLY[1]		
Rated Performance	±(14.8 to 15.3)V dc	*
Operating	±(14 to 16)V dc	*
Quiescent Current	±12mA	*
TEMPERATURE RANGE		
Rated Performance	-25°C to +85°C	*
Operating	-25°C to +85°C	*
Storage	-55°C to +125°C	*
MECHANICAL		
Weight	2 oz.	*
Socket	AC1023	*
Case Dimensions	1.5" x 1.5" x 0.62"	*

*Specifications same as Model 429A.
[1] Recommend Model 904 available from Analog Devices
Specifications subject to change without notice.

OUTLINE DIMENSIONS
Dimensions shown in inches and (mm).

1.51 MAX (38.1)
0.62 MAX (15.7)
0.04 DIA. (1.02)
0.20 MIN., 0.25 MAX. (5.0) (6.4)
1.51 MAX (38.1)
BOTTOM VIEW
0.1 GRID 2.54

PIN CONNECTIONS
Bottom View Shown in all Cases.
Optional Trim Pots
Shown are not Required
for Rated Accuracy.

MULTIPLY MODE

$e_o = \dfrac{XY}{10}$

DIVIDE MODE

$e_o = \dfrac{10Z}{X}$

SQUARE ROOT MODE

$e_o = -\sqrt{10Z}$

All trim pots 20kΩ; PN79PR20k

SPECIFICATIONS AND DATA SHEETS

Analog Devices, Inc.: Applying the Fast Multiplier

Courtesy of Analog Devices, Inc.

at 10V p-p, and the X input anywhere between ±10V dc, the break frequency is 25kHz. For corresponding X input conditions, the break occurs at 60kHz. Figure 2 is a plot of the typical nonlinearity vs. frequency for the model 429.

Gain and input to output phase shift for the model 429 are shown in Figure 2. Naturally, no multiplier will maintain accuracy at frequencies approaching the small signal bandwidth. For the model 429, the 1% amplitude error will occur at 500kHz. If input to output phase shift is a criterion, then the 1% "vector" error occurs at 50kHz.

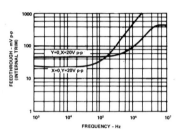

Figure 1. Feedthrough vs Frequency

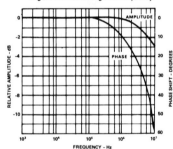

Figure 2. Typical Amplitude and Phase vs Frequency

OPTIONAL TRIM – MULTIPLY MODE
As shipped, the multiplier meets its listed specifications without use of any external trim potentiometers. Terminals are provided for optional feedthrough and offset adjustments. Using these adjustments overall static multiplication error may be reduced to only 0.2%. The 20kΩ trim potentiometers should be connected across the ± supply voltage terminals with the arm of each potentiometer connected to the desired balance terminal (see previous page).

ADJUSTMENT PROCEDURE FOR OFFSET
1. Jumper X input and Y input to ground.
2. Adjust R_0 for an output of zero volts.
3. Remove jumper from X and Y inputs.

ADJUSTMENT PROCEDURE FOR FEEDTHROUGH
1. Jumper Y input to ground and apply 20 VPP at 1kHz to X input.
2. Adjust R_Y for minimum output voltage.

3. Remove jumper from Y input.
4. Jumper X input to ground and apply 20 VPP at 1kHz to Y input.
5. Adjust R_X for minimum output voltage.
6. Remove jumper from X terminal.

DIVISION
The high bandwidth and excellent linearity of model 429 allows it to be used in divider applications achieving high performance in the dc to 8MHz region. Restrictions imposed on divide operation, and the contribution of error terms are illustrated in the error analysis below.

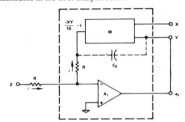

Figure 3. Divider Circuit

Shown in Figure 3 is a typical multiplier/divider which has been connected for divide operation by inserting the multiplier cell, M, in the op amp's feedback loop. Errors associated with the op amp, A_1, are incorporated in ϵ, which represents all errors. In order to insure negative feedback, the X input range is restricted to negative values.

Summing currents at the op amp's summing junction:

$$\frac{Z}{R} = \frac{\frac{XY}{10} + \epsilon}{R}$$

Solving for Y, which is also ϵ_0:

$$Y = \frac{10(Z - \epsilon)}{X}$$

or,

$$\epsilon_0 = \frac{10Z}{X} - \frac{10\epsilon}{X}$$

And now breaking ϵ into its constituents

$$\epsilon_0 = \underbrace{\frac{10Z}{X}}_{\substack{\text{ideal}\\\text{divider}}} - \underbrace{\frac{10E_{NV}}{X}}_{\substack{\text{noise}\\\text{error}}} - \underbrace{\frac{10E_{OS}}{X}}_{\substack{\text{offset}\\\text{error}}} - \underbrace{\frac{10E_{OS}/°C}{X}}_{\substack{\text{offset drift}\\\text{error}}} - \underbrace{\frac{10E_{NLX}}{X}}_{\substack{\text{X non-}\\\text{linearity}\\\text{error}}} - \underbrace{\frac{10E_{NLY}}{X}}_{\substack{\text{Y non-}\\\text{linearity}\\\text{error}}}$$

These errors can be broken down into two categories, static errors and signal dependent errors. All of the static errors associated with the divide mode are inversely proportional to the denominator signal level. The signal dependent errors are the X and Y nonlinearities. For model 429B nonlinearity errors are 0.2% for both the X and Y inputs. Substituting these values in the error terms yields:

$$-\frac{10(0.2\%)X}{X} - \frac{10(0.2\%)Y}{X}$$

Analog Devices, Inc.: Applying the Fast Multiplier (continued)

Courtesy of Analog Devices, Inc.

The importance of using the terminal with largest nonlinearity for the denominator is revealed by the above expression. Effects of X nonlinearity are virtually independent of signal level and may be trimmed out. Nonlinearities of Y typically contribute 200mV for $X = Z = 1V$ i.e., (10 [0.2%] 10V) = 200mV. This error can be reduced if external trims are used to optimize divider performance.

Bandwidth is also degraded with a decrease in denominator level, due to the increase in system gain;

i.e.) for $X = Z = 1V$, $\epsilon_0 = 10V$

and $\dfrac{\epsilon_0}{Z} = \dfrac{10}{1} = 10$

Since the gain bandwidth product is constant, a bandwidth of 1/10 of that obtained for full scale denominator levels will be obtained for division at 1V levels.

For other denominator levels, bandwidth is determined by:

$$B.W. = \dfrac{\text{Denominator Level}}{\text{Full Scale Denominator}} \times (\text{Multiplier B.W.}) \times K$$

where K is a constant having a value less than unity. It is introduced due to a combination of stray capacitance paralleling the multiplier cell and effects of feedthrough. For model 429

$$B.W. = \left(\dfrac{X}{10}\right) 8MHz$$

Before selecting a multiplier/divider for divide applications, errors resulting from the lowest anticipated denominator signal should be considered. After such considerations have been made, one can further appreciate the importance of starting with an accurate, high speed multiplier such as model 429. It is also highly recommended that the optional trim procedure for division be performed.

OPTIONAL TRIMMING – DIVIDE MODE

Connections are made as shown previously.

The suggested trim procedure is (starting with centered adjust adjustments):

*1. With $Z = 0$, trim R_0 to hold output constant, as X is varied from –10V toward –1.0V.
2. With $Z = 0$, trim R_Y for zero at $X = -10V$.
3. With $Z = X$ and/or $Z = -X$, trim R_X for minimum worst-case variation as X is varied from –10V to –1.0V.
4. Repeat 1 and 2 if step 3 required large initial adjustment.

*For best accuracy X should be allowed to vary from –10V to lowest expected denominator.

SQUARE ROOTING

When connected as shown previously, the model 429 will provide the square root of Z_{IN}.

By summing currents at the op amp's summing junction:

$$\dfrac{Z}{R} = \dfrac{XY}{10R} + \dfrac{\epsilon}{R} = \dfrac{Y^2}{10R} + \dfrac{\epsilon}{R}$$

where ϵ represents all errors associated with the multiplier. Solving for the output voltage, Y.

$$\epsilon_0 = \pm \sqrt{10(Z - \epsilon)}$$

There are two values of ϵ_0 for every value of Z. However, only negative values of ϵ_0 will provide the negative feedback necessary for circuit stability. To restrict the output from going positive, a diode is connected as shown previously. The output is then:

$$\epsilon_0 = -\sqrt{10(Z - \epsilon)}$$

Errors, ϵ, associated with the multiplier, are inside the square root and consequently their effect, for large values of Z, is

reduced. The reason for the improved performance can be seen by inspecting the circuit. The output is fed back to both the X and Z terminals, resulting in twice the feedback as would be obtained for the divide mode. An alternative method of considering error performance is to consider errors as being at the Z terminal. By differentiating the ideal transfer function with respect to Z, errors for various values of Z may be determined:

$$\dfrac{d\epsilon_0}{dZ} = \dfrac{d}{dZ} \sqrt{10Z} = \tfrac{1}{2}\sqrt{\dfrac{10}{Z}}$$

The factor of ½ has the advantage of reducing errors by a factor of 2 for $Z = 10$, but also introduces the potential problem of instability. Since the feedback gain is the reciprocal of the forward gain, the slope of the forward gain is 2. Additional phase margin is required to support the increased gain in the feedback path. Model 429 is optimized for phase margin in the multiply and divide modes producing minimum vector errors at high frequencies. To avoid the potential problem of instability, the RC network shown previously is recommended. This network restricts the bandwidth and guarantees stability for all positive values of Z.

OPTIONAL ADJUSTMENT PROCEDURE – SQUARE ROOT

1. Apply a voltage to the Z terminal equal to the lowest anticipated input voltage.
2. Adjust R_0 such that $\epsilon_0 = -\sqrt{10Z}$, where Z is the voltage applied in step 1.

DIVISION SPECIFICATIONS (TYPICAL)

OUTPUT FUNCTION	$10(Z)/X$
Numerator Range	±10V
Denominator Range, 1% Accuracy	–1 to –10V
Denominator Range, 5% Accuracy	–0.2V to –10V
Bandwidth Formula, (Hz, –3dB)	$(8MHz)(X)/10$

SQUARE ROOTING SPECIFICATIONS (TYPICAL)

OUTPUT FUNCTION	$-\sqrt{10(Z)}$
Dynamic Range	1000 to 1 (+0.010V ≤ Z ≤ +10V)
Accuracy (% of Full Scale)	0.5%
Bandwidth Formula, (Hz, –3dB)	$(5MHz)\sqrt{|X|/10}$

Table 1. Division & Square Rooting Specifications

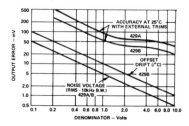

Figure 3. Typical Error Performance of Model 429 in Divide Mode for Worst Case of $|\epsilon_0| = 10V$

Analog Devices, Inc.: Low Cost, Sample/Hold Amplifier (AD582)

Courtesy of Analog Devices, Inc.

Low Cost Sample/Hold Amplifier

AD582

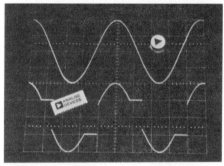

FEATURES
Low Cost
Suitable for 12-Bit Applications
High Sample/Hold Current Ratio: 10^7
Low Acquisition Time: 6µs to 0.1%
Low Charge Transfer: <2pC
High Input Impedance in Sample and Hold Modes
Connect in Any Op Amp Configuration
Differential Logic Inputs

PRODUCT DESCRIPTION

The AD582 is a low cost integrated circuit sample and hold amplifier consisting of a high performance operational amplifier, a low leakage analog switch and a JFET integrating amplifier — all fabricated on a single monolithic chip. An external holding capacitor, connected to the device, completes the sample and hold function.

With the analog switch closed, the AD582 functions like a standard op amp; any feedback network may be connected around the device to control gain and frequency response. With the switch open, the capacitor holds the output at its last level, regardless of input voltage.

Typical applications for the AD582 include sampled data systems, D/A deglitchers, analog de-multiplexers, auto null systems, strobed measurement systems and A/D speed enhancement.

The device is available in two versions: the "K" specified for operation over the 0 to +70°C commercial temperature range and the "S" specified over the full military temperature range, -55°C to +125°C. Both versions may be obtained in either the hermetically sealed, TO-100 can or the TO-116 DIP.

PRODUCT HIGHLIGHTS

1. The monolithic AD582 is the lowest cost sample and hold amplifier available. Until recently, quality sample and hold circuits could only be fabricated with costly discrete or hybrid components.

2. The specially designed input stage presents a high impedance to the signal source in both sample and hold modes (up to ±12V). Even with signal levels up to ±V_S, no undesirable signal inversion, peaking or loss of hold voltage occurs.

3. The AD582 may be connected in any standard op amp configuration to control gain or frequency response and provide signal inversion, etc.

4. The AD582 offers a high, sample-to-hold current ratio: 10^7. The ratio of the available charging current to the holding leakage current is often used as a figure of merit for a sample and hold circuit.

5. The AD582 has a typical charge transfer less than 2pC. A low charge transfer produces less offset error and permits the use of smaller hold capacitors for faster signal acquisition.

6. The AD582 provides separate analog and digital grounds, thus improving the device's immunity to ground and switching transients.

SPECIFICATIONS AND DATA SHEETS

Analog Devices, Inc.: Specifications

Courtesy of Analog Devices, Inc.

SPECIFICATIONS (typical @ +25°C, $V_S = \pm15V$ and $C_H = 1000pF$, A = +1 unless otherwise specified)

MODEL	AD582K	AD582S
SAMPLE/HOLD CHARACTERISTICS		
Acquisition Time, 10V Step to 0.1%, C_H = 100pF	6µs	*
Acquisition Time, 10V Step to 0.01%, C_H = 1000pF	25µs	*
Aperture Time, 20V p-p Input, Hold 0V	150ns	*
Aperture Jitter, 20V p-p Input, Hold 0V	15ns	*
Settling Time, 20V p-p Input, Hold 0V, to 0.01%	0.5µs	*
Droop Current, Steady State, $\pm 10V_{OUT}$	50pA max	
Droop Current, T_{min} to T_{max}	1nA	50nA max
Charge Transfer	5pC max (1.5pC typ)	*
Sample to Hold Offset	0.5mV	*
Feedthrough Capacitance 20V p-p, 10kHz Input	0.05pF	*
TRANSFER CHARACTERISTICS		
Open Loop Gain V_{OUT} = 20V p-p, R_L = 2k	25k min (50k typ)	*
Common Mode Rejection V_{CM} = 20V p-p, F = 50Hz	60dB min (70dB typ)	*
Small Signal Gain Bandwidth V_{OUT} = 100mV p-p, C_H = 200pF	1.5MHz	*
Full Power Bandwidth V_{OUT} = 20V p-p, C_H = 200pF	70kHz	*
Slew Rate V_{OUT} = 20V p-p, C_H = 200pF	3V/µs	*
Output Resistance Hold Mode, $I_{OUT} = \pm 5mA$	12Ω	*
Linearity V_{OUT} = 20V p-p, R_L = 2k	±0.01%	*
Output Short Circuit Current	±25mA	*
ANALOG INPUT CHARACTERISTICS		
Offset Voltage	6mV max (2mV typ)	8mV max (5mV typ)
Offset Voltage, T_{min} to T_{max}	4mV	
Bias Current	3µA max (1.5µA typ)	
Offset Current	300nA max (75nA typ)	
Offset Current, T_{min} to T_{max}	100nA	400nA max (100nA typ)
Input Capacitance, f = 1MHz	2pF	*
Input Resistance, Sample or Hold 20V p-p Input, A = +1	30MΩ	*
Absolute Max Diff Input Voltage	30V	*
Absolute Max Input Voltage, Either Input	$\pm V_S$	*
DIGITAL INPUT CHARACTERISTICS		
+Logic Input Voltage		
Hold Mode, T_{min} to T_{max}, -Logic @ 0V	+2V min	
Sample Mode, T_{min} to T_{max}, -Logic @ 0V	+0.8V max	
+Logic Input Current		
Hold Mode, +Logic @ +5V, -Logic @ 0V	1.5µA	
Sample Mode, +Logic @ 0V, -Logic @ 0V	1nA	
-Logic Input Current		
Hold Mode, +Logic @ +5V, -Logic @ 0V	24µA	
Sample Mode, +Logic @ 0V, -Logic @ 0V	4µA	
Absolute Max Diff Input Voltage, +L to -L	+15V/-6V	
Absolute Max Input Voltage, Either Input	$\pm V_S$	*
POWER SUPPLY CHARACTERISTICS		
Operating Voltage Range	±9V to ±18V	±9V to ±22V
Supply Current, $R_L = \infty$	4.5mA max (3mA typ)	*
Power Supply Rejection, ΔV_S = 5V, Sample Mode (see next page)	60dB min (75dB typ)	*
TEMPERATURE RANGE		
Specified Performance	0 to +70°C	-55°C to +125°C
Operating	-25°C to +85°C	-55°C to +125°C
Storage	-65°C to +150°C	*
Lead Temperature (Soldering, 15 sec)	+300°C	

*Specifications same as AD582K.

OUTLINE DIMENSIONS

Dimensions shown in inches and (mm).

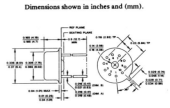

TO-100 "H"

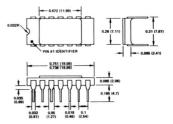

TO-116 "D"

PIN CONFIGURATIONS
TOP VIEW

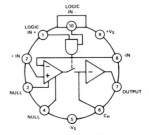

10 PIN TO-100

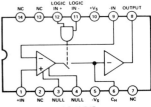

14 PIN DIP

Analog Devices, Inc.: Applying the AD582

Courtesy of Analog Devices, Inc.

APPLYING THE AD582

Both the inverting and non-inverting inputs are brought out to allow op amp type versatility in connecting and using the AD582. Figure 1 shows the basic non-inverting unity gain connection requiring only an external hold capacitor and the usual power supply bypass capacitors. An offset null pot can be added for more critical applications.

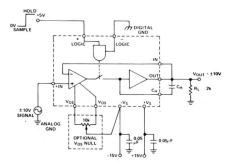

Figure 1. Sample and Hold with A = +1

Figure 2 shows a non-inverting configuration where voltage gain, A_V, is set by a pair of external resistors. Frequency shaping or non-linear networks can also be used for special applications.

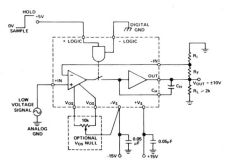

The hold capacitor, C_H, should be a high quality polystyrene (for temperatures below +85°C) or Teflon type with low dielectric absorption. For high speed, limited accuracy applications, capacitors as small as 100pF may be used. Larger values are required for accuracies of 12 bits and above in order to minimize feedthrough, sample to hold offset and droop errors (see Figure 6). Care should be taken in the circuit layout to minimize coupling between the hold capacitor and the digital or signal inputs.

In the hold mode, the output voltage will follow any change in the $-V_S$ supply. Consequently, this supply should be well regulated and filtered.

Biasing the +Logic Input anywhere between -6V to +0.8V with respect to the -Logic will set the sample mode. The hold mode will result from any bias between +2.0V and $(+V_S - 3V)$. The sample and hold modes will be controlled differentially with the absolute voltage at either logic input ranging from $-V_S$ to within 3V of $+V_S$ ($V_S - 3V$). Figure 3 illustrates some examples of the flexibility of this feature.

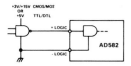

Figure 3A. Standard Logic Connection

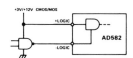

Figure 3B. Inverted Logic Sense Connection

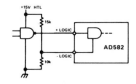

Analog Devices, Inc.: Definition of Terms

Courtesy of Analog Devices, Inc.

DEFINITION OF TERMS

Figure 4 illustrates various dynamic characteristics of the AD582.

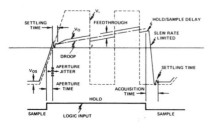

Figure 4. Pictorial Showing Various S/H Characteristics

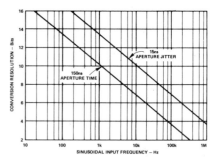

Figure 5. Maximum Frequency of Input Signal for ½LSB Sampling Accuracy

Aperture Time is the time required after the "hold" command until the switch is fully open and produces a delay in the effective sample timing. Figure 5 is a plot giving the maximum frequency at which the AD582 can sample an input with a given accuracy (lower curve).

Aperture Jitter is the uncertainty in Aperture Time. If the Aperture Time is "tuned out" by advancing the sample-to-hold command 150ns with respect to the input signal, the Aperture Jitter now determines the maximum sampling frequency (upper curve of Figure 5).

Acquisition Time is the time required by the device to reach its final value within a given error band after the sample command has been given. This includes switch delay time, slewing time and settling time for a given output voltage change.

Droop is the change in the output voltage from the "held" value as a result of device leakage. In the AD582, droop can be in either the positive or negative direction. Droop rate may be calculated from droop current using the following formula:

$$\frac{\Delta V}{\Delta T} \text{(Volts/sec)} = \frac{I(pA)}{C_H(pF)}$$

(See also Figure 6.)

Feedthrough is that component of the output which follows the input signal *after* the switch is open. As a percentage of the input, feedthrough is determined as the ratio of the feedthrough capacitance to the hold capacitance (C_F/C_H).

Charge Transfer is the charge transferred to the holding capacitor from the interelectrode capacitance of the switch when the unit is switched to the hold mode. The charge transfer generates a sample-to-hold offset where:

$$\text{S/H Offset (V)} = \frac{\text{Charge (pC)}}{C_H(pF)}$$

(See also Figure 6.)

Sample to Hold Offset is that component of D.C. offset independent of C_H (see Figure 6). This offset may be nulled using a null pot, however, the offset will then appear during the sampling mode.

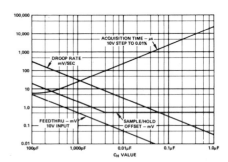

Figure 6. Sample and Hold Performance as a Function of Hold Capacitance

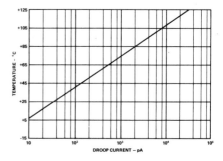

Analog Devices, Inc.: CMOS 10 & 12-Bit Monolithic Multiplying D/A Converters (AD7520, AD7521)

Courtesy of Analog Devices, Inc.

CMOS 10&12-Bit Monolithic Multiplying D/A Converters

AD7520, AD7521

FEATURES
AD7520: 10 Bit Resolution
AD7521: 12 Bit Resolution
Linearity: 8, 9 and 10 Bit
Nonlinearity Tempco: 2ppm of FSR/°C
Low Power Dissipation: 20mW
Current Settling Time: 500ns
Feedthrough Error: 1/2LSB @ 100kHz
TTL/DTL/CMOS Compatible

GENERAL DESCRIPTION

The AD7520 (AD7521) is a low cost, monolithic 10-bit (12-bit) multiplying digital-to-analog converter packaged in a 16-pin (18-pin) DIP. The devices use advanced CMOS and thin film technologies providing up to 10-bit accuracy with TTL/DTL/CMOS compatibility.

The AD7520 (AD7521) operates from +5V to +15V supply and dissipates only 20mW, including the ladder network.

Typical AD7520 (AD7521) applications include: digital/analog multiplication, CRT character generation, programmable power supplies, digitally controlled gain circuits, etc.

ORDERING INFORMATION

Nonlinearity	Temperature Range		
	0 to +70°C	-25°C to +85°C	-55°C to +125°C
0.2% (8-Bit)	AD7520JN AD7521JN	AD7520JD AD7521JD	AD7520SD AD7521SD
0.1% (9-Bit)	AD7520KN AD7521KN	AD7520KD AD7521KD	AD7520TD AD7521TD
0.05% (10-Bit)	AD7520LN AD7521LN	AD7520LD AD7521LD	AD7520UD AD7521UD

PACKAGE IDENTIFICATION
Suffix D: Ceramic DIP package
Suffix N: Plastic DIP package

FUNCTIONAL DIAGRAM

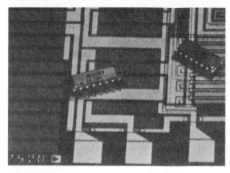

DIGITAL INPUTS (DTL/TTL/CMOS COMPATIBLE)

AD7520: N=10
AD7521: N=12
Logic: A switch is closed to I_{OUT1} for its digital input in a "HIGH" state.

PIN CONFIGURATION

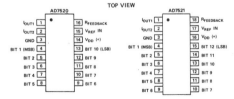

Analog Devices, Inc.: Specifications

Courtesy of Analog Devices, Inc.

SPECIFICATIONS (V_{DD} = +15, V_{REF} = +10V, T_A = +25°C unless otherwise noted)

PARAMETER	AD7520	AD7521	TEST CONDITIONS
DC ACCURACY[1]			
Resolution	10 Bits	12 Bits	
Nonlinearity (See Figure 5)			
	J, 0.2% of FSR max (8 Bit)	*	S,T,U: over −55°C to +125°C
	S, 0.2% of FSR max (8 Bit)	*	−10V ≤ V_{REF} ≤ +10V
	K, 0.1% of FSR max (9 Bit)	*	
	T, 0.1% of FSR max (9 Bit)	*	
	L, 0.05% of FSR max (10 Bit)	*	
	U, 0.05% of FSR max (10 Bit)	*	
Nonlinearity Tempco	2ppm of FSR/°C max	*	−10V ≤ V_{REF} ≤ +10V
Gain Error[2]	0.3% of FSR typ	*	−10V ≤ V_{REF} ≤ +10V
Gain Error Tempco[2]	10ppm of FSR/°C max	*	−10V ≤ V_{REF} ≤ +10V
Output Leakage Current			
(either output)	200nA max	*	Over specified temperature range
Power Supply Rejection	50ppm of FSR%/°C typ	*	
(See Figure 6)			
AC ACCURACY			To 0.05% of FSR
Output Current Settling Time	500ns typ	*	All digital inputs low to high and high to low
(See Figure 10)			
Feedthrough Error (See Figure 9)	10mV p-p max		V_{REF} = 20V p-p, 100kHz
		*	All digital inputs low
REFERENCE INPUT			
Input Resistance[4]	5kΩ min	*	
	10kΩ typ	*	
	20kΩ max	*	
ANALOG OUTPUT			
Output Capacitance I_{OUT1}	120pF typ	*	All digital inputs high
(See Figure 8) I_{OUT2}	37pF typ	*	All digital inputs high
I_{OUT1}	37pF typ	*	All digital inputs low
I_{OUT2}	120pF typ	*	All digital inputs low
Output Noise (both outputs)	Equivalent to 10kΩ typ	*	
(See Figure 7)	Johnson noise		
DIGITAL INPUTS[3]			
Low State Threshold	0.8V max	*	Over specified temperature range
High State Threshold	2.4V min	*	Over specified temperature range
Input Current (low to high state)	1μA typ	*	Over specified temperature range
Input Coding	Binary	*	See Tables 1 & 2 under Applications
POWER REQUIREMENTS			
Power Supply Voltage Range	+5V to +15V	*	
I_{DD}	5nA typ	*	All digital inputs at GND
	2mA max	*	All digital inputs high or low
Total Dissipation (Including ladder)	20mW typ	*	

NOTES:
[1] Full scale range (FSR) is 10V for unipolar mode and ±10V for bipolar mode.
[2] Using the internal $R_{FEEDBACK}$
[3] Digital input levels should not go below ground or exceed the positive supply voltage, otherwise damage may occur.
[4] Ladder and feedback resistor tempco is approximately −150ppm/°C.

Specifications subject to change without notice.

SPECIFICATIONS AND DATA SHEETS

Analog Devices, Inc.: Absolute Maximum Ratings

Courtesy of Analog Devices, Inc.

ABSOLUTE MAXIMUM RATINGS
(T_A = +25°C unless otherwise noted)

V_{DD} (to GND) .... +17V
V_{REF} (to GND) .... ±25V
Digital Input Voltage Range .... V_{DD} to GND
Output Voltage (Pin 1, Pin 2) .... -100mV to V_{DD}
Power Dissipation (package)
   up to +75°C 450mW
   derates above +75°C by .... 6mW/°C
Operating Temperature
   JN, KN, LN Versions .... 0 to +70°C
   JD, KD, LD Versions .... -25°C to +85°C
   SD, TD, UD Versions .... -55°C to +125°C
Storage Temperature .... -65°C to +150°C

CAUTION:
1. Do not apply voltages higher than V_{DD} or less than GND potential on any terminal except V_{REF}.
2. The digital control inputs are zener protected; however, permanent damage may occur on unconnected units under high energy electrostatic fields. Keep unused units in conductive foam at all times.

TYPICAL PERFORMANCE CURVES
T_A = +25°C, V_{DD} = +15V unless otherwise noted

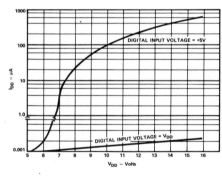

Figure 1. Supply Current vs. Supply Voltage

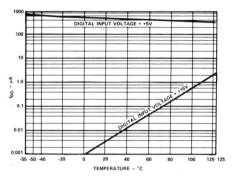

Figure 2. Supply Current vs. Temperature

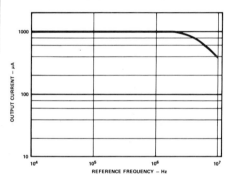

Figure 3. Output Current Bandwidth

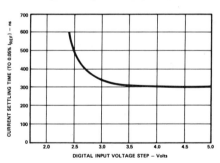

Figure 4. Output Current Settling Time vs. Digital Input Voltage

Analog Devices, Inc.: Test Circuits/DC Parameters, AC Parameters Terminology

Courtesy of Analog Devices, Inc.

TEST CIRCUITS

Note: The following test circuits apply for the AD7520. Similar circuits can be used for the AD7521.

DC PARAMETERS

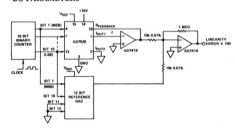

Figure 5. Nonlinearity

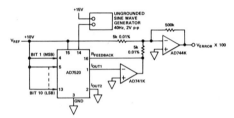

Figure 6. Power Supply Rejection

AC PARAMETERS

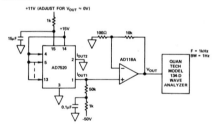

Figure 7. Noise

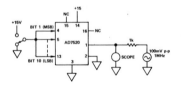

Figure 8. Output Capacitance

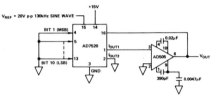

Figure 9. Feedthrough Error

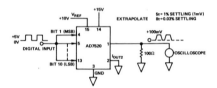

Figure 10. Output Current Settling Time

TERMINOLOGY

NONLINEARITY: Error contributed by deviation of the DAC transfer function from a best straight line function. Normally expressed as a percentage of full scale range. For a multiplying DAC, this should hold true over the entire V_{REF} range.

RESOLUTION: Value of the LSB. For example, a unipolar converter with n bits has a resolution of $(2^{-n}) (V_{REF})$. A bipolar converter of n bits has a resolution of $[2^{-(n-1)}]$ $[V_{REF}]$. Resolution in no way implies linearity.

SETTLING TIME: Time required for the output function of the DAC to settle to within 1/2 LSB for a given digital input stimulus, i.e., 0 to Full Scale.

GAIN: Ratio of the DAC's operational amplifier output voltage to the input voltage.

FEEDTHROUGH ERROR: Error caused by capacitive coupling from V_{REF} to output with all switches OFF.

OUTPUT CAPACITANCE: Capacity from I_{OUT1} and I_{OUT2} terminals to ground.

OUTPUT LEAKAGE CURRENT: Current which appears on I_{OUT1} terminal with all digital inputs LOW or on I_{OUT2} terminal when all inputs are HIGH.

Comdyna, Inc.: Comdyna GP-6

Courtesy of Comdyna, Inc.

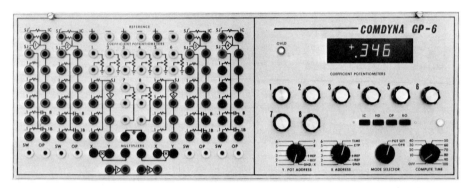

Datel-Intersil: Fast, 8 Bit Analog-to-Digital Converters (Model ADC-EH8B)

Reprinted with Permission of Datel-Intersil, Inc., Mansfield, Massachusetts.

Fast, 8 Bit Analog-to-Digital Converters Model ADC-EH8B

FEATURES
- 8 Bit Resolution
- 4.0 & 2.0 μsec. Conversion Time
- Unipolar or Bipolar Operation
- Parallel & Serial Outputs
- Low Cost

GENERAL DESCRIPTION

The model ADC-EH8B is a fast, 8 bit successive approximation type analog to digital converter in a compact 2 x 2 x .375 inch module. These converters are low cost devices with application in pulse code modulation systems and instrumentation and control systems requiring fast data conversion rates up to 500,000 per second. There are two models to choose from based on conversion speed: ADC-EH8B1 with a conversion time of 4.0 μsec. (250 kHz rate), and ADC-EH8B2 with a conversion time of 2.0 μsec. (500 kHz rate).

The high speed in a small size is made possible by the use of an MSI integrated circuit which provides all the necessary successive approximation logic, along with other new integrated circuit components. The analog input range is either unipolar 0 to +10V or bipolar –5V to +5V, determined by external pin connection. For unipolar operation no external adjustments are necessary; for bipolar operation only a bipolar offset adjustment must be made externally. Parallel output coding is straight binary for unipolar operation and offset binary or two's complement for bipolar operation. A serial output gives successive decision pulses in NRZ format with straight or offset binary coding. Other outputs are clock output for synchronization with serial data, and MSB output for two's complement coding.

Other specifications include full scale temperature coefficient of 50 ppm/°C max., long term stability of .05%/year, and linearity of ±1/2 LSB. Power requirement is ±15VDC and +5VDC.

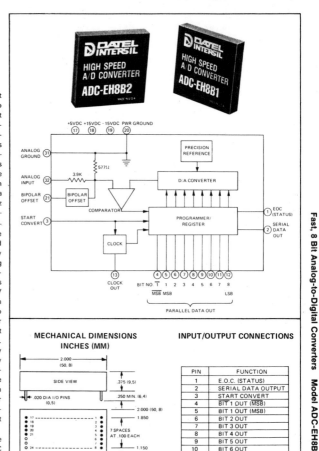

DATEL-INTERSIL INC., 11 CABOT BOULEVARD, MANSFIELD, MA 02048 / TEL. (617) 339-9341 / TWX 710-346-1953 / TLX 951340

SPECIFICATIONS AND DATA SHEETS 517

Datel-Intersil: Specifications, ADC-EH8B

Reprinted with Permission of Datel-Intersil, Inc., Mansfield, Massachusetts.

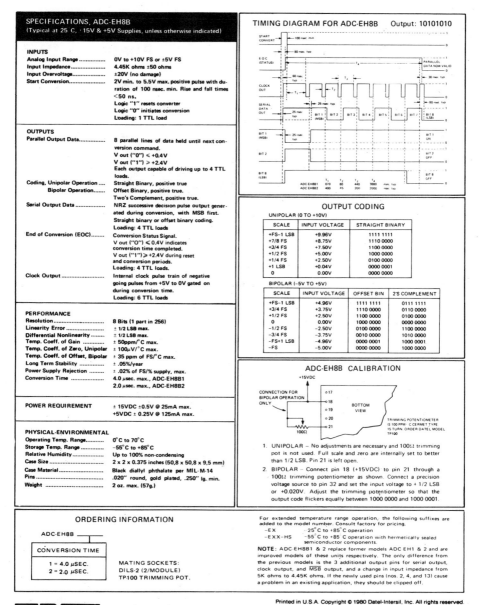

PRICES AND SPECIFICATIONS SUBJECT TO CHANGE WITHOUT NOTICE

518 SPECIFICATIONS AND DATA SHEETS

Datel-Intersil: 10 Bit, 2.0 and 4.0 μSec. Analog-to-Digital Converters (Model ADC-EH10B)

Reprinted with Permission of Datel-Intersil, Inc., Mansfield, Massachusetts.

10 Bit, 2.0 and 4.0 μSec. Analog-to-Digital Converters Model ADC-EH10B

FEATURES
- 2.0 μsec. Conversion — ADC-EH10B2
- 4.0 μsec. Conversion — ADC-EH10B1
- 10 Bit Resolution
- Compact 3" x 2" x .375" Module
- ±30ppm/°C max. Tempco

GENERAL DESCRIPTION

Model ADC-EH10B is a very fast 10 bit successive approximation type A/D converter in a compact low profile package. Low pricing makes this converter an ideal choice for many applications including fast scanning data acquisition systems, PCM systems, and fast pulse analysis. This converter is available in two versions based on conversion speed: ADC-EH10B1 with 4.0 μsec. (250kHz rate) and ADC-EH10B2 with 2.0 μsec. (500kHz rate).

High speed and moderate power consumption (1.7 watts) in a compact size (3" x 2" x .375") are made possible by use of an MSI integrated circuit successive approximation programmer/register used with 10 fast switching current sources driving a low impedance R-2R ladder network. A fast precision comparator and precision voltage reference circuit are also used.

Operating features include unipolar (0 to +10V) or bipolar (±5V) operation by external pin connection. The converter has a maximum full scale temperature coefficient of ±30ppm/°C and is monotonic over the full operating temperature range of 0°C to 70°C. External offset and gain adjustments are provided for precise calibration of zero and full scale. Parallel output coding is straight binary for unipolar operation and offset binary or two's complement for bipolar operation. A serial output gives successive decision pulses in NRZ format with straight binary or offset binary coding. Other outputs include clock output for synchronizing serial data, MSB output for two's complement coding, and end of conversion (status) signal. All outputs are DTL/TTL compatible. Power requirement is ±15VDC and +5VDC. The ADC-EH10B is also available in extended temperature range versions.

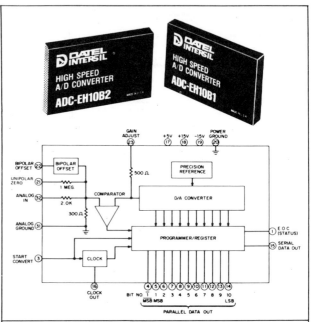

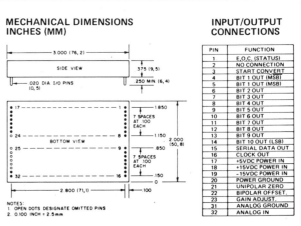

DATEL-INTERSIL INC., 11 CABOT BOULEVARD, MANSFIELD, MA 02048 / TEL. (617) 339-9341 / TWX 710-346-1953 / TLX 951340

Datel-Intersil: Specifications, ADC-EH10B
Timing Diagram for ADC-EH10B
Gain & Offset Adjustments

Reprinted with Permission of Datel-Intersil, Inc., Mansfield, Massachusetts.

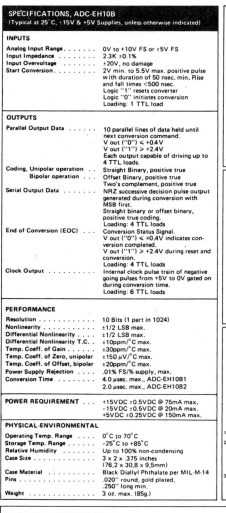

SPECIFICATIONS, ADC-EH10B
(Typical at 25°C, ±15V & +5V Supplies, unless otherwise indicated)

INPUTS

Analog Input Range	0V to +10V FS or ±5V FS
Input Impedance	2.3K ±0.1%
Input Overvoltage	±20V, no damage
Start Conversion	2V min. to 5.5V max. positive pulse with duration of 50 nsec. min. Rise and fall times <500 nsec. Logic "1" resets converter. Logic "0" initiates conversion. Loading: 1 TTL load

OUTPUTS

Parallel Output Data	10 parallel lines of data held until next conversion command. V out ("0") ≤ +0.4V. V out ("1") ≥ +2.4V. Each output capable of driving up to 4 TTL loads.
Coding, Unipolar operation	Straight Binary, positive true
Bipolar operation	Offset Binary, positive true. Two's complement, positive true
Serial Output Data	NRZ successive decision pulse output generated during conversion with MSB first. Straight binary or offset binary, positive true coding. Loading: 4 TTL loads
End of Conversion (EOC)	Conversion Status Signal. V out ("0") ≤ +0.4V indicates conversion completed. V out ("1") ≥ +2.4V during reset and conversion. Loading: 4 TTL loads
Clock Output	Internal clock pulse train of negative going pulses from +5V to 0V gated on during conversion time. Loading: 6 TTL loads

PERFORMANCE

Resolution	10 Bits (1 part in 1024)
Nonlinearity	±1/2 LSB max.
Differential Nonlinearity	±1/2 LSB max.
Differential Nonlinearity T.C.	±10ppm/°C max.
Temp. Coeff. of Gain	±30ppm/°C max.
Temp. Coeff. of Zero, unipolar	±150 μV/°C max.
Temp. Coeff. of Offset, bipolar	±20ppm/°C max.
Power Supply Rejection	.01% FS/% supply, max.
Conversion Time	4.0 μsec. max., ADC-EH10B1; 2.0 μsec. max., ADC-EH10B2

POWER REQUIREMENT

+15VDC ±0.5VDC @ 75mA max.
−15VDC ±0.5VDC @ 20mA max.
+5VDC ±0.25VDC @ 150mA max.

PHYSICAL-ENVIRONMENTAL

Operating Temp. Range	0°C to 70°C
Storage Temp. Range	−25°C to +85°C
Relative Humidity	Up to 100% non-condensing
Case Size	3 x 2 x .375 inches (76.2 x 30.8 x 9.5mm)
Case Material	Black Diallyl Phthalate per MIL-M-14
Pins	.020" round, gold plated, .250" long min.
Weight	3 oz. max. (85g.)

OUTPUT CODING

UNIPOLAR (0V TO +10V)

SCALE	INPUT VOLTAGE	STRAIGHT BINARY
+FS − 1 LSB	+9.9902V	1111 1111 11
+7/8 FS	+8.7500V	1110 0000 00
+3/4 FS	+7.5000V	1100 0000 00
+1/2 FS	+5.0000V	1000 0000 00
+1/4 FS	+2.5000V	0100 0000 00
+1 LSB	+0.0098V	0000 0000 01
0	0.0000V	0000 0000 00

BIPOLAR (−5V TO +5V)

SCALE	INPUT VOLTAGE	OFFSET BINARY	TWO'S COMPLEMENT*
+FS −1 LSB	+4.9902V	1111 1111 11	0111 1111 11
+3/4 FS	+3.7500V	1110 0000 00	0110 0000 00
+1/2 FS	+2.5000V	1100 0000 00	0100 0000 00
0	0.0000V	1000 0000 00	0000 0000 00
−1/2 FS	−2.5000V	0100 0000 00	1100 0000 00
−3/4 FS	−3.7500V	0010 0000 00	1010 0000 00
−FS + 1 LSB	−4.9902V	0000 0000 01	1000 0000 01
−FS	−5.0000V	0000 0000 00	1000 0000 00

*Using MSB output for Bit 1

GAIN & OFFSET ADJUSTMENTS

UNIPOLAR OPERATION

1. Apply START CONVERT pulses to pin 3 (see specifications and timing diagram).
2. Apply a precision reference voltage source to ANALOG IN (pin 32) and ANALOG GROUND (pin 31). Adjust the output of the voltage reference to Zero +1/2 LSB (+4.9mV). Adjust the zero trimming potentiometer so that the output code flickers equally between 0000 0000 00 and 0000 0000 01.
3. Adjust the output of the voltage reference to +FS − 1 1/2 LSB (+9.9854V). Adjust the GAIN trimming potentiometer so that the output code flickers equally between 1111 1111 10 and 1111 1111 11.

BIPOLAR OPERATION

1. Apply START CONVERT pulses to pin 3 (see specifications and timing diagram).
2. Apply a precision reference voltage source to ANALOG IN (pin 32) and ANALOG GROUND (pin 31). Adjust the output of the voltage reference to −FS +1/2 LSB (−4.9951V). Adjust the offset trimming potentiometer so that the output code flickers equally between 0000 0000 00 and 0000 0000 01.
3. Adjust the output of the voltage reference to +FS − 1 1/2 LSB (+4.9854V). Adjust the GAIN trimming potentiometer so that the output code flickers equally between 1111 1111 10 and 1111 1111 11.

ORDERING INFORMATION

ADC-EH10B ─

CONVERSION TIME	
1 = 4.0 μsec.	
2 = 2.0 μsec.	

MATING SOCKETS:
DILS-2 (2/MODULE)
TRIMMING POTENTIOMETERS:
TP20, TP200, TP20K

For extended temperature range operation, the following suffixes are added to the model number. Consult factory for pricing.
 —EX   −25°C to +85°C operation
 —EXX–HS   −55°C to +85°C operation with hermetically sealed semiconductor components

NOTE: ADC-EH10B1 replaces former Datel model ADC-EH10B and is an improved version of the model. The only differences from the previous model are the change in input impedance from 10K ohms to 2.3K ohms, and the reduction in 5V supply current from 280mA to 150mA.

THE ADC-EH10B CONVERTERS ARE COVERED BY GSA CONTRACT

Printed in U.S.A. Copyright © 1980 Datel-Intersil, Inc. All rights reserved.

11 CABOT BOULEVARD, MANSFIELD, MA 02048 / TEL. (617)339-9341 / TWX 710-346-1953 / TLX 951340
• Santa Ana, (714)835-2751, (L.A.) (213)933-7256 • Sunnyvale, CA (408)733-2424 • Gaithersburg, MD (301)840-9490
• Houston, (713)781-8886 • Dallas, TX (214)241-0651 OVERSEAS: DATEL (UK) LTD—TEL: ANDOVER (0264)51055
• DATEL SYSTEMS SARL 602-57-11 • DATELEK SYSTEMS GmbH (089)77-60-95 • DATEL KK Tokyo 793-1031

PRICES AND SPECIFICATIONS SUBJECT TO CHANGE WITHOUT NOTICE

520 SPECIFICATIONS AND DATA SHEETS

Litton Encoder Division: Photo-Optical Shaft Encoders
Theory of Operation

Courtesy of Litton Encoder Division.

PHOTO-OPTICAL SHAFT ENCODERS

In terms of High Resolution, Noise-Free performance, and in-service-life, no other type of Shaft Encoder can compare to the Optical type.

These models provide high level, unambiguous outputs directly compatible with the most popular I.C.'s—No External Signal Processing Amplifiers or Decoding Networks are Required.

SOLID-STATE ILLUMINATION

By use of a solid-state illumination source (Gallium Arsenide) these devices are most uniquely qualified for those applications requiring high MTBF (which may be calculated including the illumination source failure rate), operation in stringent environment (full Mil.-Spec. shock, vibration, and temperature capability), and the utmost in reliability.

Some of the unique features of the GaAs illumination source which enhance its application in Shaft Encoders include:

TEMPERATURE COMPENSATION
—the combination of a Silicon Photo-Transistor and a GaAs illumination source provides a self-compensating system for detector output level variations due to temperature excursions. This feature facilitates use of a much simpler Amplifier and thereby enhances reliability.

LOW POWER CONSUMPTION
—typically 50 Mw (incandescent lamp models typically require 600 Mw).

SOLID-STATE DIFFUSED CONSTRUCTION
—provides high immunity to stringent mechanical environments.

SPECTRAL RESPONSE
—the combination of a silicon photo-transistor and a GaAs illumination source constitutes an excellent marriage in terms of speed and stability.

INCANDESCENT ILLUMINATION

A long-lived, Field Replaceable, tungsten filament lamp provides illumination for photo-electric readout. This illumination source has a realistically rated useful life expectancy of up to 50,000 Hours, and is Field Replaceable with no mechanical alignment or electrical trimming required.

Available with the most popular code formats—Gray Code, Natural Binary, and 8-4-2-1 Binary Coded Decimal—with resolution up to 20,000 Counts/360° and multi-turn capacity up to 2 Million Counts; this line affords the design engineer with an application requiring Optical Encoder performance, and Direct System Compatibility, an opportunity to select the exact device required . . . . with no compromise.

THEORY OF OPERATION

1. The heart of the Encoder is the commutator disc which is mechanically connected to the input shaft.

2. A code pattern on the surface of the disc—comprised of unique combinations of opaque and transparent, segmented regions—uniquely describes each discrete shaft position (within the resolution limitations of the device).

3. The illumination source, either solid-state or incandescent, illuminates one side of the commutator disc.

4. An array of photo-detectors, located on the opposite side of the disc from the illumination source, photo-electrically responds to the presence or absence of light, as passed, or blocked by the transparent and opaque segments respectively.

5. Amplifiers process the photo-detector signals to insure bit-to-bit symmetry and broad temperature range accuracy. These amplified signals are then buffered to provide CCSL compatibility with high fan-out capability.

6. Multi-turn models utilize a mechanical speed reducer between the encoding section associated with the input shaft and additional encoding sections which provide a "turns counting" function. The code transition points for all low speed outputs (IE "turns counting") are controlled by signals generated by the high speed section, thereby eliminating code transition ambiguities which would otherwise be generated by gearing errors and/or backlash.

NOTE: The above description is general in nature only. Exact techniques depend on specific application requirements.

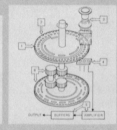

Litton Encoder Division: Photo-Optical Absolute Code Models
Single-Turn Models
Multi-Turn Models

Courtesy of Litton Encoder Division.

PHOTO-OPTICAL ABSOLUTE CODE MODELS

The models listed below represent but a small fraction of our Photo-Optical Absolute Encoder product line. Model Numbers with "L" or "H" in the dash number use incandescent illumination sources; Model Numbers with "G" in the dash number utilize solid state (Gallium Arsenide) illumination sources.

SINGLE-TURN MODELS

Model Number	Code	Counts Per Rev.	Synchro Mount Size	Model Number	Code	Counts Per Rev.	Synchro Mount Size
GCC23-10L1A	Gray	2^{10}	23	SNB23-12G1A	Nat. Bin.	2^{12}	23
GCC23-11L1A	Gray	2^{11}	23	SNB35-10L1A	Nat. Bin.	2^{10}	35
GCC23-12L1A	Gray	2^{12}	23	SNB35-11L1A	Nat. Bin.	2^{11}	35
GCC23-10G1A	Gray	2^{10}	23	SNB35-12L1A	Nat. Bin.	2^{12}	35
GCC23-11G1A	Gray	2^{11}	23	SNB35-13L1A	Nat. Bin.	2^{13}	35
GCC23-12G1A	Gray	2^{12}	23	SNB35-14L1A	Nat. Bin.	2^{14}	35
GCC35-10L1A	Gray	2^{10}	35	SNB35-10G1A	Nat. Bin.	2^{10}	35
GCC35-11L1A	Gray	2^{11}	35	SNB35-11G1A	Nat. Bin.	2^{11}	35
GCC35-12L1A	Gray	2^{12}	35	SNB35-12G1A	Nat. Bin.	2^{12}	35
GCC35-13L1A	Gray	2^{13}	35	SNB35-13G1A	Nat. Bin.	2^{13}	35
GCC35-14L1A	Gray	2^{14}	35	SBD35-102L1A	8421 BCD	1000	35
GCC35-10G1A	Gray	2^{10}	35	SBD35-182L1A	8421 BCD	1800	35
GCC35-11G1A	Gray	2^{11}	35	SBD35-202L1A	8421 BCD	2000	35
GCC35-12G1A	Gray	2^{12}	35	SBD35-362L1A	8421 BCD	3600	35
GCC35-13G1A	Gray	2^{13}	35	SBD35-642L1A	8421 BCD	6400	35
SNB23-10L1A	Nat. Bin.	2^{10}	23	SBD35-102G1A	8421 BCD	1000	35
SNB23-11L1A	Nat. Bin.	2^{11}	23	SBD35-182G1A	8421 BCD	1800	35
SNB23-12L1A	Nat. Bin.	2^{12}	23	SBD35-202G1A	8421 BCD	2000	35
SNB23-10G1A	Nat. Bin.	2^{10}	23	SBD35-362G1A	8421 BCD	3600	35
SNB23-11G1A	Nat. Bin.	2^{11}	23	SBD35-642G1A	8421 BCD	6400	35

MULTI-TURN MODELS

Model Number	Code	Counts Per Rev.	Capacity Total Counts	Turns to Full Capacity	Synchro Mount Size
SBD43-104H10	8421 BCD	1000	100,000	100	43
SBD43-105H10	8421 BCD	1000	1,000,000	1000	43

Mostek, Inc.: Z80-CPU Pin Description

Courtesy of Mostek, Inc. Copyright © Mostek, 1978. Specifications subject to change.

3.0 Z80-CPU PIN DESCRIPTION

The Z80-CPU is packaged in an industry standard 40 pin Dual In-Line Package. The I/O pins are shown in Figure 3.0-1 and the function of each is described below.

Z80 PIN CONFIGURATION

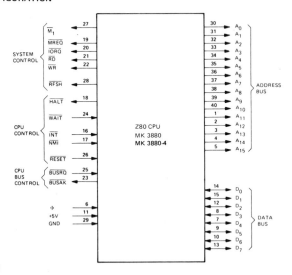

FIGURE 3.0-1

A_0-A_{15} (Address Bus)	Tri-state output, active high. A_0-A_{15} constitute a 16-bit address bus. The address bus provides the address for memory (up to 64K bytes) data exchanges and for I/O device data exchanges. I/O addressing uses the 8 lower address bits to allow the user to directly select up to 256 input or 256 output ports. A_0 is the least significant address bit. During refresh time, the lower 7 bits contain a valid refresh address.
D_0-D_7 (Data Bus)	Tri-state input/output, active high. D_0-D_7 constitute an 8-bit bidirectional data bus. The data bus is used for data exchanges with memory and I/O devices.
$\overline{M_1}$ (Machine Cycle one)	Output, active low. $\overline{M_1}$ indicates that the current machine cycle is the OP code fetch cycle of an instruction execution. Note that during execution of 2-byte op-codes, $\overline{M_1}$ is generated as each op code byte is fetched. These two byte op-codes always begin with CBH, DDH, EDH, or FDH. $\overline{M_1}$ also occurs with $\overline{IORQ}$ to indicate an interrupt acknowledge cycle.
$\overline{MREQ}$ (Memory Request)	Tri-state output, active low. The memory request signal indicates that the address bus holds a valid address for a memory read or memory write operation.

Mostek, Inc.: Z80-CPU Pin Description (continued)

Courtesy of Mostek, Inc. Copyright © Mostek, 1978. Specifications subject to change.

$\overline{IORQ}$
(Input/Output Request)
Tri-state output, active low. The $\overline{IORQ}$ signal indicates that the lower half of the address bus holds a valid I/O address for a I/O read or write operation. An $\overline{IORQ}$ signal is also generated with an $\overline{M_1}$ signal when an interrupt is being acknowledged to indicate that an interrupt response vector can be placed on the data bus. Interrupt Acknowledge operations occur during M_1 time while I/O operations never occur during M_1 time.

$\overline{RD}$
(Memory Read)
Tri-state output, active low. $\overline{RD}$ indicates that the CPU wants to read data from memory or an I/O device. The addressed I/O device or memory should use this signal to gate data onto the CPU data bus.

$\overline{WR}$
(Memory Write)
Tri-state output, active low. $\overline{WR}$ indicates that the CPU data bus holds valid data to be stored in the addressed memory or I/O device.

$\overline{RFSH}$
(Refresh)
Output, active low. $\overline{RFSH}$ indicates that the lower 7 bits of the address bus contain a refresh address for dynamic memories and current $\overline{MREQ}$ signal should be used to do a refresh read to all dynamic memories. A_7 is a logic zero and the upper 8 bits of the Address Bus contains the I Register.

$\overline{HALT}$
(Halt state)
Output, active low. $\overline{HALT}$ indicates that the CPU has executed a HALT software instruction and is awaiting either a non maskable or a maskable interrupt (with the mask enabled) before operation can resume. While halted, the CPU executes NOP's to maintain memory refresh activity.

$\overline{WAIT}$*
(Wait)
Input, active low. $\overline{WAIT}$ indicates to the Z80-CPU that the addressed memory or I/O devices are not ready for a data transfer. The CPU continues to enter wait states for as long as this signal is active. This signal allows memory or I/O devices of any speed to be synchronized to the CPU.

$\overline{INT}$
(Interrupt Request)
Input, active low. The Interrupt Request signal is generated by I/O devices. A request will be honored at the end of the current instruction if the internal software controlled interrupt enable flip-flop (IFF) is enabled and if the $\overline{BUSRQ}$ signal is not active. When the CPU accepts the interrupt, an acknowledge signal ($\overline{IORQ}$ during M_1 time) is sent out at the beginning of the next instruction cycle. The CPU can respond to an interrupt in three different modes that are described in detail in section 8.

$\overline{NMI}$
Input, negative edge triggered. The non maskable interrupt request line has a higher priority than $\overline{INT}$ and is always recognized at the end of the current instruction, independent of the status of the interrupt enable flip-flop. $\overline{NMI}$ automatically forces the Z80-CPU to restart to location 0066_H. The program counter is automatically saved in the external stack so that the user can return to the program that was interrupted. Note that continuous WAIT cycles can prevent the current instruction from ending, and that a $\overline{BUSRQ}$ will override a $\overline{NMI}$.

Mostek, Inc.: Z80-CPU Pin Description (continued)

Courtesy of Mostek, Inc. Copyright © Mostek, 1978. Specifications subject to change.

$\overline{\text{RESET}}$	Input, active low. $\overline{\text{RESET}}$ forces the program counter to zero and initializes the CPU. The CPU initialization includes: 1) Disable the interrupt enable flip-flop 2) Set Register I = 00$_H$ 3) Set Register R = 00$_H$ 4) Set Interrupt Mode 0 During reset time, the address bus and data bus go to a high impedance state and all control output signals go to the inactive state. No refresh occurs.
$\overline{\text{BUSRQ}}$ (Bus Request)	Input, active low. The bus request signal is used to request the CPU address bus, data bus and tri-state output control signals to go to a high impedance state so that other devices can control these buses. When $\overline{\text{BUSRQ}}$ is activated, the CPU will set these buses to a high impedance state as soon as the current CPU machine cycle is terminated.
$\overline{\text{BUSAK}}$* (Bus Acknowledge)	Output, active low. Bus acknowledge is used to indicate to the requesting device that the CPU address bus, data bus and tri-state control bus signals have been set to their high impedance state and the external device can now control these signals.
Φ	Single phase system clock.

*While the Z80-CPU is in either a $\overline{\text{WAIT}}$ state or a Bus Acknowledge condition, Dynamic Memory Refresh will not occur.

Mostek, Inc.: Electrical Specifications, Absolute Maximum Ratings

Courtesy of Mostek, Inc. Copyright © Mostek, 1978. Specifications subject to change.

11.0 ELECTRICAL SPECIFICATIONS
ABSOLUTE MAXIMUM RATINGS*

Temperature Under Bias	Specified Operating Range
Storage Temperature	$-65°C$ to $+150°C$
Voltage on Any Pin with Respect to Ground	$-0.3V$ to $+7V$
Power Dissipation	1.5W

D.C. CHARACTERISTICS
$T_A = 0°C$ to $70°C$, $V_{CC} = 5V \pm 5\%$ unless otherwise specified

SYMBOL	PARAMETER	MIN.	TYP.	MAX.	UNIT	TEST CONDITION
V_{ILC}	Clock Input Low Voltage	-0.3		0.8	V	
V_{IHC}	Clock Input High Voltage	$V_{CC}-.6$		$V_{CC}+.3$	V	
V_{IL}	Input Low Voltage	-0.3		0.8	V	
V_{IH}	Input High Voltage	2.0		V_{CC}	V	
V_{OL}	Output Low Voltage			0.4	V	$I_{OL} = 1.8mA$
V_{OH}	Output High Voltage	2.4			V	$I_{OH} = -250 \mu A$
I_{CC}	Power Supply Current			150*	mA	
I_{LI}	Input Leakage Current			10	μA	$V_{IN} = 0$ to V_{CC}
I_{LOH}	Tri-State Output Leakage Current in Float			10	μA	$V_{OUT} = 2.4$ to V_{CC}
I_{LOL}	Tri-State Output Leakage Current in Float			-10	μA	$V_{OUT} = 0.4V$
I_{LD}	Data Bus Leakage Current in Input Mode			± 10	μA	$0 \leq V_{IN} \leq V_{CC}$

*200mA for -4, -10 or -20 devices

CAPACITANCE
$T_A = 25°C$, $f = 1MHz$ unmeasured pins returned to ground

SYMBOL	PARAMETER	MAX.	UNIT
$C\Phi$	Clock Capacitance	35	pF
C_{IN}	Input Capacitance	5	pF
C_{OUT}	Output Capacitance	10	pF

*Comment

Stresses above those listed under "Absolute Maximum Ratings" may cause permanent damage to the device. This is a stress rating only and functional operation of the device at these or any other condition above those indicated in the operational sections of this specification is not implied. Exposure to absolute maximum rating conditions for extended periods may affect device reliability.

Mostek, Inc.: MK 3880-4/Z80A-CPU (A.C. characteristics)

Courtesy of Mostek, Inc. Copyright © Mostek, 1978. Specifications subject to change.

MK 3880-4 Z80A-CPU

A. C. CHARACTERISTICS $T_A = 0°C$ to $70°C$, $V_{cc} = +5V \pm 5\%$, Unless Otherwise Noted

SIGNAL	SYMBOL	PARAMETER	MIN.	MAX.	UNIT	TEST CONDITIONS
Φ	t_c	Clock Period	.25	[12]	μsec	
	$t_w(\Phi H)$	Clock Pulse Width, Clock High	110	(D)	nsec	
	$t_w(\Phi L)$	Clock Pulse Width, Clock Low	110	2000	nsec	
	t_r, t_f	Clock Rise and Fall Time		30	nsec	
A_{0-15}	$t_{D(AD)}$	Address Output Delay		110	nsec	
	$t_{F(AD)}$	Delay to Float		90	nsec	
	t_{acm}	Address Stable Prior to $\overline{MREQ}$ (Memory Cycle)	[1]		nsec	$C_L = 50pF$
	t_{aci}	Address Stable Prior to $\overline{IORQ}$, $\overline{RD}$ or $\overline{WR}$ (I/O Cycle)	[2]		nsec	
	t_{ca}	Address Stable From $\overline{RD}$, $\overline{WR}$, $\overline{IORQ}$ or $\overline{MREQ}$	[3]		nsec	Except T3.M1
	t_{caf}	Address Stable From $\overline{RD}$ or $\overline{WR}$ During Float	[4]		nsec	
D_{0-7}	$t_{D(D)}$	Data Output Delay		150	nsec	
	$t_{F(D)}$	Delay to Float During Write Cycle		90	nsec	
	$t_{S\Phi(D)}$	Data Setup Time to Rising Edge of Clock During M1 Cycle	50		nsec	
	$t_{S\overline{\Phi}(D)}$	Data Setup Time to Falling Edge at Clock During M2 to M5	60		nsec	$C_L = 50pF$
	t_{dcm}	Data Stable Prior to $\overline{WR}$ (Memory Cycle)	[5]		nsec	
	t_{dci}	Data Stable Prior to $\overline{WR}$ (I/O Cycle)	[6]		nsec	
	t_{cdf}	Data Stable From $\overline{WR}$	[7]		nsec	
	t_H	Input Hold Time	0		nsec	
$\overline{MREQ}$	$t_{DL\Phi(MR)}$	$\overline{MREQ}$ Delay From Falling Edge of Clock, $\overline{MREQ}$ Low	20	85	nsec	
	$t_{DH\Phi(MR)}$	$\overline{MREQ}$ Delay From Rising Edge of Clock, $\overline{MREQ}$ High		85	nsec	
	$t_{DH\overline{\Phi}(MR)}$	$\overline{MREQ}$ Delay From Falling Edge of Clock, $\overline{MREQ}$ High		85	nsec	$C_L = 50pF$
	$t_w(\overline{MRL})$	Pulse Width, $\overline{MREQ}$ Low	[8]		nsec	
	$t_w(\overline{MRH})$	Pulse Width, $\overline{MREQ}$ High	[9]		nsec	
$\overline{IORQ}$	$t_{DL\Phi(IR)}$	$\overline{IORQ}$ Delay From Rising Edge of Clock, $\overline{IORQ}$ Low		75	nsec	
	$t_{DL\overline{\Phi}(IR)}$	$\overline{IORQ}$ Delay From Falling Edge of Clock, $\overline{IORQ}$ Low		85	nsec	$C_L = 50pF$
	$t_{DH\Phi(IR)}$	$\overline{IORQ}$ Delay From Rising Edge of Clock, $\overline{IORQ}$ High		85	nsec	
	$t_{DH\overline{\Phi}(IR)}$	$\overline{IORQ}$ Delay From Falling Edge of Clock, $\overline{IORQ}$ High		85	nsec	
$\overline{RD}$	$t_{DL\Phi(RD)}$	$\overline{RD}$ Delay From Rising Edge of Clock, $\overline{RD}$ Low		85	nsec	
	$t_{DL\overline{\Phi}(RD)}$	$\overline{RD}$ Delay From Falling Edge of Clock, $\overline{RD}$ Low		95	nsec	$C_L = 50pF$
	$t_{DH\Phi(RD)}$	$\overline{RD}$ Delay From Rising Edge of Clock, $\overline{RD}$ High		85	nsec	
	$t_{DH\overline{\Phi}(RD)}$	$\overline{RD}$ Delay From Falling Edge of Clock, $\overline{RD}$ High		85	nsec	
$\overline{WR}$	$t_{DL\Phi(WR)}$	$\overline{WR}$ Delay From Rising Edge of Clock, $\overline{WR}$ Low		65	nsec	
	$t_{DL\overline{\Phi}(WR)}$	$\overline{WR}$ Delay From Falling Edge of Clock, $\overline{WR}$ Low		80	nsec	$C_L = 50pF$
	$t_{DH\Phi(WR)}$	$\overline{WR}$ Delay From Falling Edge of Clock, $\overline{WR}$ High		80	nsec	
	$t_w(\overline{WRL})$	Pulse Width, $\overline{WR}$ Low	[10]		nsec	

NOTES:
A Data should be enabled onto the CPU data bus when $\overline{RD}$ is active. During interrupt acknowledge data should be enabled when M1 and $\overline{IORQ}$ are both active.
B The $\overline{RESET}$ signal must be active for a minimum of 3 clock cycles.

(Cont'd. on page 83)

SPECIFICATIONS AND DATA SHEETS

Mostek, Inc.: MK 3880-4/Z80A-CPU (A.C. characteristics) (continued)

Courtesy of Mostek, Inc. Copyright © Mostek, 1978. Specifications subject to change.

MK 3880-4 Z80A-CPU

SIGNAL	SYMBOL	PARAMETER	MIN.	MAX.	UNIT	TEST CONDITION
$\overline{M1}$	$t_{DL(M1)}$	$\overline{M1}$ Delay From Rising Edge of Clock, $\overline{M1}$ Low		100	nsec	$C_L = 50pF$
	$t_{DH(M1)}$	$\overline{M1}$ Delay From Rising Edge of Clock, $\overline{M1}$ High		100	nsec	
$\overline{RFSH}$	$t_{DL(RF)}$	$\overline{RFSH}$ Delay From Rising Edge of Clock, $\overline{RFSH}$ Low		130	nsec	$C_L = 50pF$
	$t_{DH(RF)}$	$\overline{RFSH}$ Delay From Rising Edge of Clock $\overline{RFSH}$ High		120	nsec	
$\overline{WAIT}$	$t_{S(WT)}$	$\overline{WAIT}$ Setup Time to Falling Edge of Clock	70		nsec	
$\overline{HALT}$	$t_{D(HT)}$	$\overline{HALT}$ Delay Time From Falling Edge of Clock		300	nsec	$C_L = 50pF$
$\overline{INT}$	$t_{s(IT)}$	$\overline{INT}$ Setup Time to Rising Edge of Clock	80		nsec	
$\overline{NMI}$	$t_{w(NML)}$	Pulse Width, $\overline{NMI}$ Low	80		nsec	
$\overline{BUSRQ}$	$t_{s(BQ)}$	$\overline{BUSRQ}$ Setup Time to Rising Edge of Clock	50		nsec	
$\overline{BUSAK}$	$t_{DL(BA)}$	$\overline{BUSAK}$ Delay From Rising Edge of Clock, $\overline{BUSAK}$ Low		100	nsec	$C_L = 50pF$
	$t_{DH(BA)}$	$\overline{BUSAK}$ Delay From Falling Edge of Clock, $\overline{BUSAK}$ High		100	nsec	
$\overline{RESET}$	$t_{s(RS)}$	$\overline{RESET}$ Setup Time to Rising Edge of Clock	60		nsec	
	$t_{F(C)}$	Delay to/From Float ($\overline{MREQ}$, $\overline{IORQ}$, $\overline{RD}$ and $\overline{WR}$)		80	nsec	
	t_{mr}	$\overline{M1}$ Stable Prior to $\overline{IORQ}$ (Interrupt Ack.)	[11]		nsec	

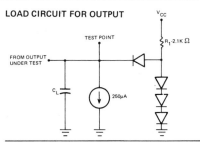

LOAD CIRCUIT FOR OUTPUT

[1] $t_{acm} = t_w (\Phi H) + t_f - 65$

[2] $t_{aci} = t_c - 70$

[3] $t_{ca} = t_w (\Phi L) + t_r - 50$

[4] $t_{caf} = t_w (\Phi L) + t_r - 45$

[5] $t_{dcm} = t_c - 170$

[6] $t_{dci} = t_w (\Phi L) + t_r - 170$

[7] $t_{cdf} = t_w (\Phi L) + t_r - 70$

[8] $t_w (\overline{MRL}) = t_c - 30$

[9] $t_w (\overline{MRH}) = t_w (\Phi H) + t_f - 20$

[10] $t_w (\overline{WR}) = t_c - 30$

[11] $t_{mr} = 2t_c + t_w (\Phi H) + t_f - 65$

[12] $t_c = t_w (\Phi H) + t_w (\Phi L) + t_r + t_f$

NOTES (Cont'd.)
C. Output Delay vs. Load Capacitance
   $T_A = 70°C$ $V_{CC} = 5V \pm 5\%$
   Add 10 nsec delay for each 50pF increase in load up to a maximum of 200pF for the data bus and 100pF for address and control lines
D. Although static by design, testing guarantees $t_w (\Phi H)$ of 200 μsec maximum.

528 SPECIFICATIONS AND DATA SHEETS

Intel Corp.: Schottky Bipolar 8212, Eight-Bit Input/Output Port

Courtesy of Intel Corporation.

Schottky Bipolar 8212
EIGHT-BIT INPUT/OUTPUT PORT

- **Fully Parallel 8-Bit Data Register and Buffer**
- **Service Request Flip-Flop for Interrupt Generation**
- **Low Input Load Current — .25 mA Max.**
- **Three State Outputs**
- **Outputs Sink 15 mA**
- **3.65V Output High Voltage for Direct Interface to 8080 CPU or 8008 CPU**
- **Asynchronous Register Clear**
- **Replaces Buffers, Latches and Multiplexers in Microcomputer Systems**
- **Reduces System Package Count**

The 8212 input/output port consists of an 8-bit latch with 3-state output buffers along with control and device selection logic. Also included is a service request flip-flop for the generation and control of interrupts to the microprocessor.

The device is multimode in nature. It can be used to implement latches, gated buffers or multiplexers. Thus, all of the principal peripheral and input/output functions of a microcomputer system can be implemented with this device.

PIN CONFIGURATION

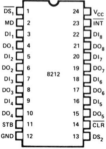

PIN NAMES

DI_1-DI_8	DATA IN
DO_1-DO_8	DATA OUT
$\overline{DS_1}$-DS_2	DEVICE SELECT
MD	MODE
STB	STROBE
$\overline{INT}$	INTERRUPT (ACTIVE LOW)
$\overline{CLR}$	CLEAR (ACTIVE LOW)

LOGIC DIAGRAM

Intel Corp.: Schottky Bipolar 8212

Courtesy of Intel Corporation.

SCHOTTKY BIPOLAR 8212

Functional Description

Data Latch

The 8 flip-flops that make up the data latch are of a "D" type design. The output (Q) of the flip-flop will follow the data input (D) while the clock input (C) is high. Latching will occur when the clock (C) returns low.

The data latch is cleared by an asynchronous reset input ($\overline{CLR}$). (Note: Clock (C) Overides Reset ($\overline{CLR}$).)

Output Buffer

The outputs of the data latch (Q) are connected to 3-state, non-inverting output buffers. These buffers have a common control line (EN); this control line either enables the buffer to trasmit the data from the outputs of the data latch (Q) or disables the buffer, forcing the output into a high impedance state. (3-state)

This high-impedance state allows the designer to connect the 8212 directly onto the microprocessor bi-directional data bus.

Control Logic

The 8212 has control inputs $\overline{DS1}$, DS2, MD and STB. These inputs are used to control device selection, data latching, output buffer state and service request flip-flop.

$\overline{DS1}$, DS2 (Device Select)

These 2 inputs are used for device selection. When $\overline{DS1}$ is low and DS2 is high ($\overline{DS1} \cdot DS2$) the device is selected. In the selected state the output buffer is enabled and the service request flip-flop (SR) is asynchronously set.

MD (Mode)

This input is used to control the state of the output buffer and to determine the source of the clock input (C) to the data latch.

When MD is high (output mode) the output buffers are enabled and the source of clock (C) to the data latch is from the device selection logic ($\overline{DS1} \cdot DS2$).

When MD is low (input mode) the output buffer state is determined by the device selection logic ($\overline{DS1} \cdot DS2$) and the source of clock (C) to the data latch is the STB (Strobe) input.

STB (Strobe)

This input is used as the clock (C) to the data latch for the input mode MD = 0) and to synchronously reset the service request flip-flop (SR).

Note that the SR flip-flop is negative edge triggered.

Service Request Flip-Flop

The (SR) flip-flop is used to generate and control interrupts in microcomputer systems. It is asynchronously set by the $\overline{CLR}$ input (active low). When the (SR) flip-flop is set it is in the non-interrupting state.

The output of the (SR) flip-flop (Q) is connected to an inverting input of a "NOR" gate. The other input to the "NOR" gate is non-inverting and is connected to the device selection logic ($\overline{DS1} \cdot DS2$). The output of the "NOR" gate ($\overline{INT}$) is active low (interrupting state) for connection to active low input priority generating circuits.

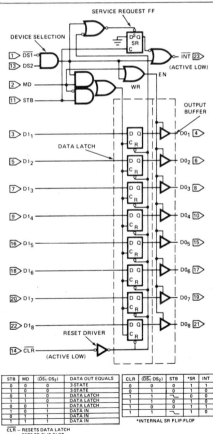

Intel Corp.: Schottky Bipolar 8214, Prior Interrupt Control Unit

Courtesy of Intel Corporation.

Schottky Bipolar 8214

PRIORITY INTERRUPT CONTROL UNIT

- Eight Priority Levels
- Current Status Register
- Priority Comparator
- Fully Expandable
- High Performance (50ns)
- 24-Pin Dual In-Line Package

The 8214 is an eight level priority interrupt control unit designed to simplify interrupt driven microcomputer systems.

The PICU can accept eight requesting levels; determine the highest priority, compare this priority to a software controlled current status register and issue an interrupt to the system along with vector information to identify the service routine.

The 8214 is fully expandable by the use of open collector interrupt output and vector information. Control signals are also provided to simplify this function.

The PICU is designed to support a wide variety of vectored interrupt structures and reduce package count in interrupt driven microcomputer systems.

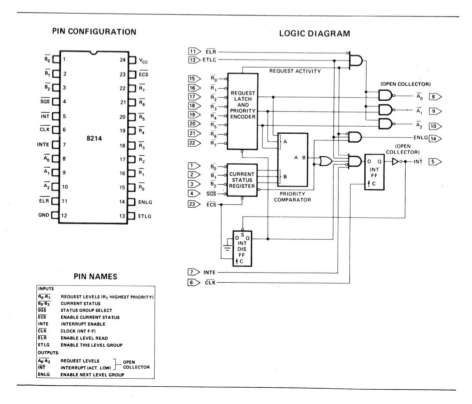

Intel Corp.: Schottky Bipolar 8214
Polled Method
Interrupt Method

Courtesy of Intel Corporation.

SCHOTTKY BIPOLAR 8214

INTERRUPTS IN MICROCOMPUTER SYSTEMS

Microcomputer system design requires that I/O devices such as keyboards, displays, sensors and other components receive servicing in an efficient method so that large amounts of the total systems tasks can be assumed by the microcomputer with little or no effect on throughput.

The most common method of servicing such devices is the **Polled** approach. This is where the processor must test each device in sequence and in effect "ask" each one if it needs servicing. It is easy to see that a large portion of the main program is looping through this continuence polling cycle and that such a method would have a serious, detrimental effect on system throughput thus limiting the tasks that could be assumed by the microcomputer and reducing the cost effectiveness of using such devices.

A more desireable method would be one that would allow the microprocessor to be executing its main program and only stop to service peripheral devices when it is told to do so by the device itself. In effect, the method would provide an external asynchronous input that would inform the processor that it should complete whatever instruction that is currently being executed and fetch a new routine that will service the requesting device. Once this servicing is complete however the processor would resume exactly where it left off.

This method is called **Interrupt**. It is easy to see that system throughput would drastically increase, and thus more tasks could be assumed by the microcomputer to further enhance its cost effectiveness.

The Priority Interrupt Control Unit (PICU) functions as an overall manager in an Interrupt-Driven system environment. It accepts requests from the peripheral equipment, determines which of the incoming requests is of the highest importance (priority), ascertains whether the incoming request has a higher priority value than the level currently being serviced and issues an Interrupt to the CPU based on this determination.

Each peripheral device or structure usually has a special program or "routine" that is associated with its specific functional or operational requirements; this is referred to as a "service routine". The PICU, after issuing an Interrupt to the CPU, must somehow input information into the CPU that can "point" the Program Counter to the service routine associated with the requesting device. The PICU encodes the requesting level into such information for use as a "vector" to the correct Interrupt Service Routine.

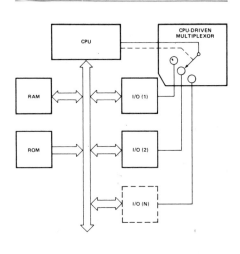

Polled Method

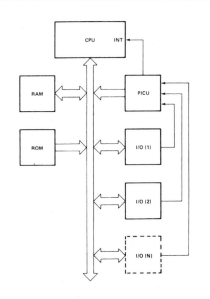

Interrupt Method

Intel Corp.: Schottky Bipolar 8214 (Functional Description)

Courtesy of Intel Corporation.

SCHOTTKY BIPOLAR 8214

FUNCTIONAL DESCRIPTION

General

The 8214 is a device specifically designed for use in real time, interrupt driven, microcomputer systems. Basically it is an eight (8) level priority control unit that can accept eight different interrupt requests, determine which has the highest priority, compare that level to a software maintained current status register and issue an interrupt to the system based on this comparison along with vector information to indicate the location of the service routine.

Priority Encoder

The eight requests inputs, which are active low, come into the Priority Encoder. This circuit determines which request input is the most important (highest priority) as preassigned by the designer. ($\overline{R7}$) is the highest priority input to the 8214 and ($\overline{R0}$) is the lowest. The logic of the Priority Encoder is such that if two or more input levels arrive at the same time then the input having the highest priority will take presidence and a three bit output, corresponding to the active level (modulo 8) will be sent out. The Priority Encoder also contains a latch to store the request input. This latch is controlled by the Interrupt Disable Flip-flop so that once an interrupt has been issued by the 8214 the request latch is no longer open. (Note that the latch does not store inactive requests. In order for a request to be monitored by the 8214 it must remain present until it has been serviced.)

Current Status Register

In an interrupt driven microcomputer system it is important to not only prioritize incoming requests but to ascertain whether such a request is a higher priority than the interrupt currently being serviced.

The Current Status Register is a simple 4-bit latch that is treated as an addressable outport port by the microcomputer system. It is loaded when the $\overline{ECS}$ input goes low.

Maintenance of the Current Status Register is performed as a portion of the service routine. Basically, when an interrupt is issued to the system the programmer outputs a binary code (modulo 8) that is the compliment of the interrupt level. This value is stored in the Current Status Register and is compared to all further prioritized incoming requests by the Priority Comparator. In essence, a copy of the current interrupt level is written into the 8214 to be used as a reference for comparison. There is no restriction to this maintenance, other level values can be written into this register as references so that groups of interrupt requests may be disallowed under complete control of the programmer.

Note that the fourth bit in the register is $\overline{SGS}$. This input is part of the value written out by the programmer and performs a special function. The Priority Comparator will only issue an output that indicates the request level is greater than the Current Status Register. If both comparator inputs are equal to zero no output will be present. The $\overline{SGS}$ input allows the programmer to, in effect, disable this comparison and allow the 8214 to issue an interrupt to the system that is based only on the logic of the priority encoder.

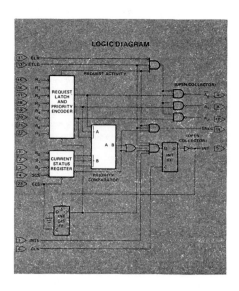

Intel Corp.: Schottky Bipolar 8214 (diagram)

Courtesy of Intel Corporation.

SCHOTTKY BIPOLAR 8214

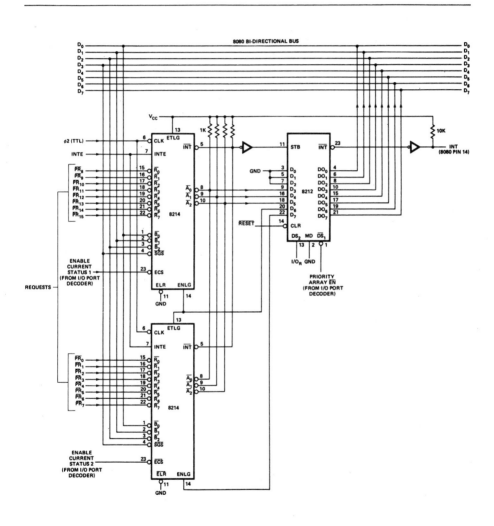

INDEX

Absolute value, of complex number, 11
Acceleration error coefficient, 205
Accumulator, microprocessor, 403
Active filters, 289
A/D converter, 4
    conversion frequency of, 387
    need for, 383
    resolution of, 387
    successive approximation method for, 386
    types of, 386
ADD instructions, microprocessor, 419
Address, memory, in microprocessor system, 289
Amplifiers, operational, 277
    important characteristics of, 281
    integration using, 286
    loading of, 280
    potentiometers used with, 286, 305
    scaling with, 280
    summation using, 283
    transfer function synthesis using, 289
Analog computer, 276
    vs. digital computer, 325
    hybrid, 313
    magnitude scaling of, 304
    modes of operation, 299
    use of, to simulate control system, 297
    to simulate mechanical system, 298
    to solve differential equations, 290
    to solve non-linear differential equations, 309
Analog/digital converter, see A/D converter
Analog/digital interface, in a microprocessor-based control system, 382
Analogs, electrical-mechanical, 148
Analog systems, vs. digital, 3, 381
Angular velocity, 182
Arbitrary constants, in solution of differential equation, 64
Asymptotic plots, 54

accuracy of, 227
Automatic iterative computation, 356

Back emf, in motor, 197, 202
Bang-bang control system, 163
BASIC language, 326
    DEF FNy(x) statement in, 328
    INT(x) function in, 337
    SGN(x) function in, 337
Bidirectional bus, 389
Block diagrams, 165
Bode plots, 46
    asymptotic, 54
    of differentiator, 51
    of integrator, 49
    for position control system, 178
    stability analysis of control system and, 227
Break frequency, 55
Bus, microprocessor, 389

Calculus, important facts from, 29
Cascade connections, 46, 51, 165
Central processing unit (CPU), 382
Characteristic equation, 224, 252
Chip-select, input to integrated circuit chip, 392
Chopper stabilization, of operational amplifiers, 281
Clock, microprocessor, 390
    periods, required for various instructions, 411
Closed loop control system, definition of, 163
Closed loop transfer function, of control system, 167
Coefficient potentiometers, 299, 304
Compensation, of control system, 260
    lag, 266
    lead, 260
    RC lag, 268
    RC lead, 264
Complex number, 7
    arithmetic of, 12

    conjugate of, 14
    exponential form of, 19
    magnitude of, 10
    polar form of, 10
    rectangular form of, 10
Complex plane, 9
Conditionally stable control systems, 246
Conjugate, complex, 14
Continuous System Modeling Program (CSMP), 363
    simulation of control system using, 370
    specification of driving functions in, 367
    specification of transfer functions in, 365
Continuous variable, 3
Control signal, in microprocessor-based control system, 389
Control system, 2, 163
    analog simulation of, 297
    compensation of, 260
    digital computations for, 341
    fully fed-back, 166
    open and closed loop, 163
    proportional, 164
    stable vs. unstable, 221
    velocity, 195
Cos (ωt), differentiation and integration of, 77
Cramer's rule, 128
Critically damped response, 142, 146, 147
    of control system, 192
Current divider rule, 110
Current sources, 105
    conversion to voltage source, 106
    in frequency domain, 119
Cutoff frequency, 46, 177
Cycles, on log-log graph paper, 47

D/A converter, 383
    output of, when driven by counter, 385

INDEX 537

use of, in A/D conversion, 386
Damped sinusoid, 94
Damping, mechanical, 69, 148
    in rotational systems, 182
Damping constant, 69
Damping in electrical networks, 143
Damping ratio, 143
Dashpot, 69
Decade, 47
Decibels, 51
    per octave and per decade, 53
    voltage gain in, 52
Decrement (DEC) instructions, microprocessor, 410
Degree of a differential equation, 64
Delayed time functions, representation of, 459
Delay line, 459
Delta, denominator determinant, 129
Delta function, 75
Determinants, 128
Differential equations, 2, 63, 65
    analog computer solution of, 290
    general and particular solutions of, 64
    order and degree of, 64
    solution of, by LaPlace transforms, 84
Differentiator, 51
Digital/Analog converter, see D/A converter
Digital-analog switch, 315
Digital computer simulation, 325
    for analysis of transfer functions, 326
    for finding time-domain response, 344
    for optimization, 356
    using CSMP, 363
Dimensionless scaling of analog computer setup, 305
Discontinuity, of function, 461, 478
Discrete valued variable, 3
Dot notation for differential equations, 64

Drift, in operational amplifiers, 282
Dynamic memory, 397

Encoder, optical shaft, 388
Equivalent circuits, for charged capacitors, 121
    for inductors with initial currents, 122
Error coefficients, 205
    improvement in, using lag compensation, 266
Exponential form of complex number, 19

Feedback, in control system, 163
    positive, 221
Filter, low pass, 43
    asymptotic Bode plot for, 54
    control system as, 177
Final value theorem, 205
FINTIM specification, in CSMP, 364
Flags, microprocessor, 409
Frequency domain, 6, 73, 85
Frequency response, 46, 51
    asymptotic, 54
    nonlinear, 47
    of operational amplifiers, 281
    of position control system, 178
Fully fed-back control system, 166
Function generators, analog, 312
    diode type, 313
Fundamental theorem of algebra, 92

Gain, 38
    frequency independent, 49
    as function of frequency, 41
    of integrator, 48
    of potentiometer, 39
    voltage, in dB, 52
Gain margin, 234
    realization of, by digital computation (example), 357
Gears, 184
    force and torque relations for, 185
General solution of a different equation, 64

538 INDEX

Hold, mode of operation of an analog computer, 303
Hybrid computer, 313

Ideal current source, 105
Ideal voltage source, 104
Imaginary axis, 9
Imaginary number, 7
Imaginary part of complex number, 8
Immediate load, op-codes, 405
Impedance, phasor form of, 24
   LaPlace transforms of, 85
   mechanical, 197
Impulse function, 75
Increment (INC) instructions, microprocessor, 410
Indirect addressing, 406
   op-codes for, 407
Inertia, 182
Infinite sum, 468
IN instruction, microprocessor, 419
Initial conditions, for differential equations, 64
     setting of, in analog computer circuits, 295, 300
   mode of operation of analog computer, 299
Initial voltages and currents in networks, 121
Input/output (I/O), 398
   memory-mapped (example), 399
   using device codes, 400
Integrators, 47, 173
   Bode plot for, 49
   construction of, using operational amplifiers, 286
Interface, analog/digital, in a microprocessor-based control system, 383
Interrupts, microprocessor, 423
   maskable and non-maskable, 428
   modes of, 432
Inverse LaPlace transform, 79
   digital computation of, 344
   by partial fractions, 87
   when F(s) has complex poles, 93

when F(s) has repeated poles, 94
I/O, *see* Input/output
Iterative computation, 356

j, 7
   properties of, 17
Jump, in function, 461, 478
Jump instructions, microprocessor, 408
   conditional, op-codes for, 409

Kirchoff's current law, 109
   using LaPlace transforms, 127
Kirchoff's voltage law, 108
   using LaPlace transforms, 123

Ladder network R-2R, 383
Lag compensation, of control system, 266
   RC network for lag compensation, 268
LaPlace transform, 1, 72
   definition of, 73
   of derivatives and integrals, 82
   table of pairs, 77
Lead network, 58
   as lead compensator in control system, 264
Logic operations, microprocessor, 412
   op-codes for, 414
Log-log graph paper, 47
Low pass filter, *see* Filter, low pass

M_m, 231, 248
Magnitude, of complex number, 11
   of transfer function, 44
Magnitude scaling, of analog computer, 304
   for second-order equation, 308
Masking, 414
Mass, 69, 148
Mass-spring combination, 150
Mechanical-electrical analogs, 148
Mechanical impedance, 197
Mechanical networks, 148

analysis of, using LaPlace transforms, 153
Memory, microprocessor, 389, 394
 -mapped I/O, 398
 pin connections for, 395
 static and dynamic, 397
Mesh analysis, 110
 using LaPlace transforms, 128
Microprocessor, 381
 architecture and programming, 403
 -based control system, 383
 design example, 435
 interrupts, 423
 system design considerations, 389
Model, mathematical, 5
Modes, operating, of analog computer, 299
Multipliers, analog, 309
 time-division type, 318

Natural frequency, 143
Network analysis using LaPlace transforms, 104
Nichols chart, 246
Nodal analysis, 112
 using LaPlace transforms, 134
Noise, in operational amplifiers, 282
Nonlinear differential equations, 309
 use of analog computer to solve, 311
Nonlinear periodic functions, 468, 473
Normalized scaling of an analog computer setup, 305
Norton's theorem, 118
 using LaPlace transforms, 140
Nyquist stability criterion, 243

Octave, 53
Op-code (operation code), 404
Open loop control system, 163
Open loop transfer function, of a control system, 166
 stability and, 222

Operational amplifiers, *see* Amplifiers, operational
Optical shaft encoder, 388
Optimization, by iterative digital computation, 356
Order:
 of differential equation, 64
 of root, 94
OUTDEL specification, in CSMP, 365
OUT instruction, microprocessor, 419
Overdamped response, 142, 147
 of control system, 191
Overshoot, as performance criterion in control system, 248

Partial fractions, 87
 for repeated roots, 94
Particular solution of differential equation, 64
Periodic functions, non-linear, 468
 LaPlace transform of, 473
Phase margin, 234
Phasors, 20
 and impedance, 24
Polar form of complex number, 10
 conversion to rectangular form, 12
Polar plots, 235
 summary of rules for sketching, 242
Pole, 91
Polling, 424
POP instructions, microprocessor, 430
Position control system, 164
Position error coefficient, 205
 compensation for, 267
Power, in rotational systems, 182
Priority interrupt control unit, Intel 8214, 450
Proportional control system, 164
Pulse, time-domain representation of, 465
 LaPlace transform of, 473
PUSH instructions, microprocessor, 429
 op-codes for, 430

540 INDEX

Quadratic, log |G| plot for, 231
   phase plot for, 232
Quarter-square multiplication, in analog computer, 309

R-2R ladder network, for D/A conversion, 383
Ramp voltage, 461
Random access memory (RAM), 394
Rationalizing, 16
Reaction force, 68
Read-only memory (ROM), 394
Real axis, 9
Real current source, 105
   voltage source equivalent of, 106
Real part of complex number, 8
Real time solutions to differential equations, 276
Real voltage source, 104
   current source equivalent of, 106
Rectangular form of complex number, 10
   conversion to polar form, 10
Registers, microprocessor, 403
   pairs, 429
   source and destination, 405
Repeated roots, 94
   partial fraction method for, 94
Repetitive operation, of analog computer, 300
Reset:
   microprocessor, 429
   mode of operation of analog computer, 299
Resolution of A/D converter, 387
Response, 3
   critically damped, 142, 146, 147
   digital computation of, 344
   overdamped, 142, 144
   steady-state, 91
   transient, 91
   underdamped, 142, 144
RET (return) instruction, microprocessor, 426

Rise, voltage, 108
Robot, design example of microprocessor control system, 435
Root locus, 252
Roots, repeated, 94
Rotate instructions, microprocessor, 415
Rotational mechanical systems, 181
RST (restart) instructions, microprocessor, 433

Sampling theorem, by Shannon, 387
Servomechanism, 165
Servomotor, 174
   in closed loop, 174
   comprehensive analysis of, in control system, 196
   as integrator, 173
   second approximation of, 189
   speed-torque curves for, 202
Shannon sampling theorem, 387
Shifting, of data bytes, in microprocessor, 415
Shifting theorem, 472
Simultaneous equations, solution of, 128
Sin (ωt), differentiation and integration of, 30
   LaPlace transform of, 76
Slew rate, of operational amplifiers, 281
Solution, of differential equation, definition of, 63
   particular and general, 64
Solving simultaneous equations using determinants, 128
Speed-torque curves, 202
Spirule, for root locus plots, 260
Spring constant, 68, 148
   in rotational systems, 182
Spring-dashpot combinations, 149
Spring-mass combination, 70
Spring-mass-dashpot combinations, 151

INDEX **541**

Springs, 68
Square wave, time-domain representation of, 470
Stability, of control system, 221
  analysis:
    of system with conjugate poles, 230
    using Bode plots, 227
    using Nichols chart, 246
    using Nyquist criterion, 243
    using root locus, 252
    and open loop transfer function, 222
Stack, microprocessor, 425
  pointer, 425
Stall torque, of motor, 203
Static memory, 397
Steady-state solution, 91
Step function, 74, 461
Stimulus, 3
Straight line, equation of, 49
SUB instructions, microprocessor, 419
Subroutine, 433
  interrupt service, 424
Subtraction, of voltages using operational amplifiers, 284
Summation, of voltages using operational amplifiers, 283
Superposition, 113
  using LaPlace transforms, 135

Table of transform pairs, 77
  use of, 79
Thevenin's theorem, 116, 481
  using LaPlace transforms, 137
Three-level logic, 391
Time constant, 74
Time-delay programs, 410
Time-division multiplication, 318
TIMER specification, in CSMP, 364
Time scaling, of analog computer, 301
Torque, 181
  relations in frequency domain, 183
Track-hold circuit, in hybrid computer, 316

Traffic controller, microprocessor example of, 416
Transducers, 40
Transfer functions, 38, 42
  analysis of, by digital computer, 326
  of cascaded devices, 46
  closed loop, 167
  in frequency domain, 84
  open loop, 166
  of passive networks, 43
Transformation, 2, 72
Transform pairs, table of, 77
Transforms of derivatives and integrals, 82
Transient solution, 91
Translational mechanical system, 68
Triangular pulse, time domain representation of, 467
Triangular wave, time-domain representation of (example), 471
Tri-state logic, 391

Underdamped response, 142, 144
  of control system, 193
Unit step function, 74, 461
Unity feedback control system, 166
Unstable control system, characterization of, 221

Velocity error coefficient, 205
Virtual ground, 280
Voltage comparator, 314
Voltage divider rule, 27
Voltage gain, in dB, 52
Voltage rise, 108
Voltage sources, 104
  conversion to current source, 106
  in frequency domain, 119

Weight, of impulse function, 75

Zero, 91
  effect of, in open loop transfer function, 234, 257
Zero-order hold, 316